AF713643

ANALYTICAL CHEMISTRY OF THE ELEMENTS

ANALYTICAL CHEMISTRY OF PLUTONIUM

Analytical Chemistry of the Elements

Series prepared by the Vernadskii Institute of Geochemistry and Analytical Chemistry
USSR Academy of Sciences
A. P. VINOGRADOV, *Editor*
English translation by ISRAEL PROGRAM FOR SCIENTIFIC TRANSLATIONS, Jerusalem

---

ANALYTICAL CHEMISTRY OF BERYLLIUM
ANALYTICAL CHEMISTRY OF BORON
ANALYTICAL CHEMISTRY OF COBALT
ANALYTICAL CHEMISTRY OF GALLIUM
ANALYTICAL CHEMISTRY OF MOLYBDENUM
ANALYTICAL CHEMISTRY OF NICKEL
ANALYTICAL CHEMISTRY OF NIOBIUM AND TANTALUM
ANALYTICAL CHEMISTRY OF PLUTONIUM
ANALYTICAL CHEMISTRY OF POTASSIUM
ANALYTICAL CHEMISTRY OF PROTACTINIUM
ANALYTICAL CHEMISTRY OF RUTHENIUM
ANALYTICAL CHEMISTRY OF TECHNETIUM, PROMETHIUM, ASTATINE AND FRANCIUM
ANALYTICAL CHEMISTRY OF THALLIUM
ANALYTICAL CHEMISTRY OF THORIUM
ANALYTICAL CHEMISTRY OF URANIUM
ANALYTICAL CHEMISTRY OF YTTRIUM AND THE LANTHANIDE ELEMENTS
ANALYTICAL CHEMISTRY OF ZIRCONIUM AND HAFNIUM

ANALYTICAL CHEMISTRY OF

PLUTONIUM

M. S. MILYUKOVA, N. I. GUSEV,

I. G. SENTYURIN, and I. S. SKLYARENKO

Translated by J. SCHMORAK

ANN ARBOR–HUMPHREY SCIENCE PUBLISHERS

ANN ARBOR · LONDON · 1969

ANN ARBOR–HUMPHREY SCIENCE PUBLISHERS, INC.
Drawer No. 1425, 600 S. Wagner Road, Ann Arbor, Michigan 48106

ANN ARBOR–HUMPHREY SCIENCE PUBLISHERS, LTD.
5 Great Russell Street London W.C. 1

© 1967 Israel Program for Scientific Translations Ltd.
© 1969 Ann Arbor–Humphrey Science Publishers, Inc.

Library of Congress Catalog Card Number 70–104367
SBN 250 39918 0

This book is a translation of
ANALITICHESKAYA KHIMIYA PLUTONIYA
Izdatel'stvo "Nauka"
Moskva 1965

FOREWORD

The Vernadskii Institute of Geochemistry and Analytical Chemistry of the USSR Academy of Sciences has begun the publication of a series of monographs on the analytical chemistry of individual elements. This series, *Analytical Chemistry of the Elements,* will include about fifty volumes, and it is planned to complete the publication during the coming five years. The necessity for such a project has been felt for a long time. We also have at our disposal the accumulated experience of numerous laboratories which now can (and should) be summarized. In this way the present series originated, the first of its kind. The analytical chemistry of any element with its different compounds is at present extremely varied; this is due to the complexity of the modern materials investigated and the wide range of concentrations determined, as well as to the variety of the methods employed.

With this in mind, a general plan has been worked out for the present series, both with respect to the contents and to the presentation of the material.

The monographs contain general information on the properties of the elements and their compounds, followed by a discussion of the chemical reactions which are the basis of the analytical work. The physical, physicochemical, and chemical methods for the quantitative determination of the element are given in the following order: first, the analysis of raw materials, next the analysis of the typical semi-manufactured products, and last, that of the finished products—metals or alloys, oxides, salts, and other compounds and materials. The underlying principles are always explained; whenever necessary, the exhaustive description of the entire analytical procedure is given. Due attention is paid to rapid analytical methods. A separate place is allotted to techniques for the determination of impurities in high purity materials.

Particular stress is placed on the accuracy and the sensitivity of the various methods, in view of the general tendency toward increased sensitivity of determination of traces of impurity elements.

The monographs contain an exhaustive and up-to-date bibliography. They are intended for a wide circle of chemists: in the first place, for the analysts of research institutes and industrial laboratories of various branches of the national economy, and also for teachers and students of chemistry in institutions of higher education.

Our most prominent experts participated in the preparation of the monographs, drawing upon their own extensive experience in the analytical chemistry of the element in question.

Each volume of the series will be published separately as soon as it has been made ready for publication. [A list of currently available volumes will be found facing the title page. Further titles are in preparation.]

We invite all our readers to send us their comments and criticisms on the monographs.

Editorial Board

CHIEF EDITOR

Academician A. P. Vinogradov

EDITORIAL BOARD

I. P. Alimarin, A. K. Babko, A. I. Busev, A. P. Vinogradov
A. N. Ermakov, V. I. Kuznetsov, P. N. Palei,
D. I. Ryabchikov, I. V. Tananaev, Yu. A. Chernikhov

EDITOR OF THIS VOLUME

P. N. Palei

ADDRESS OF THE EDITORIAL BOARD

Moskva, V-334, Vorob'evskoe shosse, 47a,
Institut Geokhimii i Analiticheskoi Khimii im. Vernadskogo
Akademii Nauk SSSR

TABLE OF CONTENTS

PREFACE

The theory and practice of analysis of one of the most important products of nuclear industry—plutonium—has not been adequately treated so far in Soviet scientific literature. Most studies on the analytical chemistry of plutonium carried out by Soviet scientists have remained unpublished, while foreign publications on the subject are not easily accessible.

The layout of the present monograph is in conformity with the general outline of all monographs on analytical chemistry published by the Vernadskii Institute of Geochemistry and Analytical Chemistry of the USSR Academy of Sciences. Considerable space has been devoted to the theoretical introduction which includes a description of the analytically important properties of plutonium and its compounds. The methods of determination and separation, described in subsequent chapters, are all based on these properties as also is the determination of impurities.

The monograph is a collective work of a group of scientists with extensive experience in developing analytical methods for plutonium. For this reason it is only possible to quote the most important contributions of each co-author to the individual chapters. Thus, the sections *Isotopes of Plutonium and their Properties*, *Behavior of Plutonium Ions in Aqueous Solutions*, *Toxic Properties of Plutonium*, *Working Techniques*, *Chromatographic Separation of Plutonium* and *Analysis of Preparations and Alloys of Plutonium*, were written by N. I. Gusev; I. G. Sentyurin wrote the sections *Valency States, Electron Configuration and Location in the Periodic System*, *Electrochemical Methods* and *Titrimetric Methods*; I. S. Sklyarenko wrote the sections *Metallic Plutonium, Its Preparation and Properties*, *Compounds of Plutonium*, *Gravimetric Methods* and *Separation by Precipitation with Inorganic and Organic Reagents*; M. S. Milyukova wrote the sections *Detection of Plutonium*, *Radiometric Methods*, *Colorimetric and Spectrophotometric Methods* and *Separation of Plutonium by Extraction*, and she also took the leading part in the compilation of the bibliographical material.

The monograph does not include the recent physical methods of analysis based on the nuclear reactions of plutonium nor the methods for the determination of the isotopic composition, as these require special treatment.

The work on the manuscript was concluded in 1963. In the two years which have since elapsed, many new studies on the analytical chemistry of plutonium were published and others were released for publication. In this connection, *Selected*

Measurement Methods for Plutonium and Uranium in Nuclear Fuel Cycle, by R.J. Jones (1963) and the chapter *Plutonium* in *Analysis of Essential Nuclear Reactor Materials* by C.J. Rodden (1964) deserve special mention. In order not to hold up the publication of the monograph, the Editorial Board decided to publish in the near future a supplementary literature survey on the analytical chemistry of plutonium.

P. Palei

CHAPTER I

General Information

In 1940 McMillan and Abelson [541] noted that neutrons interact with the nuclei of the main isotope of uranium—U^{238}—with the formation of the first transuranic element, neptunium. By using various nuclear reactions, other representatives of the actinide series were subsequently prepared, with atomic numbers higher than 93, the terminal number being lawrencium, $_{103}Lw$.

Plutonium was detected towards the end of 1940 by a group of American workers [645, 653, 660] as the product of the bombardment of uranium by deuterons:

$$_{92}U^{238}(d,2n)_{93}Np^{238} \xrightarrow{\beta^-} {}_{94}Pu^{238}. \tag{1}$$

The isotope Pu^{238} helped to clarify the principal properties of plutonium [659].

Later, Kennedy, Seaborg, Segre and Wahl [492, 493] detected a second isotope of plutonium, Pu^{239}, formed by the β-decay of the previously discovered Np^{239}:

$$U^{238}(n,\gamma)U^{239} \xrightarrow[23.5\text{ min}]{\beta^-} Np^{239} \xrightarrow[2.33\text{ days}]{\beta^-} Pu^{239}. \tag{2}$$

It was found that Pu^{239} is fissioned by slow neutrons somewhat more readily than uranium-235. This discovery stimulated an intensive study of the properties of plutonium and its compounds. During 1941–1943, prolonged irradiation of uranium by slow neutrons made it possible to prepare and isolate in the pure form microgram [367] and then milligram amounts of the isotope Pu^{239}. At present Pu^{239} is manufactured according to reaction (2) on an industrial scale.

ISOTOPES OF PLUTONIUM AND THEIR PROPERTIES

At present, 15 isotopes of plutonium are known (Table 1). The importance of the individual isotopes in nuclear technology, analytical chemistry and chemical research is highly variable. The chemist will not encounter in his work the light isotopes, which undergo electron capture (Pu^{232}, Pu^{234}, Pu^{235}, Pu^{237}). Plutonium-238 is of importance only as an isotopic power source for spacecraft. The most important and the most easily accessible isotope is undoubtedly Pu^{239}, with which the analytical chemist mostly has to work. However, the high specific activity of Pu^{239} is responsible for undesirable radiolytic effects in solutions and solid compounds, which render difficult a detailed study of the chemical properties of plutonium. From this point of view the long-lived isotopes Pu^{242} and Pu^{244} are most useful.

The preparation of Pu^{239} in nuclear reactors necessarily involves a large number of reactions resulting in the formation of other isotopes of plutonium and other transuranides, and also various fission products. This obviously affects the purity of the plutonium and raises complex technological problems of isolation of plutonium and analysis of preparations of various isotopic compositions.

Table 1

ISOTOPES OF PLUTONIUM

Isotope	Atomic weight*	Type of decay; energy, MeV	Half-life period**	Activity, disint./min · μg	Method of preparation
Pu^{232}	232.0411	e.c ($\leqq$ 98 %) α ($\geqq$ 2 %, 6.58)	36 min	—	U^{235} (α, 7*n*) U^{233} (α, 5*n*)
Pu^{233}	233.0427	e.c. (> 99 %) α (0.12 %, 6.30)	20 min	—	U^{233} (α, 4*n*)
Pu^{234}	234.0433	e.c. (94 %) α (6 %, 6.19)	9.0 hr	—	U^{235} (α, 5*n*) U^{233} (α, 3*n*) Cm^{238} α-decay
Pu^{235}	235.0453	e.c. (> 99 %) α ($3 \cdot 10^{-3}$ %, 5.85)	26 min	—	U^{235} (α, 4*n*) U^{233} (α, 2*n*)
Pu^{236}	236.0461	α (5.763; 5.716; 5.610; 5.448), β-stable	2.85 years	—	U^{235} (*d*, *n*) U^{235} (α, 3*n*) U^{238} (α, 6*n*) Cm^{240} α-decay Np^{237} (*n*, 2*n*) Np^{236} β-decay
Pu^{237}	237.0483	e.c. (> 99 %) α ($3.3 \cdot 10^{-3}$ %, 5.65, 5.36)	45.6 days	—	Np^{237} (p, *n*) U^{235} (α, 2*n*) U^{238} (α, 5*n*)

Table 1 (continued)

Isotope	Atomic weight*	Type of decay; energy, MeV	Half-life period**	Activity, disint./min · μg	Method of preparation
Pu^{238}	238.0495	α (0.1%, 5.352; 28%, 5.452; ~ 72%, 5.495) β-stable	86.4 years [466, 685]	$38.5 \cdot 10^6$	U^{238} $(d, 2n)$ Np^{238} β-decay Np^{237} (n, γ) Np^{238} β-decay Cm^{242} α-decay U^{235} (α, n) U^{238} $(\alpha, 4n)$
Pu^{239}	239.0522	α (69%, 5.150; 20%, 5.137; 11%, 5.099)	24390 years [87, 647]	$1.36 \cdot 10^5$ 87	U^{238} (n, γ) U^{239} $\xrightarrow{\beta}$ Np^{239} β-decay U^{238} (d, n) Np^{239} β-decay [478, 662] U^{238} $(\alpha, 3n)$ [478] Pu^{238} (n, γ)
Pu^{240}	240.054	α (76%, 5.162; 24%, 5.118), β-stable	6620 years [57, 647]	$5 \cdot 10^5$ 57	Pu^{239} (n, γ) U^{238} (n, γ) U^{239} (n, γ) U^{240} $\xrightarrow{\beta}$ Np^{240} β-decay
Pu^{241}	241.0567	α (1.5%, 5.036; 1.4%, 4.967; 83%, 4.893; 13%, 4.856; 11%, 4.799) [86] β^- (99%, 0.021)	$5.6 \cdot 10^5$ years [677] 13.30 years [318, 677]	5900 $2.48 \cdot 10^8$	Pu^{240} (n, γ) [426, 641, 65]
Pu^{242}	242.0587	α(4.898; 4.854) β-stable	$3.73 \cdot 10^5$ years [332, 687]	8800	Pu^{241} (n, γ) [694] Am^{242} e.c. [332]
Pu^{243}	243.062	α-stable β^- (12%, 0.37; 35%, 0.468; 53%, 0.57)	4.98 hr [393]	$5.7 \cdot 10^{12}$	Pu^{242} (n, γ) [688, 695]
Pu^{244}	—	α	$7.6 \cdot 10^7$ years [379]	42	Pu^{243} (n, γ) Am^{244} e.c. 19
Pu^{245}	—	α-stable β^-	12 hr [402]	—	Pu^{244} (n, γ) [402]
Pu^{246}	—	α-stable β^- (0.33; 0.15)	10.85 days [392, 647]	—	Pu^{245} (n, γ) [321, 392]

* Atomic weights of isotopes quoted from [119].

** Spontaneous fission half-lives are given in [19, 93, 231, 647].

Note: Data for which no literature reference is indicated have been taken from [3, 231, 647].

The relative content of Pu^{239} and the heavier isotopes Pu^{240}, Pu^{241}, Pu^{242}, and Pu^{238}, Np^{237}, Am^{241}, as well as other transplutonium elements, depends on the dose of the neutron irradiation of uranium. The capture cross sections of plutonium isotopes for the reactor neutrons are sufficiently large to produce successive (n, γ) reactions even if the concentration of Pu^{239} in uranium is small. Table 2 gives data illustrating the formation of plutonium isotopes when U^{238} is irradiated with neutrons.

Table 2

ISOTOPIC COMPOSITION OF PLUTONIUM ISOLATED FROM IRRADIATED NATURAL URANIUM [449]

Amount of plutonium formed, g Pu / kg U	Isotope content in plutonium, wt.%			
	Pu^{239}	Pu^{240}	Pu^{241}	Pu^{242}
0.5	97	2.9	0.14	—
1.0	93.6	5.6	0.60	0.02
1.5	90	8.4	1.4	0.06

Plutonium isolated from irradiated uranium contains Pu^{239} as the main component and also Pu^{240} as the main admixture; the latter isotope is formed chiefly by the reaction Pu^{239} (n, γ) Pu^{240}.

The content of Pu^{241} in irradiated uranium is low and may be calculated by the formula (Dokuchaev, 1954)

$$y = 0.0168 \cdot x^{2.05}, \tag{3}$$

where y is the content of Pu^{241} on the final day of irradiation and x is the content of Pu^{240} as determined by spectrometry, in weight percent.

The data in Table 2 are in good agreement with calculations carried out by this formula. Plutonium-241 undergoes mainly β-decay with the formation of Am^{241}, which is discarded during the chemical and technological processing of uranium slugs. (Am^{241}, however, gradually accumulates again in the purified plutonium. Thus, for instance, the α-activity of metallic plutonium containing 7.5% Pu^{240} increased by 2% per year because of the formation of Am^{241}.) Plutonium-241 has a large fission cross section for reactor neutrons—about 1,100 barns [19]—which is of importance in the utilization of plutonium as nuclear fuel.

Together with heavy plutonium isotopes, Pu^{238} is also formed in a nuclear reactor:

$$U^{238}(n,2n)\,U^{237} \xrightarrow{\beta^-} Np^{237}; \tag{4a}$$

$$U^{235}(n,\gamma)\,U^{236}(n,\gamma)\,U^{237} \xrightarrow{\beta^-} Np^{237}; \tag{4b}$$

$$Np^{237}(n,\gamma)Np^{238} \xrightarrow{\beta^-} Pu^{238}. \tag{4c}$$

The yield of Pu^{238} may be calculated from the formula of Ya. P. Dokuchaev (1954):

$$y = (1.2 + 0.59 \cdot x^{1.66}) \cdot 10^{-3}, \tag{5}$$

where y is the content of Pu^{238} in wt.% on the last day of the irradiation, and x is the content of Pu^{240} in wt. %. Equations (3) and (5) have been checked for values of x between 0 and 7.5%. Calculations by formula (5) indicate that the concentration of Pu^{238} in a sample containing 5.6% Pu^{240} (Table 2) is 0.0115%. This represents a relatively large contribution to the overall α-activity of the preparations, since the half-life of Pu^{238} is 86.4 years.

The Pu^{236} yield in uranium irradiation is $10^{-9}-10^{-8}$%. The formation reaction is

$$Np^{237}(n,2n)\,Np^{236} \xrightarrow{\beta^-} Pu^{236}.$$

The long-lived plutonium isotopes, Pu^{242} and Pu^{244}, are formed during prolonged (100 days or more) irradiation of Pu^{239} by neutrons. The yield of Pu^{242} is several tenths of one percent [19, 106], while the yield of Pu^{244} is a fraction of one percent of the yield of Pu^{242} [218, 686] (Figure 1). Americium, curium, other transplutonides and fission elements are produced at the same time.

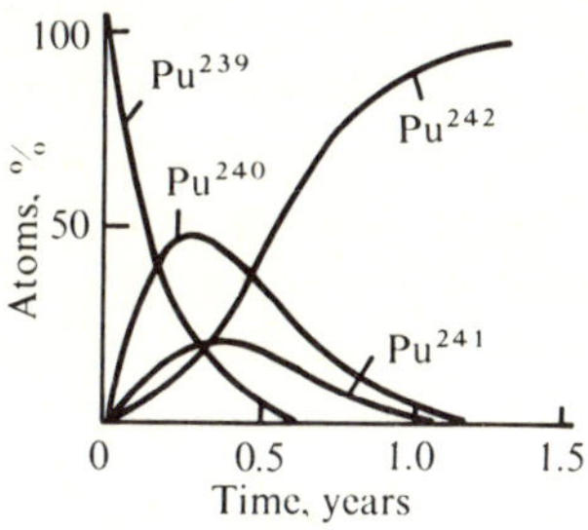

Figure 1

Variation of the relative proportion of various plutonium isotopes during prolonged irradiation of Pu^{239} by a flux of $3 \cdot 10^{14}$ neutrons/cm^2 · sec. The figure does not show the yields of transplutonides and fragment elements, which are formed in large amounts.

Butler et al. [332] made a detailed study of the irradiation of Am^{241} by neutrons, resulting in the formation of Pu^{238} and Pu^{242}.

$$\begin{array}{ccc} Cm^{242} & \xrightarrow[162.7\ \text{days}]{\alpha} & Pu^{238} \\ \uparrow \beta^-\ 64\% & & \\ Am^{241}(n,\gamma)Am^{242m} & \xrightarrow[18\%]{\text{Internal conversion}} & Am^{242}. \\ \downarrow \text{e.c.}\ 18\% & & \\ Pu^{242} & & \end{array} \tag{6}$$

These authors isolated by anion exchange a product containing 2% Pu^{238} and 98% Pu^{242} from Am^{241} which had been irradiated for two days; this indicates that it is possible to obtain plutonium preparations with a relatively low specific activity.

Plutonium-244 is the most stable isotope and is therefore of interest for chemical research. Its specific activity is 3000 times that of Pu^{239}. Unfortunately, no method is so far known for preparing isotopically pure Pu^{244} in significant amounts.

OCCURENCE IN NATURE

According to most recent data, the age of the Earth is about $5.5 \cdot 10^9$ years [201]. For this reason only those radioactive isotopes whose half-life period is more than 10^8 years are encountered in natural formations in significant amounts. Short-lived isotopes are constantly formed as a result of the decay of elements with a higher atomic number or as a result of reactions involving capture of elementary particles.*

A long time before the discovery of transuranides, numerous fruitless attempts were made to find them in nature. These studies were continued soon after the discovery of neptunium and plutonium. In the actinide series, the half-life period of α-active isotopes decreases with the increase in the atomic number, so that it was doubtful if detectable amounts of neptunium or plutonium would be found in the Earth's crust. This has been repeatedly confirmed [210, 424, 521, 590, 655, 657]. The content of Pu^{239} in pitchblende and monazite ores is in the ratio $Pu^{239} : U = (0.7-2) \cdot 10^{-11}$. It is believed that Pu^{239} is formed from U^{238} by the action of neutrons evolved as a result of the spontaneous fission of U^{238}, and also during the fission of U^{235} and through (α, n) reactions of light elements. The contribution of the (α, n) reactions seems to be the most significant.

The likelihood of discovering the long-lived isotope Pu^{244} ($T_{\frac{1}{2}} = 7.6 \cdot 10^7$ years) in nature is also small, since its half-life period is almost 100 times less than the age of the Earth. The search for this isotope in thorium ores, based on the assumed similarity in the behavior of Pu(IV) and Th(IV) in the process of ore formation involves serious difficulties. Assuming comparable initial contents of Th^{232} and Pu^{244}, 1 g of Th in an ore would correspond to about 10^{-18} g of Pu^{244}.

VALENCY STATES, ELECTRON CONFIGURATION AND LOCATION IN THE PERIODIC SYSTEM

Plutonium is located in the seventh period of the periodic system following uranium and neptunium. As regards the location of these elements in the periodic table, the

* For the formation and decay of radioactive elements in nature, see [239, 502].

theory of Seaborg [3, Ch. 17; 170, 203, Ch. 11; 646, 648] is now the most generally accepted. According to this theory, in the sequence of elements beginning formally with thorium and ending with lawrencium the inner energy level $5f$ is gradually filled with 14 electrons. Since the number of the outer valency electrons (one $6d$ and two $7s$ electrons) remains unchanged and equal to the number of valency electrons of actinium, the physical and chemical properties of all elements in the series should be similar, and the series itself received the name of actinides. A similar effect is conspicuously apparent in lanthanides, which have the electron structure $(Xe)4f^n5d6s^2$ and a principal valency of three.

The chemical properties of the elements of the transactinide series are much more variegated than those of the lanthanides. The most important property, next to the atomic number, which determines the position of an element in the periodic table is its valency. Table 3 lists the various valencies of elements ranging from actinium to mendeleevium. The valency of three, typical for actinides, becomes the principal valency beginning with americium. For thorium and protactinium this valency is quite atypical. The principal valency of the first elements in the series gradually increases to six, and is retained as the maximum valency in aqueous solutions for the transuranides up to americium.

Table 3

VALENCIES OF TRANSACTINIDE ELEMENTS

Atomic number	89	90	91	92	93	94	95	96	97	98	99	100	101
Element	Ac	Th	Pa	U	Np	Pu	Am	Cm	Bk	Cf	Es	Fm	Md
Valencies*	3	(3)		3	3	3	3	3	3	3	3	3	3
		4	4	4	4	4	(4)	(4)	4				
			5	5	5	5	5						
				6	6	6	6						

* Valencies shown in brackets are encountered in the solid states only.

Of special interest in the series is plutonium, which can exist in solution in several states of oxidation at the same time. This is explained by the disproportionation of the intermediate states of oxidation, which does not proceed to completion owing to the similar values of the redox potentials of the various ion pairs of plutonium. Plutonium ions with a charge of 4+ are the most stable.

In the absence of hydrolysis or complex formation, ions of transactinide elements, including plutonium, with a charge of 3+ and 4+ exist in aqueous solutions as strongly hydrated cations. Pentavalent and hexavalent ions exist in acid solutions as the oxygenated cations MO_2^+ and MO_2^{2+} and have a linear structure.

The nature of the redox states of the elements is closely connected with their electron configurations. Table 4 gives the structure of the electron shells of the gaseous elements from actinium to lawrencium inclusive, obtained partly from spectroscopic data and also electron configurations in the metallic state.

On inspecting the electron configurations of this series, a number of general relationships may be distinguished. In the first place, filling the 5*f* shell probably does not begin with thorium. Thorium is a typical representative of the transition series of 6*d* elements and is analogous to Zr and Hf. Secondly, subsequent elements, from protactinium to plutonium, have mixed type 5*f*6*d* electron configurations, which differ from the $4f^n5d6s^2$ configurations of the lanthanides which, according to the actinide theory, would be their analogs. It should also be noted that the electron configurations in the gaseous and in the solid state are not necessarrily identical.

Table 4

ELECTRON CONFIGURATIONS OF GASEOUS AND METALLIC TRANSACTINIDES*

Atomic number	Element	Electron configuration in the gaseous state	Ground term	Probable electron configuration in the metallic state	References
89	Ac	$6d7s^2$	$^2D_{3/2}$	$6d7s^2$	[545]
90	Th	$6d^27s^2$	$^3F_2$	$6d^27s^2$	[203, Ch. 11]
91	Pa	$5f^26d7s^2$ or $5f^16d^27s^2$	$^4K_{11/2}$ or $^4I_{9/2}$	$(5f6d)^37s^2$	[203]
92	U	$5f^36d7s^2$	$^5L_6$	$(5f6d)^47s^2$	[494]
93	Np	$5f^46d7s^2$	$^6L_{11/2}$	$(5f6d)^57s^2$	[202]
94	Pu	$5f^67s^2$ or $5f^56d7s^2$	$^7F_9$ or $^7K_4$	$(5f6d)^67s^2$	[293, 434, 542]
95	Am	$5f^77s^2$	$^8S_{7/2}$	$5f^66d7s^2$	[413]
96	Cm	$5f^76d7s^2$	$^9D_2$	$5f^76d7s^2$	[203, Ch. 11]
97	Bk	$5f^86d7s^2$ or $5f^97s^2$	$^8H_{17/2}$ or $^6H_{15/2}$	$5f^86d7s^2$	
98	Cf	$5f^{10}7s^2$	$^5I_8$	$5f^96d7s^2$	
99	Es	$5f^{11}7s^2$	$^4I_{15/2}$	$5f^{10}6d7s^2$	
100	Fm	$5f^{12}7s^2$	$^3H_6$	$5f^{11}6d7s^2$	
101	Md	$5f^{13}7s^2$	$^2F_{7/2}$	$5f^{12}6d7s^2$	
102	—	$5f^{14}7s^2$	$^1S_0$	$5f^{13}6d7s^2$	
103	Lw	$5f^{14}6d7s^2$	$^2D_{5/2}$	$5f^{14}6d7s^2$	

* Outside the radon configuration.

Note: Configurations shown in brackets vary with the physical state of the element.

In the gaseous state there is a stabilization of the *f* electrons at the expense of the *d* electrons present in the condensed state. Thirdly, with increasing atomic number there is a general tendency to an increase in the binding strength of the *f* elections. Curium has a particularly stable seven-electron configuration, corresponding to the half-filled *f* shell. The only valency displayed by curium in solution is three.

Beginning with americium, the electron configurations of the actinides seem to correspond to the configurations of the lanthanides, in full conformity with the lanthanide theory. The differences between the properties of $5f$ electrons of the actinides and those of $4f$ electrons of the lanthanides are due to the electron configurations and valency states of the heavy elements. The binding energy of $5f$ electrons is small as compared with that of $6d$ electrons. Consequently, the first few elements in the series—Th, Pa and U—may lose all their valency electrons, including the $5f$ electrons, with the formation of multiply-charged ions which are stable to reduction. In the elements which follow in the series, the binding energy of $5f$ electrons is still within the limits of the chemical bond energy, owing to which neptunium, plutonium and americium may display the high valency of six. Even curium, which has a relatively stable seven-electron configuration in the $5f$ shell, forms the tetravalent compounds CmO_2 and CmF_4 by splitting off one $5f$ electron.

The above notwithstanding, the result of the stabilization of the $5f$ electron layer is that, beginning with americium, the main valency state of the elements is three. In this state the elements of the second half of the series are very similar to the corresponding lanthanide elements. This analogy was particularly conspicuous in the ion-exchange properties, which served to make a reliable identification of all newly prepared transcurium elements up to and including mendeleevium. Even though element 102 and lawrencium have not been chemically identified because of their short half-life period, they should also have a principal valency of three, and lawrencium should be the last element of the $5f$ series. The series could not terminate, say, with element 104, since it is difficult to imagine that the last elements of the $5f$ series would continue to fill the $6d$ level before the completion of the stable 14-electron structure in the $5f$ shell. The series could possibly terminate before lawrencium, but even then element 102 would have the valency electron configuration of radium. The weaker binding strength of $5f$ as compared with $4f$ electrons also indicates that the principal valency of element 102 will be 3 and not 2.

Even though the actinide theory made an accurate forecast of the chemical properties of transcurium elements, it failed to provide a satisfactory explanation for the behavior of the first, more thoroughly studied, half of the elements in the series. The main difficulty is that the principal valency of the first five elements following actinium is higher than three. The valencies displayed by Th, Pa, U, Np, Pu and Am are no longer minor deviations from the principal valency of three, as is the case with lanthanides, but seem to vary according to an independent law. The electron configurations and the chemistry of these elements, as well as the need for a continuous placing of the elements in the periodic table in accordance with their atomic numbers, all indicate the necessity for a different approach.

One such approach is to consider the elements from thorium to lawrencium, in accordance with the true state of things, as a *mixed 5f 6d series*. The first member of the series—thorium—has the $6d^2 7s^2$ configuration. In the following few elements of the series there is a rearrangement of the electron configuration of thorium into the $5f^7 7s^2$ configuration of americium, and it is precisely the elements with the mixed configuration that display the large number of valencies up to a valency of six. Since the $5f$ sequence itself has been shortened, sudden changes in the chemical properties may occur due to the exchange of d for f electrons. Thus, for instance, there are sharp differences in the chemical properties of protactinium and uranium.

The rearrangement of the electron structure in the first few elements of the sequence leads to a concept altogether different from that of the actinide theory. The behavior of all elements in the series can be described not in terms of one precursor element, but in terms of the properties of the first few members of the series, viz., Ac, Th, Pa and U. From this point of view, the group of elements from Th to Lw might be referred to as the actinide-uranide group. It is significant that the series of $3d$, $4d$ and $5d$ transition elements, including hafnium, tantalum and tungsten, display an analogy (albeit incomplete) with the actinide-uranide series.

At present extensive experimental data is available on the chemical behavior of the heavy elements; a detailed discussion of the material from the point of view of the actinide theory will be found in the book by Seaborg and Katz [203]. It should be pointed out that most of this material is given as evidence for the existence of the $5f$-series, while evidence in favour of the actinide theory mostly concerns the trivalent elements. Nevertheless, the material presented is compatible with the existence of the actinide—uranide series.

The results of the determinations of magnetic susceptibility of ions, as well as those of paramagnetic absorption, make it possible to obtain a clear picture in the case of thorium which has no f electrons and for elements beginning with americium which have 6 or more f electrons. In the case of the intermediate elements the picture is more complex and the results can be accounted for in many cases both by the presence of f and of d electrons. This applies in particular to uranium, the behavior of which is more simply interpreted in terms of the presence of $6d$ electrons.

A large volume of data on the configuration of the electron shells of heavy elements was obtained on studying the optical absorption spectra of ions in solutions and in crystals. These spectra display sharp absorption bands in the visible and nearby regions. The corresponding electron transitions take place inside the $5f$ shell owing to the interaction with the electrostatic fields of the neighboring ions. In the middle of the series the overall absorption spectrum of the elements becomes more complex. We may distinguish isoelectronic series of ions with similar absorption spectra: U^{4+}, NpO_2^+ and PuO_2^{2+} (with two $5f$ electrons) and U^{3+}, Np^{4+} and PuO_2^+ with the configuration $5f^3$. Because of the similarity between the spectra of Eu^{3+} and Am^{3+} it was concluded that Am^{3+} has six $5f$ electrons.

It was concluded from the analysis of the optical absorption spectra of the ions [481] that in uranium and transuranic elements the $5f$ electrons have much lower energy content than the $6d$-electrons. Figure 2 shows the difference between the energies of the lowest term of the configuration $5f^{n-1}6d$ and the ground state of the $5f$ configuration as a function of the atomic number of the element. The function also shows that the stability of the f electrons as compared with the d electrons increases with increasing degree of oxidation. The main difference between the optical absorption spectra of transuranides and those of the lanthanides is that the intensity of the spectra of transuranides is 10–1000 times higher, and the bands are often much wider than those of lanthanides. In this respect uranium and transuranides occupy an intermediate position between the transition elements and the lanthanides.

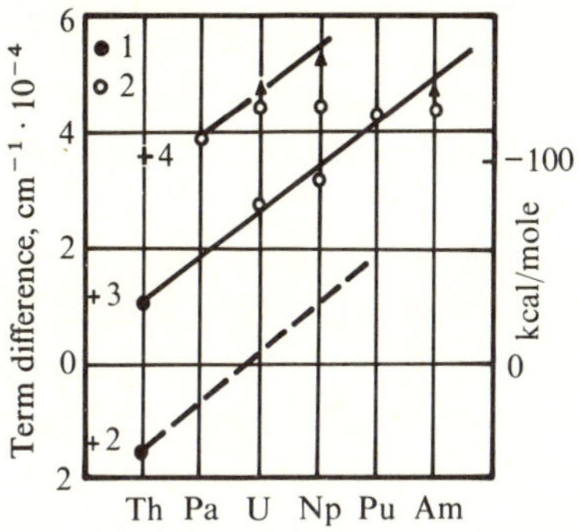

Figure 2

Difference between the lowest terms of the configurations $5f^{n-1}$ $6d$ and $5f^{n}$:

1) values obtained from atomic spectra; 2) from light absorption spectra of aqueous solutions.

From the point of view of the concept of the *mixed 5f 6d series*, valencies below three and the metallic state are of interest.

A number of bivalent transactinides are known: monoxides of the elements from thorium to americium, which are deposited as a film on the metal surface, thorium, uranium and plutonium monosulfides and diiodides. Bivalent compounds cannot exist in solution and occasionally display a distinctly metallic nature.

It was shown by Zachariasen [229] by crystallographic methods that all valency electrons in monoxides and monosulfides of elements up to and including uranium are $6d$ electrons, while in neptunium, plutonium and americium they are partly of the $5f$ type. The same applies to the sesquisulfides. Th_2S_3 and U_2S_3 have no nonbound $5f$ electrons, Np_2S_3 has no more than one, while in plutonium and americium compounds all nonbound electrons are $5f$ electrons.

Such behavior is not typical of lanthanides: Sm, Eu and Yb are normally bivalent, but have the properties of actinide-uranides. In the sesquisulfides thorium is the homolog of zirconium and uranium is the homolog of molybdenum.

Crystallographic studies make it possible to determine ionic and atomic radii, from which the types of bonding and the type of electrons may be estimated. The data given above on the lower valency states were obtained in this manner. Figure 3 gives the experimental and the calculated values of ionic (*a*) and metallic (*b*) radii of transactinide elements in various valency states [203]. The reduction of the atomic radius at equal valency states takes place irrespective of the type of bonding and is known as the actinide contraction; this is an indication that the 5*f* shell is being filled. Pentavalent and hexavalent compounds apparently have ionic type bonds only in the case of fluorides. In ions of the type MO_2^+ and MO_2^{2+} the bonding is mainly covalent and the covalent radii should be used in calculations.

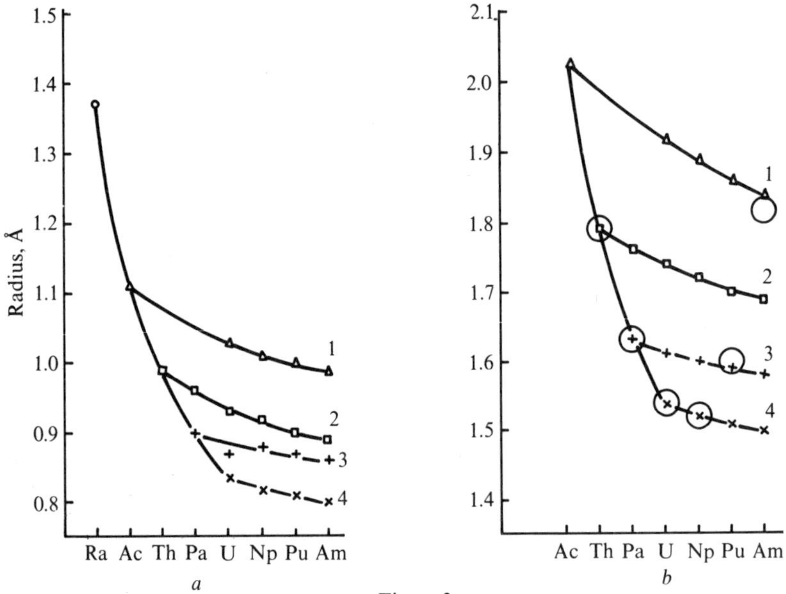

Figure 3

Ionic (*a*) and metallic (*b*) radii of actinide-uranides:

1–4) radii of tri-, tetra-, penta- and hexavalent actinide-uranides; the experimental radii are circled.

Zachariasen [203, Chap. 18, 229] suggested that tri-, tetra-, penta- and hexavalent ions of heavy elements be designated as separate series of *actinides*, *thorides*, *prot-actinides* and *uranides*. These series can be logically generalized in terms of the actinide-uranide concept. Since during ionization the first electrons to be split off are the *s* and *d* electrons and a stabilization of the *f* electrons takes place at the expense of the remaining *d* electrons, the resulting ions in their normal valency states may contain no 6*d* electrons. From this point of view the electron structures of the elements in the metallic state are the most illustrative. The values of the metallic radii (Figure 3) indicate that Th, Pa and U have no *f* electrons. The first 5*f* electron appears only in neptunium. Figure 4 [420] shows the fields of existence of crystal structures and the type of electrons at various temperatures for elements up to and including americium. Thorium and protactinium display a pure *d* orbital behavior, while americium and the following elements display an *f* behavior. The intermediate elements uranium, neptunium and plutonium display a combined (*df*) behavior and exist in numerous allotropic forms. Since the difference between the *d* and *f* energy levels is small, the distribution of the electrons over the levels may vary from one allotropic form to another and from one compound to another. All these data form a consistent picture and confirm the validity of the actinide-uranide concept.

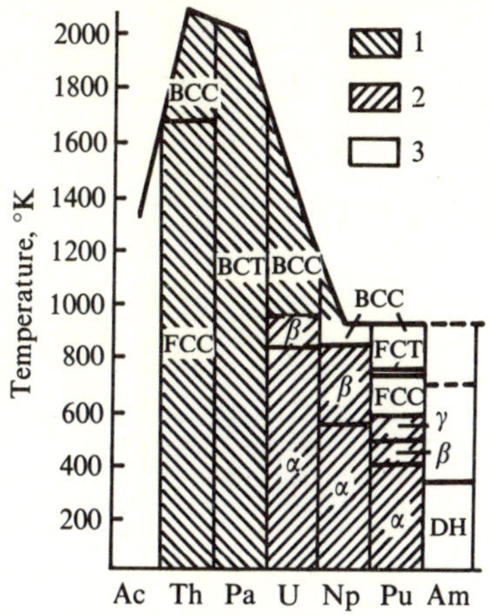

Figure 4

Fields of existence of crystalline structures of actinide-uranides.

Shaded and white areas indicate: 1) *d*-orbital; 2) *df*-hybrid and 3) *f*-orbital behavior. BCC body-centered cubic structure; FCC face-centered cubic structure; BCT body-centered tetragonal structure; FCT face-centered tetragonal structure; DH dihexagonal structure.

The last column of Table 4 gives the probable electron configurations in the metallic state which would correspond to the behavior described above. Protactinium seems to have three *d* electrons. The ratio between the number of 5*f* and 6*d* electrons in uranium, neptunium and plutonium may vary with the physical state of the element.

The continuity of the mixed 5*f* 6*d* series makes it possible to place the elements conveniently in the periodic table (p. 14). The first elements in the series are separated and located under the appropriate groups of the table in accordance with the principle of maximum valency. The remaining elements, the properties of which are within the limits of those of the first elements, are arranged in two series placed one underneath the other in order to stress the internal periodicity of the series. Such placement of actinide-uranides preserves all the internal analogies within the series as well as the analogy with the transition *d* series. This last analogy is even more striking owing to another typical property of 5*f* electrons, viz., their large orbital radii. As a result, 5*f* electron can participate in chemical bonding, and the elements themselves have an

PERIODIC TABLE OF ELEMENTS

Periods	Groups I	II	III	IV	V	VI	VII	VIII			0
1	1 H 1.00797						(H)				2 He 4.0026
2	3 Li 6.939	4 Be 9.0122	5 B 10.811	6 C 12.01115	7 N 14.0067	8 O 15.9994	9 F 18.9984				10 Ne 20.183
3	11 Na 22.9898	12 Mg 24.312	13 Al 26.9815	14 Si 28.086	15 P 30.9738	16 S 32.064	17 Cl 35.453				18 A 39.948
4	19 K 39.102	20 Ca 40.08	21 Sc 44.956	22 Ti 47.90	23 V 50.942	24 Cr 51.996	25 Mn 54.9381	26 Fe 55.847	27 Co 58.9332	28 Ni 58.71	
	29 Cu 63.54	30 Zn 65.37	31 Ga 69.72	32 Ge 72.59	33 As 74.9216	34 Se 78.96	35 Br 79.909				36 Kr 83.80
5	37 Rb 85.47	38 Sr 87.62	39 Y 88.905	40 Zr 91.22	41 Nb 92.906	42 Mo 95.94	43 Tc [99]	44 Ru 101.07	45 Rh 102.905	46 Pd 106.4	
	47 Ag 107.870	48 Cd 112.40	49 In 114.82	50 Sn 118.69	51 Sb 121.75	52 Te 127.60	53 I 126.9044				54 Xe 131.30
6	55 Cs 132.905	56 Ba 137.34	Lantha- nides	72 Hf 178.49	73 Ta 180.948	74 W 183.85	75 Re 186.2	76 Os 190.2	77 Ir 192.2	78 Pt 195.09	
	79 Au 196.967	80 Hg 200.59	81 Tl 204.37	82 Pb 207.19	83 Bi 208.980	84 Po [209]	85 At [210]				86 Rn [222]
7	87 Fr [223]	88 Ra [226]	Actinide- uranides	(104)	(105)	(106)					

Period									
6	Lanthanides	57 La 138.91	58 Ce 140.12	59 Pr 140.907	60 Nd 144.24	61 Pm [147]	62 Sm 150.35	63 Eu 151.96	64 Gd 157.25
			65 Tb 158.924	66 Dy 162.50	67 Ho 164.930	68 Er 167.26	69 Tu 168.934	70 Yb 173.04	71 Lu 174.97
7	Actinide-uranides	89 Ac [227]	90 Th 232.038	91 Pa [231]	92 U 238.03	93 Np [237]	94 Pu [242]	95 Am [243]	96 Cm [245]
			97 Bk [247]	98 Cf [249]	99 Es [254]	100 Fm [253]	101 Md [256]	102 No [254]	103 Lw [257]

increased tendency to form complexes, which is typical of the transition *d* elements.

The placement of the seventh period elements as outlined above is based on views similar to those of Haisinsky [46]. According to this author, the elements in question have three types of electron configurations. Actinium, thorium and protactinium have no 5*f* electrons, whereas curium and transcurides have 5*f* electrons in a number corresponding to the actinide hypothesis. Uranium and uranides, i.e., Np, Pu and Am, have mixed 6*d*5*f* configurations, the 6*d* configuration being more conspicuous in uranium, wheras the 5*f* configuration is prominent in americium. It can be seen from what has just been said that there are many reasons for such a view to be accepted. The electron configurations of the elements from uranium to americium may vary with the physical state and valency state of the element. Moreover, the behavior of 5*f* electrons themselves is intermediate between that of 4*f* and 4*d* electrons. According to Haisinsky, these three groups of elements are more or less independent. Curium and curides are placed with actinium; thorium and protactinium are placed under hafnium and tantalum, while uranium, neptunium, plutonium and americium are placed underneath tungsten. The disadvantage of this concept is the impairment of the principle of continuity in placing the elements in the periodic table in accordance with their atomic numbers, which has so far been preserved. Moreover, the excessive stress laid on the similarity with the elements of Groups IV–VI, which is in fact incomplete owing to the contribution of the 5*f* electrons, results in an underestimation of the internal analogies and relationships within the series itself. It would appear that this point of view is just as extreme as the hypothesis of Seaborg.

CHAPTER II

Analytical Properties of Plutonium and Its Compounds

METALLIC PLUTONIUM, ITS PREPARATION AND PROPERTIES

Preparation

Metallic plutonium must satisfy especially stringent purity requirements as regards the absence of impurity elements with a higher neutron capture cross section. Some of these impurities (Nd, Sm, Eu, Gd, Ru) are fission products in nuclear reactors [101]. After separation from uranium and fission fragments and careful purification of plutonium oxide or salt, the metal may be prepared in several ways. Most of the methods are based on the reduction of plutonium oxide or a plutonium halide by alkaline-earth metals (Ca, Mg) [222, 229, 280, 283, 419, 435, 533, 614].

Metallic plutonium was prepared for the first time in 1943 [641, p. 65]. Plutonium tetrafluoride was reduced with barium vapor at 1,400° in a high vacuum. The initial material used in the early work [229, 280, 419, 435, 533] were plutonium fluorides which are less hygroscopic than the chlorides. However, the employment of chlorides for the preparation of metallic plutonium [614] makes it possible to reduce by a factor of about 100 the neutron flux produced in the reaction of α-radiation with halogens. This makes the working conditions much more convenient. Anhydrous halides are prepared mainly from plutonium dioxide; for this reason the preparation of plutonium by direct reduction of PuO_2 [283] is of interest. The reducing agents (Ca, Mg, Ba) must meet high standards of purity. Their preparation is described in [43].

Metallic plutonium prepared from halides contains, as a rule, about 1% of impurities. One technique of supplementary purification is zone melting, by means of which exceptionally pure plutonium may be obtained [220]. The preparation of plutonium by electrolysis of molten chlorides has recently been proposed [112, 300]. The electrolytic method yields a 99.98% pure plutonium [570]. An interesting method is the high-temperature decomposition of PuC_2 with the distillation of plutonium, resulting in a product of high purity [222, 539]. This method involves, however, a number of technical difficulties.

All the methods listed above yield a metallic powder, which is smelted into ingots. Solid plutonium can be obtained directly by reducing plutonium chloride by metallic calcium in the presence of iodine.

Physical properties

Metallic plutonium is a heavy, brittle, silvery-white metal. When heated from room temperature to the melting point, plutonium undergoes five phase transformations and may consequently exist in six allotropic modifications [115, 118, 287, 524, 673]. The most reliable information on the structure and properties of the phases of plutonium will be found in [25, 116, 220, 285, 459, 482, 520], which are studies conducted on high purity metal. The most important physical properties of metallic plutonium are given below.

Property	Value	Ref.
Melting point, °C	639.5	[647]
Boiling point, °C	3235	[647]
Phase transition temperatures, °C		
$\alpha-\beta$	122	[116]
$\beta-\gamma$	211	[116]
$\gamma-\delta$	315	[116]
$\delta-\eta$	400	[116]
$\eta-\varepsilon$	460	[116]
ε – liquid	635	[116]
Density, g/cm^3 (25°C)	19.737	[647]
Vapor pressure (between 1119 and 1520°C), mm Hg	$\log P = \frac{-17.587}{T} + 7.895$	[223]
Heat of evaporation, kcal/mole	80.46 ± 0.34	[647]
Thermal conductivity at 40°C, $cal/cm \cdot sec \cdot deg$	0.008	[220]
Electrical resistivity at 25°C, $\rho \cdot 10^6$ ohm · cm	145	[482]
Hardness, kg-force / mm^2	261	[470]

Discrepancies in the individual parameters reported by different authors [116, 220, 482, 674] are due to the unequal degree of purity of the metal employed. For more details on the physical and mechanical properties of metallic plutonium, see [33, 34, 203, 520].

Chemical properties

Plutonium is very active chemically. Atmospheric oxygen oxidizes metallic plutonium. The rate of the oxidation and the composition of the products thus formed strongly depend on the atmospheric moisture and temperature. Plutonium is slowly oxidized in dry air at 50°, forming a PuO film, which protects it from further oxidation [33]. It is more rapidly oxidized in a humid atmosphere, when the dioxide PuO_2 is formed [634]. When the temperature is increased to ~ 300°, plutonium ignites spontaneously in the air.

Metallic plutonium reacts with hydrogen at a significant rate even at 25–50°. At 200° the reaction becomes very rapid and is accompanied by the formation of PuH_3 [319, 640]. At 900° plutonium slowly reacts with nitrogen, but the yield of PuN is negligible [320].

Metallic plutonium vigorously reacts with halogens and hydrogen halides with the formation of the corresponding plutonium trihalides. It forms sulfides with gaseous sulfur compounds. The reaction between metallic plutonium and CO_2 yields mainly PuO_2.

Solubility in acids. Metal, containing not less than 99% plutonium, is readily and rapidly dissolved by hydrochloric acid of various concentrations. The dissolution of plutonium in 72% perchloric acid, 85% phosphoric acid and concentrated trichloroacetic acid is somewhat slower. The metal is not attacked by concentrated sulfuric or acetic acid, but reacts slowly with dilute solutions of these acids. In nitric acid plutonium becomes passivated and does not dissolve even when left in the acid for several hours. Solutions of alkalis do not dissolve plutonium to any significant extent.

Plutonium containing 5–10% fission products (Ce, La, Mo, Nd, Nb, Ru, Zr) slowly dissolves in hydrochloric acid at 300° [159].

Reactions with other metals

The nature of the reactions of plutonium with other elements is largely conditioned by its structure and physicochemical properties, which differ considerably from those of ordinary metals. By now, a large number of binary and ternary systems of plutonium with other metals have been investigated [25–27, 33, 115–117, 435, 620, 621].

Table 5 lists the binary compounds of plutonium with various metals so far reported in the literature.

Table 5

BINARY COMPOUNDS OF PLUTONIUM [26]

Metal	Compound	Metal	Compound	Metal	Compound
Cu	$PuCu_2$, $PuCu_4$, $PuCu_6$, PuCu,* $PuCu_3$,* $PuCu_7$*	C	PuC, Pu_2C_3	Bi	PuBi, $PuBi_2$
Ag	$PuAg_3$	Si	Pu_3Si_2, PuSi, Pu_2Si_3, $PuSi_2$, Pu_5Si_3*	Te	PuTe
Be	$PuBe_{13}$	Ge	Pu_2Ge_3, $PuGe_2$, $PuGe_3$	Mn	$PuMn_2$
Mg	PuMg,* $PuMg_2$*	Sn	$PuSn_3$	Fe	Pu_6Fe, $PuFe_2$
Hg	$PuHg_3$, $PuHg_4$	Pb	$PuPb_3$	Co	Pu_6Co, Pu_3Co, Pu_2Co, $PuCo_2$, $PuCo_3$, Pu_2Co_{17}
Al	Pu_3Al, PuAl, $PuAl_2$, $PuAl_3$, $PuAl_4$	Zr	Pu_6Zr, $Pu_x Zr$ ($x > 3$), $PuZr_2$	Ni	PuNi, $PuNi_2$, $PuNi_3$, $PuNi_4$, $PuNi_5$, Pu_2Ni_{17}
In	Pu_3In, $PuIn_3$	P	PuP	Os	Pu_5Os_3, $PuOs_2$
Tl	Pu_3Tl, $PuTl_3$	As	PuAs	Th	Pu_2Th

Provisional composition.

It will be seen from Table 5 that plutonium displays a strong tendency to form intermetallic compounds. In addition, it forms solid solutions with many metals. Figure 5 represents the atomic diameters of metals [32]. According to the Hume-Rothery rule, metals with atomic diameters differing from that of plutonium by not more than 15% (area bounded by the two horizontals in Figure 5) are capable of forming solid solutions with it. These metals include Zr, Hf, Ta, Nb, Ti, W and Mo. We see that it is a matter of some difficulty to find a suitable vessel material for working with liquid plutonium. Nevertheless, these metals can be left in contact with plutonium at temperatures close to its melting point for relatively long periods of time, since the rates of diffusion of plutonium into the metals and of the metals into plutonium are quite slow. Tungsten and tantalum are the most resistant to plutonium.

Plutonium alloys have been extensively investigated [25–27, 33, 115–117, 363, 435, 446, 620, 621, 636, 640]. They may be prepared by direct interaction of the elements (e. g., Pu–Ag) or by various chemical methods. Alloys of plutonium with aluminum or beryllium are prepared by reducing plutonium trifluoride with the respective metal. Alloys of plutonium with Mn, Fe, Co and Ni are prepared by

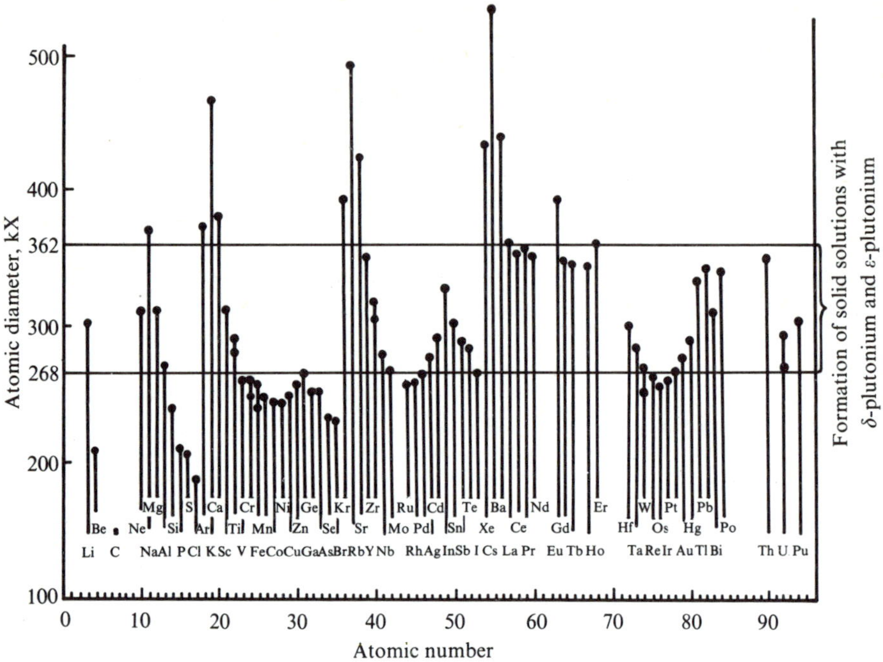

Figure 5
Atomic diameters of metals

heating a mixture of PuF_3 with the powder of the respective metal in lithium vapor at 950°. A few ternary alloys of plutonium of practical significance (U–Pu–Mo, Pu–Ce–Co, etc.) have also been studied. Complete phase diagrams for the binary and ternary systems were published by Konobeevskii [25–27, 115, 116] and other workers [33, 118, 220, 286, 363, 435, 446, 620, 621, 636, 640].

BEHAVIOR OF PLUTONIUM IONS IN AQUEOUS SOLUTIONS

If the hydrogen ion concentration is sufficiently high and if complex formation with the anions can be neglected, plutonium is present in solution as the hydrated ions $Pu(H_2O)_8^{3+}$, $Pu(H_2O)_8^{4+}$, $PuO_2(H_2O)_6^{+}$, $PuO_2(H_2O)_6^{2+}$.*

It is believed [3, Chap. 9] that plutonium is present in solutions of perchloric acid in the form of such ions. The corresponding absorption spectra [343] are represented in Figure 6. The presence of hydrated plutonium ions in major concentrations imparts a characteristic color to the solution: Pu_3^{+}—blue-violet, Pu^{4+}—yellow-brown, PuO_2^{+}—pale pink, PuO_2^{2+}—pink-orange.

* These coordination or hydration numbers have been deduced from the closest packing rule [120].

When the acidity of the solution is decreased below a certain value, hydrolysis takes place. According to modern ideas, this hydrolysis consists of a proton transfer from a coordinated water molecule to a water molecule in the outer sphere [295]. This is accompanied by the formation of hydroxyhydrated compounds. The first stage of the hydrolysis may be expressed by the equation:

$$M(H_2O)_n^{z+} + H_2O \rightleftharpoons MOH(H_2O)_{n-1}^{(z-1)+} + H_3O^+, \qquad (1)$$

where M is Pu^{3+}, Pu^{4+}, PuO_2^+ and PuO_2^{2+}.

The mechanism of formation of complex compounds with the anions is different and involves the displacement of the water molecules from the inner sphere of the central ion by the complexing anion:

$$M(H_2O)_n^{z+} + A^{y-} \rightleftharpoons MA(H_2O)_{n-1}^{(z-y)+} + H_2O. \qquad (2)$$

The effects of hydration, hydrolysis and complex formation are seldom observed separately. Depending on the relative contribution of each of these processes, mixed complexes of variable composition may be present in solution. Hydrolysis and complex formation involving plutonium ions depend on the value of the ionic potential:

$$\phi = Z/r, \qquad (3)$$

where Z is the ionic charge and r is the ionic radius.

The ionic potentials, and consequently the tendency to hydrolyze and form complexes decrease in the following sequence:

Pu^{4+} >	Pu^{3+} >	PuO_2^{2+} >	PuO_2^+
4.44	2.91	2.47	1.17

The available publications on the formation and hydrolysis of plutonium and other transuranium elements were reviewed by Gel'man et al. [60].

HYDROLYSIS

The hydrolysis reactions of plutonium ions in various states of oxidation are highly different. Reactions involving Pu(III) and Pu(V) yield monomeric products only. Pu(VI) and, to a much greater extent, Pu(IV) also participate under certain conditions in polymerization reactions which will be discussed below.

The first stage of monomeric hydrolysis can be represented, in a simplified manner, as

$$M^{z+} + H_2O \rightleftharpoons MOH^{(z-1)+} + H^+ \qquad (4)$$

with the concentration constant

$$K_1^c = \frac{[\mathrm{MOH}^{(z-1)+}]\cdot[\mathrm{H}^+]}{[\mathrm{M}^{z+}]}.$$

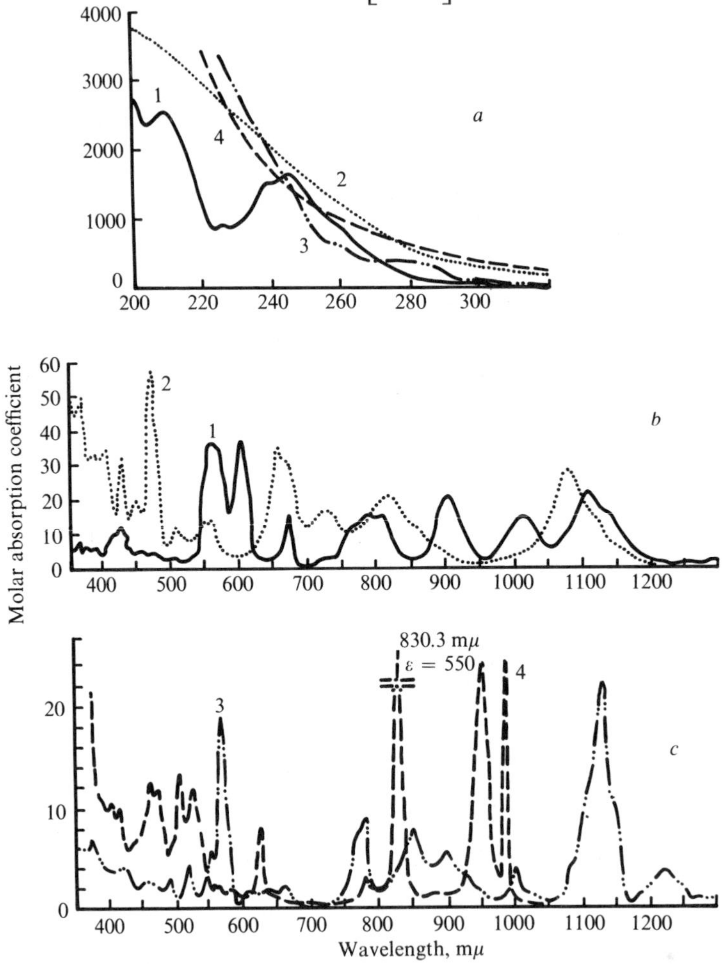

Figure 6

Absorption spectra of plutonium ions in the ultraviolet (*a*), visible (*b*) and infrared (*c*):

1) Pu(III) in 1*M* $HClO_4$; 2) Pu(IV) in 1*M* $HClO_4$; 3) Pu(V) in 0.2*M* $HClO_4$ at 10°C; 4) Pu(VI) in 1*M* $HClO_4$. Spectra 1, 2, and 4 were taken at 25°C.

The hydrolysis constants for the following stages may be written in a similar manner. The overall constant is related to the individual stage constants by

$$K_{\mathrm{tot}}^c = K_1^c \cdot K_2^c \ldots K_n^c.$$

Hydrolysis of plutonium (III) ions

The information available in the literature on the hydrolysis of trivalent plutonium is scarce. Kraus and Dam [510] analyzed the acid-base titration curves in perchloric and hydrochloric acids (ionic strength $\mu = 0.024–0.069$) and obtained the value $pK_1^c \cong 7.3$. Nothing is known about the other stages of hydrolysis, except that the final product of the hydrolysis is a hydroxide, the solubility product of which is $2 \cdot 10^{-20}$ [146, 203, p. 328].

The precipitation of $Pu(OH)_3$ is a routine analytical operation, which must be effected in an inert gas atmosphere, since Pu(III) hydroxide is rapidly oxidized in the air to the polymeric hydroxide of Pu(IV) which is sparingly soluble in acids.

Hydrolysis of plutonium (IV) ions

The hydrated Pu^{4+} ion exists in a wide range of $HClO_4$ concentrations and seems to be the predominant species in moderately acid solutions of HCl. In agreement with the value of its ionic potential, the tendency of this ion to hydrolyze is stronger than that of any other plutonium ion.

It was shown by numerous experimental observations that the hydrolysis of Pu(IV) takes place as a rule when the hydrogen ion concentration is less than 0.3 *M*. Two kinds of reaction products must be distinguished in this connection: the hydrolyzed monomeric ions of the $Pu(OH)_n^{(4-n)+}$ type, and the products of polymerization which are formed in the same acidity range, and the concentration of which will depend on the experimental conditions (temperature, time, concentration of Pu(IV)).

The constant K_1^c of the first stage of monomeric hydrolysis,

$$Pu^{4+} + H_2O \rightleftharpoons PuOH^{3+} + H^+, \tag{5}$$

in perchlorate solutions ($\mu = 0.5–2.0$) has been variously reported as 0.018 to 0.031 [460, 514, 604, 610, 611]. A rapid decrease in the acidity of a solution of Pu(IV), produced by the addition of concentrated ammonia or alkali, results in a more extensive hydrolysis and the precipitation of the monomeric hydroxide $Pu(OH)_4 \cdot xH_2O$ at pH above 3. If the precipitation is effected in this manner, there is, practically speaking, no formation of sparingly soluble polymerization products. According to Kasha [486], the solubility product of $Pu(OH)_4 \cdot xH_2O$ is $7 \cdot 10^{-56}$. The precipitate is readily dissolved in excess acid. After such dissolution the bulk of the plutonium is again present in monomeric or ionic forms, and only an insignificant proportion is present as a polymeric colloid.

The formation of polymers of plutonium (IV) during the various analytical operations may be considerable. Polymerization is enhanced if the decrease in the acidity in the hydrolysis range ($[H^+] < 0.3\ M$) is slow. Dilution of the acid solution with water produces an instantaneous local decrease in the acidity to a pH value at which polymerization tends to take place. Polymer formation is accelerated when moderately acid solutions of Pu(IV) are heated.*

Polymerization. The polymerization process may be represented in its simplest form as the polarization of a molecule of water by two molecules of plutonium. Such linking of the individual units of the polymeric chains through an oxygen bridge may well be the explanation for the stability of the polymers of Pu(IV).

Ockenden and Welch [582] studied solutions of colloidal Pu(IV), prepared by diluting acid solutions with water or with ammonium hydroxide. As soon as the acidity is reduced to 0.1 M HNO_3, disproportionation and polymerization of Pu(IV) take place (Figure 7). The equilibrium is attained within a few hours. Figure 8 shows the spectra of Pu(IV) polymers, after Pu(III), Pu(IV) and Pu(VI) ions have been separated by cation exchange. The differences in the molar extinction coefficients are probably due to the differences in preparation of the colloid. These spectra do not show the absorption band at 476 mμ, which is typical of Pu^{4+}.

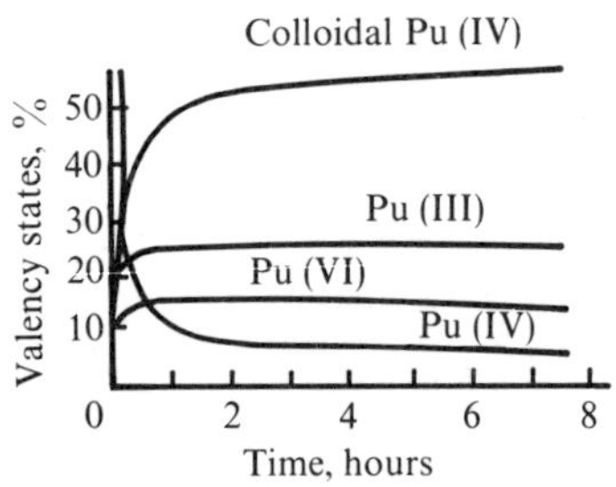

Figure 7

Equilibrium between the various valency states of plutonium in 0.1 M HNO_3 (concentration of plutonium $8 \cdot 10^{-3}$ M)

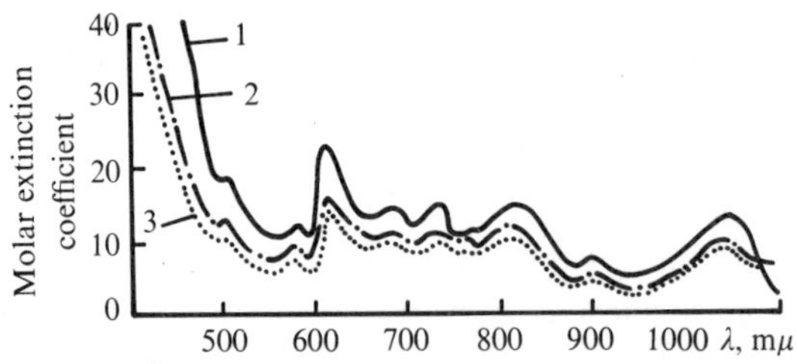

Figure 8

Absorption spectra of Pu(IV) polymers in 0.1M HNO_3:

1) polymer after aging for 10 months; 2) freshly prepared polymer, obtained by diluting a nitric acid solution of Pu(IV) with water; 3) freshly prepared polymer, obtained by diluting a nitric acid solution of Pu(IV) with ammonia.

* Slow disproportionation of Pu(V) in weakly acid solutions leads to polymerization of Pu(IV) [177].

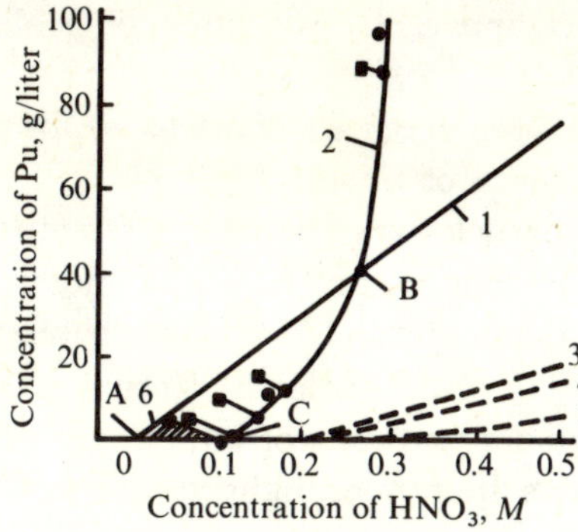

Figure 9

Polymerization of Pu(IV) as a function of the concentrations of plutonium and nitric acid and of the temperature:

1) dilution line, ■ initial composition; 2,3,4,5) stability curves at 25, 80, 90 and 100°C respectively, ● stable composition; 6) colloidal state.

Brunstad [323] carried out a spectrophotometric study of the regions of existence and conditions of formation of polymers (Figure 9). Solutions containing various concentrations of plutonium and nitric acid were prepared by diluting a solution of 400 g/liter of Pu(IV) in 2.5 M HNO_3 with water (straight line 1). It was assumed that under these conditions polymerization takes place in the solutions if the molar extinction coefficient at 415 mμ is higher than 20. The composition of the solutions obtained 10 minutes after dilution is indicated by black squares. Lines joining black squares and black circles correspond to the process of hydrolysis which is accompanied by an increase in the hydrogen ion concentration and precipitation of the solid phase with a correspondingly higher content of plutonium (area *ABC*), or by the formation of a stable colloid in solution with a low content of plutonium (area 6 in Figure 9). The line of black circles is the line of minimum acid concentrations required to preserve a true solution, i.e., the stability curve (curve 2); the area of true solutions extends to the right of this curve. With the increase in temperature the stability curves shift toward higher acidities.

The rate of polymerization depends to a great extent on the polymerization technique and on the composition of the initial solution. Pure solutions of Pu(IV) rapidly attain some stationary state (Figure 7). Solutions of mixtures of different valency forms attain equilibrium only after 24 hours or more [485].

The course of the polymerization is greatly affected by the presence of complexing anions. Dilution of nitric acid solutions of Pu(IV) with a solution of ammonium nitrate suppresses polymer formation [582]. If the concentration of the nitrate ion in solution is less than 0.5 M, the colloid is formed rapidly, if it is 1–2 M, it is formed only slowly, while if the concentration of nitrates exceeds 4 M, the colloid is not formed at all. The same effect, apparently even stronger, is observed in the presence of sulfate ions [613].

Polymers of Pu(IV) have been reported [698] to be formed in a non-aqueous medium (butex)* in the presence of traces of water.

The size of the colloidal particles may be very different depending on the conditions of preparation and the concentration of plutonium. The molecular weight of the particles may vary between 10^3 and 10^{10}. The average particle size decreases with the increase in the acidity of the solution. When solutions of alkalis or salts are made to act on the colloidal particles of Pu(IV), the polymer precipitates out. It should be noted that anions are not an indispensable part of the structure of the polymer, since they can be readily washed out of the precipitate. It has been shown [582] that the structure of the polymer is close to that of plutonium dioxide.

The chemical properties of polymerized Pu(IV) are quite different from its properties in the ionic forms. As a rule, the extractability of the colloid with solvents is very small [582]. Complete oxidation to Pu(VI), e.g., by sodium bismuthate [437], cannot be achieved within a short time.

Similar considerations apply to the action of reducing agents. The polymer does not react with ion exchange resins, but may be adsorbed on them. It is adsorbed to a significant extent on glass ($\sim 1.6 \mu g/cm^2$), steel and other materials [582].

It will be seen that work with solutions of Pu(IV) which are even partly polymerized involves difficulties which can be overcome only be effecting a total depolymerization.

Depolymerization. The depolymerization of Pu(IV) was studied in detail in [323, 582]. Some data on the subject will also be found in [120, 357, 613, 203, p. 332].

The rate of depolymerization substantially depenes on the mode of preparation of the polymer. Polymers in colloidal solutions or in the form of polymeric hydroxylated precipitate, formed when a weakly acid solution of plutonium (IV) is intensively heated or left to stand, are degraded much more slowly than a colloid which has been freshly prepared without application of heat. Thus, the depolymerization is the more rapid, the lower the degree of polymerization. Table 6 gives the depolymerization rates of polymers prepared in various manners. The addition of complexing ions (SO_4^{2-}, F^-) reduces the time of depolymerization.

There are also other techniques of converting Pu(IV) from polymeric to monomeric forms. These include changes in the valency of Pu(IV) effected with the aid of strong oxidizing or reducing agents.

Kraus [3, p. 218] oxidized colloidal Pu(IV) to Pu(VI) in 10 minutes to the extent of 90% by a solution of Ce^{4+} in 0.9 M HNO_3. The Pu(IV) residual from this treatment could be oxidized only at a much slower rate. The degradation of polymeric Pu(IV) by oxidation to Pu(VI) may be effected, for instance, with the aid of bichromate [613] and bismuthate [437] ions.

* [Diethylene glycol dibutyl ether.]

Table 6

TIME REQUIRED FOR FULL DEPOLYMERIZATION OF Pu(IV) IN NITRIC ACID SOLUTION (HOURS) [582]*
(concentration of plutonium ~ 1 mg/ml)

Description of the colloidal solution	Temperature, °C	Concentration of HNO_3		
		1 *M*	3 *M*	5 *M*
Freshly prepared	20	200	90	24
Left to stand for 10 months	20	Precipitation	Precipitation	> 200
	50	—	—	10
	70	—	—	4
	90	—	—	2.5

* Controlled by spectrophotometry.

Sklyarenko and Chubukova (1956) achieved the same result by reducing Pu(IV) by sulfur dioxide. The colloidal solution of plutonium (IV) in 1 *M* HNO_3 was treated with gaseous SO_2 for 30 minutes and the solution was left to stand in a closed vessel for one hour. The excess reducing agent was then removed by passing through a stream of air, and the trivalent plutonium was rapidly reoxidized to Pu(IV) by passing a mixture of nitrogen oxides through the solution. The oxidized solution did not contain significant amounts of Pu(IV) polymers. Hydroxylamine may be utilized instead of sulfur dioxide [613].

The formation of polymers during industrial processing results in losses of plutonium and distortion of process parameters; polymerization during analysis will lead to erroneous results. The stability curves in Figure 9 make it possible to select experimental conditions which are guaranteed to keep all Pu(IV) in a depolymerized form. It is not advisable to dilute acid solutions of Pu(IV) with water, even if the calculated resulting acidity is sufficiently high to ensure formation of a true solution, since when the two solutions are mixed, local zones may be formed in which hydrolysis and polymerization tend to occur (the coordinates of these zones are to the left of the stability curve in Figure 9). The dilution must be effected with an acid solution of at least 0.20 *M*, and the resulting solution must be stored at acidities higher than those given by the stability curves. The addition of alkali in order to precipitate $Pu(OH)_4$ should be done rapidly, in the cold and from dilute solutions. The hydroxide should also be dissolved rapidly, in a sufficient amount of acid and without heating. It will be recalled that $Pu(OH)_3$ is rapidly oxidized to the polymeric Pu(IV) hydroxide. Despite the fact that the precipitation and filtration of $Pu(OH)_3$ is carried out in an atmosphere of inert gas, some polymer is always formed, apparently due to the presence of dissolved oxygen in the reagents. $Pu(OH)_3$ is usually dissolved in concentrated HNO_3 with heating to 100° to degrade the polymer and obtain Pu(IV) exclusively in the ionic forms.

Hydrolysis of Pu (V) ions

The hydrolysis of Pu(V) has not been studied to any extent because the ion has a strong tendency to disproportionate. Kraus [120] reported a study of the acid-base titration of Pu(V) ($7.7 \cdot 10^{-3}$–$4.4 \cdot 10^{-4}$ *M*), when a buffer zone of hydrolysis was noted at pH 8.0–9.5. For the reaction $PuO_2^+ + H_2O \rightleftharpoons PuO_2OH + H^+$ in hydrochloric and perchloric acid media Kraus and Dam [511] found pK_1^c of about 9.6 ($\mu \cong 3 \cdot 10^{-3}$). This small value of K_1^c is caused by the insignificant tendency of Pu(V) to form hydroxy compounds.

Hydrolysis of Pu (VI) ions

Spectrophotometric measurements [169; 241, p. 112; 560] showed that the spectrum of the hydrated PuO_2^{2+} ion remains unchanged until a pH of 3.3–3.6 is attained. At pH above 3.5 the intensity of the absorption band at 830 mμ decreases, a new band appears at 845 mμ, and the maxima show a small shift towards the long wave region of the spectrum. Clearly, these conspicuous changes are caused by the hydrolysis of Pu(VI).

The overall mechanism of hydrolysis of hexavalent plutonium is now thought to consist of a series of successive reactions:

$$PuO_2^{2+} + H_2O \overset{K_1}{\rightleftharpoons} PuO_2(OH)^+ + H^+, \tag{6}$$

$$PuO_2(OH)^+ + H_2O \overset{K_2}{\rightleftharpoons} PuO_2(OH)_2^0 + H^+, \tag{7}$$

$$PuO_2(OH)^+ + PuO_2(OH)_2^0 \overset{K_3}{\rightleftharpoons} (PuO_2)_2(OH)_3^+, \tag{8}$$

$$PuO_2(OH)_2^0 + H_2O \overset{K_4}{\rightleftharpoons} PuO_2(OH)_3^- + H^+, \tag{9}$$

$$PuO_2(OH)_2^0 + PuO_2(OH)_3^- \overset{K_5}{\rightleftharpoons} (PuO_2)_2(OH)_5. \tag{10}$$

During the early studies [120, 512] on the acid-base titration of solutions of Pu(VI) two buffer zones at pH 5–6 and pH 10 were noted. Reactions (6) and (7) are thought to take place in the former range, and reaction (9) in the latter.

When the solution is back-titrated with acid, hysteresis is noted, indicating that the hydrolysis is to a certain extent irreversible, and that the products of hydrolysis include polymers. This effect was ignored in calculating the constants of reactions (6) and (7), given in Table 7.

Moskvin and Zaitseva [60, p. 37, 169] studied the solubility of $(NH_4)_2Pu_2O_7$ as a function of pH.*

* The precipitate is formed at pH > 13 on the addition of concentrated ammonia to ~ 0.1 *M* plutonyl nitrate. The experimentally found ratio $NH_4^+ : PuO_2^{2+} = 1:1$ corresponds to the formula $NH_4[PuO_2(OH)_3]$ or $(NH_4)_2Pu_2O_7$. The solubility of the precipitate in the experimental conditions employed is $2.28 \cdot 10^{-4}$ mole/liter plutonium [169].

According to these authors, the appearance of $PuO_2(OH)^+$ ions takes place at pH 3.3, precisely when the changes in the absorption spectrum of Pu(VI) begin. According to Krevinskaya et al. [60, p. 36, 126] the first stage of the hydrolysis takes place in a more acid range (at pH 2.74 in 0.01 *M* Pu(VI) about 20% of the plutonium is present as PuO_2OH^+).

Between pH 5.0 and 7.5 [169] reactions (7) and (8) take place, i.e., there is dimerization of the monomeric hydrolysis products. At pH above 8.3 further hydrolysis according to equations (9) and (10) takes place.

Table 7

CONSTANTS OF HYDROLYSIS OF Pu(VI) IONS

Constant	Value of constant	Reference
K_1	$1.95 \cdot 10^{-6}$	[3, p. 271]
	$5.0 \cdot 10^{-6}$	[3, p. 271]
	$4.7 \cdot 10^{-4}$	[126]
	$4.1 \cdot 10^{-4}$	[169]
K_2	$1.95 \cdot 10^{-6}$	[3, p. 271]
	$9 \cdot 10^{-5}$	[126]
	$5.6 \cdot 10^{-6}$	[169]
K_3	$1.8 \cdot 10^{-6}$	[169]
K_4	$2 \cdot 10^{-10}$	[3, p. 271]
	$3 \cdot 10^{-10}$	[169]
K_5	$2.0 \cdot 10^{-5}$	[169]

The pH ranges of the existence of hydrolyzed plutonyl ions and their respective yields are shown in Figure 10. Table 7 gives the values of the hydrolysis constants as reported by different workers.

Moskvin et al. [169] also calculated the equilibrium constants of the overall reactions of dimer formation and the concentrational instability constant of $PuO_2(OH)^+$, $PuO_2(OH)_2^0$ and $PuO_2(OH)_3^-$.

Moskvin and Zaitseva [169] represent the formation of the diplutonate ion in the form of the reactions

$$2[PuO_2(OH_3)]^- + [(PuO_2)_2(OH)_5]^- + OH^- \rightleftharpoons 2Pu_2O_7^{2-} + 6H_2O \quad (11)$$

and

$$2[PuO_2(OH)_2]^0 + 2OH^- \rightleftharpoons PuO_7^{2-} + 3H_2O\,. \quad (12)$$

The solubility of $(NH_4)_2Pu_2O_7$ has two minima: one at pH > 13 ($2.28 \cdot 10^{-4}$ mole of Pu per liter), and the other at pH 8.25–8.3 ($4.2 \cdot 10^{-4}$ mole of Pu per liter). The second minimum is connected with the conversion of $(NH_4)_2Pu_2O_7$ into the less soluble hydroxide $PuO_2(OH)_2$. The solubility product of plutonium hydroxide was found to be $1.8 \cdot 10^{-23}$ [169] and $2 \cdot 10^{-25}$ [60, p. 36]. These values are close to the solubility products of uranyl and neptunyl hydroxides.

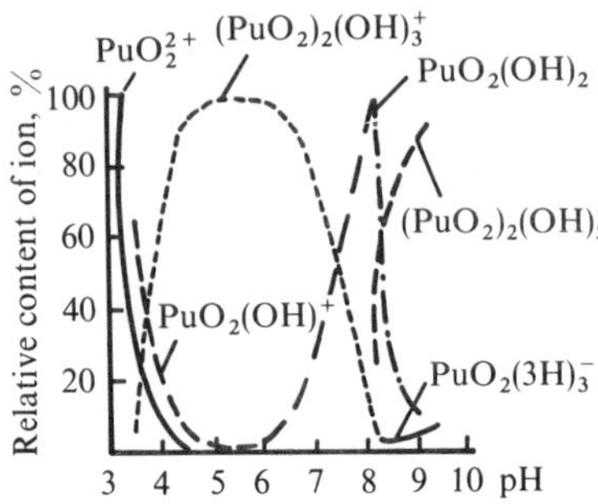

Figure 10

Effect of pH on the ionic composition of solutions of Pu(VI)

The quantitative precipitation, which is characteristic of U(VI) on the addition of excess alkali, does not take place with Pu(VI), and is rarely used in analysis.

COMPLEX FORMATION

The theoretical aspects of the chemistry of complex compounds and a discussion of the method for the determination of constants will be found in monographs by Yatsimirskii and Vasil'ev [259] and Babko [15]. A treatise on the chemistry of plutonium complexes was published in 1961 by Gel'man, Moskvin, Zaitsev and Mefod'eva [60].

Complex formation substantially affects the behavior of plutonium ions of various valencies in the process of chemical separation and determination of this element. It may accelerate or retard oxidation and reduction reactions. By a suitable choice of the complexing anion analytical purification of plutonium by precipitation, extraction and ion exchange may be carried out. Complex compounds are very important in titrimetric determinations of plutonium in the presence of interfering elements. The literature data on the complex formation by plutonium, of interest for the study of analytical chemistry of this element, will be reviewed below,

The successive reactions of complex formation may be written in a simplified form as follows:

$$M^{z+} + A^{y-} \rightleftharpoons MA^{(z-y)+}, \tag{13}$$

$$MA^{(z-y)+} + A^{y-} \rightleftharpoons MA_2^{(z-2y)+}, \text{ etc.} \tag{14}$$

or, in general,

$$M^{z+} + nA^{y-} \rightleftharpoons MA_n^{(z-ny)+}. \tag{15}$$

The stability of these complexes is characterized by means of the direct or reverse reaction constants.*

Complexes of Pu (III)

Spectrophotometric studies failed to detect any complex formation between Pu(III) and *perchlorate* ions [463].

Plutonium complexes with *chloride* ions were detected by spectrophotometry. Ward and Welch [710] showed that at HCl concentrations higher than 1 *M*, the complex ions formed are $PuCl^{2+}$ and $PuCl_2^+$. In concentrated solutions of hydrochloric acid (> 12.5 *M*) there are anionic complexes of Pu(III).

In *nitric acid* media (up to 5 *M* HNO_3) complexing of Pu(III) was also noted [203, p. 334]. The study carried out by Shevchenko, Timoshev and Volkova [252] by extraction methods resulted in the identification of the complex forms as $Pu(NO_3)^{2+}$, $Pu(NO_3)_2^+$ and $Pu(NO_3)_3$.

Complexes of both the cationic and the anionic type were found in *sulfuric acid* solutions by the electrophoretic method (the latter type is the predominant species in concentrated H_2SO_4) [3, p. 273]. Rabideau et al. [186] give the stability constants of $PuSO_4^+$ and $Pu(SO_4)_2^-$ as 10 and 50, respectively.

The fact that plutonium (III) *oxalate* is soluble in excess potassium or ammonium oxalate indicates complex formation in these media. Studies of the solubility of $Pu_2(C_2O_4)_3 \cdot 9H_2O$ [54, 56, 60, p. 44] and of the absorption spectra of Pu(III) in oxalate solutions at pH 1.4–3.0 [58, 60, p. 44] made it possible to calculate the formation constants of the ions $[Pu(C_2O_4)_2]^-$, $[Pu(C_2O_4)_3]^{3-}$, $[Pu(C_2O_4)_4]^{5-}$ ($2.0 \cdot 10^9$, $2.4 \cdot 10^9$ and $8.3 \cdot 10^9$ respectively at 20°C).

Fomin, Vorob'ev and Andreeva [227] made a polarographic study of the system Pu(III)–$C_2O_4^{2-}$ at pH 3.5–6.0. The resulting stability constants of the ions $[Pu(C_2O_4)_3]^{3-}$ and $[Pu(C_2O_4)_4]^{5-}$ are somewhat different from those given above. At $K_2C_2O_4$ concentrations above 1.5 *M* the latter ion predominates, and its relative content is directly proportional to $[K_2C_2O_4]^4$ [60, p. 53].

By the method of ion exchange [55], $[Pu(C_2O_4)_2]^-$ and $[Pu(HC_2O_4)_4]^-$ ions were detected in a solution of 1 *M* NH_4Cl + 0.005 *M* $H_2C_2O_4$.

The presence of complexes of Pu(III) was qualitatively demonstrated in *carbonate* solutions [58, 60, p. 43]; transport number data indicated that the complexes are mainly anionic.

Depending on the pH, trivalent plutonium forms a series of complexes in solutions of *ethylenediaminetetraacetic acid* [60, p. 59, 461]. These compounds were studied by electrophoretic [58] and ion exchange [59, 165] methods. At pH 1.5–2.0

* In the Appendix (p. 406) the reader will find the stability constants β of the complex plutonium ions. Only a few of these are given in the text.

the complexes $Pu\gamma^-$ and $PuH\gamma$, the latter predominating, were identified.* At pH above 2.0 the yield of the negatively charged $Pu\gamma^-$ increases. In a nearly neutral medium (pH > 5.0), the complex $Pu\gamma^-$ is partly hydrolyzed with the formation of $Pu\gamma(OH)^{2-}$.

Gryzin (1951–1955) studied the processes of complex formation by Pu(IV) and Pu(III) in certain organic acids for the purpose of a polarographic determination of plutonium. He identified the following products of the reduction of Pu(IV) complexes: in *tartaric acid* solutions (pH 5.0–6.0) $[Pu(C_4H_4O_6)_6]^{9-}$ ions with a stability constant of $5 \cdot 10^{15}$, and in *citric acid* solutions (pH 3.0–6.5) $[Pu(C_6H_5O_7)_3]^{6-}$ ions with a stability constant of $\sim 10^{11}$.

Chang-Wen-Ch'ing [241, p. 90] found that at pH > 1.5 trivalent plutonium is precipitated by *trihydroxyglutaric acid.* The precipitate, whose composition was close to $PuC_5H_5O_7 \cdot 3H_2O$, was fairly soluble in the excess of the reagent. The stability of the complex compounds formed was not estimated.

Harvey et al. [451] give in their review qualitative information on the reaction of plutonium ions, mainly Pu(III), with a number of organic acids.

The complexing ligands may be arranged in the following sequence of decreasing stability of the complex compounds formed with Pu(III):

$$\gamma^{4-} > C_4H_4O_6^{2-} > C_6H_5O_7^{3-} > C_2O_4^{2-} > SO_4^{2-} > NO_3^- > Cl^-.$$

The complexes formed by these anions with Pu(III) are considerably less stable than those formed with Pu(IV), U(IV), Th(IV), Np(IV) and certain other ions. This fact is widely used in the separation of plutonium from interfering elements by extraction and ion exchange. Thus, U(IV) and Fe(III) can be separated from Pu(III) by anion exchange on A300D resin from 5*M* HCl [600]; Pu(III) gives no anionic chloride complexes at this concentration of chloride ions.

Complexes of Pu (IV)

The absorption spectrum of Pu(IV) in perchloric acid (Figure 7) corresponds to the hydrated $Pu(H_2O)_8^{4+}$ ions.

In solutions containing *chlorides* stepwise complex formation takes place, up to the formation of anionic complexes $PuCl_6^{2-}$ at HCl concentrations above 6*M* [462]. Rabideau [603] calculated the stability constant of $PuCl^{3+}$ at $\mu = 1$ as $\beta = 2.1$. Recently published constants for the successively formed $PuCl^{3+}$, $PuCl_2^{2+}$ and $PuCl_3^+$ ions are 1.4 ± 0.2, 1.2 ± 0.2 and 0.1 ± 0.1 respectively [433].

Ryan [625] carried out spectrophotometric investigations on Pu(IV) sorbed from concentrated HCl solutions on the anion exchanger Dowex-1. The likeness of the absorption spectra of solid Pu(IV) and Pu(IV) in 12 *M* HCl to those of

* γ^{4-} is the anion of ethylenediaminetetraacetic acid.

$[(C_2H_5)_4N]_2PuCl_6$ in nitromethane indicates that most of the Pu(IV) in 12 *M* HCl is present as the hexachloride complex.

Ryan concluded from the change in the molar extinction coefficient at 638 mμ that the content of $PuCl_6^{2-}$ in 9 *M* HCl is 9%, and in 12 *M* HCl it is 75% of the total concentration of Pu(IV). In LiCl solutions, the extent of complexing of Pu(IV) is much lower than in HCl at the same concentration of Cl^- ions.

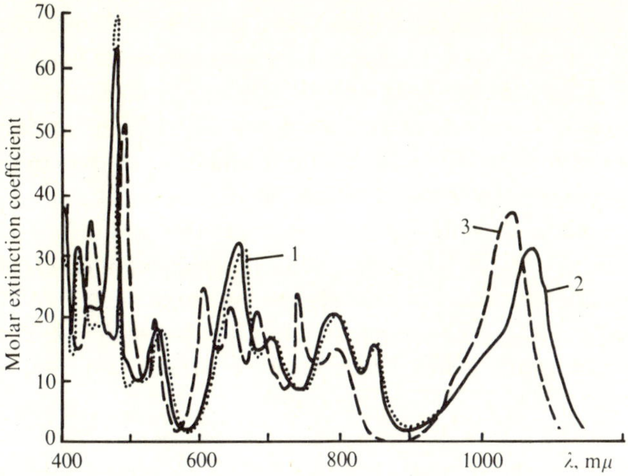

Figure 11

Absorption spectra of nitric acid solutions of Pu(IV):

1) 1 *M* HNO_3; 2) 5 *M* HNO_3; 3) 13.2–15.5 *M* HNO_3.

Pu(IV) has a greater tendency to form anionic complexes with chlorides than U(IV) or Np(IV). This is evidenced by the relatively large values of distribution coefficients of Pu(IV) during extraction with tri-n-octylamine and during anion exchange [625].

Many literature references to complex formation by Pu(IV) in solutions of *nitrates* are available. This problem is particularly interesting in view of the recent development of extraction and ion-exchange methods of nuclear fuel processing, which are, as a rule, carried out in nitrate media.

It was shown by the very first spectrophotometric investigations, carried out by Hindman [461] that a whole series of complexes, up to $Pu(NO_3)_6^{2-}$, is formed, depending on the concentration of nitrates. The absorption spectra are shown in Figure 11. The stability constants have been calculated only for the ions $Pu(NO_3)^{3+}$, $Pu(NO_3)_2^{2+}$ and $Pu(NO_3)_3^{+}$ [60, p. 66; 433, 461, 611]. Only qualitative and semi-quantitative data are available on complexes with a larger number of nitrate groups.

Lipis, Pozharskii and Fomin [149] made a thorough study of the absorption spectra of Pu(IV) in HNO_3 and found that the appearance of maxima and minima in the spectra takes place at acid concentrations of 1.5, 2.1, 3.8, 5.6 and 7.1 *M*. This may be caused by successive addition of NO_3^- groups. It follows that hexanitrate complexes appear in solutions with HNO_3 concentrations higher than 7 *M*.

The complex $Pu(NO_3)_4$ is the predominant species in 1–4 *M* HNO_3 [317], while at higher acid concentrations the complexes $HPu(NO_3)_5$ and $H_2Pu(NO_3)_6$ are formed.

The existence of such compounds has been confirmed by the isolation of complex salts of the type $R_2Pu(NO_3)_6$, where R is K^+, Rb^+, Tl^+, pyridine and tetraethylamine. Chaikhorskii, Burenina and Kondratov in 1954 prepared the acid $H_2Pu(NO_3)_6$, whose solubility in diethyl ether, calculated on the metal, was more than 1 kg/liter.

The absorption spectra of solutions of Pu(IV) in diethyl ether and methyl ethyl ketone after extraction from concentrated solutions of nitric acid are identical with the spectra of Pu(IV) in 12*M* HNO_3, while the ratio $[NO_3^-]:[Pu]$ in the complex in the organic phase is 6 ± 0.4 (Palei, Khalkin, 1953).

Moreover, the spectra of Pu(IV) in concentrated HNO_3 were observed to be similar to those of $[(C_2H_5)_4N]_2Pu(NO_3)_6$ and $[(C_4H_9)_4N]_2Pu(NO_3)_6$ in certain organic solvents (differences in the ε-values did not exceed 5%) [623].

According to Ryan [623], the absorption bands at 609 and 774 mμ are due to hexanitrate complexes only, while the peaks at 650 and 790 mμ are not specific and appear in a wide range of HNO_3 concentrations.

Figure 12 shows the yields of complexes with the ratio $[Pu]:[NO_3^-] = 1:6$ calculated from the intensity of these peaks. According to the calculation, a 100% content of the hexanitrate complex is observed at nitric acid concentrations above 13 *M*. If HNO_3 is replaced by an equivalent amount of $Ca(NO_3)_2$, the proportion of the hexanitrate complex formed is much lower; thus, it is about 91% in a solution containing 9 *M* $Ca(NO_3)_2$ and 0.25 *M* HNO_3 (i.e., 18.2 *M* NO_3^-). In the case of ammonium nitrate, the yield of Pu(IV) hexanitrate is even lower.

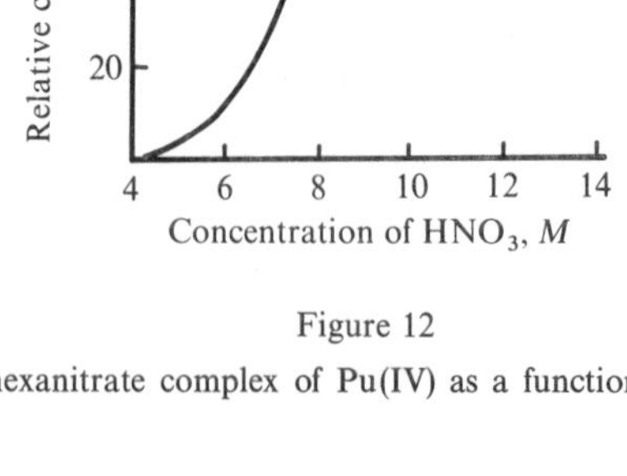

Figure 12

Formation of the hexanitrate complex of Pu(IV) as a function of HNO_3 concentration

These effects were interpreted in [149]. Hydrogen ions and cations with a high hydration energy (e. g., Ca^{2+}) favor the replacement of the water molecules in the hydrate sphere of Pu(IV) by NO_3^- ions, and the dehydration is more intense in acid than in salt solutions. This is confirmed by the identity of the spectra given by 0.0023 *M* solutions of Pu (IV) in 0.33 *M* HNO_3 + 6 *M* $HClO_4$ and in 5.7 *M* HNO_3.

It is very difficult to distinguish between individual complexes with several nitrate groups, i.e., to determine the proportion of each of the ions: $Pu(NO_3)_5^-$, $Pu(NO_3)_6^{2-}$, $HPu(NO_3)_6^-$, $H_2Pu(NO_3)_6$. In this connection studies of anion exchange by Pu(IV) [623, 626] are of interest. The aqueous phase equilibria for the last stages of complex formation may be written as:

$$Pu(NO_3)_n^{4-n} + (6-n)NO_3^- = Pu(NO_3)_6^{2-}, \tag{16}$$

where *n* is 0–5;

$$Pu(NO_3)_6^{2-} + H^+ = HPu(NO_3)_6^-, \tag{17}$$

$$HPu(NO_3)_6^- + H^+ = H_2Pu(NO_3)_6. \tag{18}$$

It was found that the maximum distribution coefficient of Pu(IV) between a solution of nitric acid and Dowex-1 is noted for a nitric acid concentration of ~ 7.7 *M*. At this acidity the proportion of the hexanitrate complex is 40%. (Figure 12). At higher concentrations, reactions (17) and (18) apparently begin to play a major part, since $H_2Pu(NO_3)_6$ is a very weak acid. The distribution coefficients K_d decrease accordingly. In dilute acid solutions (0.5 *M* HNO_3), containing $Ca(NO_3)_2$ in large concentrations, reactions (17) and (18) are almost negligible, and at the same concentrations of NO_3^- ions K_d is much higher for a solution of salt than for a solution of acid ($K_d(\text{salt})/K_d(\text{acid}) \cong 5 \cdot 10^2$ at $[NO_3^-] = 11M$) despite the relatively low yield of hexanitrate complexes.

The hexanitrate complexes of Pu(IV) are, according to Ryan, more stable than the complexes of Np(IV). The ratio of their stability constants is 3.4.

On the strength of the above data concerning the complexes in nitrate media it is possible to select the optimum conditions for the ion-exchange and extractional isolation and purification of plutonium, both in analytical and industrial practice.

Qualitative date on *acetate* complexes [3, Chap. 9] show that Pu(IV) has a considerable tendency to form negatively charged ions.

Nebel and Schwabe [573a, 663a] discovered the existence of acetate complexes of Pu(IV) with a maximum number of five ligands. Their high stability is an explanation for the considerable negative shift of the potential of Pu(IV)/Pu(III), and makes it possible to effect polarographic determinations of plutonium on a dropping mercury electrode in acetate solutions.

Spectrophotometric [3, Chap. 9] and potentiometric [471] results indicate that stable *fluoride* complexes are formed (for PuF^{3+} $\beta \cong 10^4$) [203, p. 337]. Transference number tests failed, however, to detect any anions. Several complex salts [60, p. 133] have been isolated in the solid state.

Electromigration experiments [540] definitely established that anionic complexes are formed in *sulfate* solutions. The shift of the Pu(IV)/Pu(III) potential in 1 *M* H_2SO_4 relative to that in 1 *M* $HClO_4$ (about -0.235 V) is quite large. The changes in the absorption spectrum are also conspicuous [150, 151].

Lipovskii and Chernyavskaya [151] consider that in nitric acid solutions containing sulfate ions, a gradual complex formation takes place, up to a ratio of $Pu:SO_4^{2-} = 1:4$.

According to Marov and Chmutova [157], who studied the cation exchange in 1.92–2.32 *M* solutions of $HClO_4$, containing up to 0.4 *M* $H_2SO_4([H^+] = 2.33\ M)$, only the lower complex $PuSO_4^{2+}$ is formed according to the reaction:

$$Pu^{4+} + HSO_4^- \rightleftharpoons PuSO_4^{2+} + H^+.$$

When the hydrogen ion concentration is 0.5 *M*, three complex compounds with sulfate ions are formed, in the ratios 1 : 1, 1 : 2 and 1 : 3.

Lipis, Pozharskii and Fomin [150] carried out a spectrophotometric investigation using the technique they had employed in their study of nitrate systems [149]. The extremal points on the curves giving the value of ε as a function of the concentration of H_2SO_4 for a number of absorption bands proved to coincide at H_2SO_4 concentrations of 0.5, 1.0, 2.4, 3.7 and 5.3 *M*. The authors [150] suggested that these extrema correspond to the successive replacement of water in the hydrated sphere by the SO_4^{2-} ion, up to the formation of $[Pu(SO_4)_8]^{12-}$ at acid concentrations above 5.8*M*. As distinct from the nitrate media, substitution of ammonium sulfate for the acid favors complex formation because of the diminished effect of the competing reaction $H^+ + SO_4^{2-} \rightleftharpoons HSO_4^-$. With increasing temperature the stability of sulfate complexes decreases.

A number of solid complexes of the type $Me_4Pu(SO_4)_4 \cdot H_2O$ have been isolated, where Me is Cs^+, K^+, NH_4^+, Rb^+ [60, pp. 137, 148], and of the type $Me_6Pu(SO_4)_5 \cdot \cdot nH_2O$, where Me is Na^+, NH_4^+ [148].

Many reports are available on complex formation in *oxalate* solutions [3, Ch. 9; 57, 60, 62, 94, 167, 227]. When oxalic acid is added to an acid solution of Pu(IV), the sparingly soluble salt $Pu(C_2O_4)_2 \cdot 6H_2O$ precipitates out. The solubility of the salt in excess precipitant is considerable and increases with the decrease in the acidity of the solution. Reas [615] assumed that several complex forms are present: $PuC_2O_4^{2+}$, $Pu(C_2O_4)_2$ and $Pu(C_2O_4)_3^{2-}$, and calculated the equilibrium constants of the reactions of formation as $8 \cdot 10^3$, $9.65 \cdot 10^2$ and 25, respectively. Calculations show [167] that complexes with a higher number of oxalate groups can be formed, their composition depending on the concentration of the oxalate ion. The complexes $Pu(C_2O_4)_4^{4-}$ [57, 167, 227], $Pu(C_2O_4)_5^{6-}$ [62, 94] have been identified and their stability constants determined.

Gel'man and Sokhina [60, p. 138, 62] developed a method for the isolation of solid oxalate complexes. They prepared and described compounds of the type $Me_4[Pu(C_2O_4)]_4$, where Me is K^+, Na^+, and $Me_6[Pu(C_2O_4)_5]$, where Me is K^+, NH_4^+. These salts, as well as $Pu(C_2O_4)_2 \cdot 6H_2O$, when acted upon by the α-radiation of Pu^{239} itself, are decomposed to the oxycarbonate $PuOCO_3 \cdot 2H_2O$ or to a mixture of plutonium oxycarbonate with sodium (potassium) carbonate.

The data [3, Chap. 9] obtained for the solubilities of Pu(IV) hydroxide and Pu(IV) salicylate [100] in concentrated carbonate solutions as well as spectrophotometric determinations [60, p. 73] undoubtedly indicate complex formation between Pu(IV) and *carbonate* ions. The stability constant of $Pu(CO_3)^{2+}$ ion is $9.1 \cdot 10^{46}$. Gel'man and Zaitsev [52, 53, 60, p. 140] described the preparation of the following complex salts: $Me_4[Pu(CO_3)_4]$, where Me is K^+, Na^+, NH_4^+; $Me_6[Pu(CO_3)_5]$, where Me is K^+, Na^+, NH_4^+; $Me_8[Pu(CO_3)_6]$, where Me is K^+, NH_4^+; and $K_{12}[Pu(CO_3)_8]$. In the absence of excess carbonate ions solutions of these complexes are unstable.

Many mixed carbonate-oxalate complexes have also been obtained [60, p. 153].

The observed increase in the solubility of plutonium(IV) phosphate with the increase in the phosphoric acid concentration in nitric acid solutions was explained by King [3, Chap. 9] as due to the formation of complex ions (up to five *phosphate* groups per atom of plutonium). Electromigration experiments showed that in solution of 2 *M* HNO_3 + 0.8 *M* H_3PO_4 almost all the plutonium is present as anionic complexes [84].

Denotkina, Moskvin and Shevchenko [84] measured the solubility of $Pu(HPO_4)_2$ in solutions of H_3PO_4 in order to calculate the stability constants of all phosphate complexes with Pu : HPO_4^{2-} ratio between 1 and 5. A compound whose formula is believed to be $H_6[Pu(HPO_4)_5] \cdot nH_2O$ has been isolated from 2.8–6*M* solutions of H_3PO_4 by the addition of alcohol. The complex acid $H_4[Pu(HPO_4)_4] \cdot nH_2O$ has been isolated [85] from solutions of 1.2 and 1.6 *M* in H_3PO_4 and 2 *M* in HNO_3.

Two soluble complexes [352, 353, 450] are formed by Pu(IV) with *hydrogen peroxide.* The lower, brown-colored complex $[Pu—O—O—Pu—OH]^{5+}$ is formed at low H_2O_2 concentrations and contains two plutonium atoms per peroxide group. When the H_2O_2 concentration is increased, a red coloration appears owing to the formation of a complex with two peroxide groups per two plutonium atoms $[HO—Pu—O—O—Pu—O—OH]^{4+}$. It should be remembered that, depending on the experimental conditions, hydrogen peroxide may act either as oxidizing or as reducing agent; it can oxidize Pu(III), reduce Pu(IV) and Pu(VI), and is also capable of precipitating Pu(IV) peroxide if present in sufficiently high concentrations. It follows that the true equilibria in solutions are very involved [3, p. 236].

Citrate complexes of Pu(IV) [3, Chap. 9] were detected by ion-exchange methods and by spectrophotometry.

An extensive study of the behavior of Pu(IV) in solution of *citric acid* was carried out in 1951–1954 by Vinogradov and Gryzin. The shift of the formal Pu(IV)/Pu(III) potential in these solutions with respect to solutions in $HClO_4$ is about -1.0 V and is undoubtedly due to the formation of very stable complexes. A reversible cathodic wave on mercury dropping electrode is formed at pH 3.0–6.5.

According to Vinogradov and Gryzin the ratio between the stability constants of the citrate complexes of Pu(IV) and Pu(III) is $2.2 \cdot 10^{16}$ at pH 5 and $5 \cdot 10^{17}$ at pH 6.

These authors also determined the ratio Pu : $C_6H_5O_7^{3-}$ in the complex as 4.0 ± 0.15 by means of amperometric and conductometric titrations. The overall stability constant of $Pu(C_6H_5O_7)_4^{8-}$ is $7 \cdot 10^{27}$.

Gryzin developed a method for the separation of the solid citrate complex of Pu(IV). It was shown by elementary analysis that the chemical composition of the isolated crystals is $K_8[Pu(C_6H_5O_7)_4]$. The formula was confirmed by electrical conductivity measurements conducted on dilute aqueous solutions of this salt; 0.01 *M* solution of the salt retained its pH value unchanged during 11 days (pH $\cong$ 4.8). The breakdown of the complex is accompanied by an increase in the pH and by the appearance of a yellow precipitate, apparently of basic Pu(IV) citrates.

Vinogradov and Gryzin (1951–1954) investigated the behavior of Pu(IV) in *tartaric acid* solutions. When tartaric acid is added to a solution of $Pu(NO_3)_4$ in 1.5 *M* HNO_3, a light-brown precipitate of $Pu(C_4H_4O_6)_2$ separates out which is soluble in excess of the reagent. The solution then becomes orange-colored.

Polarographic experiments in $H_2C_4H_4O_6$ demonstrated that in the pH range 3.5–5 the reversible reaction $Pu(IV) + e \rightleftharpoons Pu(III)$ takes place on the mercury dropping electrode. The composition of the complex ion was determined by the method used for citrate media. It was found that 6.0 ± 0.15 tartrate groups are present per atom of plutonium. The stability constant of the complex $Pu(C_4H_4O_6)_6^{8-}$ is $2 \cdot 10^{31}$.

Gryzin and Fionin (1952) isolated the complex salt $K_8[Pu(C_4H_4O_6)_6]$ by the technique employed in the preparation of $K_8[Pu(C_6H_5O_7)_4]$. The tartaric acid complex of Pu(IV) is a dark-brown substance, stable in the air and readily soluble in water. Aqueous solutions of the salt ($\sim$ 0.01 *M*) are stable for a week, and afterwards undergo hydrolysis with the appearance of a yellowish precipitate; the hydrolysis is caused by a radiolytic increase in the pH of the water.

Chang Wen Ch'ing [241, p. 64] carried out a potentiometric and spectrophotometric investigation of the complex formation between Pu(IV) and *trihydroxyglutaric acid* $C_5H_8O_7$. Between pH 1 and pH 13 there is a complex compound of the composition 1:1 which is responsible for the shift of the maximum at 480 mμ, typical of Pu^{4+}, to 498 mμ.

Foreman and Smith [407] studied the complexing of Pu(IV) in solutions of *ethylenediaminetetraacetic acid* (EDTA) as a function of the pH. At pH 3.3 two complexes are present—the monomer Puγ and the dimer $Pu_2\gamma^{4+}$—which are hydrolyzed at pH > 5. For the complex Puγ $\beta = 1.58 \cdot 10^{24}$, for $Pu(OH)\gamma^-$ $\beta = 4.5 \cdot 10^{17}$ (ion-exchange method) and $\beta \cong 1.3 \cdot 10^{17}$ (spectrophotometric method).

According to Klygin et al. [110], who made a precise determination of the dissociation constants of EDTA itself, the stability constant of Puγ is $1.4 \cdot 10^{26}$.

Chang Wen Ch'ing [241, p. 47] studied the complex formation in acid solutions of EDTA. He found that a change in the absorption spectrum of Pu(IV) is noted at acid concentrations less than 1 *M* HNO_3, when the absorption maxima of Pu^{4+} shift as follows: 430 → 450 mμ, 480 → 495 mμ and 665 → 678 mμ, owing to the appearance of the complex with Pu:γ^{4-} = 1:1 (Figure 13). The stability constant of Puγ is $1.38 \cdot 10^{25}$. This is important for the complexometric determination of Pu(IV) in the presence of elements which form unstable complexes with EDTA in acid solutions.

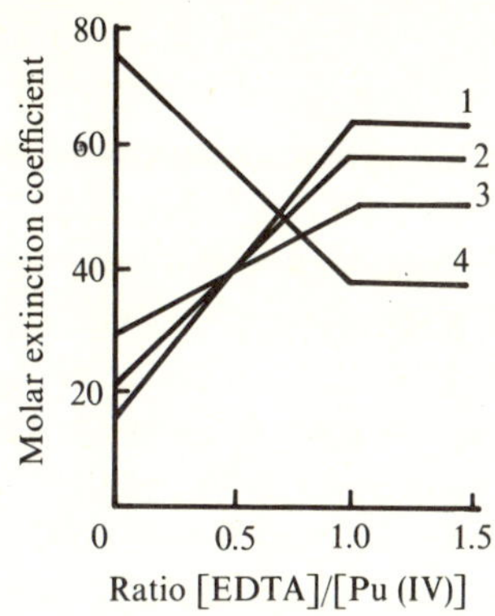

Figure 13

Molar extinction coefficients of Pu(IV) in 0.2 *M* HNO_3 as a function of the molar ratio EDTA:Pu(IV). Concentration of plutonium 0.003 *M*. Wavelengths: 1) 495 mμ; 2) 450 mμ; 3) 678 mμ; 4) 480 mμ.

Salicylate, sulfite, acetylacetonate complexes of Pu(IV) are all known [60].

In a class apart are the so-called internal complexes of plutonium ions with *colored organic reagents*, used in spectrophotometric determinations of microgram amounts of the element. These reagents include azo dyes of *arsenazo* and *thoron* Class.* The formation of complexes with these reagents was studied for uranium, thorium, rare-earth and other elements [9, p. 35, 131–133, 138, 638]. The reaction mechanism deduced from these studies is undoubtedly valid also for plutonium ions.

The analytically functional grouping responsible for complex formation is

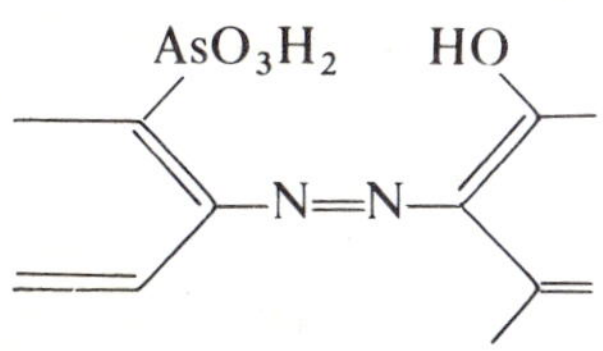

* The formulas of the reagents will be found in Chapter III, under *Spectrophotometric Methods.*

which appears twice in the reagents Thoron II, Arsenazo II and Arsenazo III. The complex formation between Pu (IV) and the functional grouping may be represented as follows:

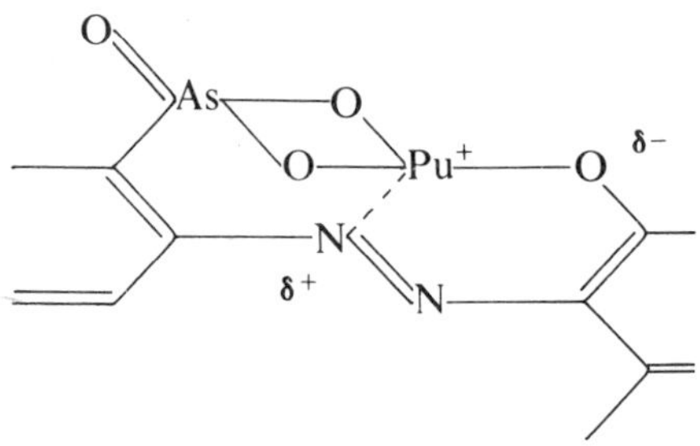

The complexing is accompanied by the formation of two six-membered rings which display aromatic properties and are very stable. The strong deepening in the color is caused, firstly, by the imposition of a positive charge on the "chromophoric center" of the molecule (the —N=N—group) and, secondly, by the imposition of a partial negative charge on the residue of the hydroxy group of the naphthalene nucleus of the reagent. The dipole formed at the ends of the conjugated bonds produces a sharp change in the color of the reagent, i. e., gives a color reaction.

A particular property of complex formation with Arsenazo III is that two functional groupings in the reagent form two practically isolated conjugated systems. The complex formation takes place with the participation of one such grouping only, the symmetry of the molecule is distorted and two absorption maxima appear. The stability of the complexes increases in the sequence Arsenazo I–Arsenazo III, owing to which the optimum pH values for complex formation show an acid shift. This produces a rapid decrease in the effect of the masking ions such as sulfates, phosphates and fluorides.

Of all plutonium ions, Pu(IV) gives complexes which are most stable in acid solutions. Complexes of Pu(IV) have been studied in greatest detail.

According to Klygin and Pavlova [109], Pu(IV) forms with Arsenazo I complexes of the composition 1 : 1, viz., $Pu(OH)H_4R^+$ and $Pu(H_3R)^+$ between pH 2 and pH 6. The maximum yield of the complexes is noted at pH 2.2, and the stability constants are $3.6 \cdot 10^6$ and $4.8 \cdot 10^7$, respectively. The complex of plutonium with Arsenazo I is relatively unstable, but its color is quite different from that of the reagent; this makes it possible to use Arsenazo I as metal indicator during the complexonometric determination of plutonium [179].

Lapshina, Mikhailov et al. (1956) studied the complex formation of Pu(IV) with Thoron I and found that the complex which is formed has a Pu : Thoron I ratio of 1 : 1 and a stability constant of $\sim 10^{14}$. According to the same authors, the complex with Thoron II has the composition 1 : 1 and is stable in 2–3 *M* HNO_3 (1958).

In 1960 Kuznetsov, Savvin, Mikhailov et al. studied the complex formation between plutonium ions and Arsenazo III. Pu(IV) gives strongly colored compounds with this reagent in strongly acid (> 1 *M* HNO_3 or HCl) solutions. The optimum pH value for the formation of the complex of Pu(III) with Arsenazo III is about 2.5.

Opinions differ as to the composition of the complexes of Pu(IV) with Arsenazo III. In analogy with other elements, compounds with the ratio Arsenazo III : Pu = 1 : 1 and 2 : 1 may be formed. A higher value of this ratio would appear to be impossible, both because the valency saturation has been attained and because of the steric hindrances preventing the coordination by Pu(IV) of more than two ligands of this size.

Under the optimum conditions of complex formation between Pu(IV) and the above reagents, Th(IV), Zr(IV), U(IV) and other elements also tend to form stable complexes.

A number of other reagents (of the thenoyltrifluoroacetone type) also give internal complexes with plutonium ions. These complexes are sparingly soluble in aqueous solutions and are employed in extraction procedures. They will be described in Chapter IV.

For tetravalent plutonium there is generally a correlation between the stability of the complex and the dissociation constant of the acid, the anion of which is the complex ligand.

The tendency of individual anions to form complexes with Pu(IV) may be represented by the following sequences. Anions of monobasic acids:

$$CH_3COO^- > F^- > NO_3^- > Cl^- > ClO_4^-.$$

Anions of di- and polybasic acids:

$$CO_3^{2-} > \gamma^{4-} > HPO_4^{2-} > C_4H_4O_6^{2-} > C_6H_5O_7^{3-} > C_2O_4^{2-} > SO_4^{2-}.$$

Complexes of Pu (V)

PuO_2^+ ions have the lowest tendency to form complex compounds.

Rabideau [603] detected the complex form PuO_2Cl in hydrochloric acid solutions.

Quantitative reports are available on the formation of *oxalate* complexes [60, p. 80], $PuO_2(C_2O_4)^-$ and $PuO_2(C_2O_4)_2^{3-}$, formed at pH 3–5.

Complex formation with *Complexone III* begins at pH > 4 [49, 59, 164], or, according to other data [104, 483], even at pH 3. The molar ratios $PuO_2^+ : \gamma^{4-}$ in the complexes were reported [104] as 1 : 1 and 1 : 2.

Complexes of Pu (VI)

In *nitrate* solutions minor changes in the spectrum of Pu(VI) are noted only at HNO_3 concentrations higher than 5 *M*. According to ion migration data [3, Chap. 9] the yield of the negatively charged ions increases with increase in the acid concentration. The complexing reaction constants are given in [126, 454].

Vdovenko, Lipovskii and Kuzina [38, 39] studied the state of Pu(VI) in acetone and aqueous solutions. The curves giving the dependence between the molar extinction coefficients for a number of absorption bands and the concentration of NO_3^- in acetone have extremal points at $[NO_3^-]$: [Pu] ratios of 2 and 3. Formation of a mononitrate complex was not noted.

In aqueous solutions the dinitrate and the trinitrate exist at high acidities. When the HNO_3 concentration is raised from 6 *M* to 12 *M*, the proportion of $PuO_2(NO_3)_2$ increases. A further increase in the acidity results in an equilibrium shift towards the formation of $PuO_2(NO_3)_3^-$. The absorption spectrum in 21.5 *M* HNO_3 is similar to the spectrum of the trinitrate complex of Pu(VI) in acetone.

According to other data [126], the dinitrate complex appears even at nitric acid concentrations below 4 *M*.

Changes in the absorption spectrum of Pu(VI) in *chloride* solutions become conspicuous in 0.5 *M* HCl [3, p. 300]. In solutions of 6 *M* HCl 70% of the plutonium moves towards the anode and in 10 *M* HCl 100% of the plutonium [3, p. 275]. The stability constants of the 1 : 1 complex were calculated by Rabideau and Masters [612], those of 1 : 1 and 1 : 2 complexes by Newton and Baker [578].

Gel'man et al. [60, p. 87] studied the solubility of $NaPuO_2(CH_3COO)_3$ in *acetate* and perchlorate solutions. They calculated the stability constants β of the complexes having the composition 1 : 1, 1 : 2 and 1 : 3 as $1.9 \cdot 10^3$, $2.0 \cdot 10^6$ and $2.3 \cdot 10^7$ respectively; these values are close to the β-values of complexes of U(IV) and Np(IV) of the corresponding compositions.

Stable *sulfate* [3, Chap. 9], *carbonate* [60, p. 88, 61, 88, 89], *oxalate* [51, 90] and EDTA [407, 483] complexes are known to exist; their stability constants are given in the Appendix.

* * *

It will be seen from the material given above that the tendency of plutonium ions to hydrolyze and form complexes decreases in the sequence

$$Pu^{4+} > Pu^{3+} > PuO_2^{2+} > PuO_2^+$$

in agreement with the values of ionic potentials.

However, the effect of the ionic potential is not as clear in the case of the ions of elements nearest plutonium in the actinide series carrying an equal charge. The increase in the atomic number in the sequence U—Np—Pu and the accompanying decrease in the ionic radii does not result, as a rule, in an increase in the stability of the complex compounds which would be large enough to enable the analyst to separate these elements, if they are present in states of equal valency.

On the other hand, the differences between the properties of complexes of elements of similar chemical behavior become much more apparent if the elements are present in various valency states.

In this connection oxidation and reduction processes are of prime importance in attaining valency combinations in which the stability of plutonium complexes will be sufficiently different from the stabilities of the interfering elements. The redox properties of plutonium ions in solution will now be discussed.

POTENTIALS

The potentials of pairs of hydrated plutonium ions are so close to one another that under certain conditions autooxidation and autoreduction reactions occur (disproportionation) in the case of tetravalent and pentavalent plutonium. As a result of these processes ionic equilibrium between all states of oxidation may be attained.

The potentials of pairs of complex ions of plutonium differ from those of the hydrated ions in $HClO_4$. For this reason, depending on the hydrogen ion concentration and the concentration and nature of the anions, the nature and the rate of disproportionation reactions will change, and so will the relative content of ions of different valencies in the system.

Just as important is the effect of the medium on the reactions of plutonium ions with redox reagents and on the stability of ions in various states of oxidation.

Literature sources give the values of formal redox potentials of pairs of plutonium ions, referred to solutions of a given ionic strength, at the ion concentration ratio of unity.

These data cannot be extrapolated to standard conditions, but are of prime importance for practical purposes.

Potentials of pairs of hydrated plutonium ions in solutions of $HClO_4$ were given in a number of publications [310, 354–356, 601, 602, 611]; they were then determined more accurately [186, 603], and the most recent determinations in 1 *M* $HClO_4$ with respect to a normal hydrogen electrode have been summarized by Rabideau, Asprey, Keenan and Newton [186]:

$$\mathrm{Pu(VI)} \xrightarrow{+0.9164\ \mathrm{V}} \mathrm{Pu(V)} \xrightarrow{+1.161\ \mathrm{V}} \mathrm{Pu(IV)} \xrightarrow{+0.9819\ \mathrm{V}} \mathrm{Pu(III)} \xrightarrow{-2.02\ \mathrm{V}} \mathrm{Pu_{met}}$$

Pu(VI)–Pu(IV): + 1.043 V

Pu(VI)–Pu(III): + 1.0228 V

It has been shown above that complex ions of Pu(IV) are much more stable than the complex ions of Pu(III), and the ratio of their stability constants is higher the lower the degree of dissociation and the higher the basicity of the acid. For this reason the negative shift of Pu(IV)/Pu(III) will be an indication of the relative degree of complexing of the ions constituting the system in the given medium.

Table 8

REDOX POTENTIALS OF PLUTONIUM IONS IN SOLUTIONS OF INORGANIC ACIDS

(volts, with respect to normal hydrogen electrode)

Acid	Pu(IV) / Pu(III)	Pu(VI) / Pu(IV)	Pu(VI) / Pu(III)	Pu(VI) / Pu(V)
1 *M* HCl	+ 0.966 [471] + 0.9701 [186] + 0.963 [105]	+ 1.053 [356] + 1.051 [606]	+ 1.025 [356] + 1.0238 [186]	+ 0.910 [12] + 0.9122 [186]
1 *M* HNO_3	+ 0.92 [471] + 0.914 [12]	+ 1.08 [583] + 1.054 [12]	+ 1.006 [12]	+ 0.917 [12] (0.1 *M* HNO_3)
1 *M* H_2SO_4	+ 0.73* + 0.74 [471] + 0.75 [3]	+ 1.2 – + 1.4 [3]	—	—
1 *M* HCl + + 0.6 *M* H_3PO_4	+ 0.59 [471]	—	—	—
1 *M* HCl + 1 *M* HF	+ 0.50 [471]	—	—	—

* Fomin, Vorob'ev and Andreeva (1951).

For pairs of ions, one of which is oxygenated, Pu(VI)/Pu(IV) and Pu(V)/Pu(IV), a considerable positive potential shift is noted, since the tendency to complex formation of the reduced form Pu(IV) is higher than that of the oxidized forms Pu(VI) and Pu(V). For the system Pu(VI)/Pu(III) the shift is usually small, since the stability constants of the complexes of Pu(VI) and Pu(III) are not much different.

Table 8 shows redox potentials of plutonium ions in solutions of inorganic acids.

It was found that in hydrochloric acid [105], nitric acid [12] and sulfuric acid (Kurbatov, 1950) solutions the potentials vary, as could have been expected, with the concentration of the acid (Figure 14).

Vinogradov and Gryzin (1951–1955) carried out a polarographic investigation of the system Pu(IV)/Pu(III) in solutions of tartaric and citric acids. The potentials in 0.5 *M* $Na_3C_6H_5O_7$ at pH 6.0 and 5.0 are -0.086 V and $+ 0.016$ V respectively; in 1.2 *M* $H_2C_4H_4O_6$ at pH 5.0 the potential is $+ 0.064$ V (all potentials n. h. e.).

The potential shifts in these media are so large that Pu(IV) can be determined polarographically on a dropping mercury electrode.

The potentials in alkaline medium are as follows [146, p. 317, 203, p. 346, 507]:

$$PuO_2(OH)^+_{aq} \xrightarrow{+0.26\,V} PuO_2(OH)_{aq} \xrightarrow{+0.76\,V} Pu(OH)_4 \cdot yH_2O \xrightarrow{-0.95\,V} Pu(OH)_3 \cdot xH_2O.$$

$PuO_2(OH)^+_{aq}$ — + 0.4 V — $Pu(OH)_4 \cdot yH_2O$

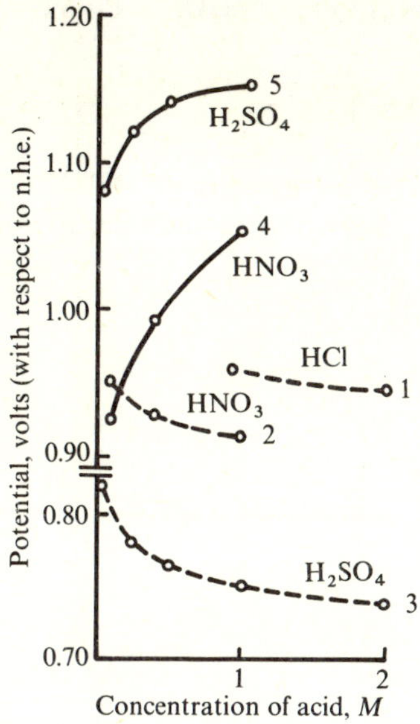

Figure 14

Potentials of Pu(IV)/Pu(III) (1–3) and Pu(VI/Pu(IV) (4–5) as a function of the concentration of HCl, HNO_3 andH_2SO_4.

1–3) Pu(IV)/Pu(III); 4, 5) Pu(VI)/Pu(IV).

EQUILIBRIA

The intermediate states of oxidation—Pu(IV) and Pu(V)—tend to undergo disproportionation in aqueous solutions. This effect is undesirable if plutonium has to be maintained in a definite valency state. Stabilization of solutions of Pu(IV) and Pu(V) is effected by a suitable choice of hydrogen ion concentration or of the concentration of a suitable complexing anion.

On the other hand, oxidation and reduction reactions involving a conversion of a simple ion into an oxygenated ion (e.g., Pu^{3+} into PuO_2^{2+}) or a reverse conversion are frequently determined by the rates of oxidation of the intermediate forms Pu^{4+} and PuO_2^+. From this point of view, disproportionation favors a faster and more complete conversion of plutonium from one valency state to another.

Disproportionation is highly important in the analytical chemistry of plutonium.

Disproportionation of Pu (IV) ions. Equilibria in the system Pu (III)—Pu (IV)—Pu (VI)

Equilibria in this system have been studied in detail [12, 13, 346, 356, 485, 486, 602, 606] and may be expressed by the overall equation

$$3\text{Pu(IV)} \underset{}{\overset{K_1}{\rightleftharpoons}} 2\text{Pu(III)} + \text{Pu(VI)} \tag{19}$$

or, in the ionic form,

$$3Pu^{4+} + 2H_2O \overset{K_2}{\rightleftharpoons} 2Pu^{3+} + PuO_2^{2+} + 4H^+. \tag{20}$$

Under certain conditions, Pu(IV), Pu(III) and Pu(VI) formed as a result of disproportionation of pure Pu(IV) may coexist in comparable concentrations (Figure 15).

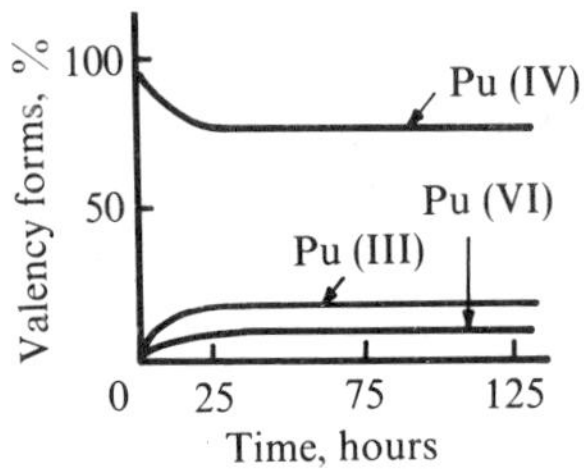

Figure 15

Disproportionation of Pu(IV) in solutions of HNO_3.

Concentration of plutonium, $7.7 \cdot 10^{-3}$ *M*; concentration of HNO_3, 0.40 *M*.

The equilibrium constants of the reactions (19) and (20) were calculated for solutions of $HClO_4$ [486, 602], HCl [485, 606] and HNO_3 [3, Chap. 8, 13]. K_1 and K_2 are interconnected by the equation $K_2 = K_1 \cdot [H^+]^4$. However, while this equality is valid in perchloric and hydrochloric acid solutions at $[H^+] = 0.2–1.0\,M$, this is not the case in nitric acid media: $K_2 \cong 1.1 \cdot 10^{-4}$ (0.4 *M* HNO_3); $K_2 \cong 4 \cdot 10^{-4}$ (0.3 *M* HNO_3); $K_2 \cong 7.2 \cdot 10^{-4}$ (0.2 *M* HNO_3) [13]. For solutions with hydrogen ion concentration of about 1 *M* at 25°, K_2 has the following values:

$HClO_4$	HCl	HNO_3
$9.0 \cdot 10^{-3}$	$1.8 \cdot 10^{-3}$	$4 \cdot 10^{-7}$

The decrease in the value of K_2 is in agreement with the tendency of Pu^{4+} to form complexes whose stabilities vary in the reverse sequence.

Artyukhin, Gel'man, and Medvedovskii [12, 13] gave the equation $K_2 = K_1 \cdot [H^+]^{5.3}$ for nitric acid solutions. It follows from equation (20) that with the increase in the concentration of hydrogen ions and of the complexing anions the equilibrium is shifted to the left and the disproportionation of Pu(IV) decreases. In analytical practice the stabilization of tetravalent plutonium in concentrated solutions of nitric and sulfuric acids is widely employed.

The equilibrium constant in HCl solutions increases by 2–3 orders of magnitude as the temperature increases from 25 to 70° [3, Chap. 8].

Disproportionation has been represented as a two-stage process [12, 13, 485, 486]:

$$2Pu^{4+} + 2H_2O \rightleftharpoons Pu^{3+} + PuO_2^+ + 4H^+, \tag{21}$$

$$PuO_2^+ + Pu^{4+} \rightleftharpoons PuO_2^{2+} + Pu^{3+}. \qquad (22)$$

In moderately acid solutions ($\sim 0.5\ M\ H^+$) the relative content of Pu(V) does not exceed 1% [3, Chap. 8] and increases to 10% with the decrease in the acid concentration to 0.1–0.2 *M* [13]. The maximum stability range of Pu(V) is at pH 1.5–3.5. A further increase in pH results in an intense hydrolysis of Pu(IV), an equilibrium shift towards the formation of Pu(IV) and its stabilization as the hydroxide, whose tendency to disproportionation is insignificant [177].

Reaction (21) is much slower than reaction (22), since it consists in the formation of a Pu–O bond. It is thought [602] that the slow reaction (21) takes place with the participation of hydrolyzed plutonium ions:

$$Pu^{4+} + H_2O \rightleftharpoons PuOH^{3+} + H^+, \qquad (23)$$

$$PuOH^{3+} + H_2O \rightleftharpoons Pu(OH)_2^{2+} + H^+, \qquad (24)$$

$$PuOH^{3+} + Pu(OH)_2^{2+} \rightleftharpoons Pu^{3+} + PuO_2^+ + H_2O + H^+, \qquad (25)$$

since these ions are apparently more reactive. In agreement with the mechanism of the slow stage here given, the disproportionation rate constant is directly proportional to $[Pu(IV)]^2$ and inversely proportional to $[H^+]^3$.

The rate constants in solutions of $HClO_4$ and HNO_3 are almost equal, while in HCl solutions the rate constant is several times higher [13, 602, 606]. It would appear that in solutions in which complex formation takes place, the complex ions of plutonium participate in the disproportionation reaction.

Disproportionation of Pu (V) ions. Equilibria in the system Pu (III)—Pu (IV)—Pu (V)—Pu (VI)

The behavior of Pu(V) in moderately acid solutions, at $[H^+] = 0.1–0.5\ M$, not containing plutonium in lower valency states, is of interest. Under these conditions, Pu(V) disproportionates with the formation of Pu(III), Pu(IV) and Pu(VI) (Figure 16). This process in hydrochloric acid solutions was studied in [345], and in nitric acid solutions in [13] and [177]. It was found that the reaction

$$2PuO_2^+ + 4H^+ \rightleftharpoons Pu^{4+} + PuO_2^{2+} + 2H_2O \qquad (26)$$

is the only one possible at the initial instant of time. However, Pu(IV) does not immediately accumulate in the solution. With the appearance of small amounts of Pu(IV), the most probable reaction is (22), which does not involve formation or scission of Pu—O bonds. With the appearance of Pu(III) reaction (26) loses its importance and the predominant reaction is

$$PuO_2^+ + Pu^{3+} + 4H^+ \rightleftharpoons 2Pu^{4+} + 2H_2O, \qquad (27)$$

the rate of which in a HCl solution is 35 times higher [345] and in HNO_3 solutions

30–37 times higher than the rate of reaction (26). The competition of this relatively fast reaction determines the autocatalytic nature of the process (Figure 17) which terminates in reaction (22), bringing the system of four valency states to equilibrium. For the rate constants of reaction (22) in $HClO_4$, HCl and HNO_3, see [486], [347] and [13], respectively.

In accordance with reactions (26) and (27), the stability of pentavalent plutonium increases with the decrease in the acidity. Solutions of Pu(V) not containing other valency forms are usually prepared by reducing Pu(VI) at pH 1.5–3.5.

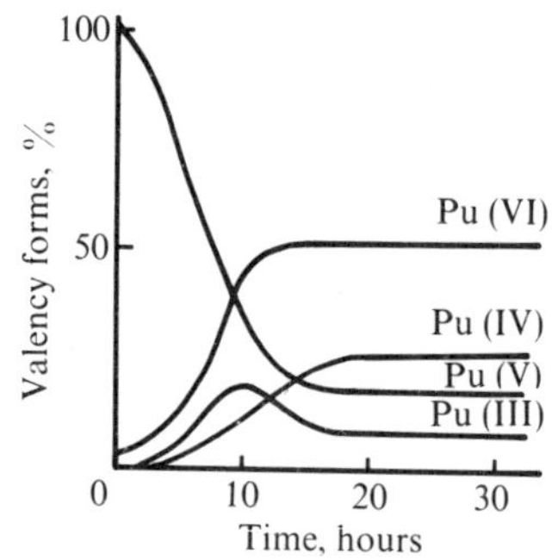

Figure 16
Disproportionation of Pu(V) in 0.2 *M* HNO_3 and 0.2 *M* $NaNO_3$.
Concentration of Pu(V), $(2–7) \cdot 10^{-3}$ *M*.

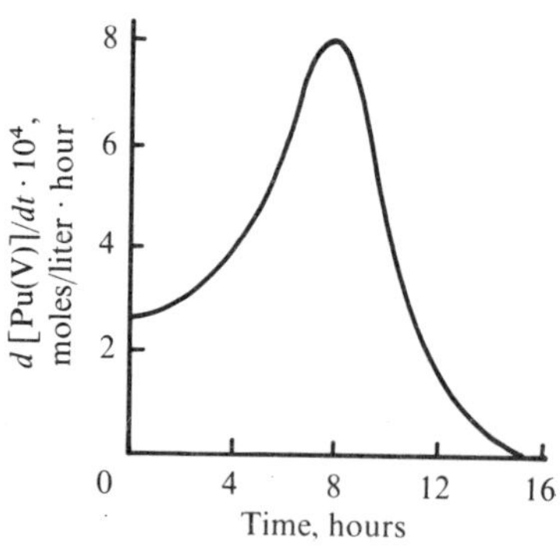

Figure 17
Dependence of the rate of disproportionation of Pu(V) in 0.2 *M* HNO_3 and 0.2 *M* $NaNO_3$ on time

At low acidity (pH 3.5–6) Pu(V) is in equilibrium with Pu(IV) hydroxide [425]:*

* In this pH range Pu(VI) partly becomes hydrolyzed. For this reason it is more correct to represent this process as

$$2PuO_2^+ + 3H_2O \rightleftharpoons PuO_2(OH)^+ + Pu(OH)_4 + H^+.$$

$$2PuO_2^+ + 2H_2O \rightleftharpoons PuO_2^{2+} + Pu(OH)_4. \quad (28)$$

At pH > 6 plutonyl hydroxide is formed:

$$2PuO_2^+ + 6H_2O \rightleftharpoons Pu(OH)_4 + PuO_2(OH)_2 + 2H_3O^+, \quad (29)$$

and pentavalent plutonium ceases to exist in significant amounts. The increase in the temperature reduces the stability of Pu(V).

For general information on the kinetics of disproportionation of Pu(IV) and Pu(V), see the review by Newton and Rabideau [580].

ACTION OF OXIDIZING AND REDUCING AGENTS

The similar redox potentials of pairs of plutonium ions, and the tendency of Pu(IV) and Pu(V) to disproportionate complicate the preparation of plutonium in one given valency state.

The transformations $Pu^{3+} \rightleftharpoons Pu^{4+}$ and $PuO_2^+ \rightleftharpoons PuO_2^{2+}$ do not involve formation or scission of Pu–O bonds and are therefore easily reversible. On the contrary, conversion of plutonium from the lower valency form Pu(III) to the higher valency form Pu(VI) and back, and also the transformations $Pu^{4+} \rightleftharpoons PuO_2^{2+}$, $Pu^{4+} \rightleftharpoons PuO_2^+$ and $Pu^{3+} \rightleftharpoons PuO_2^+$, which are accompanied by addition or detachment of oxygen, are more difficult to carry out.

The mechanism of these irreversible transformations, which take place under the action of oxidizing or reducing agents, is not always clear. Thus, for instance, the conversion $Pu^{4+} \rightleftharpoons PuO_2^{2+}$ may be represented as either a one- or two-stage process:

$$Pu^{4+} - 2e^- + 2H_2O = PuO_2^{2+} + 4H^+, \quad (30)$$

or

$$Pu^{4+} - e^- + 2H_2O = PuO_2^+ + 4H^+, \quad (31a)$$

$$PuO_2^+ - e^- = PuO_2^{2+}. \quad (31b)$$

Similar considerations apply to other irreversible conversions.

The one-electron reaction (31b) may be expected to proceed at a fast rate in both directions; the overall rate of the process would then be determined by the slow irreversible stage (31a).

However, a large proportion of the experimental data indicate that the true mechanism of the redox process is more complicated and that it involves, as one of the stages, disproportionation of the intermediate valency states. The following reactions then compete with reactions (30), (31a) and (31b):

$$\text{Oxidation process} \begin{cases} 3Pu^{4+} + 2H_2O = PuO_2^{2+} + 2Pu^{3+} + 4H^+, \\ Pu^{3+} - e^- = Pu^{4+}; \end{cases} \quad (32)$$

$$\text{Reduction process} \begin{cases} PuO_2^{2+} + e^- = PuO_2^+, \\ 2PuO_2^+ + 4H^+ = Pu^{4+} + PuO_2^{2+} + 2H_2O. \end{cases} \tag{33}$$

The final products of these reactions are the same: PuO_2^{2+} and Pu^{4+} Whether it is the direct oxidation (reduction) or the oxidation (reduction) through disproportionation which is the predominant process will depend on the experimental conditions: concentrations of plutonium and reagent, acidity of solution, complex formation both by plutonium ions and by electron donor ions.

At high concentrations of plutonium the probability of disproportionation is high. In strongly acid solutions and in the presence of complexing anions Pu(IV) will become stabilized, while the stability of Pu(V) will decrease. Complex formation between the ion of the reagent and the anion of the medium is also important. Thus, the reduction of Pu(IV) by Fe(II) or Sn(II) is accelerated in the presence of chloride ions. The mechanism of the one-electron reaction $Pu^{3+} \rightleftharpoons Pu^{4+}$ becomes more involved if an Me—O bond is formed or broken in the oxidizing or the reducing agent.

At present much empirical information is available on the redox behavior of plutonium ions in solutions. The most complete survey on the subject is that of Connick [3, Chap. 8]. Kinetic studies were carried out on only a limited number of reactions.

In what follows, the methods for the oxidation and reduction of plutonium will be considered from the point of view of obtaining the metal in a definite valency state.

Reactions Pu (IV) ⇌ Pu (III)

These reactions are easily reversible. Conversion of plutonium from tetravalent to trivalent state presents no special difficulty. The reverse reaction may be complicated by the disproportionation of Pu(IV) or by oxidation of this form in a higher valency state.

Reduction of Pu(IV)

Connick et al. [357] were the first to demonstrate the existence of trivalent plutonium by reducing Pu(IV) with *hydroxylamine.* To the green solution of Pu(IV) in concentrated nitric acid enough hydroxylamine hydrochloride was added for the final mixture to contain 0.14 *M* Pu(IV), 1.7 *M* H^+ and 0.3 *M* NH_3OH^+. After a few minutes standing at room temperature the solution assumed a blue tinge, and became deep blue after a few hours. The reduction in hydrochloric and sulfuric acid solutions was conducted in a similar manner, but in the latter case the rate of reduction was much slower because of complex formation. The reaction is accelerated by raising the temperature.

According to the results obtained on trace amounts of plutonium [3, p. 233], the expression for the reduction rate is

$$-\frac{d[\mathrm{Pu(IV)}]}{dt} = k[\mathrm{Pu(IV)}];$$

the rate constant is 1.2 min^{-1} at 18°, NH_3OH^+ concentration 0.1 *M* and HNO_3 concentration 0.5 *M*.

The hydroxylamine reduction is utilized in various analytical operations in the course of isolation and determination of plutonium: to transfer Pu(IV) from an organic to an aqueous phase, to desorb plutonium sorbed on an ion exchanger as the anionic Pu(IV) complex, in stabilization and storage of Pu(III) solutions, etc.

The reduction of Pu(IV) by *sulfurous acid* was described in early publications [3, p. 244, 464]. In a solution of 1 *M* HNO_3, saturated with H_2SO_3 and containing 0.0085 *M* Pu(IV), the reduction is complete within one minute. In hydrochloric acid solutions the reaction rate is also rapid.

Palei and Milyukova (1953) investigated the completeness of the reduction of gravimetric amounts of plutonium (0.0012–0.042 *M*) when sulfur dioxide is passed through the solution. They found that the reaction rate decreases with the increase in the hydrogen ion concentration. The time required for a full reduction at $[H^+] =$ 3 *N* is 5–10 minutes in HCl and HNO_3 and several hours in sulfuric acid solutions. Clearly, the more stable the Pu(IV) complex, the slower the rate of reaction.

A convenient method of preparation of solutions of Pu(III) free of extraneous ions is to reduce with gaseous *hydrogen* in the presence of platinized platinum as a catalyst [3, p. 222, 349].

Through a hydrochloric acid solution containing 10 mg/ml of Pu(IV) and about 2.4 *M* HCl, hydrogen is passed for 45 minutes at the rate of 5 cm^3 of hydrogen per minute per cm^3 of the liquid. The gas stream is directed against the platinum foil (2 cm^2 foil per 1 cm^3 solution). The extent of the reduction is more than 99% if precautions have been taken against the entry of oxygen which reacts with hydrogen in contact with platinum with the formation of hydrogen peroxide. For this reason the foil must be kept out of contact with the atmosphere. For the same reason, after washing the foil, the solution should be collected separately unless the oxygen has been previously removed.

In the presence of bright platinum the reduction is slow. In the absence of a catalyst the rate of reduction is negligible [349].

In acid solutions Pu(IV) may be fairly rapidly reduced by *hydrazine* and *hydroquinone* [3, p. 217]. Both reagents are utilized in the reductive reextraction of Pu(IV) from the organic phase [37; 66, p. 701]. In [66] it is pointed out that the reaction becomes much slower when the concentration of Pu(IV) increases above 2 μg/ml during the reextraction with a solution of 0.3 *M* N_2H_4 + 0.2 *M* HNO_3.

The formal redox potential of Fe^{3+}/Fe^{2+} (+ 0.771 V relative to n. h. e.) indicates that *ferrous iron* may act as reducing agent for tetravalent plutonium. Newton and Cowan [579] studied the reduction kinetics of Pu(IV) in 2 *M* $HClO_4$. The rate constants in the equation

$$d[\mathrm{Pu(IV)}]/dt = k_1[\mathrm{Pu}^{4+}][\mathrm{Fe}^{2+}] + k_2[\mathrm{Pu}^{4+}][\mathrm{Fe}^{2+}][\mathrm{H}^+]^{-1}$$

are $k_1 = 10.6\,M^{-1}\,\mathrm{min}^{-1}$; $k_2 = 187\,\mathrm{min}^{-1}$ at 2.5°. Small amounts of sulfuric acid ($9 \cdot 10^{-4}\,M$ H_2SO_4 in 2 *M* $HClO_4$) increase the rate of the process almost fivefold. Chloride ions also bring about a considerable increase in the reaction rate. In nitric acid media Fe(II) rapidly reacts with Pu(IV). This reagent is most frequently employed in mixture with sulfamic acid HNH_2SO_3. The ferrous ions and the sulfamic acid ions act independently: ferrous ions reduce Pu^{4+}, while the sulfamate decomposes nitrites and nitrogen oxides, which may oxidize ferrous iron and interfere with the formation of Pu^{3+}.

An aqueous solution of iron (II) sulfamate in mixture with hydrazine in 0.2 *M* HNO_3 is employed during reductive reextraction of Pu(IV) from solutions of tributyl phosphate in inert solvents [66]. The stability of iron (II) sulfamate decreases with the increase in the concentration of nitric acid. In 2 *M* HNO_3 Fe(II) is oxidized within two weeks, and in 0.2 *M* HNO_3 the solutions will keep for a year [337]. Iron (II) sulfamate is considered to be the best reagent for the reductive reextraction of Pu(IV) [337], since its reducing action on Pu(IV) is fast, and it is readily oxidized if it is desired, say, to oxidize Pu(III).

In sulfuric acid solutions the potentials Pu(IV)/Pu(III) and Fe(III)/Fe(II) are so near that equilibrium between the ions is easily attained [3, p. 225] and the reaction between Pu(IV) and Fe(II) does not proceed to completion.

A rapidly acting reducing agent is the *iodide ion* [3, p. 223], since the potential I_2/I^- is + 0.535 V. The time required for a 50% reduction to occur in a solution containing $10^{-3}\,M$ Pu(IV), 0.1 *M* KI and 0.4 *M* HCl at room temperature is 2 minutes. The rate of the reduction in dilute sulfuric acid is probably somewhat slower than in HCl and HNO_3 solutions. The reagent is utilized for the desorption of Pu(IV) sorbed on ion exchangers.

The value of the Sn(IV)/Sn(II) potential (+ 0.15 V) ensures a rapid reduction of Pu(IV) to Pu(III) by *stannous tin* [3, p. 244]. Rabideau [605] found that the overall reaction between Pu(IV) and Sn(II) is a resultant of two simultaneous reactions and strongly depends on the concentration of chlorides:

$$-\frac{d[\mathrm{Pu(IV)}]}{dt} = k_1[\mathrm{Pu}^{4+}][\mathrm{Sn}^{2+}][\mathrm{Cl}^-]^4 + k_2[\mathrm{Pu}^{4+}][\mathrm{Sn}^{2+}][\mathrm{Cl}^-]^5.$$

At 25°, $k_1 = 720\,M^{-5} \cdot \mathrm{sec}^{-1}$ and $k_2 = 1636\,M^{-6} \cdot \mathrm{sec}^{-1}$. It was found that the rate of the reaction is independent of the hydrogen ion concentration between 0.445 and 2.0 *M*, but increases with the temperature.

The reported value [146, p. 270] of the potential of V(IV)/V(III) is + 0.361 V. Rabideau and Kline [607] investigated the kinetics of the reaction between Pu(IV) and *trivalent vanadium* in perchloric acid solutions ($\mu = 2$):

$$-\frac{d[\mathrm{Pu(IV)}]}{dt} = k_1[\mathrm{Pu}4^+][\mathrm{V}^{3+}][\mathrm{H}^+]^{-1} + k_2[\mathrm{Pu}^{4+}][\mathrm{V}^{3+}][\mathrm{H}^+]^{-2}.$$

At 2.4° $k_1 = 1.70\ \mathrm{sec}^{-1}$ and $k_2 = 1.71\ M \cdot \mathrm{sec}^{-1}$. At 20° these values increase to $11.9\ \mathrm{sec}^{-1}$ and $20.1\ M \cdot \mathrm{sec}^{-1}$, respectively.

According to Latimer [146, p. 277], the potential of Ti(IV)/Ti(III) is about + 0.1 V. Metz [547] reported the use of the reduction of Pu(IV) by *trivalent titanium* in potentiometric analysis.

Rabideau and Kline [609] studied the kinetics of this reaction in solutions of perchloric acid:

$$-\frac{d[\mathrm{Pu(IV)}]}{dt} = k[\mathrm{Pu}^{4+}][\mathrm{Ti}^{3+}][\mathrm{H}^+]^{-1}.$$

The value of the rate constant at 25° in the acidity range 0.2–2.0 M is $65.5\ \mathrm{sec}^{-1}$. The presence of small amounts of chlorides is immaterial.

Tetravalent uranium has been used as reducing agent in the reextraction of Pu(IV) and Pu(VI) from the organic phase into the aqueous phase. The U(VI)/U(IV) potential is + 0.334 V.

According to Newton [576], the process in perchloric acid comprises two stages:

$$\mathrm{Pu}^{4+} + \mathrm{U}^{4+} + 2\mathrm{H_2O} = \mathrm{Pu}^{3+} + \mathrm{UO_2^+} + 4\mathrm{H}^+.\} \quad \text{Rate-determining stage} \qquad (34a)$$

or

$$\left.\begin{aligned} \mathrm{Pu}^{4+} + \mathrm{UO_2^+} &= \mathrm{Pu}^{3+} + \mathrm{UO_2^{2+}} \\ 2\mathrm{UO_2^+} + 4\mathrm{H}^+ &= \mathrm{UO_2^{2+}} + \mathrm{U}^{4+} + 2\mathrm{H_2O}. \end{aligned}\right\} \quad \text{Rapid stages} \qquad (34b)$$

The constant k in the expression $-\frac{d[\mathrm{Pu(IV)}]}{dt} = k[\mathrm{Pu}^{4+}][\mathrm{U}^{4+}][\mathrm{H}^+]^{-2}$ is $2175\ M \cdot \mathrm{min}^{-1}$ at 20°.

Rydberg [631] and Jenkins [479] studied this reaction in nitric acid media. It was shown by Rydberg [631] that Pu(IV) at concentrations of $10^{-5}\ M$ is reduced in wide ranges of acidity and U(IV) concentration (0.1–10 M $\mathrm{HNO_3}$; 0.01–0.1 M U(IV)). The best results were obtained in a solution of 1 M $\mathrm{HNO_3}$ at the maximum uranium content: the reduction is 99% complete within a few minutes. It was found by Rydberg that the reextraction of plutonium from methyl isobutyl ketone by uranium (IV) is more rapid and more complete than by hydroxylamine and ferrous sulfamate.

According to Jenkins [479], Pu(IV) in a solution of 3 M $\mathrm{HNO_3}$ containing 10^{-6}–$10^{-7}\ M$ Pu(IV) in the presence of 0.1 M sulfamic acid is 50% reduced by the excess of U(IV) nitrate within 1–5 minutes. It was pointed out [479, 631] that oxygen interferes with and decreases the extent of the reduction. Jenkins showed that the oxidation of Pu(III) by air in the presence of U(IV) increases with the decrease in the concentration of nitric acid and that U(IV) acts as a catalyst of the process. The reaction rate is rapid in an inert gas atmosphere. Solutions of tetravalent uranium in 1 *M* perchloric and hydrochloric acids containing 0.1 mole of U(IV) per liter are stable in the dark under nitrogen for several months. When the concentration of the reducing agent decreases to 0.001 *M*, it becomes rapidly oxidized (~ 10% per hour). The nitric acid solution of $U(NO_3)_4$ is prepared by electrolysis of uranyl nitrate on a mercury cathode in the presence of sulfamic acid, which is introduced in order to suppress the formation of nitrites. In this case the stability of U(IV) is satisfactory. In solutions containing $2 \cdot 10^{-2}$ *M* U(IV) in 3 *M* HNO_3 + 0.1 *M* HNH_2SO_3, from which oxygen has not been removed and which are not protected from light, uranium is oxidized at the rate of 15% in 5 days. In the absence of sulfamic acid the oxidation rate is several dozen percent per hour.

During the extractional purification of plutonium the utilization of U(IV) as reducing agent is expedient in stages prior to the complete separation of uranium.

The mechanism of the reaction between solutions of plutonium and *hydrogen peroxide* is not yet clear. The rate of reduction of Pu(IV) depends on the concentration of plutonium, hydrogen peroxide and the nature of the acid [3, p. 239, 353]. The possibility of the formation of peroxide complexes and of the occurrence of the reverse reaction cannot be ignored. When Pu(IV) is present in macro concentrations, a mobile equilibrium between Pu(III) and Pu(IV) is established. In 0.5 *M* HCl, the equilibrium is shifted towards formation of Pu(III), and only a few percent of Pu(IV) is present in the solution. In sulfuric acid media, owing to the stabilization of Pu(IV) by sulfate ions, the equilibrium is shifted in the opposite direction: dilute hydrogen peroxide oxidizes Pu(III) rapidly and practically quantitatively. Under the effect of a large excess of H_2O_2 trace amounts of Pu(IV) appear to pass into higher valency states.

For a survey of the experimental results on the action of hydrogen peroxide, see Connick [3, p. 238].

Brief literature references are available on the reduction of Pu(IV) to Pu(III) by *formaldehyde* [282], *chromium (II) sulfate* [423], *ascorbic acid* [315, 414] and *hydrogen sulfide* [3, p. 244]. Most of these methods are of no practical value.

Complexonometric and redox titrimetric determination of plutonium are carried out after the Pu(IV) in sulfuric acid solution has been reduced by certain *metals* and their amalgams. Koch [3, p. 245] and Pietri and Baglio [592] used zinc amalgam, Kharlamov (1954), Palei and Kochetkova (1956) carried out the reduction with bismuth, and Barker [288] used silver. These metals are commonly employed to prepare tetravalent uranium.*

* The electrolytic methods of oxidation and reduction will be discussed below, p. 67.

Oxidation of Pu(III) *to* Pu(IV)

The oxidants MnO_4^-, VO_3^-, $Cr_2O_7^{2-}$, BiO_3^- and Ce^{4+} in acid medium rapidly oxidize trivalent plutonium to the tetravalent state. The oxidation to a higher state is much slower, especially at room temperature. This is the principle behind the use of certain of these reagents (VO_3^-, $Cr_2O_7^{2-}$, MnO_4^-, Ce^{4+} in the oxidimetric titration of Pu(III) to the stage of Pu(IV).

The oxidation of Pu(III) to Pu(IV) in the various stages of purification of plutonium and in preparing solutions of Pu(IV) free from extraneous ions is usually carried out by milder reagents.

Connick and McVey [350] studied the oxidation of Pu(III) by *oxygen* in hydrochloric acid solutions. Passing oxygen through a solution of 0.0052 *M* Pu(III) in 0.5 *M* HCl during 20 minutes, after which the oxygen-saturated solution was left to stand for two days, produced no significant changes in the valency state. Oxidation to the extent of a few percent occurs only on heating at 97° for four hours in an atmosphere of oxygen. Similar results were obtained in solutions of perchloric acid.

The oxidation of Pu(III) in sulfuric acid solutions proceeds at a measurable rate [3, p. 235]. The reaction kinetics were studied by Newton and Baker [577]. When hydrochloric or nitric acid solutions were boiled with concentrated sulfuric acid, plutonium could be prepared in the tetravalent state only; the product obtained by evaporation of these solutions to dryness had the formula $Pu(SO_4)_2$.

Reactions with *nitrates* and *nitrites* are widely employed in analytical and preparative work to prepare solutions of tetravalent plutonium; the mechanism of the reactions is unclear. The potentials of the system

$$NO_{gas} \xrightarrow{+0.99V} HNO_2 \xrightarrow{+0.94V} NO_3^- \qquad (NO_{gas} \text{ — } NO_3^-: +0.96\ V)$$

are close to that of Pu(IV)/Pu(III) system, which is + 0.92 V in 1 *M* HNO_3. It was experimentally found that the NO_3^- ion oxidized Pu(III) to Pu(IV) in acid media, and under certain conditions the oxidation may even proceed to Pu(VI). In a solution containing 0.5 *M* HCl, 0.5 *M* $NaNO_3$ and $5.7 \cdot 10^{-3}$ *M* Pu(III), plutonium becomes 87% oxidized to a mixture of Pu(IV) and Pu(VI) within 15 minutes at 100° [351]. At room temperature the extent of the oxidation is only a fraction of one percent even after several hours.

The oxidation by nitrous acid is much more rapid; the authors [351] consider that the oxidation by nitrate proceeds slowly until significant amounts of its reduction products—HNO_2 and NO—have been formed. Nitric oxide in turn reacts with nitrate ions to form nitrous acid, which accelerates the reaction, so that the process is actually autocatalytic.

In a later study, the rate constants of the reactions in HCl and $HClO_4$ (35a) and in HNO_3 (35b) were determined [390]:

$$-\frac{d[\,Pu(III)\,]}{dt} = k_1[\,Pu(III)\,][\,H^+\,][\,HNO_2\,];\; k_1 = 18 \pm 3.5\,M^{-2}\cdot min^{-1}, \quad (35a)$$

$$-\frac{d[\,Pu(III)\,]}{dt} = k_2[\,Pu(III)\,][\,H^+\,][\,NO_3^-\,][\,HNO_2\,];$$

$$k_2 = 90 \pm 20\ M^{-3}\cdot min^{-1}. \quad (35b)$$

It will be seen that chloride and perchlorate ions have no effect on the reaction rate, whereas nitrate ions have a strong accelerating effect. The authors give a discussion of the possible oxidation mechanisms.

The treatment of solutions of Pu(III) or precipitates of $Pu(OH)_3$ by concentrated nitric acid with heating results in an instantaneous oxidation of Pu(III) to Pu(IV). The concentration of such solutions by evaporation should be carried out in 14–16 *M* HNO_3 because there is some danger of Pu(IV) being oxidized to Pu(VI) at lower acidities (Nikol'skii, Rudenko, 1956).

The oxidation of Pu(III) by concentrated nitric acid and by nitrous acid is a widely employed method for the preparation of tetravalent plutonium.*

Iodate and *bromate ions* rapidly oxidize Pu(III) to Pu(IV). The rate of oxidation of Pu(IV) to hexavalent plutonium depends on a number of factors, in the first place on the temperature and the identity of the acid [3, p. 214]. In moderately concentrated (1–1.5 *N*) sulfuric acid the reaction is slow even at elevated temperatures.

In the first tentative separations of plutonium and neptunium [531], the reduced sulfuric acid solutions were treated with bromate to oxidize Pu(III) to Pu(IV), and Np(III) and Np(IV) to Np(VI), after which the elements were separated by co-precipitating Pu(IV) on lanthanum fluoride.

The oxidation of Pu(III) to Pu(IV) by $KBrO_3$ in dilute nitric acid seems to be difficult to effect quantitatively because of the danger of the Pu(IV) undergoing further oxidation or disproportionation.

Reaction Pu (VI) ⇌ Pu (V)

This reaction is of minor practical importance. It is very easy to oxidize Pu(V) to Pu(VI); the reverse reaction must be carried out with certain precautions owing to the instability of pentavalent plutonium in strongly acid and neutral solutions.

Reduction of Pu(VI)

Pu(V) was first successfully prepared by Kraus et al. [425, 513]. They effected the reduction of Pu(VI) using *iodide ions.* The reaction is very fast at pH 3.2–3.5 and may be followed potentiometrically. The extent of reduction is not increased in the

* The action of nitrate ions on other valency states will be discussed below (p. 62).

presence of excess iodide. The liberated iodine is removed by extraction with carbon tetrachloride.

Hydrogen peroxide is known [3, p. 235, 353] to reduce Pu(VI) to Pu(V) at a measurable rate in 0.5 *M* HCl. Under these conditions lower valencies also appear, owing to the disproportionation of Pu(V). The rate of oxidation of Pu(V) by hydrogen peroxide is much lower than the rate of reduction of Pu(VI), and at equilibrium Pu(V) is the predominant species. The oxidation of Pu(V) by hydrogen peroxide is catalyzed by ferric ions.

In perchloric acid solutions containing $4.2 \cdot 10^{-4}$ mole of Pu(VI) per liter, at pH 3, plutonium is quantitatively reduced to Pu(V) in 20 minutes in the presence of 0.05 *M* H_2O_2. In the optimum pH range (2–4) rapid reduction or disproportionation of Pu(V) did not take place [3, p. 238] when the concentration of plutonium was increased to $7.6 \cdot 10^{-3}$ *M*, that of hydrogen peroxide to 1.0 *M* and the temperature to 50°.

Artyukhin, Medvedovskii and Gel'man [13] used H_2O_2 to obtain nitric acid solutions of Pu(V). According to their results, the reduction of Pu(VI) at pH 3–4 is 95–97% complete.

The action of *nitrous acid* on Pu(VI) and Pu(V) solutions was studied in [3, p. 231]. In a nitric acid solution containing $1.7 \cdot 10^{-3}$ *M* Pu(VI), 0.1 *M* HNO_3, 0.1 *M* $NaNO_3$ and 0.1 *M* HNO_2, 93.8% Pu(VI) were rapidly reduced to Pu(V); the reaction was 50% complete in less than 25 seconds. The subsequent appearance of lower valencies was due to the disproportionation of Pu(V).

Attempts have been made to reduce Pu(VI) to Pu(V) by *hydroxylamine* [3, p. 232] and *hydrazine* [3, p. 233] in moderately acid (about 0.5 *M*) solutions. The appearance of Pu(III) was noted until the disappearance of all Pu(VI); this was clearly caused by the disproportionation of Pu(V).

Kraus and Moore [3, p. 343] quantitatively converted Pu(VI) ($5 \cdot 10^{-4}$ *M*) to Pu(V) in 5 minutes at pH 1.9 by treatment with $3.3 \cdot 10^{-3}$ *M* solution of sulfurous acid. Apparently, for this reaction to proceed to completion, the pH must be rigidly maintained in the optimum 1.5–3.5 range.

The method of preparation of Pu(V), proposed by Markin and McKay [536], is very interesting and is based on a rapid establishment of the equilibrium between plutonium ions according to reaction (22) (p. 47). Equal volumes of solutions of Pu(VI) and Pu(III) in 0.2 *M* HNO_3, with the concentration of each component at 2 mg/ml, are mixed in the presence of 0.1% solution of dibutyl phosphate in benzene. The tetravalent plutonium, formed in the reaction between Pu(VI) and Pu(III), is extracted into the organic phase, while Pu(V) remains in the aqueous layer. Pu(VI) and Pu(III) are fully converted into Pu(IV) and Pu(V) owing to the Pu(IV) being removed from the system. It was shown by spectrophotometric determinations that after 15 minutes extraction Pu(V) alone remains in the aqueous phase.

It is not possible to obtain a quantitative yield of Pu(V) by direct oxidation of lower valencies of plutonium, since the strong oxidants used for the purpose rapidly oxidize Pu(V) to Pu(VI) even before all Pu(IV) has been oxidized to Pu(V).

Oxidation of Pu(V).

Pu(V) may be readily oxidized to Pu(VI) by practically any reagent which can be used to oxidize Pu(III) to Pu(IV). During the oxidation of Pu(III) and Pu(IV) to Pu(VI) any pentavalent plutonium which may be formed as an intermediate product is rapidly converted to Pu(VI).

Reactions Pu (VI) $\rightleftharpoons$ Pu (IV) and Pu (VI) $\rightleftharpoons$ Pu (III)

These reactions involve formation or scission of a Pu—O bond. They are accordingly effected with strong oxidizing or reducing agents.

Oxidation of Pu(IV) *and* Pu(III) *to* Pu(VI)

Bivalent silver is a strong oxidant, since the potential Ag(II)/Ag(I) is about + 1.93 V in 4 *M* HNO_3 [146, p. 195]. This is one of the first reagents which were employed to oxidize trace amounts of Pu(III) and Pu(IV) to Pu(VI) [3, p. 241].

The reaction takes less than a minute* if conducted in 1–11 *M* HNO_3 with Ag(II) concentration between 0.01 and 0.1 *M*. The oxidation seems to be somewhat slower in sulfuric acid medium.

The conversion of microgram quantities of plutonium ions into the hexavalent state by *persulfate ions* in the presence of silver (I) ions [3, p. 241, 659] is rapid. Persulfate ions oxidize Ag(I) to Ag(II) or Ag(III), and these ions react with Pu(III) and Pu(IV). At $(NH_4)_2S_2O_8$ concentration of about 0.1 *M*, and $AgNO_3$ concentration of 0.01 *M*, the oxidation in 1 *M* sulfuric or nitric acid takes less than one minute, even at room temperature. The reaction rate in 10 *M* HNO_3 is somewhat slower. The oxidation of macro amounts of Pu(IV) also takes only a short time.

Yakovlev and Gorbenko-Germanov [256] took two hours to oxidize Pu(IV) to Pu(VI) by ammonium persulfate (about 20 g/liter) in concentrated solutions of potassium carbonate containing about 4 g/liter of plutonium at about 80°.

Earlier studies of Seaborg and Wahl [659] showed that trace amounts of Pu(IV) may be converted into a higher state of valency by treatment with *hexavalent chromium* (oxidation potential + 1.33 V) at elevated temperatures.

The rate of this process is proportional to the concentration of plutonium

$$-\frac{d[\mathrm{Pu(IV)}]}{dt} = k[\mathrm{Pu(IV)}]$$

* Solutions of Ag(II) are prepared by ozone oxidation of solutions of Ag(I) [146, p. 195] or by dissolving AgO or Ag(II) peroxynitrate in cold 9 *M* HNO_3 + 0.4 *M* $AgNO_3$[312].

and decreases by two orders of magnitude when the $HClO_4$ concentration is increased from 0.05 *M* ($k = 4\ min^{-1}$) to 0.1 *M* ($k = 0.02\ min^{-1}$) [3, p. 221]. The rate of oxidation decreases particularly rapidly with the increase in the concentration of nitric and sulfuric acids, nitrate ions and uranyl nitrate [3, p. 221; Zakolupin, 1957].

It was found [3, p. 220] that the times required to oxidize macro and micro amounts of Pu(IV) are not very different. In the presence of excess Cr(VI) and at about 95° the reaction is completed within one hour.

The potential of Bi(V)/Bi(III) is about + 1.6 V; accordingly, *pentavalent bismuth* in the form of solid sodium bismuthate rapidly oxidizes plutonium from lower valency states to Pu(VI) [3, p. 214]. The oxidation of $5 \cdot 10^{-3}$ *M* Pu(IV) in 5 *M* HNO_3 takes 5 minutes when the concentration of $NaBiO_3$ is 0.84 g/liter. According to Moiseev et al. (1951) the oxidation takes a longer time in sulfuric acid medium: 20–40 ml of a solution containing 1 mg Pu(IV) per ml and 150 mg sodium bismuthate in 1 *N* H_2SO_4 were boiled for 15 minutes.

The reaction with trace amounts of plutonium is slower, but the reaction rate can always be increased by stirring the solid phase in the solution.

The potentials in the system Mn(VII)–Mn(IV)–Mn(II) indicate that *potassium permanganate* and *manganese dioxide* may oxidize plutonium from lower valencies to Pu(VI):

$$MnO_4^+ \xrightarrow{+1.695V} MnO_2 \xrightarrow{+1.23V} Mn^{2+}$$

The rate of oxidation of trace amounts at room temperature is, however, slow. In a solution of 1 *M* HNO_3 + 0.001 *M* $KMnO_4$ the oxidation was 50% complete after 50 minutes. With the increase in the concentration of HNO_3 to 5–6 *M*, and more particularly of H_2SO_4 to 3–4 *N*, the reaction rate becomes markedly slower. It is considerably accelerated at higher temperatures.

Potassium permanganate was tested as titrant for potentiometric determination of macro amounts of plutonium. The reactions tested were Pu(III)→Pu(IV) at room temperature and Pu(IV)→Pu(VI) at 60°. Two-stage titration makes it possible to determine plutonium in the presence of reducing impurities, e. g. Fe(II), which become oxidized during the first stage.

In 45% solution of K_2CO_3, containing $5 \cdot 10^{-3}$ *M* Pu(IV), the permanganate oxidation of plutonium to Pu(VI) takes a few minutes at room temperature [3, p. 226].

As compared to oxidation by potassium permanganate, the oxidation of plutonium to Pu(VI) by manganese dioxide is slow. Pu(IV) or a mixture of different valencies in 10^{-3}–10^{-2} *M* solutions of plutonium in 0.5 *M* HNO_3 are oxidized by heating on a water bath at 90° for one hour [39, 90, 177]. The excess MnO_2 is removed by centrifugation. To obtain solutions of Pu(VI) not containing Mn(II), the former ion is extracted with ether or with another suitable solvent.

The formal oxidation potential of *tetravalent cerium* in 1 *M* $HClO_4$ is + 1.70 V; in 1 *M* HNO_3 it is + 1.61 V, i. e., sufficient to oxidize plutonium to the hexavalent state. In fact, in perchloric and nitric acid media the reaction is rapid even without heating: in 0.5 *M* HNO_3 containing $1–5 \cdot 10^{-3}$ *M* Pu the oxidation of plutonium takes a few minutes at 25° [3, p. 218].

In 1 *M* H_2SO_4 the potential of Ce(IV)/Ce(III) decreases to + 1.44 V, while that of Pu(VI)/Pu(IV) increases from +1.04 to +1.3 V. In this connection the oxidation rate in sulfuric acid media considerably decreases: a complete oxidation of $7 \cdot 10^{-3}$ *M* Pu(IV) in 1 *M* H_2SO_4 takes 6 hours at room temperature and a twofold excess of ceric ions. The slow rate of the oxidation of Pu(IV) in sulfuric acid makes it possible to effect a quantitative redox titration of Pu(III) to Pu(IV) in 3–4 *N* H_2SO_4.

Ozone is an oxidant with the aid of which it is possible to prepare solutions of Pu(VI) free from extraneous ions. The O_3/O_2 oxidation potential is + 2.07 V.

Trials carried out on microgram amounts of plutonium in sulfuric acid medium showed that the rate of oxidation decreases rapidly with the increase in the concentration of sulfuric acid [3, p. 234]. The reaction is considerably accelerated in the presence of Ce(III) and Ag(I) as catalysts. Thus, in a solution containing 0.03 *M* H_2SO_4, $1.45 \cdot 10^{-3}$ *M* Ce(IV), the oxidation of indicator amounts of Pu(IV) to Pu(VI) with the aid of ozonized oxygen (7% ozone) takes place at 0°.

The possibility of oxidation of plutonium by ozone was pointed out as early as 1948 by Grinberg et al. Tarakanov and Skarbov (1954–1955) made a detailed study of the oxidation of milligram amounts of the element in nitric acid solutions. The oxidations were carried out with ozonized air (0.6–1.1% O_3) or ozonized oxygen (1.5–2.2% O_3), which were bubbled through the solutions at a rate of about 5 liters per hour. The reaction rate is a function of the concentration of plutonium. Thus, in 1 *M* nitric acid containing 0.06 g Pu per liter, complete oxidation by ozonized air at 40° was attained in 60 minutes, and when the plutonium content was 0.6 grams per liter—in 100 minutes. The effect of nitric acid concentration and temperature is shown in Figure 18.

It was found by Tarakanov and Skarbov that the rate of oxidation by ozonized oxygen is 2–3 times higher than that by ozonized air, i. e., it would appear to be proportional to the concentration of ozone. The presence of about 100 g/liter of uranyl nitrate has little effect on the reaction rate, while small amounts of added Mn(II) produce a marked acceleration of the reaction rate.

Macro amounts of Pu(IV) have been oxidized in concentrated potassium carbonate solution by passing ozone at 90–100° for 15–20 minutes [256].

In dilute solutions of *perchloric acid* at room temperature, oxidation of lower valency forms to Pu(VI) is not noted [3, p. 219]. The dissolution of metallic plutonium in 72% $HClO_4$ yields solutions containing trivalent plutonium only. A prolonged treatment of Pu(IV) nitrate or Pu(IV) chloride with concentrated perchloric acid with heating results in complete conversion of plutonium to the hexavalent

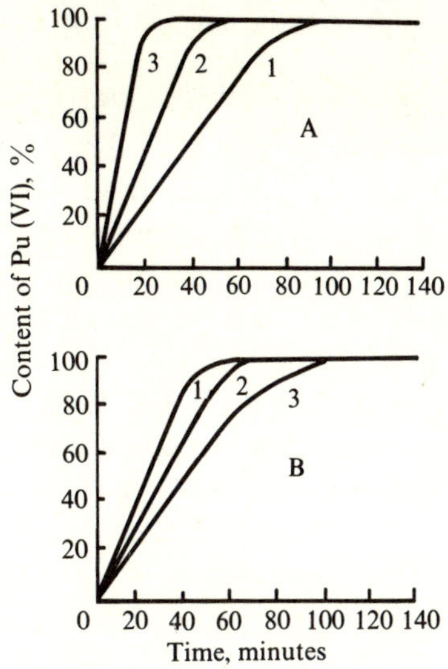

Figure 18

Rate of oxidation of Pu(IV) to Pu(VI) by ozonized air as a function of temperature (*a*) and concentration of nitric acid (*b*).

A Concentration 1 *M* HNO_3; temperatures: 1) 40°C, 2) 60°C, 3) 80°C.
B Temperature 60°C; concentrations of HNO_3: 1) 1 *M*, 2) 2 *M*, 3) 4 *M*. Pu concentration 220 mg/liter.

state. Evaporation of the product solutions yields the salt $PuO_2(ClO_4)_2$ [483].

Helbig [457] converted plutonium to the Pu(VI) form by treating micro volumes (of the order of 1 microliter) of solutions containing about 1 gram of plutonium per liter with concentrated perchloric acid containing 10% HNO_3. Evaporation to a volume of 0.1 microliter during not more than 20 seconds at a temperature close to the boiling point will ensure that Pu(VI) is obtained as the final product. Slow evaporation results in incomplete oxidation.

Potassium bromate [659] was also tried out as oxidant. In 0.1–3 *M* sulfuric acid the rate of oxidation of Pu(IV) to Pu(VI) is very slow. At room temperature the oxidation of 1 mg per ml of Pu(IV) by 0.2 *M* bromate in 1 *M* H_2SO_4 was only 1% complete in 48 hours [245]. It was found [3, p. 214] that indicator amounts of plutonium are oxidized at 95° in 0.05 *M* H_2SO_4 to the extent of over 90% in one hour. In nitric acid the reaction rate is much higher [3, p. 215, 245]. Plutonium in the form of 0.2 *M* Pu(IV) in 1 *M* HNO_3, oxidized by excess potassium bromate at 85°, is fully converted to the hexavalent form in four hours. The oxidation of macro amounts of plutonium takes place by the disproportionation mechanism.

Bromate ions oxidize neptunium to Np(VI) at a fairly rapid rate. Neptunium ($6 \cdot 10^{-3}$ *M*) in 1 *M* H_2SO_4 + 0.1 *M* $KBrO_3$ at 35° becomes oxidized within one hour while plutonium remains almost totally present as Pu(IV) [245]. Subsequent separation of plutonium from neptunium may be effected by extraction, ion-exchange and precipitation techniques.

The oxidation by *lead dioxide* [Pb(IV)$\xrightarrow{+1.455\text{ V}}$ Pb(II)] at room temperature is slow, but can be accelerated by heating [3, p. 225]. The reaction rate increases with the increase in the amount of the reagent. Lead tetraacetate oxidizes trace amounts of Pu(IV) to Pu(VI) in 5 *M* HNO_3 within one hour.

Excess of *hypochlorite* converts both gravimetric and trace amounts of Pu(IV) to Pu(VI) during 5–15 minutes at 90° in solutions at pH 4.5–8.2 and in concentrated carbonate media [3, p. 219, 245].

The reactions of macro amounts of Pu(IV) and Pu(VI) with *nitrate* and *nitrite ions* in acid media are very complex, and their mechanism is not yet clear. It is known that Pu(IV) is readily preserved in 1–2 *M* HNO_3 at room temperature for several months, but is oxidized when the solution is heated. In a solution of 0.5 *M* HNO_3 containing $3.93 \cdot 10^{-3}$ *M* Pu(IV), plutonium is 92% oxidized in $4\frac{1}{2}$ hours when heated on a water bath [508]. With the increase in the content of HNO_3 the rate of reaction shows a rapid drop. At an acidity of 7.7 *M* there is no significant oxidation of Pu(IV).

It is thought [3, p. 227] that the reaction mechanism involves the disproportionation of Pu(IV) to Pu(III) and Pu(VI) and the oxidation of Pu(III) to Pu(IV) by nitrate ions at elevated temperatures. Clearly, an increase in the concentration of nitric acid stabilizes Pu(IV). The oxidation of trace amounts of Pu(IV) takes place very slowly. This is undoubtedly an indication that the overall reaction rate is determined by the rate of disproportionation. During the oxidation of macro amounts, the reaction rate is probably determined by the stage Pu(III) → Pu(IV).

In selecting the conditions of the oxidation of plutonium by nitrate ions it must be remembered that polymers of Pu(IV) may be formed at elevated temperatures and in moderately acid solutions (0.5–0.7 *M* HNO_3). The concentration of plutonium should most probably be not more than 3–4 g/liter, in order to ensure that the element remains in the ionic forms (curves 4 and 5 in Figure 9).

In dilute solutions of sulfuric acid the rate of oxidation of Pu(IV) by nitrate ions is very low [3, p. 229].

It has been shown in the discussion of the oxidation of Pu(III) that nitrous acid rapidly oxidizes Pu(III) to Pu(IV). With respect to Pu(VI) the nitrite ion acts as a reducing agent. In moderately acid solutions Pu(VI) is not quantitatively reduced to Pu(V) because of the fairly fast disproportionation of Pu(V), which is considerably accelerated at elevated temperatures. The Pu(III) formed by the disproportionation is in turn oxidized to Pu(IV) by the nitrite ions. In this way, nitrite treatment of nitric acid solutions containing plutonium in several states of valency results in the entire plutonium being converted to the tetravalent state. This technique is utilized to prepare Pu(IV) both in analytical practice and in technology.

Reduction of Pu(VI) *to* Pu(IV) *and* Pu(III)

These reactions are frequently used during reductive reextraction of Pu(VI) from the organic into the aqueous phase. Their other application is the conversion of a mixture of plutonium ions in various valency states including Pu(VI) to one single valency state—Pu(III) or Pu(IV). The reactions with nitrite ions described above are an example of such an application.

Ferrous iron is used in potentiometric [547, 592] and amperometric [457] titration of hexavalent plutonium.

This reagent is also employed in the reductive reextraction of Pu(VI) from the organic into the aqueous phase. The addition of sulfamic acid to the solution of ferrous iron ensures a complete reduction of Pu(VI) to Pu(III). Nitrite ions accumulating in nitric acid solutions oxidize ferrous iron and thus prevent the reduction of Pu(VI). This interference is removed by the amino group of the sulfamate which reacts with the nitrite. The system Pu(VI)–Fe(II) was studied in 1956 by Volkova. In order to reduce Pu(VI) to Pu(III) it is enough to have a twofold excess of Fe(II) when working with macro amounts, and a fivefold excess when working with trace amounts. According to the author, the content of sulfamic acid may be as low as one-sixtieth of the concentration of plutonium. The reduction proceeds rapidly in solutions less acid than 2 *M* HNO_3 [337]. According to [335] (abstract), the reduction of major amounts of Pu(VI) (250 g/liter) proceeds only to Pu(IV) at HNO_3 concentration of 2–6 *M*, temperature 23–50°, and iron sulfamate concentration of about 1 *M*.

It is known that Pu(VI) is instantaneously reduced by *ferrocyanide* with the formation of a black precipitate, probably Pu(IV) ferrocyanide [3, p. 225].

The reduction of Pu(VI) to Pu(V) by *hydroxylamine* in 0.5 *M* HCl, discussed above, is accompanied by the disproportionation of Pu(V) with the appearance of Pu(III). Consequently, the preparation of Pu(III) with the aid of this reagent in moderately acid medium (0.2–1.5 *M* HNO_3 or HCl) should be the faster the higher the concentration of plutonium and of the acid, and the higher the temperature. However, according to Vdovenko, Lipovskii and Kuzina (1955), increases in the concentration of plutonium from 0.1 to 1 g/liter, of acid from 0.2 to 1.5 *M* and of the temperature from 25 to 50° do not result in the expected acceleration of the reaction. They also found that the reaction is accelerated when the nitrate ion concentration is increased. In a solution containing 54% $Ca(NO_3)_2$, 0.03 *M* HNO_3, 0.3 *M* $NH_2OH \cdot HNO_3$ and $4.1 \cdot 10^{-4}$ *M* Pu(VI), plutonium is reduced to Pu(III) within 2–3 minutes at room temperature. The increase in the nitric acid concentration to 0.4 *M* under otherwise identical conditions results in the reduction of plutonium to Pu(IV) stage only. The Pu(IV) thus obtained is rapidly (within 5 minutes) converted into Pu(III) when the temperature is raised to 50°. The transfer of Pu(VI) from the dibutyl ether phase to the aqueous phase of the above composition is just as fast.

The mechanism of the reduction of Pu(IV) to lower valency states by *hydrazine* would also seem to involve a disproportionation stage. It is thought [337] that the reagent is not very efficient in the reextraction of Pu(VI) from the organic into the aqueous layer.

According to Khalkin (1954), if 0.25 ml of a concentrated solution of hydrazine and 2 ml 13.2 *M* HNO_3 is added to 1 ml solution of hexavalent plutonium containing 15 ml plutonium per liter, plutonium will be completely converted to the tetravalent state within 30 minutes.

The reduction of the trace and macro amounts of Pu(VI) to Pu(III) in nitric acid medium is completed within a few minutes under the action of *formaldehyde sodium sulfoxylate* (Rongalite), $NaHSO_2 \cdot CH_2O \cdot 2H_2O$, employed as solid or as 50% solution in water. This reagent also rapidly reduces U(VI) to U(IV).

Solutions of Pu(VI) treated with *potassium iodide* at pH $\sim$ 2 rapidly yield Pu(V) (cf. p. 56). In more acid media, the large amounts of Pu(V) formed begin to disproportionate, and, in all probability, it is this very process that determines the rate of reduction of Pu(VI) to Pu(III), since the reaction Pu(VI) $\rightarrow$ Pu(V) is very fast [3, p. 223].

Hexavalent plutonium may also be reduced by *oxalate ions* [3, p. 216]. In a solution containing 0.018 *M* Pu(VI), 0.14 *M* $H_2C_2O_4$ and 0.85 *M* HCl, the time required for a 50% conversion of Pu(VI) into Pu(IV) is about 20 minutes at room temperature. In nitric acid solutions the rate of reaction increases with the decrease in acidity [60, 90, p. 94]. In hydrochloric acid solutions the increase in the acidity in the presence of the oxalate results in an increased rate of reduction of Pu(VI).

Kabanova, Danyushchenkova and Palei [483] noted a rapid reduction of $PuO_2(ClO_4)_2$ in 1 *M* solution of $NaClO_4$ at pH 3–5 in the presence of a large excess of the sodium salt of *ethylenediaminetetraacetic acid* (Complexone III). After a few minutes the color of the solution changed from pale pink to lemon-yellow and then the color rapidly disappeared. It was found from spectrophotometric measurements that the initial color change is produced by the reduction of Pu(VI) to Pu(V) according to the reaction

$$6PuO_2^{2+} + \left[\begin{matrix} HOOC{-}CH_2 \\ OOC{-}CH_2 \end{matrix}\!\!>\!N{-}CH_2{-}CH_2{-}N\!<\!\!\begin{matrix} CH_2{-}COO \\ CH_2{-}COOH \end{matrix}\right]^{2-} + 3H_2O \longrightarrow$$

$$6PuO_2^{+} + \begin{matrix} HOCH_2 \\ HOCH_2 \end{matrix}\!\!>\!N{-}CH_2{-}CH_2{-}N\!<\!\!\begin{matrix} CH_2OH \\ CH_2COOH \end{matrix} + 4H^+ + 3CO_2. \quad (36)$$

The process does not stop at this stage. After a few days the solution became permanently orange-yellow (this stage may be accelerated by heating). The spectrum of the solution became similar to that of the Complexone III complex with tetravalent plutonium.

These reactions, as well as the reactions between Pu(VI) and oxalate ions, are not utilized in the preparation of plutonium in lower valency states. They must, however, be borne in mind in the course of analytical and research work.

Rabideau and Kline [608] studied the kinetics of the reactions between Pu(VI) and *trivalent titanium* in perchlorate medium. The reaction rate of the slow stage

$$PuO_2^{2+} + TiOH^{2+} = PuO_2^{+} + TiO^{2+} + H^{+} \qquad (37)$$

is given by the equation

$$-\frac{d\mathrm{Pu(VI)}}{dt} = k[\mathrm{Pu(VI)}][\mathrm{Ti(III)}][H^{+}]^{-1}.$$

The rate constant k in 1 M $HClO_4$ at 25° is 108 sec^{-1}.

More advanced reduction takes place according to the reactions

$$\mathrm{Pu(V)} + \mathrm{Ti(III)} = \mathrm{Pu(IV)} + \mathrm{Ti(IV)} \qquad (38)$$

and

$$\mathrm{Pu(IV)} + \mathrm{Ti(III)} = \mathrm{Pu(III)} + \mathrm{Ti(IV)}.^{*} \qquad (39)$$

The rate of reaction (38) was too fast to be measured. The overall rate of the process is determined by the rates of reactions (37) and (39).

Rabideau and Masters [612] studied the reduction of Pu(VI) by *stannous tin* in chloride-perchlorate solutions. It was found that the reaction with Pu(VI) is much faster than with Pu(V). The former reaction apparently involves two simultaneous charge transfers:

$$\mathrm{Pu(VI)} + \mathrm{Sn(II)} = \mathrm{Pu(IV)} + \mathrm{Sn(IV)}. \qquad (40)$$

In the presence of chloride ions (~ 1.5–2.0 M) 1 g-mole of Sn(II) reduces less than 1 g-mole of Pu(VI), due to the competing reaction between Pu(IV) and Sn(II), which in the presence of Cl^- ions is accelerated to a much greater extent than reaction (40). In the kinetic equation of the process

$$-\frac{d[\mathrm{Pu(VI)}]}{dt} = k[\mathrm{Pu(VI)}]\cdot[\mathrm{Sn(II)}]$$

the constant k is 18.5 $M^{-1}\cdot sec^{-1}$ at 20° at the ionic strength of 2 and 0.5 M concentration of chlorides. Variation in the hydrogen ion concentration does not affect the reduction rate. The effect of the chloride ions is considerable: a decrease in their concentration from 1.75 to 0.25 M results in a decrease of k by a factor of 20 (the functional dependence is given in [612]).

Reducing properties of *tetravalent uranium* in perchlorate [575] and nitric acid [479] media have been studied.

* The reaction between Pu(IV) and Ti(III) is described above, p. 53.

If Pu(VI) and U(IV) have been introduced in equal amounts (each in the concentration 0.002 *M* in 1 *M* $HClO_4$), the main product of the reduction is Pu(V):

$$2Pu(VI) + U(VI) = 2Pu(V) + U(VI). \quad (41)$$

It was found that under these conditions the reaction

$$P(V) + U(IV) = Pu(III) + U(VI) \quad (42)$$

is also of some importance.

In the presence of a large excess of U(IV) in 1 *M* HNO_3 (concentrations of uranium and plutonium 0.1 *M* and 10^{-5} *M*, respectively), hexavalent plutonium is reduced to Pu(III) within 5–10 minutes. The detailed mechanism of the process is unclear; the overall reaction may be represented as follows:

$$3U^{4+} + 2PuO_2^{2+} + 2H_2O = 3UO_2^{2+} + 2Pu^{3+} + 4H^+. \quad (43)$$

In the reaction between Pu(VI) and *trivalent vanadium* [186] the first product is Pu(V):

$$Pu(VI) + V(III) = Pu(V) + V(IV). \quad (44)$$

The reaction rate is expressed by the equation:

$$-\frac{d[Pu(VI]}{dt} = k_1[PuO_2^{2+}][V^{3+}][H^+]^{-1} + k_2[PuO_2^{2+}][V^{3+}][H^+]^{-2},$$

where $k_1 = 2.12\ \text{sec}^{-1}$ and $k_2 = 0.23\ M \cdot \text{sec}^{-1}$. The course of further reduction is unknown, but the reaction between Pu(IV) and V(III) has been described [607].

Of the metallic reducing agents, studies were made on *zinc* [3, p. 245]. The solution in 0.5 *M* HCl after reduction was green; the color was probably caused by the polymerization of Pu(IV) owing to the local deficiencies of hydrogen ions near the surface of the metal. If the work were to be carried out under suitable acidity conditions, quantitative reduction of plutonium to Pu(III) would most probably be attained.

Conversion of mixtures of various valency states into Pu (IV)

The working solutions sometimes contain a mixture of plutonium forms in various states of valency. With the aid of the methods described above, solutions in which it is present exclusively in the highest (Pu(VI)) or in the lowest (Pu(III)) valency state can be readily prepared. The intermediate forms Pu(V) and Pu(IV) are obtained by treating the solution in two stages. Pu(V) may be obtained by oxidizing all the plutonium into Pu(VI) and reducing the product to Pu(V) at certain definite pH values.

Pu(IV) is obtained in an analogous manner, viz., by reducing all the plutonium to Pu(III) and oxidizing the product to Pu(IV). Certain reagents, e. g., nitrous acid and nitrites, convert plutonium from any valency state to Pu(IV).

The treatment with nitric acid in the presence of hydrogen peroxide is employed for the same purpose. It is known that nitric acid may oxidize plutonium to Pu(VI) in the heat. In order to prevent this reaction, and also in order to reduce the Pu(VI) already present in the solution, hydrogen peroxide is added. Thus a combination of HNO_3 in its capacity as oxidizing agent for Pu(III) with H_2O_2 in its capacity as a reducing agent for Pu(VI) makes it possible to obtain pure solutions of Pu(IV). The reaction is carried out in concentrated solutions of HNO_3 ($\geqq 10\,M$) near the boiling point. The hydrogen peroxide treatment is repeated 2–3 times in gradually decreasing volumes of solution. The last stage is evaporation with concentrated nitric acid.

Cunningham and Miles [365] prepared a solution of pure Pu(IV) in nitric acid by first reducing all the plutonium to Pu(III) with hydrazine, and then oxidizing Pu(III) to Pu(IV) with sodium nitrite. The latter reaction proceeds in the presence of 10–25% excess of nitrite. According to their data, the Pu(IV) : Pu(III) ratio in the final solution should be about 40.

A very popular method for the preparation of solutions of Pu(IV) is to reduce all the plutonium to Pu(III) by any suitable method and precipitate Pu(III) hydroxide. By dissolving $Pu(OH)_3$ in concentrated HNO_3 with heating, a solution of plutonium (IV) nitrate is obtained.

Plutonium(IV) sulfate may be prepared by threefold evaporation of a solution containing a mixture of various valency forms with concentrated sulfuric acid, since under these conditions Pu(III) is rapidly oxidized to Pu(IV), while Pu(VI) is reduced to this valency [3, p. 243].

The transfer of Pu(IV) from one medium to another is readily achieved by precipitating $Pu(OH)_4$ and dissolving the hydroxide in the corresponding acid.

Electrolytic methods of oxidation and reduction

The convenience of the electrolytic methods for preparing the solutions is the fact that plutonium is obtained in the desired state of valency without the addition of foreign materials.

The oxidation and reduction reactions in the reversible systems Pu(IV)—Pu(III) and Pu(VI)—Pu(V) take place easily, since they do not involve scission or formation of Pu—O bonds. For such reactions to proceed, it is enough for the potential of the working electrode (cathode for reduction processes, anode for the oxidation processes) to differ by 120–150 mV from the formal potential of the given system. It is much more difficult to effect the reactions Pu(VI) $\rightleftharpoons$ Pu(III), which amounts to effecting the irreversible reactions Pu(V) $\rightleftharpoons$ Pu(IV). The overvoltage on the working electrode must then be at least of the order of 500–600 mV. For this reason, in the coulometric analysis of preparations of plutonium contaminated with elements whose redox potentials are close to that of plutonium the reaction Pu(III) $\rightleftharpoons$ Pu(IV) is exclusively employed.

The time required for a quantitative* conversion of an element from one form into another is shorter, the larger the surface of the working electrode and the more intense the stirring rate. Under otherwise identical conditions this time is directly proportional to the volume of the electrolyte.

A simple cell and apparatus for electrolysis were described by Healy and Gardner [454]. The cell consists of two compartments divided by a partition of finely porous glass; one partition accommodates the working electrode, while the other accommodates the auxiliary electrode. Cells with a third compartment for a reference electrode are most frequently employed [342]. The working electrode and the auxiliary electrode usually consist of a platinum (or gold) spiral or gauze. The stirring of the solution is best effected by passing a stream of purified nitrogen or argon. A rotating electrode may also be used for the purpose [499].

In order to minimize the migration of the electrolyte between the anode and the cathode compartments, the composition of the electrolyte is made identical in the two compartments and the liquid levels are made equal.

Methods for the electrolytic preparation of plutonium in various states of valency in perchloric and nitric acids will be found in [3, p. 245, 347, 454, 460, 499, 512, 574, 641, 720]. Cohen [342] carried out a major study of the nature of the oxidation and reduction of plutonium ions in solutions of perchloric acid.

Figure 19, taken from Cohen [342], shows potential vs. current intensity curves, obtained by measuring the current resulting from a gradual increase in the potential of the anode (for oxidation) or a gradual decrease in the cathode potential (in the case of reduction).* Even though the reproducibility of the i vs. E curves, which have the aspect of ordinary polarographic curves for reversible systems, is within 10–15%, the data make it possible to select the optimum conditions for conversion of plutonium to any desired valency state.

Reduction to Pu(III)

The reduction in $HClO_4$ solutions [342, 574] may be carried out irrespective of the original valency state of the solution. For the reduction of Pu(IV) at a potential $\leqq +0.70$ V (relative to n. h. e.) a conversion better than 99% will be rapidly obtained. If it is desired to reduce Pu(V), Pu(VI) or a mixture of different valency forms, the potential employed should be about + 0.2 V because of the large overvoltage of the reaction Pu(V) $\rightarrow$ Pu(IV), but even so the reaction is very slow. The

* A quantitative conversion is usually understood to mean a 99–99.5% oxidation or reduction.

** The electrode surface area (a platinum spiral, 1.25 mm dia.) was 20 cm^2, while the solution volume was ~ 5 ml.

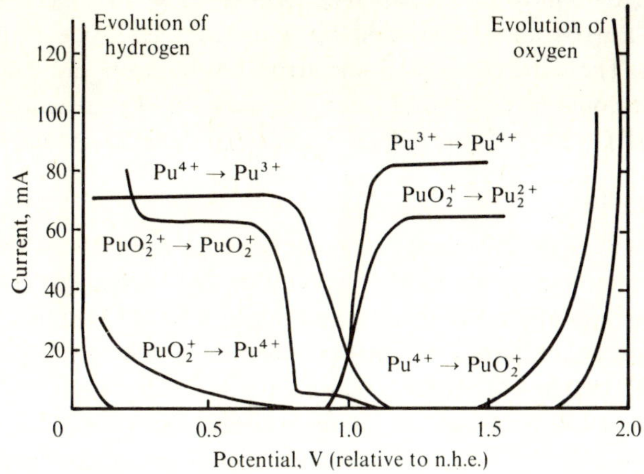

Figure 19

Volt-ampere curves for plutonium ions in 1 *M* $HClO_4$ at 25°.

The concentration of plutonium was 0.016 *M* for the reactions Pu(IV) → Pu(V), Pu(V) → Pu(VI), Pu(V) → Pu(IV) and 0.025 *M* for the reactions Pu(IV) → Pu(V), Pu(III) → Pu(IV) and Pu(IV) → Pu(III).

evolution of hydrogen may be prevented if the cathodic potential applied is somewhat above + 0.2 V (cf. Figure 19).

Scott and Peekema [641] studied the system Pu(III)–Pu(IV) in nitric and hydrochloric acid solutions for the purpose of coulometric determination of plutonium. The conditions for the reduction of Pu(IV) to Pu(III) are similar to those described above for perchloric acid solutions. The first stage of the analytical procedure proposed by these authors is the conversion of a mixture of all valency forms into Pu(III). At + 0.80 V this operation took 5–30 minutes, depending on the amount of higher valency forms in the initial sample.

Preparation of Pu(IV)

Tetravalent plutonium is obtained by the oxidation of Pu(III). That the reaction Pu(V) → Pu(IV) could never be carried out quantitatively is explained by the fact that the reduction of Pu(IV) to Pu(III) invariably takes place at the same time. The oxidation of Pu(III) in 1 *M* $HClO_4$ proceeds smoothly and rapidly at about + 1.35 V, and the Pu(IV) obtained contains less than 1% of other valency forms, since a much higher overvoltage is required for oxidation to the pentavalent state. Nebel and Nebel [547] recommend that the bulk of Pu(III) be oxidized at + 1.1 V and the remainder at + 1.16 V. This totally prevents any oxidation of Pu(IV).

The conditions for the oxidation of Pu(III) to Pu(IV) in solutions of nitric and hydrochloric acid are practically identical (see pp. 211–212).

It was shown by Vinogradov and Gryzin (1951–1954) that electrolysis in the system Pu(IV)–Pu(III) may be effected on a mercury electrode in tartrate and citrate solutions. The same conclusion was arrived at by Scott and Peekema [641] who carried out coulometric studies in citrate solution. The reactions Pu(IV) → Pu(III) and Pu(III) → Pu(IV) were carried out at −0.05 V and + 0.19 V, respectively.

Preparation of Pu(V)

The reaction Pu(V) ⇌ Pu(IV) cannot be made to go to completion. The oxidation of Pu(IV) on a platinum anode yields Pu(VI). For this reason Pu(V) can only be obtained by reduction of Pu(VI). It is pointed out by Cohen [342] that the cathode potential must not be too negative if further reduction is to be prevented. In a solution of 0.025 *M* Pu(VI) in 1 *M* $HClO_4$, plutonium was reduced to Pu(V) to an extent of 97 ± 3 % at + 0.9 V. However, at this acidity value Pu(V) is quite unstable and disproportionates. For this reason spectrophotometric investigations were carried out on a solution of Pu(V) in 0.2 *M* $HClO_4$.

The final product of the electrolytic oxidation of plutonium ions on a platinum anode is Pu(VI). This reaction (cf. Figure 19) requires a high overvoltage and is very slow. When the applied potential is + 1.95 V in 1 *M* $HClO_4$ and at Pu(IV) concentration of 0.025 *M*, plutonium becomes oxidized in three hours [342]. Similar results were obtained by Kabanova (1954) who oxidized plutonium in ~ 0.07 *M* concentrations. A large (surface area ~ 20 cm^2) rotating anode was employed in the oxidation of Pu(IV) in 0.5 *M* $HClO_4$ [499]. It may be noted that pure Pu(VI) cannot be prepared without evolution of oxygen.

Oxidation in HNO_3 solutions should be carried out at controlled acidity [512]. It is considered that the optimum concentration of HNO_3 is about 0.5 *M*. A decrease in the hydrogen ion concentration involves the danger of polymerization of Pu(IV). Healy and Gardner [454] oxidized plutonium (2–20 mg/ml in a 5–10 ml cell for a few hours in 0.4–1.0 *M* HNO_3. Increasing the concentration of nitric acid adversely affects the rate of the reaction.

Electrolytic preparation of solutions of plutonium in a given valency state is now increasingly employed, both in preparative chemistry and in electrometric analysis.

REACTIONS PRODUCED BY α-ACTIVITY AND EXTERNAL γ-RADIATION

The action of α- and γ-radiations on the valency states of plutonium has been studied by a number of workers [14, 17, 90, 177, 181, 183, 209, 485, 606].

The high specific α-activity of Pu^{239} preparations (~ 136,000 α-disintegrations/min · μg) produces a radiolytic decomposition of water accompanied by the formation of oxidizing agents (free radical HO and HO_2) and also H_2O_2 and reducing agents (atomic hydrogen and H_2O_2).

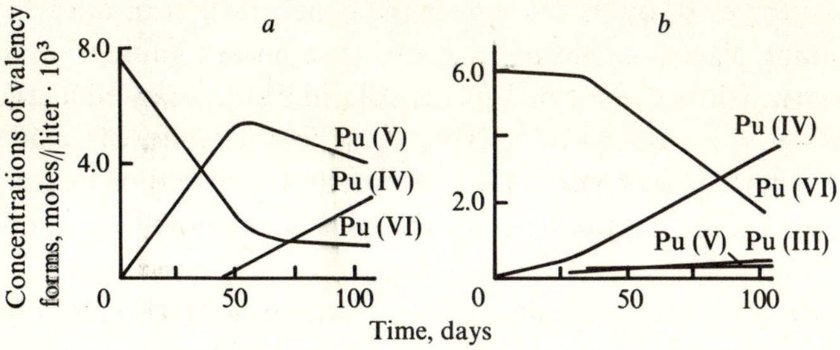

Figure 20

Reduction of Pu(VI) by α-activity as a function of time:

a at pH 2 and NO_3^- concentration of 0.05 *M*; *b* in 0.4 *M* HNO_3.

It was found that Pu(VI) in nitric acid solutions at pH 2.5–3, containing $(2–7) \cdot 10^{-3}$ *M* of plutonium, is partially converted to Pu(V) a few days after preparation, and that the rate of reduction, at constant hydrogen ion concentration, increases with increasing concentration of plutonium [14, 177]. In 0.3–8 *M* HNO_3 Pu(VI) is stable for at least one or two months (Figure 20). The appearance of Pu(V) is accompanied by its disproportionation, as a result of which Pu(IV) is formed, which is rapidly reduced to Pu(III) by hydrogen peroxide. The formation of Pu(III) initiates reactions (22) and (27) (p. 47), with the resulting accumulation of Pu(IV) [14, 177]. A solution, originally containing 0.0045 moles/liter Pu(VI) in 0.3 *M* HNO_3, was found to contain 83.3% Pu(IV) after ten months' standing.

It has been shown [14] that the average valency of plutonium is the lower, the higher the nitrate ion concentration. This is thought to be due to the formation of nitrite ions as a result of α-radiation.

When solutions of plutonium (IV) are held in 1.5 *M* $HClO_4$ (at such hydrogen ion concentration disproportionation is suppressed), practically all the plutonium is converted into Pu(III) within three months. The stability of Pu(IV) increases in the presence of complexing inorganic anions. Thus, the reduction of plutonium by the radiolysis products may be very extensive and may depend on the initial valency of plutonium and the time of holding. α-Radiation may also result in a partial oxidation of Pu(III) to Pu(IV) in hydrochloric acid solutions [203, p. 359].

Changes in valency states were also noted under the action of intense beams of γ-rays (up to $4 \cdot 10^{21}$ eV/cm³) in nitric acid and perchloric acid solutions [181, 182]. It was found by Popov [181] that in 0.3–2.3 *M* $HClO_4$ plutonium may become oxidized or reduced, depending on the relative proportions of the different valency forms in the initial solution ($3.4 \cdot 10^{-4}$ moles of plutonium per liter). The irradiation produces a stationary state, in which about 47% of the plutonium is present in the oxidized forms Pu(V) and Pu(VI). In the presence of U(VI) the relative content of the oxidized Pu forms is lower, the higher the concentration of uranyl perchlorate.

Processes taking place in nitric acid solutions are more complicated, since nitrous acid and nitrogen oxides are formed in the solution [182] as well as the radiolysis products of water. In mixtures of $HNO_3 + NaNO_3$ (Figure 21) oxidation alone takes place; it is retarded when the concentration of nitrate ions increases. The authors [182] explain the retardation by complex formation between Pu(IV) and nitrate ions.

Processes taking place as a result of X-ray irradiation of a solution of $HNO_3 + UO_2(NO_3)_2 + K_2Cr_2O_7$ are more complex (Figure 22). After the reduction of all the bichromate which maintains the plutonium in the oxidized state, reduction of plutonium takes place and is more rapid, the higher the concentration of $UO_2(NO_3)_2$.

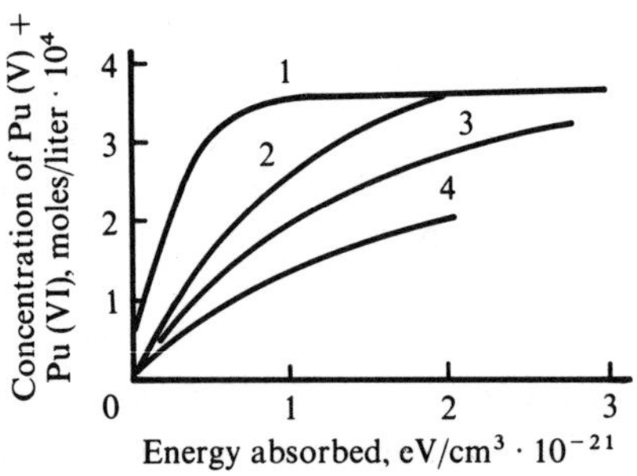

Figure 21

Oxidation of plutonium by X-ray irradiation in 3 M HNO_3 and $NaNO_3$ solutions:

1) in the absence of $NaNO_3$; 2) 0.35 M $NaNO_3$; 3) 0.88 M $NaNO_3$; 4) 1.76 M $NaNO_3$. Concentration of Pu(IV) $3.4 \cdot 10^{-4}$ M.

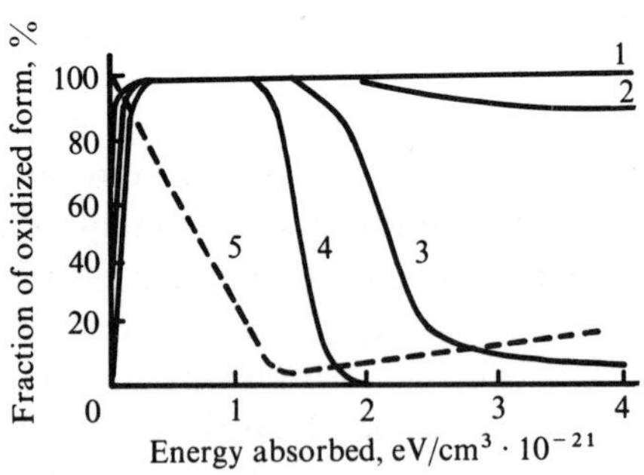

Figure 22

Effect of X-ray irradiation on the valency states of plutonium in 0.3 N HNO_3 and 0.1 N $K_2Cr_2O_7$ at various concentrations of $UO_2(NO_3)_2$:

1) 0.3 M; 2) 0.45 M; 3) 0.6 M; 4) 0.88 M; 5) decomposition curve of $K_2Cr_2O_7$.

The reducing effect of uranyl nitrate was explained [181, 182] by the appearance of UO_2^+ ions as a result of radiochemical transformation of UO_2^{2+}. It is, however, unclear if the species reacting with PuO_2^{2+} is the UO_2^+ ion or its disproportionation product U^{4+}.

The stability of plutonium to γ-radiation in the presence of organic solvents, in particular, diethyl ether, is discussed in [17].

The available data on the action of α- and γ-radiation on the state of the ions in aqueous solutions are not comparable, since the yields of the radiolysis products in both cases are not given. Changes in valency state due to α-activity are rapid only in weakly acid media. In moderately acid solutions (acidity about $1\,M\,H^+$) the changes take place slowly and are significant only if the plutonium solutions are stored for a long time. Large doses of γ-rays ($5 \cdot 10^{16}$ eV/cm^3 · sec) produce fairly rapid oxidation or reduction reactions, depending on the nature of the medium, and this must be taken into consideration when working with highly γ-active preparations.

STABILITY OF VARIOUS VALENCY FORMS IN SOLUTIONS

In the preceding paragraphs the reactions of hydrolysis, complex formation, disproportionation, oxidation and reduction brought about by chemical reagents and ionizing radiations have been discussed. We shall now give a brief summary of this material in terms of the stabilities of individual valency forms when present in solution in macro amounts.

The stability of *plutonium* (III) in inorganic acid solutions depends on their nature and concentration and on the temperature.

Its stability in hydrochloric acid media in the absence of reducing agents would not seem to be as high as reported by Connick and McVey [350] who failed to observe any oxidation of Pu(III) in oxygen-saturated 0.5 *M* HCl during two days. According to Caldwell et al. [333], the amount of Pu(III) gradually decreases from 100 to 99.7%, when the holding time is increased from 5 to 60 minutes.

In nitric acid solutions Pu(III) is less stable than in HCl, due to oxidation by nitrate ions. At nitric acid concentrations above 5 *M*, Pu(III) is fairly rapidly oxidized even at room temperature. The oxidation is accelerated in the presence of even small amounts of nitrite ions and nitrogen oxides. In sulfuric acid solutions plutonium (III) is gradually oxidized by atmospheric oxygen.

Plutonium (III) becomes less stable when its solutions are heated. This effect is particularly noticeable in solutions of nitric and sulfuric acid. In oxalate and carbonate solutions plutonium (III) is unstable and becomes oxidized after a few hours.

Solutions of trivalent plutonium should be stored in the presence of excess reducing agent (hydroxylamine, sulfurous acid, Rongalite, etc.) in an inert gas atmosphere.

The stability of solutions of *plutonium* (IV) depends on the complexing tendency of the anion. In 0.5 *M* sulfuric acid Pu(IV) retains its valency for a long time. It is also very stable in fairly concentrated (5–10 *M*) nitric acid. On the other hand, as the hydrogen ion concentration is decreased, disproportionation reactions begin to play an increasing part and, at equal acidities, the amounts of other valency forms increase in the sequence HNO_3–HCl–$HClO_4$. It was calculated from the values of the rate constants of the reaction of disproportionation of Pu(IV) (see p. 45) that in 1 *M* $HClO_4$ about 30% of Pu(IV) is converted into Pu(III) and Pu(VI) whereas in 1 *M* HNO_3 only about 3% of Pu(IV) undergoes disproportionation. Equilibrium is attained at a rate which is directly proportional to $[Pu(IV)]^2$ and inversely proportional to $[H^+]^3$. Nitrous acid and its salts stabilize tetravalent plutonium in aqueous solutions.

In solutions of certain organic acids (citric, tartaric) which form with plutonium stable complexes in fairly narrow pH ranges, the α-radiation emitted by Pu^{239} increases the pH of the medium and within a few days results in a changed ionic and valency state of plutonium.

Solutions of *plutonium* (V), which usually contain a few percent of Pu(VI), may be stored at pH ~ 3–3.5 for at least one month. In this pH range they are stable to nitrate and chloride ions (up to, respectively, 10 *M* and 8 *M*) at room temperature for several days. With the increase in the concentration of plutonium to more than $4 \cdot 10^{-4}$ *M* and at higher temperatures the stability of Pu(V) decreases considerably [177, 425]. The stability of Pu(V) decreases both when the acidity exceeds 0.01 *M* and when the pH is raised above 6.

Plutonium (VI) retains its valency unchanged for several months in solutions more concentrated than 1 *M* HNO_3; the same probably applies to other mineral acids. The lower the acidity, the sooner Pu(V) is formed by the reduction of Pu(VI) as a result of the radiolysis of water [177]. Acid solutions of plutonium (VI) may keep for a long time in the presence of an excess of a strong oxidant.

COMPOUNDS ISOLATED FROM AQUEOUS SOLUTIONS

The preparation and properties of certain compounds of plutonium utilized in analytical chemistry is described below. Sparingly soluble compounds are isolated by precipitation; soluble salts are isolated in the solid state by evaporation or crystallization.

Plutonium nitrates

Plutonium (IV) *nitrate hydrate* $Pu(NO_3)_4 \cdot 5H_2O$

Plutonium (IV) nitrate hydrate crystallizes out when nitric acid solutions are slowly evaporated [387]. Chemical and spectrographic analysis of the crystals showed that its formula is $Pu(NO_3)_4 \cdot 5H_2O$. Plutonium (IV) nitrate with a smaller number of molecules of water of crystallization may be obtained if nitric acid solutions are evaporated at a faster rate on a water bath [716]. The crystals of the nitrate, if larger than 1 mm, are lustrous black prisms; small crystals are green. Dry $Pu(NO_3)_4 \cdot 5H_2O$ may be stored unchanged for several months, so that it is suitable for use as an analytical standard. It is readily soluble in water, forming a brown aqueous solution.

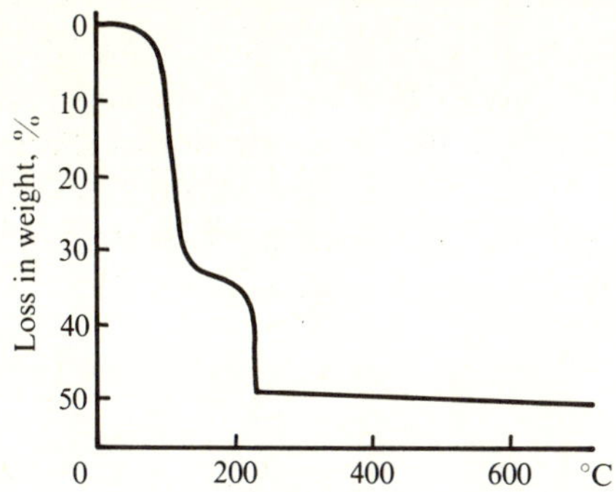

Figure 23
Thermogravigram of $Pu(NO_3)_4 \cdot 5H_2O$

Thermal decomposition of plutonium nitrate is represented in Figure 23 [387]. The crystals are stable up to 40°, then an insignificant loss of weight occurs, and at 60° the crystals begin to melt in their own water of crystallization and a considerable decomposition occurs. This process is particularly fast above 100°. The less steep part of the thermogravigram between 150 and 220° corresponds to the formation of the unstable plutonyl nitrate, which eventually decomposes into plutonium dioxide.

Plutonyl nitrate $PuO_2(NO_3)_2 \cdot 6H_2O$

Hydrated plutonyl nitrate crystallizes out when nitric acid solutions of Pu(VI) are evaporated over phosphoric anhydride [127]. It crystallizes as rectangular pink or brown plates, depending on the size of the crystals. Plutonyl nitrate is hygroscopic. When kept in moist atmosphere, Pu(VI) is slowly reduced to Pu(IV) or even to Pu(III). Plutonyl nitrate is readily soluble in water, nitric acid and alcohol and is insoluble in chloroform. Ethyl ether, nitromethane and other organic solvents extract plutonyl nitrate from ammonium nitrate-saturated aqueous solutions.

In addition to plutonium nitrate and plutonyl nitrate, a number of complex nitrates of the general formula $Me_2[Pu(NO_3)_6]$, where Me is Cs^+, Rb^+, K^+, Tl^+, $C_9H_7NH^+$, $C_5H_5NH^+$, NH_4^+ have also been prepared [278].

Plutonium sulfates

Plutonium (IV) *sulfate* $Pu(SO_4)_2$

Plutonium (IV) sulfate is obtained by evaporation of sulfuric acid solutions of tetravalent plutonium, with subsequent cautious calcination at about 300° to remove the excess acid [48, 389]. The composition was determined by chemical analysis.

Two methods for the preparation of hydrated plutonium sulfate have been described. The first method consists in crystallizing the salt out of dilute sulfuric acid solutions [48]. The second method [272] is salting out $Pu(SO_4)_2 \cdot 4H_2O$ from sulfuric acid solutions by methanol. The color of the compound may vary from pale pink to red-brown. The effect of the concentration of sulfuric acid on the solubility of plutonium sulfate is shown in Figure 24. Plutonium sulfates are readily soluble in dilute mineral acids. The anhydrous sulfate is stable in the air and may be used as a gravimetric form and analytical standard [389].

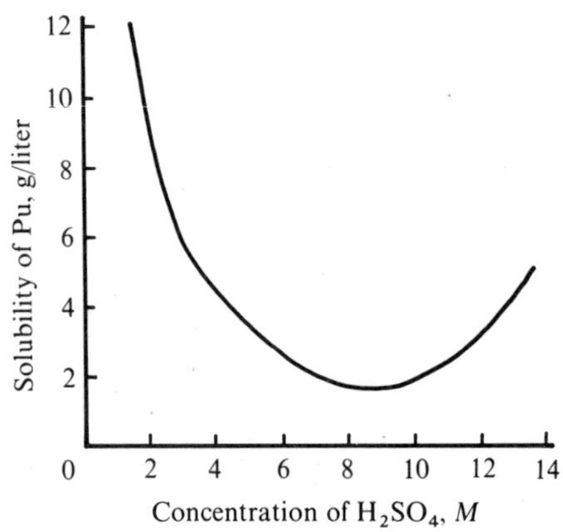

Figure 24

Solubility of $Pu(SO_4)_2 \cdot 4H_2O$ in sulfuric acid solutions

Thermogravimetric investigations of sulfates of tetravalent plutonium showed that the volatilization of water from the tetrahydrate terminates at 270–280°. Between 280 and 450° there is a slightly inclined plateau, indicating that the decomposition of the sulfate is insignificant. The decomposition of the anhydrous sulfate to the dioxide begins at ~ 500° and ends at ~ 700° (Figure 25).

Double plutonium (III) *sulfates*

Double sulfates are precipitated from acid solutions by the addition of an excess of alkali metal sulfate and ethanol or methanol [274, 275]. Depending on the excess

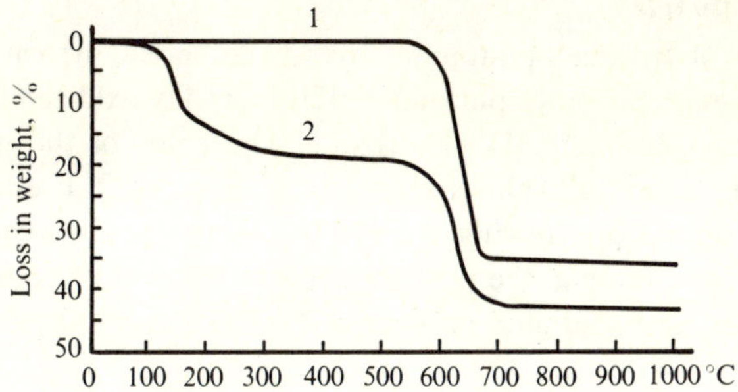

Figure 25

Thermogravigram of plutonium (IV) sulfate:

1) $Pu(SO_4)_2$; 2) $Pu(SO_4)_2 \cdot 4H_2O$.

of the alkali metal sulfate, the double sulfates which precipitate out have the composition $MePu(SO_4)_2 \cdot xH_2O$ or $Me_5Pu(SO_4)_4$, where Me is the alkali metal ion.

Anderson [274–276] isolated and studied the following double sulfates of plutonium (III): $KPu(SO_4)_2 \cdot 5H_2O$, $K_5Pu(SO_4)_4$, $NaPu(SO_4)_2 \cdot 4H_2O$, $TlPu(SO_4)_2 \cdot 4H_2O$, $RbPu(SO_4)_2 \cdot 4H_2O$, $CsPu(SO_4)_2 \cdot 4H_2O$ and $(NH_4)Pu(SO_4)_2 \cdot 4H_2O$. The solubilities of $K_5Pu(SO_4)_4$ and $NaPu(SO_4)_2 \cdot 4H_2O$ in the presence of a fivefold excess of the alkali metal sulfate and an alcohol content of about 50% is 10 mg/liter. These compounds are utilized to separate plutonium from the accompanying elements.

Double plutonium (IV) *sulfates*

Compounds of the type $Me_4Pu(SO_4)_4 \cdot (1\text{–}2)H_2O$ were isolated by Anderson [270]. The compound $K_4Pu(SO_4)_4 \cdot H_2O$ was obtained by treating a solution of plutonium (IV) sulfate in 0.5 *M* H_2SO_4 with a double volume of 10% K_2SO_4 and 20% by volume of methanol. After the addition of methanol, the pink precipitate of $Pu(SO_4)_2 \cdot 4H_2O$ was first formed; on standing for half an hour it was converted to the green $K_4Pu(SO_4)_4 \cdot H_2O$. The concentration of plutonium in the solution over the precipitate is 20 mg/liter. Anderson also prepared the readily soluble $(NH_4)_4Pu(SO_4)_4 \cdot 2H_2O$ and the sparingly soluble $Rb_4Pu(SO_4)_4 \cdot 2H_2O$.

Plutonium hydroxides

When an excess of alkali or ammonia is added to solutions of Pu (III) or Pu (IV) salts, sparingly soluble precipitates of the hydroxides are formed. Pentavalent plutonium is not precipitated as the hydroxide. In alkaline solutions Pu (VI) gives the dark-brown precipitate of plutonate.

Plutonium (III) *hydroxide*

The precipitation of trivalent plutonium hydroxide should be carried out in an atmosphere of an inert gas, since plutonium (III) is readily oxidized in the air to the insoluble polymeric plutonium (IV) hydroxide. The color of the precipitate thus obtained may vary from light blue to dirty blue [3, p. 328]. Freshly precipitated plutonium (III) hydroxide is readily soluble in mineral acids. When $Pu(OH)_3$ is precipitated by 5 *M* ammonia, the residual concentration of plutonium in solution is about 90 mg/liter. The solubility product of $Pu(OH)_3$ according to Latimer [146, p. 316] is $2.7 \cdot 10^{-20}$.

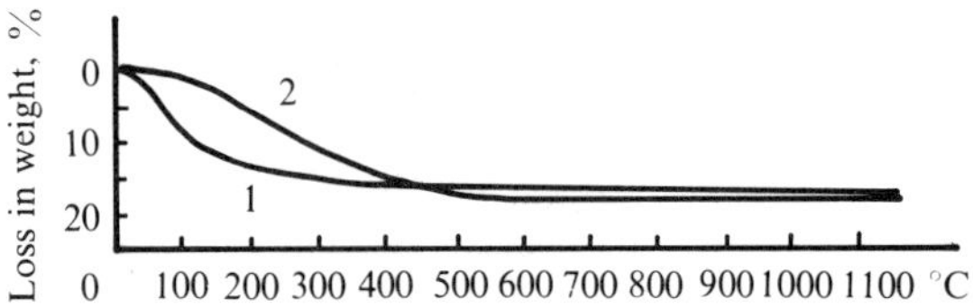

Figure 26

Thermogravigram of plutonium (IV) hydroxide:

1) hydroxide precipitated by a solution of ammonia; 2) hydroxide precipitated by pyridine.

Plutonium (IV) *hydroxide*

The action of aqueous solutions of ammonia and alkalis on acid solutions of tetravalent plutonium results in the precipitation of $Pu(OH)_4$. In 0.001 *M* solution of Pu(IV) the precipitate appears even when the concentration of hydrogen ions is as high as 0.1 *M* [48]. Plutonium (IV) hydroxide is a dark-green or olive-green amorphous powder. The precipitate is stable in the air. Basic salts were reported [509] to be formed when plutonium (IV) is precipitated from solutions of sulfates or nitrates. The hydroxide is readily soluble in strong acids. The hydroxide, when dried at 100°, rapidly dissolves in 1 *N* mineral acids, best of all in nitric acid, somewhat less readily in sulfuric and hydrochloric acids. The solubility of plutonium (IV) hydroxide at 25° in 1 *M* sodium sulfate at pH 6.2 is 1.1 mg/liter, while in 1 *M* sodium carbonate it is 1572 mg/liter. Its solubility in saturated solutions of KCl at an ionic strength of 3.5 is insignificant, $6.92 \cdot 10^{-6}$ moles/liter [57]. The solubility product $[Pu^{4+}][OH^-]^4$ is $7 \cdot 10^{-56}$ [146, p. 316].

The investigation of the thermal stability of plutonium (IV) hydroxide, obtained in various manners, is represented in Figure 26 (Sklyarenko and Chubukova, 1958). The decomposition of the hydroxide which has been precipitated by ammonia is more intense and the weight of the dioxide becomes constant at ~ 500°. The decomposition of the hydroxide precipitated by pyridine is complete at 700°. The air-dry precipitates of the plutonium (IV) hydroxide retain about two molecules of water.

On prolonged storage or heating of weakly acid solutions (0.1 *M*) containing more than 0.001 *M* Pu(IV), polymeric particles of a hydroxide of a high molecular weight are formed [3, p. 331]. The polymerized hydroxide is emerald-green.

Plutonium (IV) peroxide

When a large excess of hydrogen peroxide is added to an acid solution of plutonium in any valency state, plutonium (IV) peroxide precipitates out. The precipitation of the peroxide is preceded by the formation of two soluble peroxy complexes of plutonium [352].

Depending on the initial acidity of the plutonium solution, plutonium peroxide may be obtained in two different crystal forms (pseudocubic and hexagonal). When precipitated from solutions of acids more dilute than 2 *M*, the peroxide has, as a rule, a pseudocubical structure. In nitric acid solutions, pseudocubical precipitates are formed only when the acid concentration is below 1 *M*. In more acid solutions the hexagonal peroxide is usually formed [444, 519]. In weakly acid solutions (e. g., 0.4 *M* $HClO_4$) amorphous peroxide precipitates are obtained. It is usually preferred to precipitate the peroxide in the hexagonal form, since the precipitation then takes place in a wider acidity range and is more complete. The hexagonal peroxide lattice has a layered structure. The gravimetric factor experimentally obtained for the cubic peroxide is 0.640 (theoretical value 0.637), while that for the hexagonal peroxide is 0.467 [519]. The density of the dry pressed precipitate of the cubic peroxide is 0.7 g/ml, while that of the hexagonal peroxide is 0.17 g/ml [519].

The peroxide precipitate captures sulfate, nitrate and chloride anions from solution. The peroxide is olive-green when precipitated in the presence of sulfate ions, otherwise it is light green [504].

A readily filterable precipitate is obtained when the precipitation has been carried out from sulfate media (Kondrashova, 1953). The solubility of the peroxide in water is a fraction of a milligram of plutonium per liter of solution, while the solubility in weakly acid solutions is 5–10 mg/liter of plutonium. Plutonium peroxide is readily soluble in acid solutions in the presence of oxidizing or reducing agents which decompose hydrogen peroxide. On calcination the peroxide is converted into plutonium dioxide.

According to Deberdeeva (1960), plutonium (IV) peroxide is quantitatively precipitated from carbonate media.

Plutonium iodates

Tetravalent and trivalent plutonium iodates are widely employed in analytical work for the quantitative precipitation of plutonium and also to separate it from a number of elements which interfere with the determination.

$Pu(IO_3)_3$

When potassium iodate is added to sulfuric or hydrochloric acid solutions of trivalent plutonium, the light-brown precipitate of plutonium (III) iodate is obtained [3, p. 324, 277]. In a solution 0.17 *M* in sulfuric acid and 0.017 *M* in KIO_3, the solubility of the compound is 1.5 mg of plutonium per liter.

$Pu(IO_3)_4$

When iodic acid or an iodate is added to nitric acid solutions of tetravalent plutonium, a bulky amorphous pink precipitate of $Pu(IO_3)_4$ is formed [48, 367]. Quantitative precipitation of plutonium tetraiodate also takes place from strongly acid media (6 *M* HNO_3). The effect of the concentration of nitric acid and potassium iodate on the solubility of $Pu(IO_3)_4$ is shown in Figure 27. When the precipitate of $Pu(IO_3)_4$ is treated with ammonia, plutonium (IV) hydroxide is obtained.

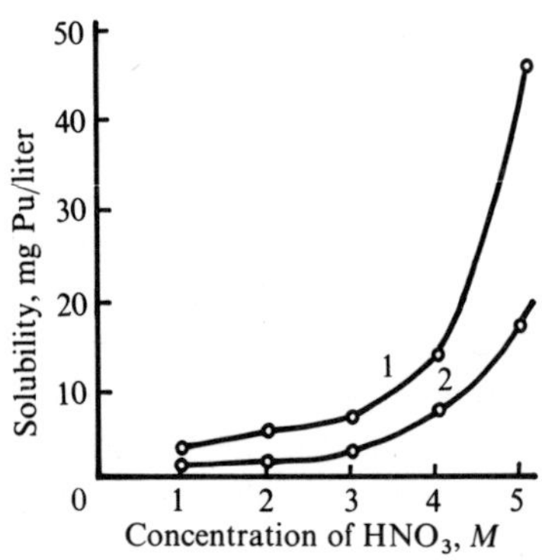

Figure 27

Solubility of $Pu(IO_3)_4$ in nitric acid solutions containing potassium iodate:

1) 0.1 *M* KIO_3; 2) 0.15 *M* KIO_3.

Plutonium tetraiodate readily dissolves in a 6% solution of sulfurous acid, when the plutonium is reduced to the trivalent state.

Plutonium tetraiodate is not hygroscopic and is stable when heated in the air at 70–80°. According to Sklyarenko and Chubukova [206], the thermal decomposition of plutonium (IV) iodate involves three stages (Figure 28). The first drop on the weight vs. temperature curve is caused by the elimination of the adsorbed water, since the precipitate was not washed with alcohol. The plateau between 150 and 250° corresponds to the anhydrous iodate. At 250–350° there is a partial decomposition of the iodate and the elimination of oxygen. This process terminates in the formation of the mixed compound $Pu(IO)_2(IO_3)_2$, which is stable between 350 and 450° and decomposes to PuO_2 above 500°. It was shown that anhydrous

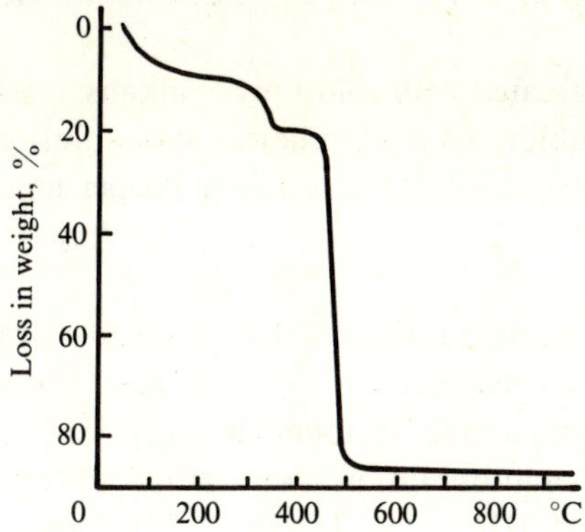

Figure 28
Thermogravigram of plutonium (IV) iodate

plutonium (IV) iodate may be employed as a gravimetric form if all the precipitant has been carefully removed. It was also found that when plutonium iodate is calcined at 900° to yield plutonium dioxide a small excess of the precipitant in the precipitate will not affect the gravimetric results, since the decomposition of potassium iodate is complete at 850°.

$PuO_2(IO_3)_2 \cdot 6KIO_3 \cdot xH_2O$

When potassium iodate is made to act on a solution of hexavalent plutonium a white precipitate of the above composition is formed [277]. The solubility of the precipitate in weakly acid solutions (~ 0.1 *M*) is about 0.5 g of plutonium per liter. This compound is of no interest to the analyst, since its solubility, as compared with those of trivalent and tetravalent plutonium iodates, is high.

Plutonium fluorides

$PuF_3 \cdot H_2O$

When hydrofluoric acid is made to act on an aqueous solution of trivalent plutonium, the salt $PuF_3 \cdot H_2O$ precipitates out [3, p. 312]. Dawson, Elliott, Hurst and Truswell [376] described the separation of $PuF_3 \cdot H_2O$ from a solution in which Pu(IV) is preliminarily reduced to Pu(III) with hydroxylamine. Freshly prepared plutonium trifluoride is an amorphous, pale purple-colored precipitate, which becomes crystalline on being left to stand for a few hours at room temperature.

Anhydrous plutonium trifluoride was obtained by heating $PuF_3 \cdot H_2O$ at 200–300° in a stream of gaseous hydrogen fluoride. When heated in the air to 300°, plutonium trifluoride is decomposed to the dioxide.

Plutonium trifluoride is rapidly hydrolyzed in aqueous solutions at 70°. It is soluble in solutions of Ce(IV), Zr(IV) and other cations which form complexes with fluoride ions.

When the trifluoride is treated with solutions of alkalis, it is converted into plutonium hydroxide. The solubility of the trifluoride in a solution saturated with SO_2 and containing 0.5 *M* HF and 1 *M* HCl is 37 mg Pu per liter [3, p. 312].

$PuF_4 \cdot 2.5H_2O$

When hydrofluoric acid is added to acid solutions of tetravalent plutonium, a crystalline pink precipitate separates out [48]. Data on the solubilities of the hydrated plutonium tetrafluoride will be found in [3, p. 314]. The solubility of the tetrafluoride (calculated as plutonium) is: in water 0.1 g/liter; in 0.8 *M* HCl + 2 *M* HNO_3, 0.62 g/liter; in 2 *M* HF + 2 *M* HNO_3, 0.57 g/liter.

The dehydration of the aqueous tetrafluoride can be effected at 350° in a stream of anhydrous hydrogen fluoride. For a more detailed description of the preparation and properties of anhydrous PuF_4, see p. 102.

$PuO_2F_2 \cdot xH_2O$

Plutonyl fluoride is readily soluble even in 19 *M* HF (its solubility is about 40 g of plutonium per liter). It is isolated from solution by the addition of methanol [277].

Double fluorides of plutonium (IV)

These compounds are prepared by adding a solution of plutonium (IV) nitrate to a solution of hydrofluoric acid containing an excess of an alkali metal fluoride [271].

$NaPuF_5$

When an excess of NaF is added to a nitric acid solution of plutonium nitrate, $NaPuF_5$ is obtained. The dense green precipitate, which separates out initially, becomes pink on standing. A similar effect is noted when solutions of plutonium (IV) in hydrochloric and sulfuric acids are employed. It was found [5] that the green precipitate has the composition $NaPuF_5$ and has a pseudocubic structure [736]. The pink precipitate has the formula Na_2PuF_6 and a hexagonal lattice. Preparations dried at room temperature contain 1–3 molecules of water. Deichman and Tananaev [81] determined the solubility of $NaPuF_5$ as $3 \cdot 10^{-5}\%$ Pu. The authors recommended the use of $NaPuF_5$ in the presence of NaF in analytical work because of its low solubility under these conditions.

Double compounds of Pu^{4+} with K^+, NH_4^+ and Rb^+ were obtained in a similar manner [5, 27]. Cesium forms a compound of the composition $CsPu_2F_9 \cdot 3H_2O$. The solubilities of double plutonium fluorides are given in Table 9.

Table 9

SOLUBILITY OF DOUBLE PLUTONIUM FLUORIDES IN SOLUTIONS OF ALKALI METAL FLUORIDES [170, 271]

Cation	Concentration of fluoride, *N*	Concentration of HF, *N*	Solubility of the double fluoride, mg Pu / liter
Li^+	0.5	3	120
NH_4^+	0.5	3	70
Na^+	0.5	3	9
K^+	0.5	3	5
Rb^+	0.5	3	30
Cs^+	0.5	3	45
Cu^{2+}	0.4	0.2	500
Co^{2+}	0.4	0.2	400

Plutonium phosphates

Plutonium phosphates are sparingly soluble compounds.

$PuPO_4 \cdot xH_2O$

Plutonium (III) phosphate was isolated by Zhagin (1959) by the addition of phosphoric acid or a phosphate to an acid solution of Pu(III). The amorphous gray-green precipitate is readily isolated from the mother liquor. Its composition corresponds to the formula $PuPO_4 \cdot xH_2O$. Its solubility, even in strongly acid solutions, is very low: in 1 *M* HCl it does not exceed 4 mg of plutonium per liter.

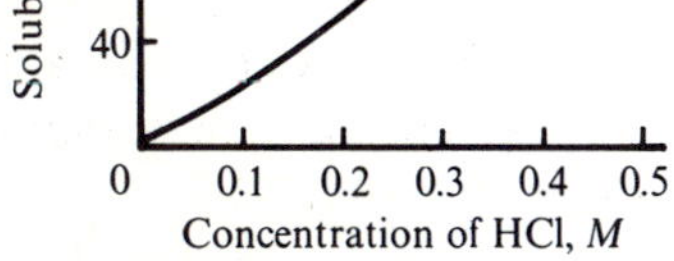

Figure 29

Solubility of plutonium (III) phosphate in hydrochloric acid solutions

Smith [3, p. 342] described the preparation of $PuPO_4 \cdot 0.5H_2O$. A 0.008 *M* solution of Pu(IV) in 0.5 *M* H_2SO_4 and 0.8 *M* H_3PO_4, saturated with SO_2, were placed in a sealed ampule and held for $2\frac{1}{2}$ hours at 75°. When plutonium was reduced to Pu(III), light purple crystals of $PuPO_4 \cdot 0.5H_2O$ precipitated out.

The effect of the acidity on the solubility of $PuPO_4 \cdot 0.5H_2O$ is illustrated in Figure 29. The differences in the properties of the compounds obtained by the authors quoted above are caused by the different water contents.

$Pu(HPO_4)_2 \cdot xH_2O$

When phosphoric acid is added to acid solutions of plutonium (IV), a white, amorphous precipitate of the above composition separates out [495].

Denotkina, Moskvin and Shevchenko [83] determined the solubility of $Pu(HPO_4)_2 \cdot xH_2O$ in solutions of nitric acid (Figure 30). It was found that the solubility increases with the increase in hydrogen ion concentration and that it is higher in nitric acid due to the formation of complexes between plutonium and nitrate ions. These authors also calculated the solubility product of bisubstituted plutonium (IV) phosphate as $2 \cdot 10^{-28}$ [83]. At hydrogen ion concentrations below 0.3 *M* the bisubstituted phosphate is hydrolyzed with the formation of aquohydroxyphosphate $[Pu(HPO_4)(H_2O)_{n-1}\,OH]^+$, where $n = 1, 2, \ldots, i$.

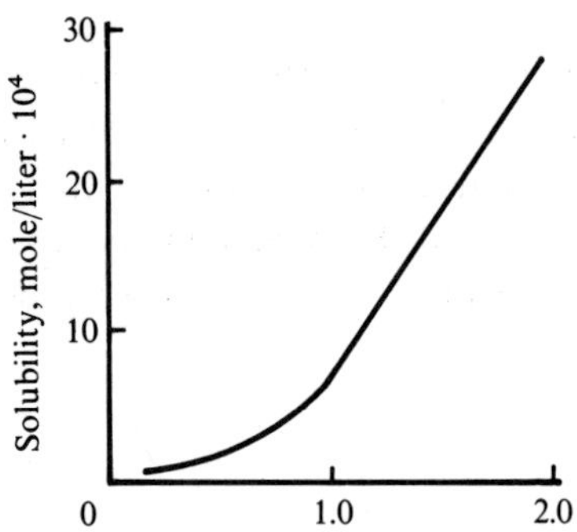

Figure 30

Solubility of plutonium (IV) phosphate in nitric acid solutions

The amorphous precipitate may be converted into a dense, brown, crystalline precipitate by heating at 100° for several days in a sealed tube in the presence of excess nitric and phosphoric acids [3, p. 343].

$Pu_2(HPO_4)(PO_4)_2 \cdot xH_2O$

When the precipitate $PuPO_4$ is treated with mild oxidizing agents ($KBrO_3$, Cl_2)

in the presence of phosphoric acid, a compound of the composition $Pu(HPO_4)\cdot(PO_4)_2\cdot xH_2O$ is obtained (Zhagin, 1959). The compound was isolated as fine red crystals after prolonged (20 days) heating of the gelatinous compound in a sealed ampule and holding for two months at room temperature. The solubility of the amorphous form in 1 *N* HCl at 18° is 6.9 mg Pu/liter. The salt is sparingly soluble in water and in most mineral acids. Its solubility in a solution of sodium carbonate was shown by Zhagin to increase with the increase in the concentration of sodium carbonate because of the formation of soluble carbonate complexes.

$Pu_3(PO_4)_4\cdot xH_2O$

This compound was obtained by heating $Pu(HPO_4)_2\cdot xH_2O$ at 110° for several days in a sealed tube with a solution containing 1.8 *M* HNO_3 and 0.4 *M* H_3PO_4 [495].

Pu (VI) is not precipitated from acid solutions of phosphoric acid [3, p. 344].

Plutonium ferrocyanides and ferricyanides

Anderson [273] identified five compounds formed between plutonium and potassium ferrocyanide or ferricyanide. The sparingly soluble compounds of trivalent plutonium alone are of analytical importance.

$KPu[Fe(CN)_6]\cdot 7H_2O$

When potassium ferrocyanide is made to act on a solution of trivalent plutonium in 0.8 *M* HCl containing 7 mg plutonium per ml, a light blue precipitate of the above composition separates out. Its solubility is 9 mg plutonium per liter. The precipitate dissolves in concentrated sulfuric acid only with difficulty [273].

$Pu[Fe(CN)_6]\cdot 7H_2O$

A compound of the composition corresponding to this formula precipitates out when a solution of trivalent plutonium in 0.5 *M* HCl is acted upon by potassium ferricyanide. The precipitate is black-colored. The residual concentration of plutonium in solution after precipitation is only a few mg per liter. The precipitate dissolves in 9 *M* H_2SO_4 with difficulty.

Plutonium (IV) and plutonium (VI) ferrocyanides and ferricyanides are readily soluble in water.

Plutonium (IV) molybdate

When excess ammonium molybdate is added to a solution of plutonium (IV) nitrate at pH 2–4, a precipitate of $Pu(MoO_4)_2$ separates out. Under these conditions its solubility calculated as plutonium is about 16 mg/liter [3, p. 354].

Plutonium oxalates

When oxalic acid is made to act on acid solutions of Pu(III), Pu(IV) and Pu(VI), the corresponding insoluble plutonium oxalates are formed.

$Pu_2(C_2O_4)_3 \cdot 9H_2O$

When crystalline oxalic acid or an oxalate is added to weakly acid solutions (0.3–0.8 *M*) of trivalent plutonium, a crystalline, bright-green precipitate of the formula $Pu_2(C_2O_4)_3 \cdot 9H_2O$ separates out. The precipitant is added in an amount sufficient to precipitate the plutonium and to render the solution 0.2 *M* in oxalate. To ensure complete precipitation, the precipitate is left to stand for some 30 minutes [587; 3, p. 347]. The compound is stable in an inert atmosphere. Its solubility in water at 20° is $4.32 \cdot 10^{-6}$ moles/liter and the solubility product is $1.62 \cdot 10^{-25}$ [54, 56]. With the increase in the concentration of potassium oxalate and ammonium oxalate, the solubility of the compound increases owing to the formation of soluble oxalate complexes of plutonium (III). Variations in the acidity of the medium substantially affect the solubility of plutonium (III) oxalate [3, p. 347]. Figure 31 represents the dependence of the solubility of $Pu_2(C_2O_4)_3 \cdot 9H_2O$ on the hydrogen ion concentration.

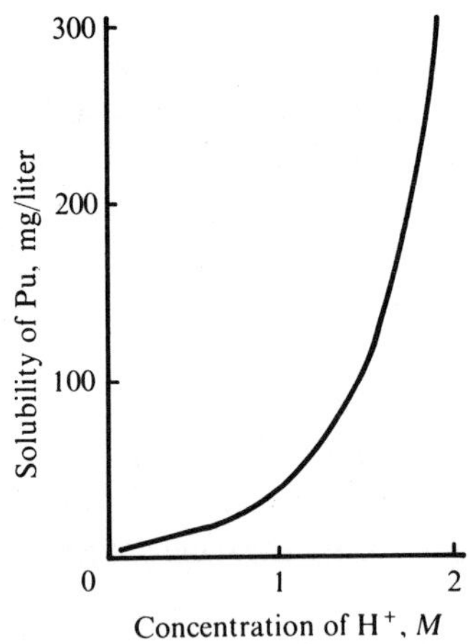

Figure 31

Effect of hydrogen ion concentration on the solubility of $Pu_2(C_2O_4)_3 \cdot 9H_2O$

The thermal decomposition of plutonium (III) oxalate was investigated with the aid of a Kurnakov pyrometer [108]. The results obtained are shown in Figure 32. The curve of thermal decomposition of plutonium (III) oxalate in the air (Figure 32a) has a maximum and a minimum, corresponding to an endothermal effect at 140°

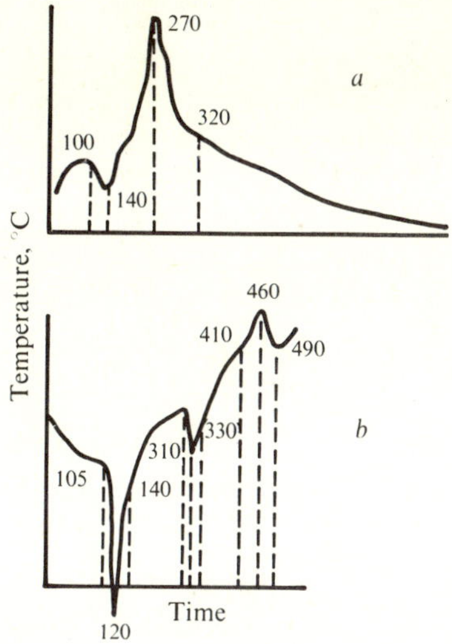

Figure 32
Differential heating curves of Pu(III) oxalate in the air (*a*) and in an inert gas atmosphere (*b*).

and an exothermal effect at 270°. The dehydration in an inert atmosphere also takes place at 140°, but the anhydrous oxalate decomposes at 330° with the formation of a mixed oxalate–carbonate. At 460° this compound decomposes and at the same time Pu(III) is oxidized to Pu(IV) with dioxide formation. At 140° the anhydrous Pu(III) oxalate is stable and at 270° it is rapidly decomposed with the formation of PuO_2. The anhydrous oxalate $Pu_2(C_2O_4)_3$ is obtained by heating the hydrated oxalate in vacuo at 225° [3, p. 347].

$Pu(C_2O_4)_2 \cdot 6H_2O$

When oxalic acid is made to act on acid solutions of tetravalent plutonium a crystalline, yellow-green precipitate separates out [3, p. 348]. The composition of the precipitate is $Pu(C_2O_4)_2 \cdot 6H_2O$ [615]. Gel'man and Moskvin [57, 166] determined its solubility in water as $10.3 \cdot 10^{-5}$ moles/liter. The concentration of ammonium oxalate in solution affects the solubility of $Pu(C_2O_4)_2 \cdot 6H_2O$ (Figure 33). When the concentration of ammonium oxalate is increased from 0.001 to 0.005 *M*, the solubility decreases, passes through the minimum value of $3.55 \cdot 10^{-5}$ moles/liter Pu(IV) and then increases because of the complex formation between Pu(IV) and oxalate ions.

Moskvin and Gel'man [166] studied the effect of acidity on the solubility of plutonium(IV) oxalate in a number of mineral acids. The results are shown in Figure 34. The solubility of $Pu(C_2O_4)_2 \cdot 6H_2O$ in solutions of HNO_3 and $HClO_4$

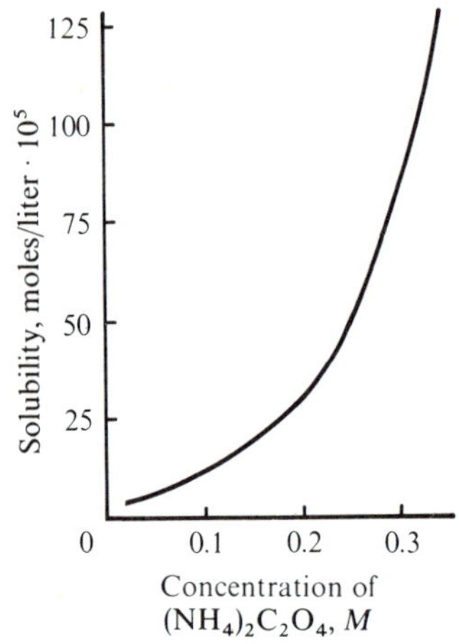

Figure 33

Effect of ammonium oxalate concentration on the solubility of $Pu(C_2O_4)_2 \cdot 6H_2O$ in 1 *M* HNO_3

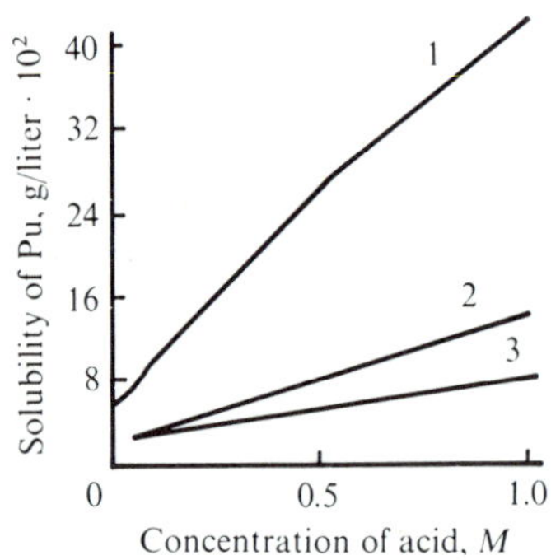

Figure 34

Effect of acid concentration on the solubility of $Pu(C_2O_4)_2 \cdot 6H_2O$:

1) H_2SO_4; 2) HNO_3; 3) $HClO_4$.

first decreases and then, when the acid concentration becomes higher than 0.1 *N*, increases again. The decrease in the solubility is due to the acid properties of the oxalate. The solubility product of the oxalate was determined from its solubility in solutions of HNO_3 containing $(NH_4)_2C_2O_4$; it is $4 \cdot 10^{-22}$ [166]. This value is in agreement with that obtained by determining the solubility of $Pu(C_2O_4)_2 \cdot 6H_2O$ in perchloric acid [166]. Plutonium oxalate decomposes under the effect of its own α-radiation with the formation of plutonium oxycarbonate, CO and CO_2 [209, 240].

Thermal decomposition of plutonium (IV) oxalate was studied in [108, 205, 571]. Figure 35 shows the thermogravigram obtained by Sklyarenko and Chubukova

[207]. When heated in the air, plutonium oxalate loses 6 molecules of water between 30 and 110°. The small horizontal part of the curve at 120–150° corresponds to the anhydrous plutonium (IV) oxalate. Further decrease in weight is caused by the decomposition of $Pu(C_2O_4)_2$, during which an unstable intermediate compound of unknown composition is formed. The final product of the thermal decomposition is plutonium dioxide.

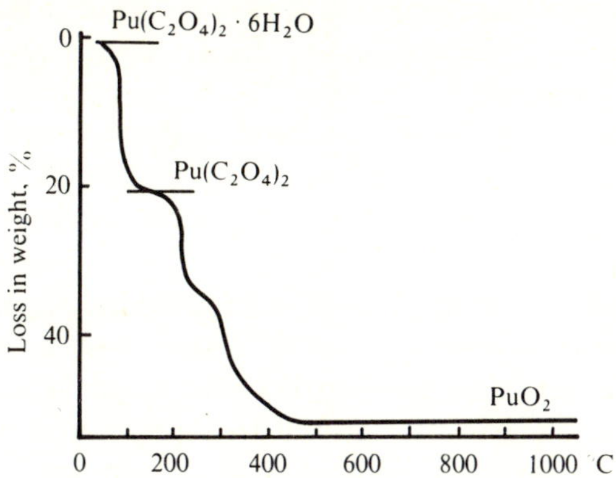

Figure 35
Thermogravigram of plutonium (IV) oxalate

A somewhat different mechanism of the decomposition of plutonium (IV) oxalate has also been proposed [108]. The discrepancies may be due to different experimental conditions and techniques.

$PuO_2(C_2O_4) \cdot 3H_2O$

This compound was prepared in 1950 by Gel'man and Drabkina [50]. Pu(IV) was oxidized to Pu(VI) by heating at 90° with potassium dichromate or MnO_2 in 1.5–2 *M* HNO_3, and plutonyl oxalate was precipitated by the addition of crystalline oxalic acid. A crystalline, pink precipitate separates out. When stored dry, plutonyl oxalate gradually changes color to green due to the reduction of Pu(VI) by its own α-radiation. $PuO_2(C_2O_4) \cdot 3H_2O$ is sparingly soluble in water and dilute mineral acids, and is readily soluble in ammonium carbonate and ammonium oxalate with formation of complex compounds. Its solubility product was determined as $6.0 \cdot 10^{-10}$ [91].

On being heated to 180°, plutonyl oxalate decomposes explosively with the formation of PuO_2.

Organic plutonium compounds

Plutonium (IV) *8-hydroxyquinolate* $Pu(C_9H_6NO)_4$

Plutonium 8-hydroxyquinolate was first obtained by the action of 8-hydroxyquinoline on an acetate solution of plutonium (IV) at pH 4–8 and plutonium concentration of 0.004 *M* [588]. The residual concentration of plutonium in solution after precipitation was 77 mg/liter at pH 4 and 17 mg/liter at pH 8.

Moiseev (1954) showed that tetravalent plutonium is quantitatively precipitated by 8-hydroxyquinoline from ammoniacal and bicarbonate solutions at pH 4.5–12 in the presence of tartrate ions.

Trivalent plutonium is oxidized by 8-hydroxyquinoline, and the precipitate consists of 8-hydroxyquinolate of tetravalent plutonium $Pu(C_9H_6NO)_4$.

In the absence of tartrate ions plutonium 8-hydroxyquinolate precipitated from sulfate solutions has a variable composition.

$Pu(C_9H_6NO)_4$ separates out as dark-red or brown-purple crystalline precipitate [588, 589]. The solubility product is $5.6 \cdot 10^{-39}$. It was shown by thermogravimetric investigations (Figure 36) that plutonium (IV) 8-hydroxyquinolate is stable up to 250° and is suitable as a gravimetric form. It is decomposed at 300–500° with the formation of plutonium dioxide. If the precipitate has not been washed completely free of the precipitant, the thermogravimetric curve 1 in Figure 36 has a supplementary drop at about 200°, due to the sublimation of the excess 8-hydroxyquinoline. The small plateau at 250° corresponds to $Pu(C_9H_6NO)_4$.

When 8-hydroxyquinoline reacts with solutions of hexavalent plutonium at pH 4–8, an orange-red precipitate separates out. Its approximate composition [3, p. 352] is $PuO_2(C_9H_6NO)_2 \cdot xH_2O$. Plutonyl 8-hydroxyquinolate is readily extracted with amyl acetate [48].

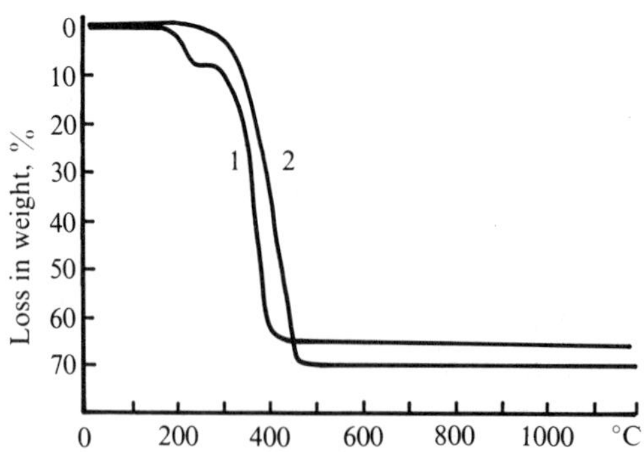

Figure 36

Thermogravigram of plutonium (IV) 8-hydroxyquinolate:

1) Pu (IV) 8-hydroxyquinolate with some 8-hydroxyquinoline; 2) pure Pu (IV) 8-hydroxyquinolate.

Plutonium (IV) *cupferronate* $Pu(C_6H_5N_2O_2)_4$

When sulfuric acid solutions of Pu(IV) interact with a solution of cupferron, an amorphous orange-yellow precipitate is formed, corresponding to the formula $Pu(C_6H_5N_2O_2)_4$ [161].

When trivalent plutonium is added to the solution, the composition of the precipitate, which is formed immediately, is one atom of plutonium to four organic radicals $C_6H_5N_2O_2^-$. The oxidation mechanism is not yet clear, but it was found [161] that no cupferron is consumed in the oxidation. In acid solutions cupferron reduces Pu(VI) to Pu(IV), which is precipitated by the excess of the reagent. Thus, plutonium in any valency state will invariably be precipitated as the tetravalent cupferronate.

The solubility product of plutonium (IV) cupferronate is $1.2 \cdot 10^{-31}$. Table 10 gives the solubility data for plutonium (IV) cupferronate.

Table 10

SOLUBILITY OF PLUTONIUM(IV) CUPFERRONATE AS A FUNCTION OF THE HYDROGEN ION CONCENTRATION OF THE REAGENT (HR) (21 ± 1°C) [161]

Concentration of	Equilibrium concentrations, *M*		Solubility of
H_2SO_4, *M*	$[H^+]$	$[HR] \cdot 10^3$	PuR_4, $M \cdot 10^6$
0.25	0.278	1.89	7.9
0.25	0.278	3.47	2.07
0.5	0.538	3.48	4.93
0.5	0.538	5.07	2.53
1.0	1.055	10.1	1.76
1.5	1.586	15.2	1.38

The average value of the equilibrium constant for the precipitation of Pu(IV) by cupferron is $6.6 \cdot 10^{-13}$.

The thermal decomposition of plutonium cupferronate was shown by Sklyarenko and Chubukova (1959) to begin at ~ 40° (Figure 37). Plutonium cupferronate cannot serve as a gravimetric form. Above 600° plutonium exists as PuO_2.

Plutonium (IV) *benzenesulfinate* $Pu(C_6H_5SO_2)_4$

When benzenesulfinic acid or its salts are made to act on weakly acid solutions of Pu(IV), an amorphous, orange-colored precipitate separates out [48]. Its composition corresponds to the formula $Pu(C_6H_5SO_2)_4$. Trivalent plutonium is not precipitated by this reagent (Milyukova, 1953). Hexavalent plutonium is reduced to Pu(IV) and is then precipitated as the benzenesulfinate.

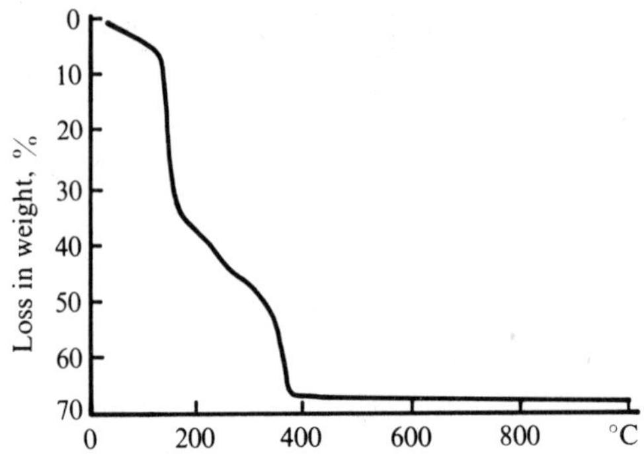

Figure 37
Thermogravigram of plutonium (IV) cupferronate

Plutonium (IV) benzenesulfinate is soluble in concentrated hydrochloric and nitric acids. It is decomposed by alkalis with the formation of plutonium hydroxide. The absorption spectrum of $Pu(C_6H_5SO_2)_4$ in concentrated nitric acid is identical with the absorption spectrum of the hexanitrate complex.

The solubility of plutonium (IV) benzenesulfinate under conditions of quantitative precipitation (0.15 M HNO_3 and about 2.5% excess precipitant) is $1.2 \cdot 10^{-5}$ moles/liter (Milyukova, 1959). Plutonium (IV) benzenesulfinate can be extracted from acid solutions by a number of organic solvents: benzene, isovaleric and isobutyric acids.

Figure 38 shows the thermogravigram of plutonium (IV) benzenesulfinate (Milyukova, 1958). When heated slightly above room temperature, the weight of

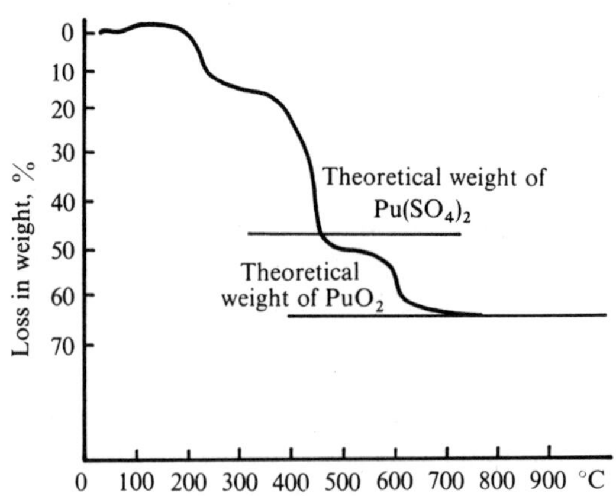

Figure 38
Thermogravigram of plutonium (IV) benzenesulfinate

the precipitate begins to drop. A certain increase in weight when the temperature is further increased may be explained by a partial oxidation of sulfinic to sulfonic group. The shape of the descending part of the pyrolytic curve indicates that unstable intermediate products are formed. The weight of the precipitate on the horizontal section between 450 and 660° approximately corresponds to plutonium sulfate.

Plutonium salicylates

Salicylates of trivalent and tetravalent plutonium have been described [48, 100, 588]. These compounds were thoroughly investigated by Zvyagintsev and Sudarikov [100].

Plutonium (III) *salicylate* $Pu(Sal)_3 \cdot 1.5H_2O$

The formation of the blue precipitate of plutonium (III) salicylate begins at pH 0–1, depending on the concentration of plutonium in the initial solution. Quantitative (99.8 %) precipitation of Pu (III) is attained in the presence of a large excess of salicylic acid and at pH 3.5. The precipitate is allowed to settle fully and the solution is decanted. The solubility of the compound in water is 0.12 g Pu/kg, in alcohol 0.50 g Pu/kg and in acetone 3.50 g Pu per kg of solution. Figure 39 illustrates the dependence of the completeness of precipitation on the excess precipitant present [100].

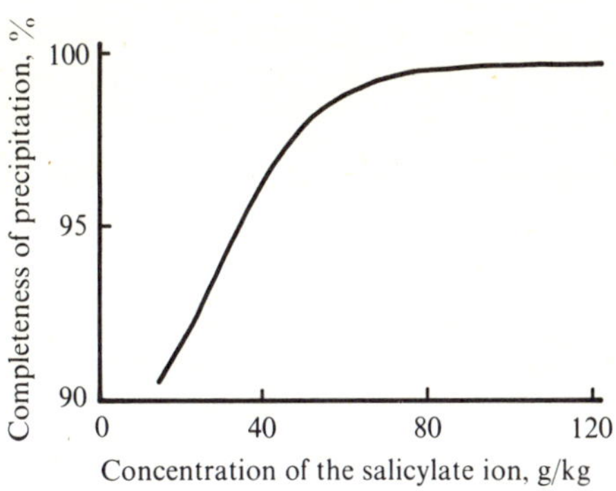

Figure 39

Precipitation of plutonium (III) salicylate as a function of the concentration of precipitant

Plutonium (IV) *oxysalicylate* $PuO(Sal)_2$

When salicylic acid is added to a solution of Pu (IV) in ~ 1.0 *N* HNO_3, the solution becomes dark brown. If the acidity is decreased to pH 0–1, the flocculent precipitate of plutonium (IV) oxysalicylate separates out. Pu (IV) is quantitatively precipitated from solution as the oxysalicylate at pH 4–5 and a minimum excess of precipitant. The solubility of plutonium (IV) oxysalicylate in water at 25° is 0.3 mg Pu/kg. The effect of the pH value on the solubility of $PuO(Sal)_2$ is represented in Figure 40.

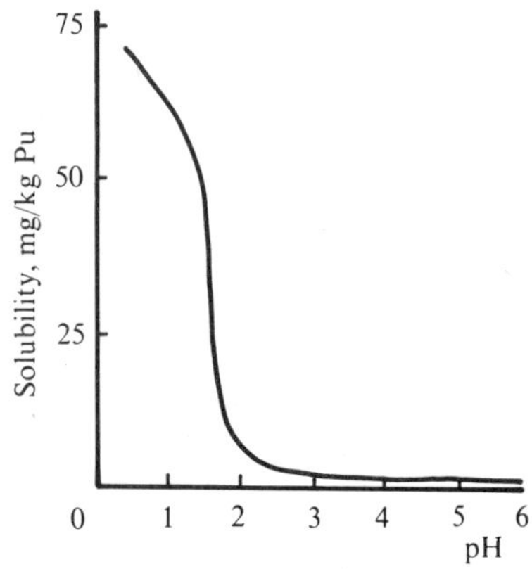

Figure 40

Solubility of plutonium (IV) oxysalicylate as a function of the pH of the solution

Solutions of complex plutonium (III) and plutonium (IV) salicylates are not stable on being heated, and the degree of hydrolysis of plutonium depends both on the temperature and on the final concentration of ammonium or sodium salicylate. The product of hydrolysis is an amorphous gray-yellow and sometimes brown precipitate. The compound has not been obtained in crystalline form. The composition of the isolated precipitate is $PuO_2 \cdot PuO \cdot Sal \cdot xH_2O$, irrespective of the initial valency state.

The minimum solubility of the compound under optimum conditions (25°, concentration of salicylate ion 5 g/kg, pH 7.4) is 0.02 mg Pu/kg of solution.

Plutonium (IV) *phytate* $Pu_3(C_6P_6O_{27}H_{12})_2 \cdot xH_2O$

The reaction between phytic acid and acid solutions of tetravalent plutonium results in the formation of a light-violet amorphous precipitate [194]. Sklyarenko [205] showed that the composition of the precipitate thus formed depends on the excess of the precipitant and approximately corresponds to the formula $Pu_3(C_6P_6O_{27}H_{12})_2 \cdot xH_2O$. The precipitate is rapidly formed in acid media, easily compacted and readily filtered.

The solubility of plutonium (IV) phytate in water at $20 \pm 0.1°$ is $5 \cdot 10^{-5}$ moles/liter. Its solubility in mineral acids is insignificant. It is readily soluble in ammonium and sodium citrates.

Figure 41 shows the thermogravigram of plutonium (IV) phytate. The gradual decrease in weight throughout the curve indicates that the compound has no gravimetric forms. The plateau between 700 and 900° approximately corresponds to the formula $2PuO_2 \cdot 3P_2O_5$. Further increase in temperature produces a decrease in weight due to the gradual elimination of P_2O_5 from the precipitate.

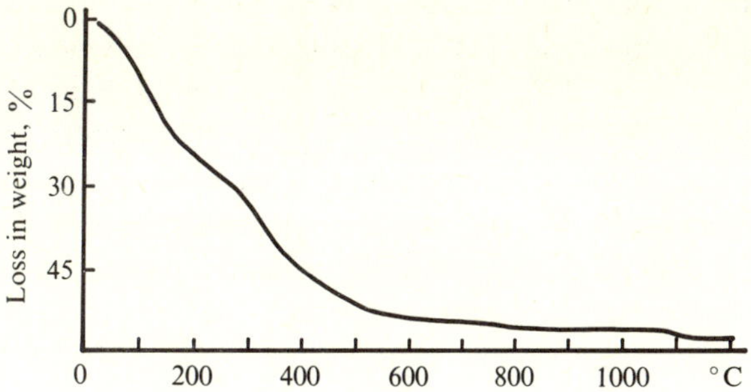

Figure 41
Thermogravigram of plutonium (IV) phytate

Plutonium (IV) *phenylarsonate* $Pu(C_6H_5AsO_3)_2$

A crystalline precipitate of this composition is formed when phenylarsonic acid is made to react with acid solutions of plutonium (IV) [588]. The solubility of the compound in water is 1.1 mg Pu/liter. The solubility product is $4.7 \cdot 10^{-6}$. The thermogravimetric curve (Figure 42) determined by Mikhailov (1958) has three plateaus. The first (below 150°) corresponds to plutonium (IV) phenylarsonate and

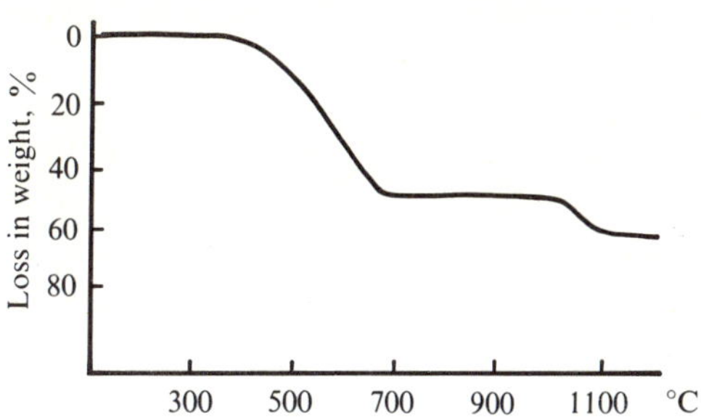

Figure 42
Thermogravigram of plutonium (IV) phenylarsonate

depends on the conditions of precipitation. The horizontal section at 600–900° corresponds most closely to the compound $3PuO_2 \cdot PuAs_2O_7$. Neither of these two compounds can be employed for the gravimetric determination of plutonium. The calcination of plutonium (IV) phenylarsonate is carried out at 1100°; at this temperature plutonium dioxide is formed.

Sodium plutonyl triacetate $NaPuO_2(CH_3COO)_3$

When sodium nitrate and sodium acetate are added to a nitric acid solution of tetravalent plutonium, containing 1 *M* CH_3COOH, the compound of this composition precipitates out [3, p. 351]. The crystals which precipitate under these conditions are light pink [738]. The solubility of sodium plutonyl triacetate is strongly affected by the concentration of sodium nitrate (Figure 43). Variation in the acetate ion concentration between 0.12 and 1.0 *M* has practically no effect on the completeness of the precipitation of sodium plutonyl triacetate. Its solubility in water at 5° is $6.1 \cdot 10^{-3}$ mg Pu per liter and at 95° it is $18.7 \cdot 10^{-3}$ mg Pu per liter.

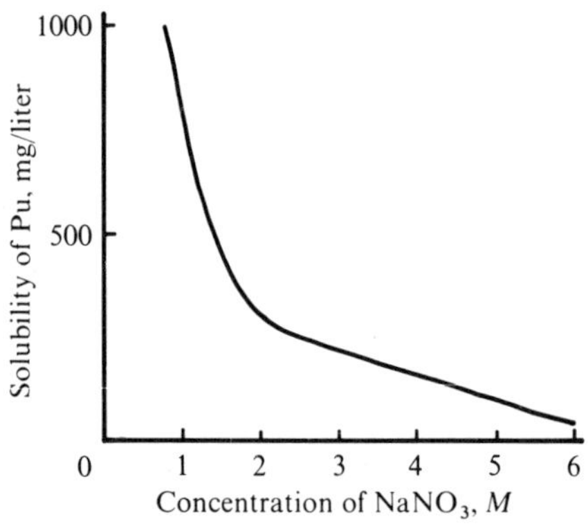

Figure 43

Effect of the $NaNO_3$ concentration on the solubility of $NaPuO_2(CH_3COO)_3$ in 0.6 *M* CH_3COOH and 0.2 *M* CH_3COONa

Plutonium (IV) *acetylacetonate* $Pu(C_5H_7O_2)_4$

The preparation of plutonium (IV) acetylacetonate described in the literature [381] is as follows. To a weakly acid solution of plutonium sulfate an alkaline solution of acetylacetone is added in the amount corresponding to 25 mg of the reagent per mg of plutonium. A red-brown crystalline precipitate separates out, which is isolated by centrifugation and dried over P_2O_5. The dried product is purified by sublimation at 170° and 1 mm Hg. The compound is not decomposed to an appreciable extent on being exposed to the atmosphere. M. p. 170°. $Pu(C_5H_7O_2)_4$ is readily soluble in benzine, benzene and chloroform.

The sublimation of plutonium acetylacetonate is employed to deposit plutonium as a thin uniform layer on various supports, and also for purposes of separation.

Plutonium (IV) *m-nitrobenzoate* $Pu(C_6H_4NO_2COO)_4$

The reaction between a solution of m-nitrobenzoic acid and a 0.004 *M* solution of plutonium (IV) in acetic acid–acetate medium at pH 3–5 results in the formation of an amorphous white precipitate [588]. The residual concentration of plutonium in the solution is 4 mg/liter at pH 5.

It was shown by Milyukova (1956) that when 20% alcoholic solution of m-nitrobenzoic acid is allowed to react with solutions of Pu(III), Pu(IV) and Pu(VI) in 0.1 *N* HNO_3, only Pu(IV) separates out as an amorphous precipitate, light-cream to yellow-green in color. The addition of acetone, aniline or ethanol improves the completeness of precipitation of plutonium. Nevertheless, the solubility of plutonium (IV) m-nitrobenzoate in 0.1 *N* HNO_3, even in the presence of ethanol, is much higher than during precipitation from acetic acid–acetate medium at pH 3–5, and is 20 mg Pu per liter.

Plutonium (IV) *pyrogallate*

When pyrogallol is made to react with a solution of plutonium acetate (0.004 *M* Pu) at pH 3–5, a red-brown crystalline precipitate separates out [588].

COMPOUNDS OBTAINED IN THE DRY WAY

Plutonium oxides

The following compounds of plutonium with oxygen have been described [554, 726, 732]: PuO_2, Pu_2O_3 and PuO.

An attempt to obtain higher plutonium oxides by oxidizing PuO_2 with nitrogen dioxide was unsuccessful [487]. It was shown by the thermodynamic calculations of Brewer [309] that stable plutonium oxides higher than PuO_2 cannot be prepared.

Plutonium dioxide PuO_2

Plutonium dioxide is widely employed in analytical work as the most stable weighing form, and in technology as a starting material in the preparation of halides and other compounds.

Plutonium dioxide may be obtained by calcining in the air any plutonium compound except plutonium phosphates. Drummond and Welch [388] found that the composition of the dioxide may vary, depending on the initial compound, from $PuO_{2.00}$ to $PuO_{2.09}$ due to additional absorption of oxygen. Calcination of PuO_2 in the air at 1200° results in the formation of the oxide $PuO_{2.002}$. These results were subsequently confirmed [189].

The most reactive dioxide with a practically stoichiometric composition is obtained by calcination of Pu(IV) peroxide and Pu(III) and Pu(IV) oxalates [60, 237]. For the dioxide to be pure, it must be crystalline [519].

Plutonium dioxide is a fine-crystalline powder; its color may vary from greenish-gray to dark brown, depending on the initial compound and temperature of calcination. Plutonium dioxide is not hygroscopic (Milyukova, 1953). The X-ray density of the compound has been calculated [554] as 11.44 g/cm^3. The dioxide has a face-centered cubic lattice (structure of the fluorite type, isomorphous with UO_2 and ThO_2) [732], with $a = 5.936 \pm 0.001$ Å.

The heat of formation of plutonium dioxide, calculated by various authors from the heat of combustion of metallic plutonium, is 252.4 $\pm$ 1.1 kcal/mole [184, 237]. The experimentally found melting points are around 2240°. X-ray diagrams after melting showed a two-phase structure because of incongruent (loss of oxygen) evaporation of PuO_2 near the melting point. The observed melting point corresponds to the two-phase system PuO_2–cubic Pu_2O_3. For this reason the melting point of stoichiometric PuO_2 should be determined under equilibrium pressure of oxygen [237]. Such data are not available in the literature.

Plutonium dioxide is stable up to 1200°, but its volatility increases when the temperature is raised further. Investigations of the volatility of PuO_2 at high temperatures were carried out by Phipps et al. [597]. They noted that at high temperatures the volatility of PuO_2 decreases; they explain this effect by the elimination of oxygen and conversion of the dioxide to the less volatile Pu_2O_3 at these temperatures. The composition of the volatile fraction is not known.

Chemical properties. It was noted [451] that the reactivity of plutonium dioxide varies with the temperature at which it has been calcined. This effect was ascribed to a rearrangement of the crystal lattice of the dioxide. Plutonium dioxide prepared at temperatures below 800° has a strongly distorted lattice. The lattice becomes perfect at 1,000–1,200°. Dioxide calcined at this temperature has a composition nearest to the stoichiometric, but is not very reactive.

Plutonium dioxide is practically insoluble in water and dilute mineral acids. However, the dioxide calcined at low temperatures (below 275°) is soluble in hydrochloric acid. Dioxide calcined at up to 600° is soluble in concentrated sulfuric and nitric acids [451]. The reagents employed to solubilize strongly calcined plutonium dioxide include 85%–100% phosphoric acid at 200°, or a solution of 10 M HNO_3 containing 0.05 M HF (6 M HI) [237]. A prolonged treatment with concentrated sulfuric acid [60] or HBr [726, Sklyarenko, 1953] also results in the solubilization of the dioxide. This process may be accelerated by treating the dioxide with a mixture of sulfuric and hydrobromic acids (1 : 10). The precipitate is first wetted with sulfuric acid, hydrobromic acid is then cautiously added and the mixture evaporated to dryness. This operation is then repeated once more. One hundred milligrams of PuO_2 can be converted into sulfate in about two hours (Kochetkova, 1955).

Rubinshtein, Zaitsev and Shubochkin (1953) proposed a method in which the dioxide is treated with a mixture of concentrated HCl and potassium iodide. The dissolution is quite fast: 100 mg PuO_2 are dissolved in $1\frac{1}{2}$ hours. Ten parts of potassium iodide are taken per one part of the dioxide. Concentrated HCl is added until the solution is 8–10% in KI. The authors consider that plutonium is reduced to the trivalent state in the process. To remove iodine, nitric and sulfuric acids are added and the solution is heated.

Formerly [388, 451] strongly calcined PuO_2 was dissolved by fusion with potassium pyrosulfate or sodium bisulfate. Plutonium dioxide may be converted into PuF_3 by heating in a stream of hydrogen mixed with hydrogen fluoride at 400–600°, and into PuF_4 by heating at the same temperature in a stream of a mixture of oxygen with hydrogen fluoride, or in a stream of HF alone [3, p. 327]. At 250–280° PuO_2 reacts with $NH_4F \cdot HF$ with the formation of PuF_4 [154]. The reaction of PuO_2 with CCl_4 at 800° results in the formation of $PuCl_3$ [32, 260]. Plutonium dioxide reacts with H_2S in a graphite crucible at 1,225–1,300° with the formation of Pu_2O_2S. When the temperature is raised to 1,340–1,400°, Pu_2S_3 is obtained [262]. When acted upon by excess graphite, plutonium dioxide is converted into a carbide at 1,200–1,500° [3, p. 327].

Plutonium oxide PuO

A number of X-ray powder patterns have been attributed to PuO [3, 554, 726, 732]. Most probably, however, this oxide exists only as a surface film on metallic plutonium. It was shown [726] that PuO is obtained together with other reaction products when PuOCl is reduced by barium vapors. These results were later shown to be reproducible [237], but no substance giving the X-ray patterns of PuO could be isolated from the mixture. The properties of PuO have been described. Plutonium oxide has a color similar to that of coke [726]; its calculated X-ray density is 13.89 ± 0.02 g/cm³ [554]. The predicted heat of formation is 130–140 kcal/mole [309]. PuO is readily dissolved by hydrochloric acid in concentrations above 1 *M*.

Plutonium sesquioxide Pu_2O_3

The sesquioxide exists in two crystal forms. The hexagonal sesquioxide is obtained by reducing PuO_2 with metallic plutonium in the form of shavings or filings in a covered tantalum crucible at 1,500°; the plutonium is used in 20% excess. The hexagonal Pu_2O_3 thus obtained contains metallic plutonium as a second phase; it is eliminated by evaporation in vacuo at 1,800–1,900°. The sesquioxide may also be prepared from plutonium hydride.

The hexagonal Pu_2O_3 is pyrophoric and may catch fire when triturated or milled, but undamaged oxide is stable in the air at room temperature for several days.

The cubic oxide was prepared by heating PuO_2 in vacuo at 1,650–1,800° [726]. Since the cubic sesquioxide cannot be prepared by thermal treatment or quenching of the hexagonal sesquioxide, it cannot be considered as its polymorphic modification. The cubic structure appears in the sesquioxide only at an oxygen content somewhat higher than the stoichiometric. The exact composition of cubic Pu_2O_3 is not known, but is assumed to lie between $PuO_{1.5}$ and $PuO_{1.6}$.

Plutonium halides and oxyhalides

Plutonium halides and oxyhalides are employed in various stages during the industrial manufacture of plutonium. Halides were also reported to have been used in the separation of actinides. The separation of the volatile uranium hexafluoride from plutonium hexafluoride has been described [314]. A patent has been taken out for a "dry" separation of neptunium from plutonium, based on the volatility of $NpCl_4$, which is much higher than that of $PuCl_3$ [415].

The decreasing stability of the higher valency states in the sequence uranium–neptunium–plutonium is particularly conspicuous for the halides. Except for fluorine, the halogens form only trihalides with plutonium. Compounds between plutonium halides and oxygen—oxyhalides—are characteristic of the trivalent state only. Table 11 gives certain physical properties of plutonium halides and oxyhalides.

Table 11

PLUTONIUM HALIDES AND OXYHALIDES [231, p. 82; 728, 737]

Compound	Color	Crystal structure	Density, g/cm^3	M.P.	ΔH, kcal / mole
PuF_6	Red or reddish-brown	Orthorhombic	—	50.75	—
PuF_4	Light brown	Monoclinic	7.0	1037	−424
PuF_3	Violet	Hexagonal	9.32	1425	−375
$PuCl_3$	Emerald green	Hexagonal	5.70	760	−230
$PuBr_3$	Green	Orthorhombic	6.69	681	−198.8
PuI_3	Light green	Orthorhombic	6.92	770	−155
PuOF	Metallic	Tetragonal	9.76	> 1635	—
PuOCl	Blue-green	Tetragonal	8.81	—	−223.6
PuOBr	Dark green	Tetragonal	9.07	—	−210
PuOI	Green	Tetragonal	8.46	—	−190

Plutonium trifluoride PuF_3

Plutonium trifluoride is usually prepared by the reaction between plutonium dioxide and hydrogen fluoride in the presence of hydrogen [376]:

$$PuO_2 + 3HF + \tfrac{1}{2}H_2 \xrightarrow{600°C} PuF_3 + 2H_2O.$$

Different compounds may be used as the starting material: Pu(IV) and Pu(VI) nitrates, Pu(IV) and Pu(III) oxalates, etc. The aqueous fluoride described in the preceding section may be rendered anhydrous by heating at 250–300° [376]. Figure 44 gives the thermal decomposition curves of aqueous (1) and anhydrous (2, 3) PuF_3. The final product of the calcination is plutonium dioxide. The anhydrous trifluoride is formed in a narrow temperature range (~ 250–300°). A comparison

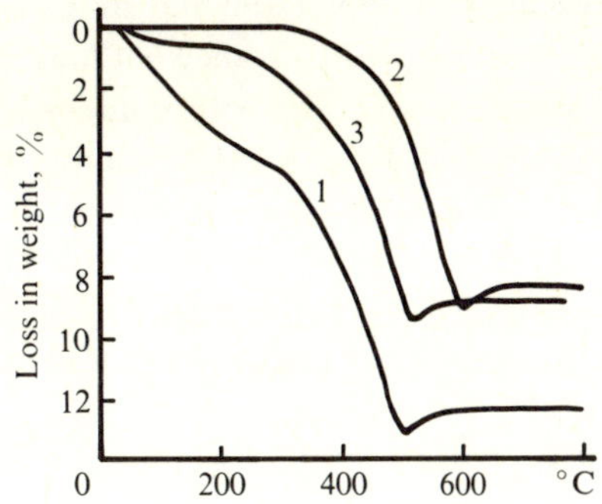

Figure 44
Thermogravigram of Pu(III) flourides:

1) PuF_3 precipitated from aqueous solution; 2) anhydrous PuF_3 prepared by hydrofluorination of PuO_2; 3) anhydrous PuF_3 after 12 days standing in the air.

of curves 2 and 3 indicates that PuF_3 is hygroscopic. Plutonium trifluoride is oxidized by dry oxygen at 600° (in vacuum the reaction is reversed [376, 416]):

$$4PuF_3 + O_2 = 3PuF_4 + PuO_2.$$

Plutonium trichloride $PuCl_3$

This compound is obtained by reacting plutonium dioxide with strong chlorinating agents such as CCl_4, HCl, SCl_2 or $POCl_3$, above 700°. Phosphorus pentachloride reacts with plutonium dioxide at 280° [3, p. 318, 260]. Plutonium (III) oxalate and plutonium carbonate can also be chlorinated, but the product is contaminated with carbon [260]. Sublimation of $PuCl_3$ at 800° yields the pure crystalline product.

The trichloride is quantitatively converted into the oxychloride PuOCl at 650° when treated with a mixture of hydrogen chloride, water vapor and hydrogen. When heated in the air at 900° it is converted to the dioxide [260]. The volatility of $PuCl_3$ is utilized in mass-spectrographic separations.

Plutonium tribromide $PuBr_3$

The preparation of plutonium tribromide is based on the reaction of various brominating agents with plutonium dioxide. As in the case of the trichloride, the preparation of $PuBr_3$ from the dioxide is faster if the oxide employed has been calcined at moderate temperatures [372]. The reaction with hydrogen bromide takes place in the absence of hydrogen, as well as in presence of hydrogen.

The tribromide may also be prepared by the reaction between metallic plutonium and bromine at 300–325° and under a pressure of bromine of 200 mm Hg and also by the reaction between $PuCl_3$ and dry HBr [372].

At 750–800° the tribromide is quite volatile and may be purified by sublimation in vacuo. In oxygen at 100° or in the air at a somewhat higher temperature plutonium tribromide is converted to the dioxide. The tribromide is hygroscopic and in moist air it forms $PuBr_3 \cdot 6H_2O$. Pure $PuBr_3$ is easily soluble in water [372].

Plutonium triiodide PuI_3

This compound is formed by the reaction between anhydrous gaseous hydrogen iodide and metallic plutonium at 450° according to [441]:

$$Pu + 3HI \xrightleftharpoons{450°C} PuI_3 + 1\tfrac{1}{2}H_2.$$

Plutonium (III) *oxyhalides*

The stability of plutonium oxyhalides increases from fluorine to iodine.

Oxyfluoride PuOF

This compound has been identified as one of the phases obtained by reducing plutonium tetrafluoride by atomic hydrogen [3, p. 317].

Oxychloride PuOCl

Plutonium oxychloride was obtained by treating plutonium dioxide at 650° by a mixture of gaseous hydrogen chloride and hydrogen, according to [261]:

$$PuO_2 + \tfrac{1}{2}H_2 + HCl \rightleftharpoons PuOCl + H_2O.$$

The oxychloride reacts with gaseous hydrogen chloride with the formation of plutonium trichloride. Plutonium oxychloride is insoluble in water, but is soluble in dilute acids [261].

Oxybromide PuOBr

The oxybromide is formed by the reaction between plutonium dioxide and a mixture of hydrogen with hydrogen bromide saturated with water vapor.

Oxyiodide PuOI

This compound is obtained by treating metallic plutonium with iodine vapor. When hydrogen iodide is made to react with plutonium dioxide at 750°, PuOI is formed rather than the triiodide. The oxyiodide can also be obtained by prolonged reaction at 750° between a mixture of gaseous hydrogen iodide and hydrogen and plutonium hydroxide dried at 70° [441].

Plutonium tetrafluoride PuF_4

This is the only stable halide of tetravalent plutonium. Plutonium tetrafluoride is utilized to prepare metallic plutonium. The tetrafluoride is prepared by fluorination of plutonium dioxide with a mixture of hydrogen fluoride and oxygen at 350–600° [3, 376, 417, 418]:

$$PuO_2 + 4HF + O_2 \overset{550°C}{\rightleftharpoons} PuF_4 + 2H_2O + O_2.$$

The oxygen prevents PuF_4 from being reduced to PuF_3.

Maly et al. [154] proposed a method of fluorination of plutonium dioxide by $NH_4F \cdot HF$ in the heat. When PuO_2 is heated with $NH_4F \cdot HF$ to 300°, three compounds are formed. The first plateau on the thermogravigram (Figure 45) corresponds to the dark green compound $NH_4PuF_5 \cdot 2H_2O$. When the temperature is further

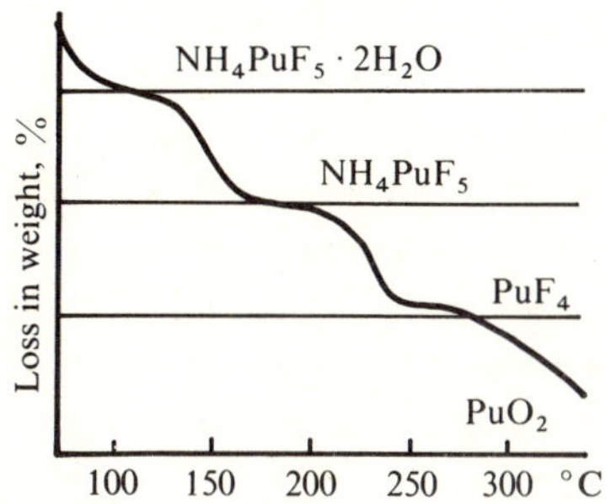

Figure 45

Thermogravigram of $NH_4PuF_5 \cdot 2H_2O$

increased, the compound loses its water of crystallization and is converted, at 160–200°, into the light green compound NH_4PuF_5, which in turn decomposes to PuF_4. Plutonium tetrafluoride is stable at 250–270°, and above 280° in the air it is converted into plutonium dioxide [376].

Plutonium tetrafluoride is stable to oxygen up to 600°.

The tetrafluoride is sparingly soluble in water, but is readily soluble in aqueous solutions containing Ce^{4+}, Fe^{3+}, Al^{3+} and other cations which form stable complexes with fluorine ions. Calcined tetrafluoride is much less soluble [34].

Plutonium hexafluoride PuF_6

Plutonium hexafluoride is most often prepared by reactions between fluorine and plutonium (III) and plutonium (IV) fluorides or with plutonium dioxide [163, 221, 405, 535, 681, 721]:

$$PuF_4 + F_2 = PuF_6,$$

$$PuO_2 + 3F_2 = PuF_6 + O_2.$$

The reaction becomes rapid above 700°. The apparatus and the experimental conditions have been described [47, 163, 221, 405, 535, 681, 721].

The similar volatilities and methods of preparation of the hexafluorides of uranium and plutonium render these compounds very much alike.

The formation of plutonium hexafluoride is accompanied by the formation of small amounts of hydrogen fluoride. Work with PuF_6 in glass vessels is only permissible after the compound has been thoroughly purified from HF in vacuo [721]. PuF_6 is stored in special quartz vessels. At 180° plutonium hexafluoride is a white crystalline substance. At room temperature the crystals become light brown, and when the temperature is increased, they darken and become red-brown [405, 535]. A number of workers [405, 535, 722] proposed equations giving the vapor pressure of PuF_6 as a function of temperature. Under a pressure of 511 mm Hg the melting point of plutonium hexafluoride is 50.75°. The normal boiling point of the compound is 62.3°. Thus, under atmospheric pressure PuF_6 may be liquid; the compound is not associated in the liquid state.

Plutonium hexafluoride is thermodynamically unstable and dissociates to give fluorine and lower fluorides of plutonium. The extent of the decomposition is, however, only 0.1% per day for gaseous plutonium hexafluoride. The rate of the radiolytic decomposition of solid PuF_6 is about 1.5% per day.

Plutonium hexafluoride vigorously reacts with water, forming plutonyl fluoride PuO_2F_2, which is isostructural with UO_2F_2. When the gaseous hexafluoride is passed through sulfuric acid, a pink crystalline precipitate of plutonium sulfate separates out. The mechanism of the reaction is unclear.

Plutonium hexafluoride is an even stronger fluorinating agent than BrF_3. The high volatility of PuF_6 accounts for its considerable toxicity, and special safety precautions must be taken when working with the compound. The differences in the respective volatilities of fluorides of uranium, neptunium and certain fission products are utilized in their separation [314, 415].

Other plutonium compounds

Plutonium hydrides and deuterides

Metallic plutonium reacts with hydrogen with the formation of PuH_2 and PuH_3. The solid metal reacts with hydrogen at 150–200°, while the powdered element reacts at 25–50° [33, 319, 537]. Mulford and Sturdy [568] prepared and studied in detail plutonium dihydride. Plutonium hydride was identified by analysis of the pressure vs. composition curves for the system metal—hydrogen [203]. The system has not so far been studied in detail. In the presence of excess hydrogen, a gray, metal-like substance of the composition $PuH_{2.7}$ is formed. Mulford and Sturdy subsequently confirmed [569] that a uniform solid phase exists between the compounds PuH_2 and PuH_3. Between PuH_2 and $PuH_{2.7}$ the fluorite structure of PuH_2 is retained; beginning with the composition $PuH_{2.75}$ the hexagonal phase appears. The solid phase of both hydrides, often obtained in practical work, has the same chemical reactivity as metallic plutonium. This solid solution is slowly oxidized in the air below 100° and gradually reacts with water and with acids; it may be converted into PuN by heating in nitrogen at 230°, and into plutonium carbide by heating at 800° in an atmosphere of CO_2. For this reason plutonium hydride is often employed as initial substance for the preparation of other plutonium compounds. The vapor pressure of the mixture of hydrides is low below 500° and the compounds themselves remain stable at 1000° under a hydrogen pressure of 1 atm [33]. Similar compounds of plutonium with deuterium have also been obtained [537, 568].

Plutonium nitride

The preparation and properties of plutonium nitride were described in detail by Brown, Ockenden and Welch [320]. The nitride was obtained by the action of anhydrous ammonia vapors or nitrogen on metallic Pu at 1000° and by the reaction between plutonium trichloride and ammonia at 800–900°. The best method is the action of ammonia or nitrogen on plutonium hydride at 600° under 250 mm Hg pressure. Pure PuN is a black brittle substance, with a face-centered cubic lattice. Its heat of formation is 95 kcal/mole, and its density is 14 g/cm^3. Plutonium nitride is easily hydrolyzed in moist warm air; the reaction becomes more rapid at elevated temperatures. Plutonium nitride readily dissolves in 3 *M* HCl and 3 *M* H_2SO_4 in the cold, with the formation of the corresponding solutions of trivalent plutonium.

Plutonium sulfides

Plutonium sulfides may be prepared only in the dry way. Depending on the conditions of preparation (temperature, rate, time) various compounds of plutonium are formed. The compound PuS was obtained when plutonium trifluoride was reduced by barium vapor in a barium sulfide crucible [262]. This compound is golden-brown and has a face-centered cubic lattice. Its density is 10.60 g/cm^3 [735].

A mixture of Pu_2S_3 and Pu_3S_4 may be prepared by heating the dry hydroxide in a graphite crucible in an atmosphere of anhydrous hydrogen sulfide at 1,340°, or by heating the trichloride in a quartz crucible in a stream of hydrogen sulfide at 840–916° [262]. The product is black or black purple. Body-centered cubic lattice.

Plutonium oxysulfide Pu_2O_2S

This compound was prepared by passing hydrogen sulfide over dry hydroxide in a graphite crucible at 1,225–1,300°. The compound has a metallic luster and has a hexagonal lattice. The density of the compound is 9.95 g/cm^3 [262].

Plutonium carbides

Zachariasen [733–735] used X-ray diffraction analysis to identify two carbides:

the monocarbide PuC and the sesquicarbide Pu_2C_3. The monocarbide was obtained in an inert atmosphere by the reaction of graphite powder with PuO_2 at 1,800° or with $PuH_{2.7}$ at 800°, or with metallic plutonium at 1000°. Pu_2C_3 is formed by heating a mixture of graphite with plutonium dioxide at 1,850°. The density of PuC is 13.99 g/cm^3, while that of Pu_2C_3 is 12.70 g/cm^3. Plutonium monocarbide is easily hydrolyzed by dilute acids; the hydrolysis of the sesquicarbide is much more difficult [368]. No other information is available in the literature on the properties of these compounds.

TOXIC PROPERTIES OF PLUTONIUM, AND WORKING TECHNIQUES

The analyst will usually work with the α-active isotopes of plutonium, mainly Pu^{239}.

Alpha-particles have a relatively low penetrating power; for Pu^{239} the path of an α-particle in air is 3.7 cm, and in soft biological tissue about 43 μ [155]. In conjunction with the high total ionization ($1.47 \cdot 10^7$ pairs of ions per α-particle), this short range results in a considerable density of ionization.

The ionization processes are accompanied by various chemical transformations in the biological tissue which have a harmful effect on the vital activity of the organism. The higher the density of ionization, the larger the so-called biological damage.

External radiation is dangerous only when the worker is in actual contact with the radioactive substance. Since the range of α-particles in a solid is short, rubber gloves afford sufficient protection.

Penetration of radioactive substances into the organism, on the other hand, is especially dangerous. The effects of the ionizing radiation will then be felt by the worker until the radioactive material is eliminated from the organism by physiological processes or until it decomposes.

In view of the fact that α-irradiation by plutonium produces major irreversible changes in the bones, liver, spleen and kidneys [255], all isotopes of plutonium are highly radiotoxic [200]. The pathological changes are difficult to diagnose; they do not become apparent soon enough for artificial elimination of plutonium with solutions of complexing agents to be practicable. The maximum permissible amount of plutonium-239 in the human body is a fraction of a microgram [548, 644].* For this reason any operations conducted on plutonium must be effected with utmost care and accuracy. A careful control must be maintained to detect any contamination on bench tops, floors and walls of laboratory accommodations and in particular in the atmosphere. The maximum permissible content of Pu^{239} in the atmosphere is $1 \cdot 10^{-13}$ curie/liter (~ 20 disintegrations per minute per cubic meter)

* Regular medical checks must be maintained to determine the concentration of plutonium in the urine [644] (and see p. 124).

[200, 548]. Accordingly analytical operations must be performed in special boxes, the design of which will vary with the nature of the sample and type of operation. The boxes should be accommodated in spacious lighted rooms, the floors of which are covered with highly wettable plastic material, and the walls and the ceiling are coated with oil paint.

Even when working with boxes, the various chemical operations on solutions of plutonium and solid plutonium should comprise safety measures aimed at reducing the contamination inside the box to a minimum.

A few types of boxes for working with plutonium preparations have been described [75, 77, 360, 548]. For analyses which are carried out in a certain sequence for long periods of time, the boxes are mounted in line. The transfer of the sample is effected through openings in the side walls of the boxes, which are closed by flaps actuated by compressed air [548].

In research laboratories and for nonroutine analyses, mounting the boxes in fixed rows is not always efficient. The transfer of the preparations from one box to another may be effected with the aid of standard portable cassettes and gates provided in the boxes [75, 77].

A very simple technique is transfer in bags (for schematic diagram see Figure 46). The bottom of the box is equipped with a sleeve, over which is passed a bellows-

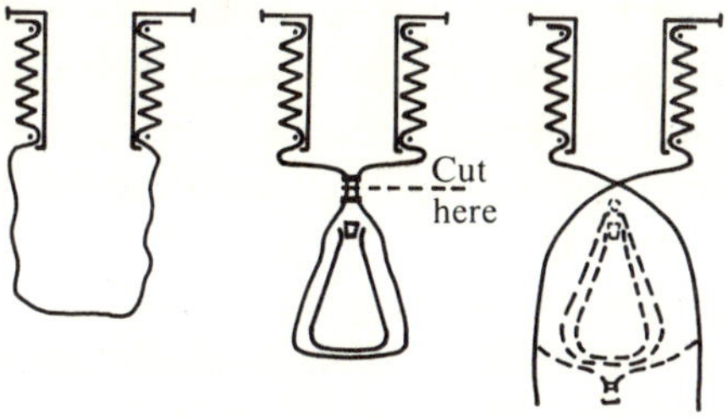

a *b* *c*

Figure 46

Transfer of radioactive preparations

shaped tube made of a thin plastic material, e. g., polyethylene. The lower end of the tube is sealed or tied (position *a*). The object to be transferred is shifted from the box through the sleeve into the bag, and the bag is then tied in two places (position *b*). The bag is cut off and carried to another box, while the first box can be used again after the bellows-shaped tube has been pulled down into position a. The bag with the object is introduced into the box in the same manner (position *c*) if for any reason it cannot be introduced through the front wall of the box.

The dimensions of the boxes may vary between 800 and 1,400 mm in width, about 600 mm in depth, 600–900 mm in height. The front wall of the box with holes and flanges for the gloves is usually inclined, and is made of glass or Plexiglas. The top wall is also made transparent and the source of illumination is placed on it. Bottom, side and back walls are usually made of noncorroding materials (stainless steel, vinyl plastic). The bottom of the box contains a funnel, to drain nonregenerated liquids into a special drainage system. If the preparations are weakly β- or γ-active, the front wall is equipped with lead glass.*

The side walls carry inlets for gas, water, compressed air, vacuum line and electric terminals. The inner surface of the box, except for the transparent parts, is covered with a resistant white paint to reduce corrosion by acid fumes and improve the illumination.

All boxes are equipped with an exhaust ventilation system. The boxes are connected to the system by pipes, which are best made of vinyl plastic. Before being vented to the atmosphere, the air is made to pass through a filter, one variant of which was described in detail in [75, 77]. The filtering element is a strip of acid-resistant fabric FPP-1 or FPP-15, folded in the shape of a bellows. In order to check the purity of the air passing through the filter, the air duct behind the filter carries a flange with a hole which is usually covered. During the check the end cap is removed and the flange is connected to another flange with an absorber filter. The air from the air duct is aspired for a certain time and the filter checked for radioactivity.

In order to prolong the useful life of the fabric, it is recommended that water be sprayed from nozzles into the duct leading to the filter in order to condense acid fumes [548]. The water should be introduced so as to prevent the fabric from being splashed. The condensate should be periodically drained through a special duct.

The design of the boxes may be classified under three headings, depending on the type of operation to be effected.

1. Hermetic glove boxes for working with solid preparations of plutonium (metal, alloys, salts). Such operations include comminution, drying, calcination and weighing of samples. Spectroscopic analyses should also be carried out in such boxes. For most of these operations the boxes need not be perfectly tight. Boxes described in [75, 77] have a removable front wall, which is not tightly secured. Small openings are not sealed. A round-the-clock exhaust system, to which the box is connected, ensures that the air inside the box is rarefied (20–30 mm water column), as a result of which the pure air in the laboratory is vigorously aspired through the small slits into the box, while the radioactive dust in the box is thus prevented from penetrating into the atmosphere of the laboratory.

It is often required to break up a plate of metallic plutonium and to distribute the pieces over a number of separate hermetically sealed vessels for storage. The plate may be broken in a mortar, which is diagrammatically shown in Figure 47.

* This description does not include the safety precautions to be maintained when working with preparations which are highly β- or γ-active. These have been described elsewhere [155, 257, 516.]

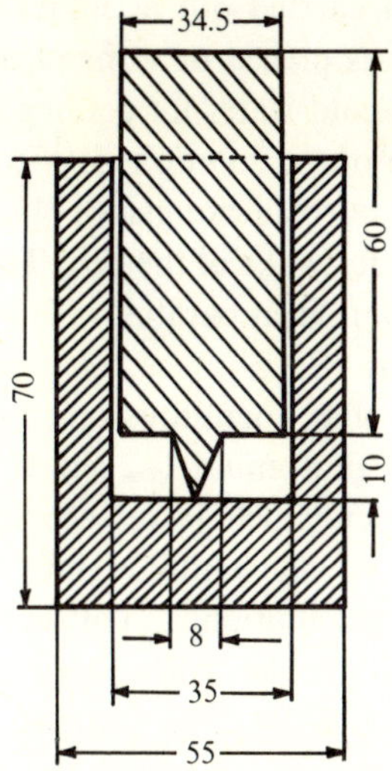

Figure 47
Steel mortar for breaking up metallic plutonium

The metal splinter is carefully placed in a thick-walled test tube by means of tweezers, and the tube is then sealed in vacuo or in an atmosphere of an inert gas (argon). Storing plutonium in sealed tubes protects the metal from oxidation.

To prepare solutions of plutonium with an accurate titer it is usual to employ metallic plutonium covered with a thin film of oxide. The oxide is removed from the surface by filing or electropolishing the surface in a solution of potash [717]. This "wet" treatment prevents the atomization of the highly toxic oxide particles.

Plutonium compounds are calcined in a furnace, which is usually accommodated in a box cabinet, equipped with a gate carrying a standard cassette serving to transfer the crucibles to the weighing box. The weighing is done on an ADV-200 balance. The control grips of the balance are extended and led outside the weighing box through felt seals, affixed by metal flanges to the outside of the front wall [75, 77].

2. Boxes with partly opened front wall (the aperture allows the operator to move his hands freely in the horizontal direction, and over a limited distance vertically). This group also includes boxes from which the gloves have been removed. These serve for the solubilization of solid materials, evaporations, sampling, precipitations, extractions, chromatography, etc. Operations of this sort need not be carried out in airtight boxes, since the danger of formation of radioactive aerosols is small. Such boxes must not be used unless air is aspired across the open surface of the front wall at a linear velocity of 1.5 m/sec [200]. According to American data [548], a velocity of 0.5 m/sec is sufficient for such work.

In order to prevent spattering during heating, the dissolution of plutonium and its compounds in acids is best carried out in sufficiently large long-necked flasks. To meet emergencies, the flask is placed on a metal tray; when solubilizing metals, the flask is held in a beaker of cold water for cooling purposes. The vessel in which the heating or the evaporation of the solutions is effected should be placed under a special exhaust hood made of borosilicate glass [717] in order to prevent condensation of acid vapors on the inner walls of the box. The best method is to evaporate acid solutions in a glass distillation apparatus with water cooling; the condensate is then run into special drains.

3. It has been shown [548] that operations on small samples of dilute solutions of plutonium (total amount of plutonium 1–2 μg) may be effected in box cabinets with the front wall largely open. Such boxes are suitable for sampling solution volumes of a few hundredths of one milliliter into flasks and onto supports for α-count, their evaporation and calcination, and other operations with microamounts of the element. According to sanitation rules [200], the operations can be carried out in the usual way (as for nonradioactive isotopes) if the concentration of plutonium in solution is at most 0.03 μg/ml, and its total amount is at most 1.5 μg.

If plutonium solutions are stored in tightly closed vessels, it must be remembered that there may be gas formation due to the radiolysis of water. To prevent blow-ups, the vessels should be periodically opened; safety valves are occasionally used.

Sampling of solutions for various operations, including application to targets for radiometric analysis, should be effected by means of pipets connected to syringes by rubber tubes or other pliable tubes. Several pipets of various volumes may be affixed to the same stand. The most convenient procedure is to fix the syringes horizontally underneath the stand base or on a separate panel.

The evaporation and combustion of the materials on the target is usually effected in the box in which the solutions have been applied. Targets for α-count are transferred in a can (Figure 48) to prevent losses of the applied material.

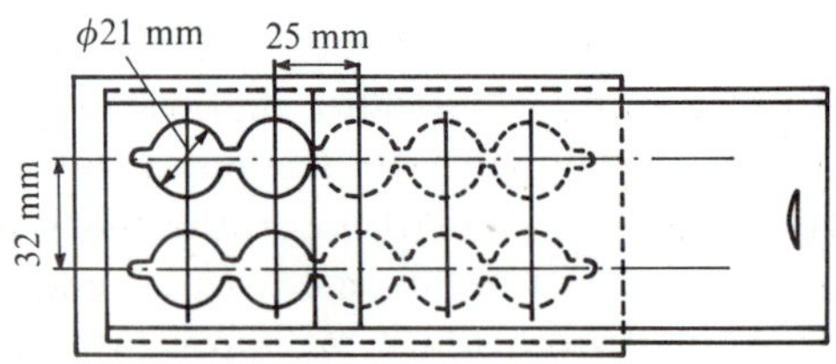

Figure 48
Target can

Whatever the precautions taken to prevent contamination of boxes and laboratory accommodations, they will still have to be occasionally deactivated. The deactivating solutions employed include mineral acids, sodium carbonate, citrates, oxalates, complexones [200] and other materials [255].

CHAPTER III

Detection of Plutonium

Certain properties of plutonium, such as α-activity, tendency to complex formation, etc., may be used for a rapid detection of plutonium. The most suitable methods for the purpose are radiometric and chemical analysis.

Detection of plutonium in chemically complex media is required only rarely.

The most important method in the detection of plutonium is the radiometric technique, based on the measurement of the total α-radiation of plutonium and its energy. There is then no need to conduct complicated and time-consuming chemical operations and the method itself is quite sensitive. The amount of Pu^{239} which can be detected is 0.0001 μg. If other α-emitters are also present in the sample, plutonium may be identified by measuring the energy of α-particles with the aid of α-analyzers.*

A number of physicochemical and chemical methods for the detection of plutonium are based on the differences in the properties of the various valency states of plutonium. In fairly concentrated aqueous solutions, the trivalent plutonium ion may be identified by its bright blue coloration, as opposed to the yellow-brown color of aqueous solutions containing ions of tetravalent plutonium.

The adsorption spectra of salts of plutonium in various stages of oxidation have characteristic, narrow adsorption bands, which makes it possible to identify the valency form and detect them in the presence of each other. The most typical adsorption maxima are 600 and 900 mμ for Pu (III), 480 and 660 mμ for Pu (IV), 569 mμ for Pu (V) and 830–835 mμ for Pu (VI) [348].

* For radiometric determination of plutonium see Chapter IV.

Plutonium may also be detected by its reactions with the intensely colored reagents of Arsenazo and Thoron types (Table 12).

Plutonium in various valency states forms a large number of insoluble compounds and extractable complexes with both inorganic and organic reagents. Thus, for instance, trivalent plutonium is precipitated by ammonia or alkali as the blue hydroxide, while tetravalent plutonium precipitates out as the brown hydroxide from nitrate, chloride, sulfate and perchlorate solutions. In addition, plutonium forms a large number of sparingly soluble compounds, such as iodates, fluorides, phosphates, oxalates and many others, including compounds formed with organic reagents.

However, none of these reactions, including the color reactions, are sufficiently specific since many elements exhibit similar behavior, in particular, uranium, neptunium, thorium, and zirconium.

Table 12

COLOR REACTIONS OF Pu(IV) WITH ORGANIC REAGENTS

Reagent	Color of reagent	Color of Pu(IV) complex	Sensitivity of reaction, $\mu g/ml$	Authors
Thoron I	Orange-yellow	Raspberry red	0.2	Kuznetsov, Mikhailov, Lapshina, 1956
Thoron II	Orange	Red	0.6	Kuznetsov, Mikhailov, 1959
Chlorphostoron	Orange	Red	~ 0.5	Palei, Kochetkova, Nemodruk, 1961
Aminothoron	Pink	Violet	0.5	Pal'shin, 1955
Arsenazo I	Pink	Blue	0.1	Klygin, Pavlova, 1955 [109]
Arsenazo II	Pink	Violet	~ 0.2	Pavlova, 1955
Arsenazo III	Pink	Bluish-green	0.01	Kuznetsov, Savvin et al., 1960
Chlorophosphonazo I	Red	Blue	~ 0.5	Palei, Kochetkova, Nemodruk, 1961
Chlorophosphonazo III	Red-violet	Violet	< 0.01	Palei, Kochetkova, Nemodruk, 1961
Alizarin S	Lemon-yellow	Red-violet	2.0	[547]
Rhodamine 3B	Pink	Blue	0.5	Kuznetsov, Pal'shin, 1957

CHAPTER IV

Quantitative Determination of Plutonium

RADIOMETRIC METHOD

The radiometric determination of plutonium is now widely used. The reason for this is the fact that the main isotope of plutonium, Pu^{239}, has a very high α-activity, much higher than that of other elements usually obtained in a nuclear reactor. Small amounts of plutonium—0.1 μg or even less—may be determined by this method. Its high sensitivity, the relatively simple apparatus required and the uncomplicated experimental procedure make it possible to employ the radiometric method both in production control and in research work.

The specific activity of Pu^{239} is 136,200 disintegrations per minute per microgram.* The energy spectrum of its α-particles contains two main groups, with energies of 5.147 and 5.134 MeV, and another group with energies ranging from 5.10 to 4.66 MeV [3, Chap. 7; 657]. The radiations of other isotopes of plutonium are given in Table 1 (pp. 2–3).

α-Radiation is characterized by the range of the particles and by their energies. During absorption, the bulk of the energy is expended in ionizing the medium. The specific ionization of air by α-particles varies between 2200 and 7000 ion pairs per 1 mm for the energy range 7.9–0.95 MeV. The specific ionization of air by β-

* The most accurate value of the specific activity of Pu^{239} found by Soviet workers [87] is 136,200 ± 200 disintegrations per minute per microgram. This value is in agreement with those found by other workers [411, 467, 656, 727].

particles is only 5–20 ion pairs per 1 mm path between 1.5 MeV and 60 KeV. The specific ionization by γ-rays is almost a hundred times lower. Thus, it is possible to determine the α-activity of a preparation against the background of a dominant β- and γ-activity of the accompanying elements, which is specially important in the analysis of reactor materials. The permissible β-activity levels in the samples will depend on the nature of the radiation and the type of counter employed. Thus, for instance, if an electron pulse chamber is employed, α-radiation may be measured in the presence of high energy β-activity, e.g., Au^{198}, of more than $2.5 \cdot 10^{10}$ disintegrations per minute. The permissible level of softer β-radiation, e.g., Np^{239}, is about $5 \cdot 10^8$ disintegrations per minute. When proportional counters are employed, the limiting β-activity level for hard β-radiation is about $2 \cdot 10^9$ disintegrations per minute [3, Chap. 16].

Other alpha-emitters strongly affect the results of radiometric measurements made on preparations of Pu^{239}. The main interfering emitter is the isotope Pu^{240}, whose radiation includes α-particles with an energy of $\sim$5.15 MeV, similar to that of the α-particles of Pu^{239}.

When uranium is irradiated in a nuclear reactor, a number of plutonium isotopes other than Pu^{239} are formed: Pu^{238}, Pu^{240}, Pu^{241} and Pu^{242}, which are not separated during the chemical treatment of the irradiated material. The content of these isotopes in the irradiated material is variable and depends on the isotopic composition of uranium, on the duration and intensity of the irradiation, and the holding time after the irradiation. The content of the isotope Pu^{240} in Pu^{239} may vary from 1 to 7% by weight [339, 411, 663], while the content of other isotopes is usually not more than a fraction of one percent.

The knowledge of the isotopic composition and of the specific activity of the isotopes of plutonium makes it possible to introduce suitable corrections in the measured values of α-activities of the preparations. The percentage content of Pu^{240} is calculated from the empirical formula (Dokuchaev, 1955):

$$y = 136{,}200 \pm 4211x, \tag{1}$$

where y is the number of α-disintegrations per minute per microgram of plutonium isotope mixture, x is the percentage content of Pu^{240} and 136,200 is the specific activity of Pu^{239}, in disintegrations per minute per microgram.

The content of Pu^{241} in plutonium is directly proportional to that of Pu^{240}. As a result of β-decomposition of Pu^{241} (half-life period 13 years), the intensive alpha-emitter Am^{241} is formed. For this reason the activity of plutonium preparations increases with time, which is especially undesirable in the case of standard preparations. In order that the correction to be introduced one year after the preparation of standards might not exceed 0.2%, the plutonium employed must not contain more than 2% by weight of Pu^{240}. In recent times alpha-analyzers are increasingly used, by means of which individual α-radiations can be determined in the presence of each other.

Another factor in the absorption of α-radiation by matter is the path length (range) of α-particles. The ranges of all α-particles with the same initial energy are nearly the same [16]. The range of α-particles originating from different emitters in the air at NTP varies from 2.6 to 8.6 cm. For Pu^{239} the α-particle range in the air is about 3.7 cm, while in water it is only 35 μ [46].

An approximate (to within $\pm 15\%$) estimate of the α-particle range in other media may be made using the formula

$$R = 3 \cdot 10^{-4} \frac{R_0}{p} \sqrt{A} , \tag{2}$$

where R_0 is the range of the α-particles in the air, p is the density of the absorbing medium and A is the atomic weight of the absorbing element.

The low penetrating power of α-radiation as compared to other kinds of radiation means that the radiometric methods employed must meet certain specific requirements; this particularly applies to the preparation of the sample for the α-count. The usual technique is to apply a thin layer of the α-emitter onto a support (planchet). In order to prevent absorption of the α-radiation by the preparation itself, the layer thickness must constitute only a small fraction of the range of α-particles. Significant self-absorption has not been noted when 50 μg of Pu^{239} in the form of a salt or the dioxide were spread over an area of about 1 cm^2.

The determination of plutonium is greatly affected by the presence of extraneous salts, which are present in the sample to be analyzed. Palei and Milyukova (1951) investigated the effect of a number of elements on the radiometric determination of plutonium. The starting solutions contained lithium, sodium, lanthanum and uranium salts in concentrations from 1 to 8 g/liter, calculated on the element, at a constant concentration of plutonium of 0.7 mg/liter. The solution (0.25 ml) was applied to a support with a surface area of about 2 cm^2, after which the preparation was evaporated, calcined and its α-activity determined. The results of these experiments show (Figure 49) that light elements give lower results than the heavy elements at equal concentrations by weight. The approximate permissible amount of salts on the support is not more than 0.1 mg per cm^2 of surface.

Major absorption of α-particles in the preparation may also take place owing to the uneven distribution of material over the surface. The precision requirements are particularly high when the α-spectra are measured with a spectrometer.

Most radiometric determinations are carried out on a certain volume of the solution of plutonium, applied to a metallic planchet, and evaporated, after the thin layer of solid residue has been calcined. Activity measurements carried out directly on the surface of the solution or on a thick layer of the solid substance are less frequently employed. In what follows we shall describe the preparation of samples for

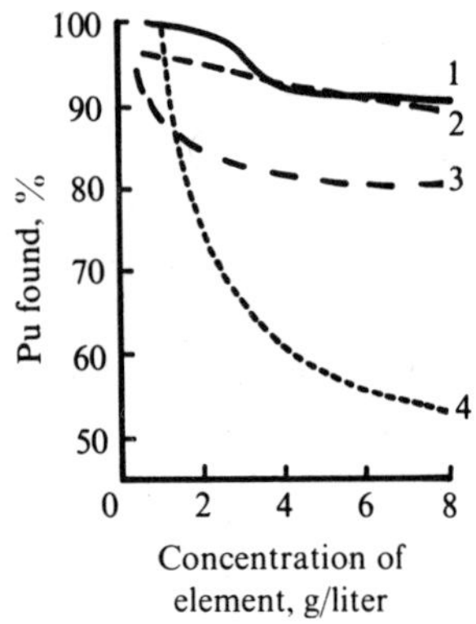

Figure 49

Effect of elements on the radiometric determination of plutonium:

1) U(VI); 2) La; 3) Na; 4) Li.

radiometry, techniques of quantitative application of the solution, standardization of α-measurements and the techniques of the measurement itself.

Preliminary treatment of α-preparations

The preparation to be analyzed is applied to the planchet as a thin layer. The planchet is usually a disc or a small plate with raised edges. Depending on the kind of work and the type of pick-up employed, the planchets used are of various kinds and dimensions. The planchet should be smooth and flat, without depressions or folds which might bring about absorption of α-particles by the preparation itself. The material of which the planchet is made is also important. It must be chemically stable to media with which it is brought into contact during the preparation of the sample. Also, if the sample is in contact with the planchet for a long time, a possible diffusion of the substance into the planchet is to be expected.

Dokuchaev and Lebedeva (1952) studied the stability of the radioactive preparations of plutonium with time in order to find the effect of diffusion on the accuracy of the count. The studies were carried out during $1\frac{1}{2}$ years on preparations containing plutonium dioxide on planchets made of platinum, EYa-IT stainless steel, nickel, copper, brass, zinc and other metals. It was found that there is no diffusion of plutonium into platinum at 20 and 200°. Samples prepared on stainless steel planchets proved stable at 20°, but at 200° their activity decreased by 5.1 % per year. A considerable diffusion of plutonium into aluminum, nickel and tin was observed not only at 200°, but even at 20°. It will be seen that platinum and stainless steel are the best planchet material. Quartz can also be used in certain cases [64].

Platinum planchets are not particularly popular, since their decontamination before reuse involves considerable difficulties. Plutonium dioxide is removed from the support by one of the methods described above (p. 98) and the support is then repeatedly treated with concentrated HNO_3.

In analytical work stainless steel planchets are most often employed. Before application of the sample solution, the planchet is washed with an organic solvent, e.g., benzene, to remove fat and oil stains which interfere with the uniform distribution of the active substance. The planchet is then boiled in 1.5 *N* HNO_3, carefully washed with water and dried. In order to prevent the radioactive solution from spreading out to the edges of the planchet, the edges are varnished with a mixture of 0.4 g polystyrene, 0.1 g paraffin and 5 g benzene.

In sample preparation it is important to know its optimum α-activity. From the point of view of an accurate count and its contrast with the background, it is advisable to have high activity preparations, of the order of 1–10 μg Pu per cm^2. The top limit of the content of plutonium in the preparation is determined by the self-absorption of α-radiation. The maximum amount of plutonium and salts on the planchet should not exceed 50 $\mu g/cm^2$. Another limit on the activity level of the sample is also imposed by the apparatus employed (type of chamber, amplifier and counting unit). Thus, for instance, when using a DA type installation with MK-ZM chamber, the optimum amount of plutonium in the preparation is 0.05–0.1 μg Pu/cm^2. In certain cases the upper limit of activity is restricted by the size of the sample. The lower activity limit is determined by the level of the background, the time which can be allowed for the measurement and the accuracy requirements.

In the radiometric method of analysis standard and sample preparations must be distinguished. The standard preparations are divided in turn into primary and secondary preparations. The primary standards serve only to check the secondary standard preparations, to control the count accuracy and to check the geometry of the chambers. Such standards are prepared on platinum planchets and are utilized for one year.

Secondary standards, usually prepared on stainless steel planchets, are used in everyday work. They remain essentially unchanged for 3–4 months.

The preparation of α-active samples of plutonium may be effected in several ways: evaporation of the solution to be analyzed, electrolytic deposition, sublimation or deposition of plutonium on a carrier.

Evaporation of solutions

Because of the simplicity and ease in handling, plutonium is most often applied to the planchet by evaporation of solutions. The accuracy of this operation is determined by the metering technique; a certain volume or a certain weight of the solution may be applied, depending on the requirements. The sampling is effected with a graduated pipet connected to a syringe [472], or with a weight buret. Weighing the solution directly on the support should be done in a closed weighing glass. The volume of the solution on the support may be 0.1–0.2 ml/cm^2. The accuracy of volume sampling is 0.2–0.5 %, that of weight sampling is 0.02 %.

The solution on the support is evaporated to dryness with the aid of a 500 W infrared lamp. The distance between the preparation and the lamp is so adjusted as to prevent the solution from boiling and spattering, with consequent loss of material. The distance is usually 25–30 cm. After drying the sample is calcined on an electric hotplate until the varnish on the edges of the support has burnt away. The volatile salts in the sample solution are eliminated at the same time. During the drying and the calcination safety measures must be taken against contamination of the premises. The storage and transfer of preparations must be effected in a can.

To prepare standards by the *evaporation method*, a standard solution of plutonium with an accurately known titer is employed. The preparation of primary standards of plutonium by the evaporation method is described below.

1. From one weighed sample of pure metallic plutonium two solutions of plutonium in 1.5 *M* HNO_3 are prepared by the gravimetric method. The concentration of the solutions should be 300–400 mg Pu per liter.
2. One of these solutions is used to determine the isotopic composition of plutonium by the mass-spectrometric method. The content of Pu^{240} is expressed as weight percent of the combined content of $Pu^{239} + Pu^{240}$.
3. With the aid of a weight buret a few 0.1 gram portions of the second solution are applied onto platinum supports, evaporated and calcined. The preparations thus obtained are measured in parallel in two counter units, three times for 30 minutes each time. From the results the number of pulses per minute per microgram of plutonium of the given isotopic composition is calculated. The value obtained is then compared with that calculated from formula (1).
4. From the second solution two samples are withdrawn, weighed, placed in a weighed volumetric flask and made up to the mark with 1.5 *M* HNO_3. The flask with the solution is again weighed. The extent of the dilution should be such as to have 0.1–0.15 μg Pu per 0.1–0.2 g of the solution applied to the support. The diluted solutions are the starting material for the preparation of primary standards. About 0.1–0.2 g of the solution is applied to the support, evaporated and calcined until the varnish ring is completely eliminated.

To prepare secondary standards, the above procedure may be employed except that the isotopic composition of plutonium is not determined.

In order that the evaporation of the solution on the support might yield a homogeneous film, special additives are introduced. Hufford and Scott [472] described the preparation of plutonium samples with the use of tetraethylene glycol. The glycol, in the amount of 50 μg/cm^2, was added to the solution already on the support and the solution was dried. In this case the evaporation takes place from the film and not from the drop. During the heating tetraethylene glycol is polymerized and then burns up. Plutonium is then distributed very evenly over the surface of the support, but the films are not strong. Sikkeland and Haaland use lactic acid as an additive [670]. The samples were prepared on 0.1 mm thick aluminum discs. Before applying the solution to the support, it was freed from chlorides by evaporation with nitric acid, and the excess nitric acid was removed with formaldehyde. A portion of the solution was placed on the support with the aid of a micropipet and some lactic acid was added. In order to prevent self-absorption, the amount of the substance on the support should be less than 50 $\mu g/cm^2$. Samples prepared in this way were used by the authors for α-analyzer determinations of plutonium. The measurements were carried out with a typical ionization chamber with a good geometry (about 50%), connected with a 63-channel amplitude analyzer. Plutonium present in uranium at concentrations as low as 1 μg Pu per gram of U could be determined to within a few percent. Figure 50 represents the α-spectrum of a source containing 6 μg Pu per 1 gram of U.

Similar methods for the preparation of α-active samples were also reported by other authors [476, 635].

Collodion [704] may be used to obtain preparations with a high mechanical strength. About 20 mg of a standard solution, containing 10 μg Pu/ml, are applied from a weight buret to the center of the planchet and cautiously evaporated under the lamp. A little acetone, containing 10 μg/ml of collodion, is then added to cover the active spot and the mixture is dried at room temperature. The absorption of α-particles in the film of collodion is very small.

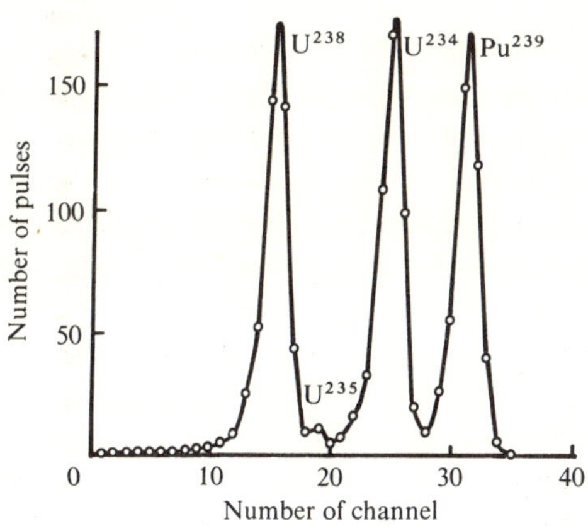

Figure 50

Alpha spectrum of a sample of plutonium-containing uranium

Standards may also be prepared by evaporation of alcoholic solutions [476]. The alcoholic solution of plutonium is applied in small portions to the support. The alcohol is evaporated after the application of each portion. The resulting films are thin and homogeneous; if necessary, their density can be made as high as 1 mg/cm^2.

To prepare working samples by the *evaporation method*, portions of the sample solution are withdrawn, applied to the support and dried under the lamp. If the sample contains large amounts of plutonium and relatively small amounts of admixtures, it is sufficient to make the appropriate dilution and bring an aliquot of the solution onto the support. It is not recommended, however, to dilute the solutions too much in order to obtain thin films, since the experimental error in the count increases with the decrease in the α-activity of the preparation. If the solution to be analyzed contains large amounts of extraneous elements, these must first be separated from plutonium. Trace amounts of plutonium are separated by coprecipitation, while semimicro- and macroamounts are separated by precipitation. For instance, precipitation of plutonium hydroxide by ammonia effects a separation of the element from large amounts of salts of sodium, potassium, lithium, copper, etc., which form soluble compounds under these conditions.

In certain cases quantitative separation of plutonium from extraneous salts is effected by extraction and chromatography [684, 692, 693]. When the solution of plutonium in the organic solvent is evaporated, the residual film will contain a lower amount of impurities. When an organic solution is evaporated, the size of the drop should be smaller to prevent creep towards the edges of the support. The varnish should then be applied not only to the edges of the disc, but also to the circumference of the disc itself.

Electrolytic deposition of plutonium from solutions

During electrolysis of plutonium solutions, the acidity of the solution near the cathode decreases and as a consequence plutonium is deposited on the cathode as hydroxide of variable composition.

Electrolytic separation of plutonium is effected in electrolytic cells of various shapes, with a cathode made of plutonium, stainless steel or nickel [204, 258, 549]. In order to prevent self-absorption of α-particles by the sample, the amount of the material deposited on the cathode, including impurities, should be of the order of 50 $\mu g/cm^2$.

The design of the electrolyzer, rate of mixing and ratio between the surface of the cathode and the volume of the solution determine the duration of the electrolysis, which may last for any time between 15 minutes and several hours. The main factors which determine the course of the electrodeposition and the completeness of separation of plutonium are the acidity of the solution and the current density on the cathode [198, 204, 562]. Clearly, the higher the current density, the greater the decrease in the acidity around the cathode, which is responsible for the deposition of plutonium.

Electrolysis has been conducted in nitric and hydrochloric acid [235, 552], bicarbonate [472], acetate [358], alkaline [549], formic acid [498] and alcoholic and alcohol–acetone [258, 659] media. In order to isolate plutonium from acid solutions, large current densities must be employed.

Not all the published methods of electrolytic separation of plutonium are quantitative. Methods in which the separation of plutonium is not quantitative may be

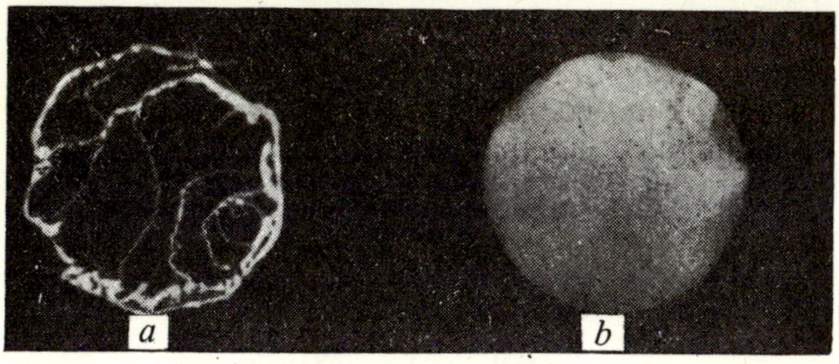

Figure 51

Autoradiograms of films obtained by evaporation of solution (*a*) and by electrolytic deposition (*b*)

used to determine the ratio between α-emitters of various energies, if the extent of the deposition of plutonium is known.

The electrolytic method is distinguished by a number of advantages. An important advantage is the homogeneity of the films. Figure 51 shows autoradiograms of films obtained, respectively, by ordinary evaporation of a solution of plutonium and by electrodeposition from alkaline solution (Kabanova and Danyushenkova, 1954). Moreover, during electrolysis plutonium may be separated from many interfering elements. These include weakly hydrolyzed elements, e.g., K, Na and Ca, and elements which do not separate out at the cathode as metals at the separation potential of hydrogen, e. g., Cr, Mn and La. A number of elements may be kept in solution with the aid of complexants [198]. The strength of the resulting films is usually low. The Pu standards and samples are prepared in the same way, except that the standards are usually prepared from chemically and radiochemically pure plutonium solutions of a known titer.

Muller and Brouns [549] proposed a method for a quantitative electrolytic deposition of Pu(VI) from solutions in $\geq 1.6\,N$ KOH. The diagram of the electrolyzer is seen in Figure 52. The electrolysis was conducted during 2 hours at a current density of 40 mA/cm^2. The solutions were stirred at 200 revolutions per minute. The yield of plutonium when fully converted to the hexavalent state was 100%. The resulting films were homogeneous, and their density was 0.4 μg Pu/cm^2.

The main difficulty of the method is the quantitative conversion of plutonium into Pu(VI) prior to the electrolysis. Tetravalent plutonium could be oxidized by ozone during a few hours to the extent of 98%.

The method was modified by Kabanova and Danyushenkova (1954). The oxidation of plutonium to hexavalent plutonium was carried out by evaporating solutions of Pu(IV) with perchloric acid or by electrolysis. The electrolysis was conducted in 1.7–1.9 N KOH at a current density of 25 mA/cm^2. An increase in the current density resulted in nonhomogeneous precipitates and a reduced yield of plutonium. The electrolysis was conducted for $3\frac{1}{2}$–4 hours, the solutions being stirred at 100 rpm. The

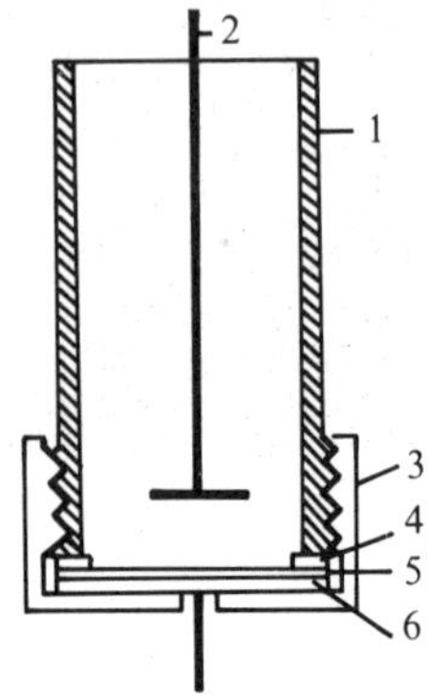

Figure 52
Electrolyzer:

1) glass cell; 2) rotating anode; 3) screw-on cap; 4) neoprene gasket; 5) platinum disc, 22 mm in dia.; 6) cathode (copper disc).

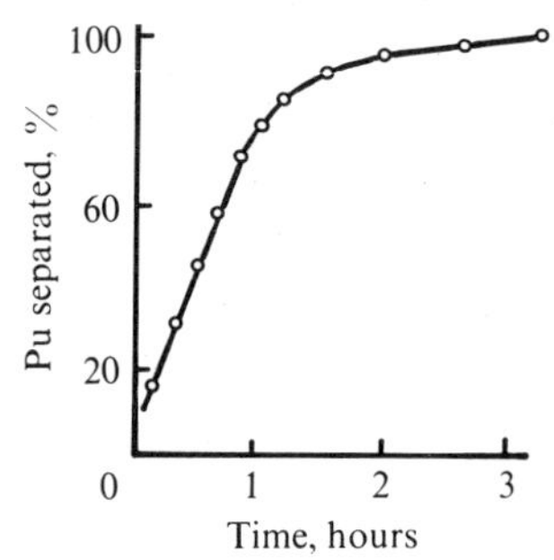

Figure 53
Effect of the duration of electrolysis on the completeness of separation of plutonium.

Composition of electrolyte: 5% water, 50% ethanol and 45% acetone (pH 1.5–2).

cathode was a platinum or a stainless steel disc. The preparations obtained contained 0.01, 0.03, 0.1 and 0.4 μg Pu/cm^2 on platinum, and 0.1 μg Pu/cm^2 on stainless steel. The completeness of separation of plutonium was 96% on platinum discs and about 89% on stainless steel discs. Strong homogeneous films were obtained (Figure 51).

Yakovlev et al. [79, 258] proposed a method in which plutonium, americium and curium are electrodeposited from neutral and weakly acid alcohol–acetone solutions of Pu(III), Am(III) and Cm(III) chlorides. Platinum, nickel or aluminum leaves are used as cathodes. When the current density is 0.2 mA/cm^2, plutonium is deposited from neutral solutions within 10–15 minutes. The stirring was effected with a rotating platinum anode. The electrolyte was a mixture of 50% ethanol, 45% acetone and 5%

water. When the electrolysis is prolonged, the hydroxide of tetravalent plutonium may precipitate out if the concentration of plutonium in the solution is more than 0.03 mg/ml. The density of the resulting films, which are perfectly satisfactory, may be as high as 0.5 mg/cm^2. The yield of plutonium is 94%.

When Pu(III) is deposited out of weakly acid alcohol–acetone solutions of the above composition, the following conditions of electrolysis must be rigorously maintained: pH 1.5–2, current density 5–10 mA/cm^2. Plutonium separates out almost quantitatively (film density up to 0.25 mg/cm^2). Figure 53 shows the relative yield of plutonium as a function of the duration of electrolysis.

Very strong, thin films of plutonium were obtained by Khlebnikov and Dergunov [235]. The cathodes employed were of platinum, nickel, and aluminum and the anodes were platinum, nickel and graphite. Electrolysis was carried out from weakly alkaline oxalate solutions. Ammonium oxalate solution (1.5–2 ml) was introduced into the electrolyzer and 0.2 ml of a hydrochloric or nitric acid solution of plutonium (IV) was added, with vigorous stirring. The pH was adjusted to 8–9 with the aid of a concentrated solution of ammonia. The current density was 100–150 mA/cm^2. In 5–6 hours 95–98% of plutonium separated out on the cathode; after 12–14 hours 99–99.5% were separated. When the electrolysis was terminated, the preparations were washed with CO_2-free water, made alkaline with ammonia, and were first air-dried at room temperature, and then in a drying oven at 80°. The films thus obtained were strong. The authors checked the mechanical strength of their preparations (0.2 mg/cm^2) by rubbing the active surface with a dry cotton pad. The resulting loss in weight was only 5–10%.

Samartseva [198] proposed a quantitative electrolytic separation of plutonium from nitric acid solutions. The electrolytic cell consisted of a platinum cup which served as the anode. The cathode consisted of an elongated platinum leaf with a working area of 1 cm^2. This cell design prevents any adsorption of plutonium, so that the electrolysis may be effected on solutions at all pH values. To deposit plutonium from nitric acid solutions at pH 3 the required current density is 100 mA/cm^2. The electrolysis lasted for two hours; the solution volume was 20 ml. The yield of plutonium was 100%.

When plutonium is isolated from 0.1 M HNO_3, the current density must be raised to 750–1000 mA/cm^2. Quantitative deposition from 30–40 ml of solution is attained within 2–3 hours.

The method can be used to isolate plutonium from solutions containing Fe^{3+}, Al^{3+}, La^{3+}, Ba^{2+}, Cr^{3+}, Mn^{2+}, Ca^{2+}, Mg^{2+}, K^+ and Na^+ ions. The interference by iron is removed by adding oxalic acid. The concentration of HNO_3 in the electrolyte is 0.125 M and the concentration of oxalic acid is 0.002 M (pH 1.3).

0.1–0.5 ml of the initial solution are placed into the platinum cup and the solution is evaporated to dryness on a water bath. The residue is treated 2–3 times with a small amount of concentrated HNO_3. To the dry residue are added 0.1 ml 1 M $H_2C_2O_4$, 10 ml 0.5 M HNO_3, and the mixture is heated on a water bath for

10–15 minutes in order to dissolve the residue. Thirty ml of water are then added to the cup, and the solution is electrolyzed at cathode current density of 1000 mA/cm^2. The electrolysis takes $1\frac{1}{2}$–2 hours. The average yield of plutonium, when precipitated in the presence of the ions listed above, is 99 ± 0.7%.

The electrolytic separation method makes it possible to concentrate the plutonium on a very small surface of the cathode from a relatively large volume of solution and in this way to increase the sensitivity of the determination. This is particularly valuable in the analysis of low-activity materials, e.g., products of human biological activity.

Schwendiman and Healy [644] give a method of determination of plutonium in a sample containing $(1–2) \cdot 10^{-10}$ mg of plutonium, which is based on the electrolytic deposition of plutonium. The method is intended for routine check-ups on personnel working with plutonium.

A sample of urine is evaporated to dryness, calcined, and the residue is dissolved in HNO_3. Plutonium is then coprecipitated with lanthanum fluoride. After the fluoride precipitate has been dissolved and plutonium oxidized to Pu (IV), the latter is extracted with thenoyltrifluoroacetone. It is then reextracted with a solution of HCl and the aqueous phase is evaporated to a small volume. The electrolytic separation of plutonium, which is almost quantitative, is effected in 1–2 *N* KOH in the presence of NaClO. The hypochlorite oxidizes Pu (IV) to Pu (VI), which is then reduced to the tetravalent state at the cathode and is deposited as the hydroxide. The duration of the electrolysis is 5 hours at current intensity of 200 mA.

The electrolytic cells consist of organic glass cartridges, to the bottom of which are affixed cathode discs, 12.5 mm in diameter, made of stainless steel. Plutonium is deposited on a surface 7 mm in diameter. The source of the electrolytic current is designed to feed 20 cells simultaneously, which are affixed directly to the front wall of the instrument.

The prepared samples are then made to act for a week on a Kodak NTA nuclear emulsion with a layer thickness of 25 μ, applied to 25 × 75 mm microscope object glasses. One object glass is exposed to eight preparations. The plates are developed, the tracks are counted under the microscope and the activity of the samples estimated.

Together with the samples, blank and standard determinations are run in parallel to allow for the background and incompleteness of deposition of plutonium. The standard deviation for $4 \cdot 10^{-6}$ μg Pu in a sample was 12%, most of which was accounted for by track variations.

A similar determination of plutonium in urine was described in detail by Dalton [371], who carried out electrolytic isolation of plutonium by the above method.

The hydrochloric acid solution, obtained after isolation and purification of plutonium from the urine sample by coprecipitation, extraction and ion exchange, is evaporated to 1–2 ml and neutralized to phenolphthalein. Four drops of concentrated HCl are then added to the solution to prevent plutonium from coprecipitating with the hydroxide of iron, which may be present in trace amounts, and heated gently. To the solution 1 ml of a NaClO solution (10–14 wt. % active chlorine) are added and immediately afterwards (to prevent decomposition of hypochlorite) 5 ml of 2 *M* KOH. The solution is then evaporated to half the volume, transferred quantitatively to the electrolytic cell and diluted with water to 10 ml.

The electrolytic cell (about 12 ml in volume) consists of a Perspex tube, which narrows down towards the base. The diameter of the base opening is 7 mm. The opening is covered by a disc-shaped cathode, which is clamped to the base by a screw-on cap. A polythene ring gasket is inserted between the base and the disc. A thick metallic supporting disc is inserted between the cathode and the cap. The cell has a cover lid which is perforated to vent the gases and prevent spattering. Cathode discs made of stainless die steel are 0.125 mm thick and 12.5 mm in diameter. Before use the discs are electropolished for 20 seconds at anodic current intensity of 3–5 amperes in a solution containing 200 g citric acid, 200 g water and 57.5 ml concentrated H_2SO_4. The polished discs are then washed with a mixture of ethanol and water (1:1). The electrolysis is carried out at 400 mA current density. Figure 54 shows the separation of plutonium as a function of the time of electrolysis. A 100% separation of Pu is obtained within 7–8 hours. For routine analyses it proved more convenient to electrolyze for $5\frac{1}{2}$ hours, even though the yield of plutonium was only 92%.

When the electrolysis is terminated, the cells are disconnected from the source, without interrupting the current. The discs with the plutonium are washed with distilled water, dried and heated to a red glow in a Bunsen flame to convert plutonium hydroxide to the dioxide, which adheres more strongly to the disc.

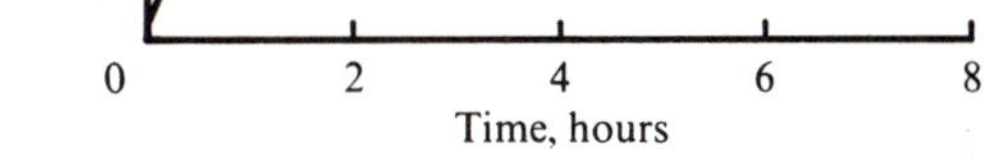

Figure 54
Electrolytic separation of plutonium from a solution of KOH and NaClO

Determinations on plutonium preparations thus obtained are carried out by autoradiography on nuclear emulsions. The amount of plutonium detectable by this method is 10^{-6} μg.

In routine analysis, the current feed ensures a voltage of 10–15 V in twenty channels, each of which is equipped with an amperometer and an individual current adjustment unit. To each channel an electrolytic cell is connected. The anodes, which are located on the front panel above the cell clamps, consist of fixed platinum spirals. The solutions are kept stirred by the evolution of gas during electrolysis.

Weiss and Shipman [723] also used the electrolytic separation of plutonium [644] for a radiometric determination of plutonium in urine.

Sublimation method

Very homogeneous and thin films of preparations of plutonium are obtained by sublimation of its volatile compounds. Plutonium is usually sublimed at high temperatures in a high vacuum and is condensed on various materials: platinum, stainless steel, aluminum, quartz, plastic materials, etc.

Phipps et al. [597, 598] used the sublimation method to study the vapor pressure of plutonium compounds (effusion method). Plutonium halides proved to be most convenient. Plutonium oxide may also be employed, but the sublimation must then be conducted at a higher temperature. The plutonium compound was evaporated in a high-frequency furnace and deposited on a planchet cooled with liquid nitrogen.

Dawson [375] described the sublimation of plutonium nitrate $Pu(NO_3)_4$ from a tantalum thread, heated to 2000° in a 10^{-5} mm Hg vacuum. The duration of heating required for the sublimation of plutonium is 5 seconds.

Gol'din and Tret'yakov [64] obtained α-preparations of plutonium in a similar manner.

Plutonium chlorides were applied to a tantalum strip, heated by an alternating current. A glass disc, on which the sublimate was deposited, was located 3–5 cm away from the strip. Plutonium may be sublimed in several ways onto different glass receivers to separate plutonium from certain admixtures if required.

The sublimation of plutonium to obtain an α-source may be carried out on highly volatile plutonium acetylacetonate [366, 472]. Plutonium acetylacetonate is easily sublimed at about 150° under a pressure of 10^{-5} mm Hg and is deposited on a cooled metal plate. After the sublimation the compound is converted into a nonvolatile form by bromination and is slowly calcined to the dioxide. The disadvantage of the method consists in the fact that at least 50 μg plutonium are required to prepare the acetylacetonate, when the yield of the compound is 50%. When smaller amounts of plutonium were employed, the yield proved to be even lower.

The sublimation method is mainly employed to obtain smooth, uniform preparations but not in quantitative analysis.

Precipitation of plutonium on a carrier

Samples are prepared by precipitation on a carrier when the amounts of plutonium to be determined are too small to be separated from the interfering admixtures and concentrated by extraction or chromatography. Compounds which yield crystalline precipitates are best used as coprecipitants. The samples are prepared as in the evaporation method; the precipitate is transferred onto the support as a suspension, with the aid of a micropipet, spread out evenly, dried and calcined. When the coprecipitation method is employed, the precipitation of plutonium on the carrier is often more than 99% complete [3, Chap. 16].

Hunter and Chenley [475] described a method for the determination of trace amounts of Pu^{239} by coprecipitation of plutonium on lanthanum fluoride for the α-count. The plutonium in solution must be in the trivalent or tetravalent form.

An aliquot of the sample solution (containing Pu in an amount corresponding to about 5000 cpm) is placed in a 10 ml centrifuge tube, and 3 *M* NH_4OH solution is added drop by drop until the precipitate separates out (pH > 5). If uranium is absent or is present in small amounts, the pH of the solution is adjusted to bromophenol blue indicator. Nitric acid (1 *M*) is then added until the precipitate redissolves, with another 0.2 ml in excess, followed by 0.25 ml of 1 % solution of hydroxylamine hydrochloride and 0.5 ml of a solution of $La(NO_3)_3$ (3.1 g $La(NO_3)_3 \cdot 6H_2O$ in one liter of 1 *M* HNO_3). The solution is left to stand for 5 minutes and 1 ml of buffer solution is added (400 g ammonium acetate are dissolved in 600 ml water, concentrated HCl is added to pH 3.3 and the solution is diluted to one liter). To the solution thus obtained 0.25 ml of a 10% solution of ammonium fluoride are added, the solution is left to stand for 15 minutes, then centrifuged and decanted. To the precipitate are added 0.2 ml 1 *M* HNO_3, 0.25 ml of NH_4F solution and 3 ml water. The mixture is stirred up and centrifuged. The mother liquor is removed by decantation, and the precipitation is repeated one or two more times if the aliquot portion of the sample contains more than 25 mg of uranium. A little water is added to the precipitate and the precipitate is transferred onto a stainless steel disc. The mixture is evaporated to dryness and the residue calcined. The activity of the preparation thus obtained is determined with a scintillation or a proportional counter.

To obtain a uniform coating, the precipitation is carried out in a cylinder with a disc [472]. The films obtained by precipitation on carriers are weak and do not adhere firmly to the planchet. The quality of the film may be improved by covering the preparation with a layer of collodion (25–50 $\mu g/cm^2$); a 5% solution of collodium in methanol is used for the purpose [3, Chap. 16].

The method of coprecipitation on a carrier makes it possible to concentrate small amounts of plutonium; it is easy to carry out, but is not as accurate as other methods of preparing α-active samples.

Comparison of the different methods

The selection of the method to be employed in preparing an α-active sample will depend on the task in hand, concentration of plutonium, the salt composition of the sample solution and the presence of extraneous α- and β-emitters. The preparation of samples by evaporation of a weight or volume aliquot is now widely used in analytical practice. With this method, an accurate amount of the sample can be taken, but the resulting films are not homogeneous owing to the nonhomogeneous evaporation of the solutions and particularly to the crystallization towards the end of the evaporation. The quality of the films may be improved by adding tetraethylene glycol. The evaporation method gives satisfactory results only if the sample solution does not contain large amounts of extraneous salts.

Sample films obtained by electrolytic deposition are thin and very homogeneous. The method is most often used to obtain standard preparations from pure solutions of plutonium which are free from foreign salts. The method is time-consuming and special apparatus is required.

Sublimation methods furnish thin, smooth and homogeneous films, but the sublimation method is only semiquantitative. The films are weak and must be stored under vacuum. Complicated apparatus is required, including a vacuum installation, and high temperatures must be employed. Samples thus prepared are used in α-spectrophotometry.

The method of coprecipitation of plutonium on a carrier is inaccurate.

Activity measurements

The determination of the amount of plutonium in samples amounts to the determination of the number of particles emitted by the preparation in unit time within a given solid angle, which is determined by the geometrical location of the preparation and that of the pick-up unit. To obtain more accurate results, it is recommended that for each sample solution several parallel determinations be run on multiplicate preparations. The shape and the size of the planchet will depend on the type of counter employed, but should be identical for sample and standard preparations.

The measurements are begun with the determination of the background, i.e., the number of pulses in the chamber in the absence of sample. The standard and the sample are then measured. Each preparation is measured several times, 4–8 minutes each time, depending on the value of the activity and on the accuracy desired. The results are corrected for the background radiation and the average value for each sample is found. The concentration of plutonium in the preparation is measured in two ways: a) by comparing the radiation intensities emitted by the standard and the sample respectively; b) by calculation, allowing for the counting efficiency of the chamber and the specific activity of plutonium. The amount of plutonium on the support is given by the formula:

$$\text{Pu}\ (\mu\text{g}) = \frac{N}{P \cdot N_0}, \tag{3}$$

where N is the count in cpm; N_0 is the specific activity of plutonium, corrected for the isotopic composition, in disintegrations per minute per microgram; and P is the count efficiency.

A technique which has recently become popular is the determination of α-activity directly on the surface of liquid preparations of plutonium.

The typical feature of the determination of α-activity of liquid preparations is that the layer thickness in this case is larger than the range of the α-particles in the sample. The amount of the solution in the layer should be more than 5 mg/cm^2. In such preparations it is the surface concentration of plutonium, and not the overall content of plutonium in the sample, which is of prime importance. The α-count of the surface of a liquid preparation is considerably affected by the density of the solution. Variations in the density of the solution not exceeding 0.4% are immaterial, but larger density fluctuations are not permissible (Chvankin and Ternovskaya, 1956).

The determination of the plutonium concentration on the surface of liquid preparations may be employed with aqueous (Chvankin, Ternovskaya et al., 1956) and nonaqueous solutions (Osipov and Samoilova, 1957; Zemlyanukhina, Yarkin et al., 1958) at plutonium concentrations between 0.01 and 250 mg/liter.

The activity is measured with the aid of a scintillation counter. The surface of the preparation is placed at a rigorously fixed distance from the working surface of the counter (screen), but not less than 5 mm. In order to protect the luminophore from acid fumes, special films are employed (Pavlov, 1960); these may be made of quartz, gold, or nitrocellulose varnish and are applied directly to the screen.

The concentration of plutonium in solution is determined from the formula:

$$C = k \cdot N, \tag{4}$$

where C is the concentration of plutonium in the sample, mg/liter, N is the count after deduction of the background, cpm, and k is the count factor.

The larger the count factor, the higher the sensitivity of the apparatus and the lower the amounts of plutonium in solution which can be measured. Its value depends on the density and salt composition of the solution; it is determined by measuring the activities of standard solutions of a density equal to that of the sample solution, using the formula

$$k = \frac{C_0}{N_0} \text{ mg} \cdot \text{min/liter} \cdot \text{count}, \tag{5}$$

where C_0 is the concentration of plutonium in the standard solution in mg/liter, and N_0 is the count of the standard solution less background, in cpm.

The error in the determination of plutonium in solutions with concentrations exceeding 1 mg/liter is not more than 10%, while for solutions more dilute than 1 mg/liter it is about 20%. Since the method does not involve any special preparation of the sample, the time of one determination may be as short as 15–20 minutes.

The errors in radiometric determinations are due to several factors: a) inaccuracy in the preparation and dilution of the sample solutions, and in the application of the aliquot to the support; b) absorption of α-particles owing to the uneven distribution and thickness of the solid sample on the support; c) statistical count error.

Statistical error of count

When using a counter with a given count efficiency (see below), the measured number of disintegrations per minute is never equal to the average rate as given by the decay law, but fluctuates around that value within a certain random error. This is because each disintegration is a separate random event. The count rate, as experimentally determined, approaches the average value only for a very large number of pulses. In order to estimate the accuracy with which a limited number of pulses can be determined, statistical methods are employed.

If the average experimental count is N particles during a time t, the probability that x particles will be counted during that time is

$$W_x = \frac{N^x}{x!} e^{-N}. \tag{6}$$

This expression is known as Poisson's distribution law. Figure 55 shows the relative distribution of the particles for $N = 10$ and $N = 100$. The mean square or standard deviation σ from the average number of pulses N may be found from the expression

$$\sigma = \sqrt{N} \tag{7}$$

and the relative deviation is

$$\frac{\sqrt{N}}{N} = \frac{1}{\sqrt{N}}.$$

It will be seen that the accuracy of the measurements can be increased by increasing the number of pulses counted, obviously provided that other errors do not exceed the value of the random error.

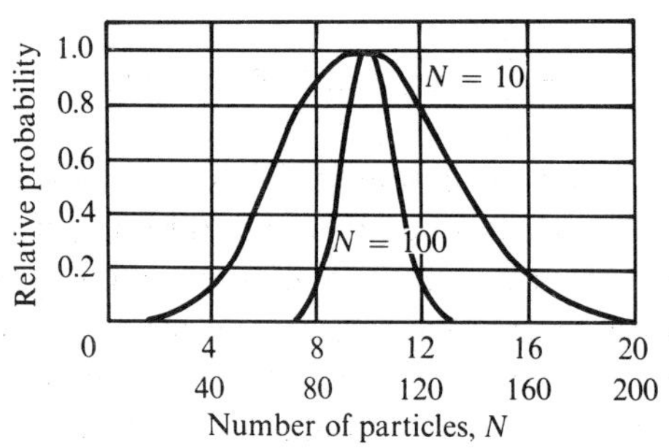

Figure 55
Poisson's distribution law

The probability for the experimental result to differ from the true result by a value lower than the standard deviation is approximately 68.3%. The relationships between the standard deviation and other reliability levels are given in Table 13.

When determining the activity of the preparations, the average number of background pulses during the same period of time is usually deducted. The presence of the background introduces a supplementary error in the activity determinations; the overall statistical error is the square root of the sum of the squares of individual errors, which for the standard deviation is

$$\sigma = \sqrt{\sigma_1^2 + \sigma_2^2}. \tag{8}$$

Table 13

RELATIONSHIP BETWEEN THE AVERAGE VALUES OF DEVIATIONS

Error	Ratio: error / σ	Probability of lower error
Probable, ζ	0.6745	0.500
Average, η	0.7979	0.575
Standard (mean square), σ	1	0.683
Error for different confidence levels:		
90%	1.645	0.900
95%	1.960	0.950
99%	2.575	0.990
99.9%	3.30	0.999

The same law applies to other kinds of average deviation (ζ, η, etc.). When calculating the composite error, the number of pulses of the preparation and the background and their deviations should refer to equal periods of time, best of all unit time, since the durations of counting may be different for the sample and for the standard. Let us say, as an example, that in the determination of the background 100 pulses with a standard deviation of 10 pulses were counted in 10 minutes, i.e., 10 ± 1 cpm. The sample yielded a value of 600 pulses with a standard deviation of 24.5 pulses in 20 minutes, i.e., 30 ± 1.23 cpm. Then the activity of the sample alone will be 20 cpm with a standard deviation of $\sqrt{1^2 + 1.23^2} = 1.58$ cpm, or 7.9%. In accordance with Table 13, the error for a confidence level of 90%, which is equal to $1.645\,\sigma$, will be 2.5 cpm or 13%.

The standard deviation may be expressed in terms of the rate and duration of count by the formula

$$\sigma_y = \sqrt{\frac{N_y + F_y}{t_N} + \frac{F_y}{t_F}}, \tag{9}$$

where σ_y is the standard deviation, cpm, N_y is the net count rate of the sample, cpm, F_y is the background noise, cpm, and t_N and t_F are the duration in minutes of the counts for the preparation and background, respectively.

The calculation of our former example by this formula yields the same result in a more simple manner.

Other problems in the measurement technique of the preparations, including instrument errors, will be briefly discussed in the following section.

Apparatus

The most extensive description of the instruments used in measuring α-radiation will be found in the work of Jaffee [3, Chap. 16], which, although written some time ago, is valuable even now. We shall give only a brief description of the main types of instruments employed in technology and in analytical chemistry.

When determining the total activity of α-plutonium, other α-emitters must be eliminated, and the isotopic composition of the plutonium must be corrected for. The use of α-analyzers will often make it possible to determine the intensity of α-emission in the presence of a background of another α-emission.

Electronic and recording instruments used with α-detectors are described in the specialized literature.

The typical parameters of α-detectors are: 1) counting efficiency, which is determined in the first instance by the geometry of the chamber; 2) background level, which determines the minimum determinable amount of an α-emitter; 3) maximum measurable activity; 4) allowable level of β- and γ-radiation; 5) response to the distribution of the sample over the surface and in depth; 6) convenience in use, response to interfering effects, including vibrations and noises (microphonic effect). The reliability and the accuracy of an instrument are determined by all these factors. When making an analysis of the energies of α-particles, the detector should also be characterized by the dependence of the pulse amplitude on the energy of the α-particle and by the scatter of the pulse amplitudes for a given energy of the particles.

The minimum determinable activity will depend on the size of the background of the instrument, which is mainly due to the presence of radioactive impurities in counter and chamber materials. It may be taken that the minimum determinable activity is one-half of the background activity. Since the size of the background activity does not significantly depend on the geometry of the counter, low activities are difficult to measure with instruments of low count efficiency. The background activity usually increases with time, owing to the contamination of its inner surfaces, which cannot be altogether prevented however careful the handling of the instrument; accordingly, the inner surfaces must be periodically cleaned and washed.

The maximum measurable activity and the permissible level of β- and γ-radiation will mainly depend on the width of the pulses, or on the resolving time of the counter, which is defined as the time during which two successive pulses merge into one. The probability of such error will increase with the rate of the count. Similarly, if the β-radiation is very intense, several β-pulses succeeding each other during the resolving time may be mistaken for one α-particle. The resolving time should thus be reduced.

Pulse ionization chambers and counters

Instruments of this type utilize the electric charge resulting from the ionization of a gas by α-particles. The ions of the gas thus formed are collected on the electrodes by the electric field in the chamber and produce current pulses in the external circuit of the chamber. At a certain value of the electric field, or, in other words, of the voltage applied to the electrodes, all ions formed in the chamber may be collected. The result is the so-called saturation current, when the pulse amplitude is not significantly dependent on the external voltage (voltage plateau) and is proportional to the ionization produced by the particles being determined. These chambers may be filled with air or with special gas mixtures.

At higher voltages, the ions accelerated by the electric field induce secondary ionization, owing to which the total charge collected increases. This is equivalent to using additional amplification. Instruments of this type are called proportional counters. The amplification factor chosen is usually 10^2–10^4.

At amplifications of about 10^5 a state is attained in which any ion pair in the counter produces the maximum possible pulse, irrespective of the identity of the ionizing particles. This is the principle of the Geiger-Muller counter, which is only rarely used in α-radiometry owing to its poor selectivity with respect to other kinds of radiation.

Plane-parallel air ionization α-chambers

These chambers are the simplest type of gas ionization instruments, since neither the use of special gas mixtures nor a vacuum is necessary. The sample is placed on the lower electrode. The geometric factor is about 50%. To reduce the background and the resolving time of the chamber, the size of the chamber is made shorter than the α-particle range. The average background may be 0.5–2 cpm. The resolving time imposes an upper limit of about 10,000 cpm on the samples. If the counting rate is 1000 cpm, the resolving error may be as high as 1%. The permissible level of β-radiation in the sample is about 10^6 disintegrations per minute. Like all instruments with a geometric factor of 50%, air chambers are not very sensitive to the distribution of material on the surface of the source, but are sensitive to the thickness of the sample and to the planchet material (back-scattering). These chambers are very susceptible to interferences of all kinds, including vibrations, noise, and low-frequency coincidences, since their signal : noise ratio is low. Nevertheless, with a very careful working technique, the results may be made reproducible to within 0.1%. These chambers are now being displaced by more sophisticated instruments.

Gas-filled electron-pulse α-chambers

The design of gas-filled chambers is much like that of air-filled chambers, and their geometric coefficient is 50%. The chambers may be filled with nitrogen, helium, argon, mixtures of argon with carbon dioxide and hydrogen, which have a low affinity for electrons. The pulses are produced by collection of electrons and not of heavy ions. As a result, the dead time becomes 100 times shorter. The use of a special gas atmosphere makes it possible to increase the inter-electrode distance (at the same voltage) and the size of the electrodes. In this way the paths of the α-particles are more fully utilized, the pulse amplitude is increased, which means that the signal: noise ratio increases as well. In some chambers the sample planchets are as large as 60 mm in diameter. After the sample has been charged, the chamber is sealed, evacuated if necessary, and filled with the working gas.

The background of electron pulse chambers is somewhat higher than that of air-filled chambers and is about 1–3 pulses per minute. Because of the decreased dead time, the maximum α-activity of samples is about 10^6 disintegrations per minute. The permissible soft β-activity in the sample (Np^{239}) is $5 \cdot 10^8$ disintegrations per minute, while the hard activity is $2.5 \cdot 10^{10}$ disintegrations per minute. At the rate of 1000 cpm the count error may be less than 0.01%. The chambers, together with high frequency amplifiers, are resistant to mechanical vibrations and noises. High-frequency electrical disturbances interfere, e.g., spark disturbances. The reproducibility error of the count may be as low as 0.01% during a prolonged period of time.

Proportional counters

We have said above that proportional counters utilize the amplification in the interior of the gas, due to the secondary ionization of the gas produced by the electrons being collected. A typical chamber consists of a metal cylinder with a thread along the axis of the cylinder. The thread acts as a positive electrode; its small diameter ensures that there is a high field potential around the thread, which is necessary to produce secondary ionization.

The main characteristics of proportional counters are similar to those of electron pulse chambers. The most reproducible results are obtained when the counter is filled with methane or gas mixtures containing methane. The geometrical coefficient is about 50%. The average value of the background is 1–2 cpm. The range of the measurable α-activities is 2–800,000 cpm. At a level of 1000 cpm the count error is less than 0.01%. The permissible β-activity of the samples is more than 10^8–10^9 disintegrations per minute.

The advantages of this type of counter is the larger amplitude of the pulses and a larger signal:noise ratio. For this reason the electronic amplifiers required are of simpler design, and they are less affected by electrical interference.

Their disadvantage is that the count efficiency is greatly affected by the feed voltage and by the external and internal amplification. The count rate is influenced by the composition, purity and pressure of the gas. In addition, the count efficiency is very sensitive to the location of the source in the counter. The reproducibility is 0.5–1.0% under the usual working conditions, but may become better than 0.1% for long periods of time if the operator is very careful.

The types of chambers and counters described above may be utilized in measurements with a low geometric factor. For this reason the samples are placed outside the counter in a special accessory situated a given distance from the chamber. α-Particles enter the chamber through a window, mostly covered with thin mica or a plastic film. The decrease in the count efficiency makes it possible to measure higher activities, and also makes absolute determinations easier. Moreover, the requirements regarding the uniform distribution of the sample are less exacting, the self-absorption by the sample takes place only along the vertical, there are no back-scattered α-particles and, in the case of proportional counters, the internal amplification for the various α-particles is more constant.

Ionization chambers and, more rarely, proportional counters are also used for energy counts of α-particles. To do this, the size of the chamber is increased so as to have the entire path of the α-particles inside the chamber. The full absorption of α-particles inside the chamber produces an ionization which is proportional to their energies; the design of the chamber should be such as to ensure the minimum scatter of pulse amplitudes for each energy value of the particles. The amplified pulses are sorted by amplitudes and counted with the aid of multichannel electronic pulse analyzers. To determine the energies of α-particles the instrument is calibrated using emitters of α-particles of known energies. Ionization chambers are more accurate and have better energy resolution than proportional counters. The resolving capacity in ionization chambers may be about 0.5% (half-width of a peak) while the energy values in the middle range (about 0.01 MeV) may be measured to within 0.2%.

Scintillation counters

Scintillation counters are at present the most frequently used instruments in measuring α-activities. As regards their characteristics they occupy a place intermediate between the plane-parallel ionization chambers and electron pulse chambers, as well as proportional counters.

A scintillation type detector consists of a scintillator, in which the excitement produced by charged particles is converted into light quanta, and a photomultiplier, which transforms the light quanta into large electrical pulses. Most often the scintillator for α-counting is a screen coated with silver-activated zinc sulfide. The layer thickness of ZnS(Ag) is chosen somewhat larger than the range of α-particles, about 10 mg/cm^2. Under these conditions the maximum light yield and the maximum β-particle discrimination is obtained.

The highest geometrical coefficient of scintillation counters is 30–40%. The background of zinc sulfide scintillators is very low, owing to the small surface area and intrinsically low specific activity. The usual background value is not more than a few pulses per hour. The duration of one scintillation of the ZnS(Ag) screen, which is of the order of 10^{-5} second, puts the maximum measurable α-activity at about 100,000 cpm, and the permissible β-activity at 10^8–10^9 disintegrations per minute. The count error at 1000 cpm is less than 0.1%. Scintillation counters with large and medium geometrical coefficients are very sensitive to the location and the size of the sample.

The counters must be protected from light. Accordingly, samples are exchanged in light-proof cassettes, or else the scintillator is covered with an opaque film, e.g., a 0.2 mg/cm^2 aluminum foil.

Small gain amplifiers are required with scintillation counters, since a considerable amplification is effected by photomultipliers. The voltage-intensity plateau of the amplifier is inclined, which is explained by the presence of a large number of small pulses emitted by the scintillator. These pulses are caused by the polycrystallinity of ZnS and its inadequate transparency to its own radiation. The slope of the plateau requires efficient stabilization of the feed voltage and amplification.

Various interferences and photomultiplier noise emit small pulses, which are easily separated in the electronic unit.

Zinc sulfide gives a large amplitude scatter with α-particles and consequently the energy resolution is poor. In spectrometry better results can be obtained with thin layers of monocrystalline scintillators NaI(Tl) and CsI(Tl). In the latter case an energy resolution of about 4% was obtained for α-particles, and the energetic yield was linear. For more information on scintillation spectrometry, see [44].

Nuclear emulsion method

A few examples of the utilization of nuclear emulsions were described above in connection with the preparation of samples. It follows that nuclear emulsions may have substantial advantages over other methods whenever very low α-activities are to be determined.

The nuclear emulsion method involves exposing a plate with nuclear emulsion to the action of the sample for a given period of time, development of the plate and counting the number of tracks under the microscope. The exposure may be effected either by the contact method, in which discs with the α-active preparation is put over the emulsion, or by impregnating the emulsion with the solution of the α-active substance. In the former, more frequent case the count geometry is nearly 50%, in the latter it is nearly 100%.

The range of the α-particles in the emulsion is 16–40 μ, and for Pu^{239} α-particles with an energy of 5.15 MeV it is about 22 μ. Thus, the emulsion for the contact exposure need not be thicker than 25 μ. The plates are practically insensitive to light and to weakly ionizing particles (electrons). For this reason, α-activity can be determined against an intense background of β- and γ-radiation, provided the presence of such radiation does not result in a general blurring of the emulsion.

Nuclear emulsions have a very weak intrinsic background, which is mainly due to radioactive impurities in glass and in gelatine. The size of the background may be of the order less than 10 disintegrations per day per cm^2 of surface.

The background tracks in the emulsions are erased by holding the emulsion prior to use over saturated vapor of 0.3% solution of hydrogen peroxide or simple saturated water vapor at room temperature. The emulsion is then dried for 3–4 hours when its original sensitivity is restored.

The detection threshold of plutonium is about 0.5 μ/μg at the exposure time of one week, with 99% degree of confidence.

The tracks must be counted under a high quality microscope giving a high magnification. The accuracy is 3% at best. The selectivity of the method may be improved by measuring the range of α-particles, but the energy resolution is poor—3–4%.

Count corrections

The count correction may be introduced empirically, from the curve giving the instrument readings (n) as a function of the radiation intensity (N). If the discrimination time is denoted as τ, the relationship between n and N will be:

$$n = Ne^{-N\tau}, \tag{10}$$

or, very approximately, if n is not too different from N

$$N = n + n^2\tau. \tag{11}$$

The maximum of this function corresponds to

$$n_{\max} = \frac{1}{e\tau} \quad \text{and} \quad N = \frac{1}{\tau}. \tag{12}$$

To find τ, a high intensity source is gradually brought closer to the counter until the maximum count rate $n_{\max}$ is obtained; from this value τ is found with the aid of equation (12).

The value of the dead time makes it possible to calculate the error from (10) or (11). In particular, the approximate equation (11) shows that the count error, relative to the count of the instrument, is proportional to the count:

$$\frac{N-n}{n} = n\tau.$$

Thus, if at a count of 1000 cpm the error is, say, 0.1%, it will be about 0.5% at a count level of 5000 cpm.

Other methods of determination of the dead time are also employed, e.g., two sources measured separately and together.

Detailed descriptions of the apparatus are usually furnished with each instrument and will not be discussed here. A few types of instruments have been described in the literature. Thus, for instance, the popular α-counting installation of the DA type has been described by Baranov [16, Chap. 5]. For a description of a few modern instruments, including PS-5M and ASU counters, the reader is referred to [95].

COLORIMETRIC AND SPECTROPHOTOMETRIC METHODS

Both aqueous and organic solutions of plutonium in all valency states have typical colorations.

All known photometric methods for the determination of plutonium may be classified into three groups.

1. Methods based on color reactions of plutonium ions with inorganic anions and with simple colorless and weakly colored organic reagents. Plutonium ions in solutions have a number of absorption bands, some of which are specific for a given valency state only. The optical absorption of solutions of plutonium remains unchanged for a long time, provided, of course, the valency state remains unchanged.

The sensitivity of the methods in the first group is low and they are used to determine milligram amounts of plutonium.

2. Methods based on the formation of chelate type complexes of plutonium with intensely colored reagents. The sharp difference between the color of the reagent and that of the complex compound formed is due to a charge redistribution in the molecule of the reagent brought about by the complex-forming cation. These methods are the most sensitive of all chemical methods of determination of plutonium and make it possible to determine plutonium concentrations of a fraction of a milligram per liter of solution. Reagents such as Arsenazo III have a fairly high selectivity, since they form stable complexes, particularly with Pu(IV) in strongly acid media. As a rule, complexes between plutonium and azo dyes of the arsenazo and thoron type have one absorption maximum between 550 and 600 mμ.

3. Methods based on solid phase color reactions between plutonium and certain organic dyes. The suspensions of the compound formed are of a different color from the solutions of the dyes themselves. The reaction mechanism involves the precipitation of the cation of the heavy hydrophobic organic reagent by the anionic complex of plutonium. These reactions are very sensitive and quite specific, since the reagent precipitates only those elements which give anionic complexes under the given experimental conditions. An important drawback is the poor reproducibility of these methods.

The absorption of plutonium solutions is usually determined with the aid of photoelectric instruments.

The connection between the light absorption of a solution, the concentration of the substance in solution and the thickness of the absorbing layer is given by

$$D = \log \frac{I_0}{I} e \cdot C \cdot l, \tag{1}$$

where D is the optical density at wavelength λ, I_0 is the intensity of radiation entering the cell, I is the intensity of the light after passage through the cell, C is the concentration of the absorbing substance in moles per liter, l is the thickness of the absorbing layer, cm; ε is the molar extinction coefficient, which is numerically equal to the optical density of a one-molar solution of the substance 1 cm thick.

METHODS BASED ON ABSORPTION IN AQUEOUS SOLUTIONS OF INORGANIC ACIDS

The presence of sharp bands in the absorption spectrum is characteristic for solutions of plutonium in all degrees of oxidation. Each oxidation state of plutonium has its own characteristic absorption minima and maxima (Figure 7). This is the base for the spectrophotometric identification of plutonium, and for the determination of the concentration of one valency form against the background of another.

The complexing anions, depending on their nature and concentration, have different effects on the nature of the absorption spectrum of solutions of plutonium [3, Chap. 9]. The literature contains numerous data on the effect of concentration of inorganic acids—perchloric, hydrochloric, nitric and sulfuric—on the absorption spectra of solutions of tri-, tetra- and hexavalent plutonium [3, 460–462, 493; Palei and Milyukova, 1953; Tarakanov, 1951].

The analytically important absorption bands of plutonium ions are listed in Table 14.

The molar extinction coefficients of solutions of trivalent plutonium attain their maximum values at 560 and 600 mμ. The advantage of the band at 600 mμ consists in the fact that the molar extinction coefficients of solutions of tetravalent and hexavalent plutonium have minimum values at this wavelength. Consequently, trivalent plutonium may be determined with satisfactory accuracy in the presence of other valency forms of plutonium. As compared with the band at 600 mμ, the band at 560 mμ does not depend to the same extent on the variations in the concentration of acid in solution [460] and on the temperature [348]. For this reason the choice of the particular absorption band for the determination of trivalent plutonium will depend on the nature of the task in hand.

The most important absorption band of tetravalent plutonium is the 470 mμ band, while that of hexavalent plutonium is at 833 mμ. Both bands are very narrow, which renders accurate determination of optical densities difficult. On the other hand, the high molar extinction coefficient of the band at 833 mμ enhances its practical importance in analysis; a fraction of one mg of Pu in 1 ml of solution may be determined at this wavelength.

Determination of plutonium (III) by self-absorption

The absorption spectra of solutions of trivalent plutonium do not vary much with the nature and the concentration of the acid in solution. Thus, for instance, in per-

Table 14

ABSORPTION BANDS OF PLUTONIUM IONS [203, 348]

Degree of oxidation	Main absorption bands, mμ	Molar extinction coefficient			
		Pu(III)	Pu(IV)	Pu(V)	Pu(VI)
Pu(III) (0.1 M $HClO_4$)	900	19.30*	4.00	5.16	0.52
	665	14.65	30.09	0.43	0.55
	603	35.40	0.96	0.60	1.20
	600	35.30	0.91	0.50	1.35
	560	36.10	11.64	3.62	2.50
Pu(IV) (0.5 M HCl)	815	14.63	19.61	1.55	2.30
	730	1.35	14.60	1.03	0.50
	700	0.75	10.88	0.44	0.25
	655	3.10	34.40	1.15	0.90
	470**	3.46	49.60	1.82	11.25
Pu(V) (0.5 M HCl)	775**	12.40	11.90	9.87	2.90
	569**	34.30	5.60	17.10	1.75
Pu(VI) (0.1 M $HClO_4$)	983	3.15	1.76	1.18	8.90
	953	1.20	0.40	1.76	19.90
	833**	5.25	15.50	4.00	~ 300.0

* Absorption bands generally employed in the determination of plutonium are given in heavy type.

** Absorption band is very narrow.

chlorate solutions of plutonium no change in the absorption spectrum is noted when the concentration of perchloric acid is increased from 1 to 5 M. Further increase in the concentration of perchloric acid does not affect the spectrum to any significant extent. Also insignificant are the changes in the absorption spectrum of Pu(III) in solutions less than 4.4 M in HCl and in solutions less than 16 M in HCl. The effect of acid concentration between 1 M and 5 M HNO_3 is more significant, but Pu(III) is not very stable in solutions of nitric acid, especially at acidities above 5 M.

These effects are small, but in accurate spectrophotometric determinations, the acidity of the solution must be kept within a narrower range.

In the absorption spectrum of Pu(III) the absorption bands between 560 and 1100 mμ are of greatest practical interest.

The optical absorption of the solutions obeys Beer's law between 0 and 11 grams Pu per liter at 600 mμ and between 0 and 15 grams/liter at 560 mμ (Tarakanov, 1951).

Palei and Milyukova (1953) proposed a spectrophotometric method for the determination of 0.5–10 mg Pu(III) per ml in pure solutions and also in the presence of other elements. Pu(III) in pure solutions is determined with the aid of the bands at 560 and 600 mμ.

Through the sample solution in hydrochloric or nitric acid, containing 0.5–10 mg Pu/ml, sulfur dioxide is passed during 5–10 minutes. The solution is left to stand for 15–20 minutes to complete the reduction of plutonium to the trivalent state. The acid concentration in solution is kept at 0.5–1.5 *M*. 2.5–3 ml of the reduced solution are placed in a 10 mm cell and the optical density is measured at 560 and 600 mμ. The concentration of plutonium is calculated by formula (1). The relative error of the determination is $\pm$ 0.5%.

The determination of Pu(III) in the presence of U(VI), Cr(III), Fe(III), La, Pb(II) and Mn(II) is carried out in a special region in which these impurities do not absorb. The optical density is measured at 915, 1020 and 1110 mμ. For samples containing 0.5–10 mg Pu/ml, the error does not exceed $\pm$ 1 rel. %.

A similar method was proposed by Allison (quoted from [547]), who used the 600 mμ band to determine plutonium (III) in solution of pure Pu.

The most accurate is the method of differential spectrophotometry, developed by Phillips [593]. The method consists in the measurement of the optical density at 565 mμ of the solution being analyzed relative to that of a comparison solution (standard solution) containing plutonium in a somewhat lower concentration. In order that this difference be small, the content of plutonium in the sample is first determined by the direct spectrophotometric method. The optimum concentrations of plutonium were 8–13 mg/ml. Temperature variations within $\pm$ 1° change the absolute value of the optical density by 0.1% (at an absolute optical density of 1.5).

Increasing the concentration of hydrochloric acid from 1.0 to 2.0 *M* and that of nitric acid from 0.0002 to 3.2 *M* has practically no effect on the optical density. Careful adjustment of the instrument, thermostatic control of the temperature of the sample solution, and controlled acidity make it possible to determine plutonium with an error not exceeding $\pm$ 0.05%.

The accuracy of the method is not impaired by the presence of equal amounts of aluminum, fourfold amounts of thorium and eightfold amounts of U(VI), Ce(III) and Ca. Ferric iron strongly interferes; the interference is removed by reducing it to the ferrous state by $SnCl_2$ (5 ml 5% $SnCl_2$ in 25 ml of solution).

The results of [593] were used by Barker who developed analytical methods for the determination of plutonium in the products of technological purification of plutonium [703] and in the metal [702]. The latter method is quoted below.

Preparation of solution for analysis. An accurately weighed amount (300–320 mg) of metallic plutonium (for use with a 40 mm cell or 240–260 mg for a 50 mm cell) is placed in a 100 ml volumetric flask and dissolved in 10 ml of a mixture containing 0.025 *M* NH_4F in 11 *M* HCl (the so-called *dissolving mixture*). The flask is then heated on a water bath at 70–80° for 2–3 minutes. After cooling, 5 ml of 0.125 *M* aluminum nitrate $Al(NO_3)_3$ are added to the solution to bind fluoride ions. Five drops of 16.3 *M* HNO_3 are then added to oxidize Pu (III) to Pu (IV); the oxidation takes only a few minutes. After the solution has turned from blue to brown, ~ 60 ml 0.1 *M* HCl and 10 ml of 50% hydroxylamine hydrochloride are added.

The solution is now left overnight to ensure a quantitative reduction of plutonium to the trivalent state. The flask is then filled to the mark with 0.1 *M* HCl and the contents are mixed.

Preparation of standard solutions. For sample weights of 300–320 mg, standard solutions with the following concentrations of plutonium are prepared: $C_1 = 3.0$ mg/ml, $C_2 = 3.2$ mg/ml and $C_3 = 3.4$ mg/ml. For 240–260 mg samples standards should have Pu concentrations of 2.4, 2.6 and 2.8 mg/ml, respectively.

The initial solution for the preparation of standards is a solution of plutonium in 1 *M* HNO_3 of an approximate concentration of 50 mg/ml. The titer of the solution is determined gravimetrically. Portions of 4.8, 5.2, 5.6, 6.0, 6.4 and 6.8 ml of the solution are withdrawn, placed in tared 100 ml volumetric flasks and the flasks are again weighed. Into each flask are added, in this order, 10 ml of 0.025 *M* NH_4F in 11 *M* HCl (dissolving solution), 5 ml of 0.125 *M* $Al(NO_3)_3$, about 60 ml 0.1 *M* HCl, and 10 ml of 50% solution of hydroxylamine hydrochloride. The solutions are left overnight for the plutonium to be quantitatively reduced to Pu (III), and the solutions are made up to the mark with 0.1 *M* HCl.

The concentration *C* of plutonium in the standard solutions is calculated by the formula

$$C\ (\text{mg/ml}) = \frac{W \cdot a}{v} \cdot 1000, \tag{2}$$

where *W* is the weight of the initial solution in grams, *a* is the concentration of plutonium in grams per gram of solution, and *v* is the volume of the flask in ml.

Determination of the concentration of plutonium. The solution to be analyzed is placed in the cell and its optical density at 565 mμ is measured relative to the standard C_1 with a similar, but somewhat lower concentration. The next step is to determine the optical densities of the more concentrated standard solutions C_2 and C_3, again relative to the standard solution C_1. The results are corrected for the different charactetistics of individual cells. To do this, both cells are filled with C_1 and the difference in the optical densities determined.

The corrected data are now used to construct the curve giving the dependence between the optical densities of standards C_2 and C_3 with respect to standard C_1 and the concentration of plutonium. The calibration curve then serves to find the accurate content of plutonium in the sample. The concentration of plutonium in the metal is calculated from the formula ,

$$\%\,\text{Pu} = 100 \cdot \frac{C \cdot v \cdot F}{W}, \tag{3}$$

where *C* is the concentration of plutonium in the sample found from the calibration curve, mg/ml; *v* is the volume of the sample solution, ml; *W* is the sample weight, mg; *F* is the ratio between the atomic weight of plutonium in the sample and in the standard.

The accuracy of the determination of plutonium in the metal is ± 0.19%.

The procedure employed to analyze the solutions obtained in the course of the technological purification of plutonium is a little different from that just described.

A 15 ml aliquot of the sample solution is placed in a 25 ml flask, 2.5 ml of 50% solution of hydroxylamine hydrochloride are added and the solution obtained is left to stand overnight. 2.5 ml of 11 *M* HCl are then added and the solution is made up to the mark with 0.1 *M* HCl. Henceforth the procedure is the same as for the analysis of the metal.

Milligram amounts of trivalent plutonium may also be determined by spectrophotometry in oxalate, carbonate and other complexing media.

Dissolution of plutonium (III) oxalate $Pu_2(C_2O_4)_3 \cdot 9H_2O$ in 20% $K_2C_2O_4$ results in the formation of the complex plutonium (III) oxalate, which is stable in the presence of excess of the reducing agent—hydrazine hydrate or sodium formaldehydesulfoxylate (Rongalite) [58, 60]. The absorption spectrum of the solution has maxima at 563, 605, 665, 780, 910 and 1015 mμ (Figure 56).

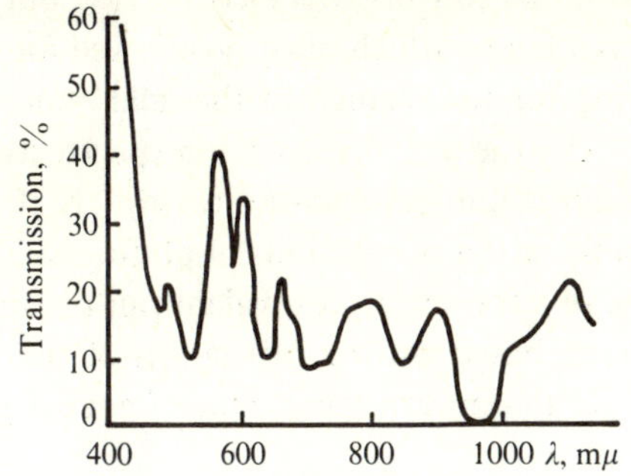

Figure 56
Absorption spectrum of Pu(III) in 20% $K_2C_2O_4$.
Concentration of Pu(III) 1.07 mg/ml.

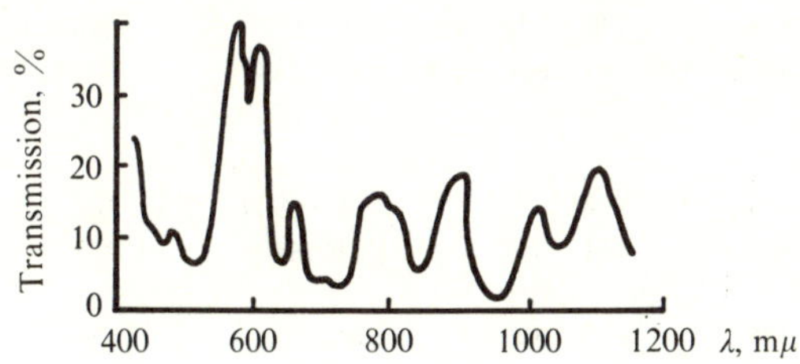

Figure 57
Absorption spectrum of Pu(III) in 20% solution of $(NH_4)_2CO_3$.
Concentration of Pu(III) 1 mg/ml.

The absorption spectrum of a solution obtained by treating trivalent plutonium hydroxide with 45% solution of K_2CO_3 or by treating plutonium (III) oxalate with 20% solution of $(NH_4)_2CO_3$ has typical absorption maxima at 565, 600, 835 and 920 mμ (Figure 57) [3, Chap. 9].

The values of the molar extinction coefficients for these maxima are similar to those of nitric acid solutions of Pu(III) at the corresponding wavelengths. However, the difficulties involved in the preparation and stabilization of solutions of plutonium just described limits their utilization in spectrophotometric methods.

Determination of plutonium (IV) by self-absorption

Tetravalent plutonium tends to form complexes much more than plutonium in any other valency state. With the change in the acid concentration in solution, the composition of the complex compound may change as well, which will in turn bring about changes in the absorption spectra. Major changes in the absorption spectra are observed for nitric and sulfuric acid solutions of Pu(IV). For this reason, when a spectrophotometric determination of Pu(IV) is carried out, the nature and the concentration of the acid is a factor which must be allowed for.

The difficulty in allowing for the change in the acid concentrations during the work is one of the reasons why the use of absorption spectroscopy in the determination of tetravalent plutonium is limited. Another reason are the difficulties involved in the preparation and stabilization of solutions of plutonium (IV). It is much easier to prepare stable solutions of trivalent or hexavalent plutonium, and for this reason spectrophotometric methods based on the utilization of the absorption bands of trivalent and hexavalent plutonium are much more frequently employed.

Table 15

ABSORPTION BANDS OF Pu(IV) IN VARIOUS MEDIA

Medium	Wavelength, mμ								References
1.0 M $HClO_4$	**426** (~ 27)	447	**470** (~ 53)	508	557	**656** (~ 33)	730	815	[3, 460]
2 M HNO_3	**420** (~ 25)	447	**476** (79.4)	501	544	660 (~ 30)	703	795	[3, 461]
1.1 M CH_3COOH, 0.06 M CH_3COONa, 1.1 M $NaNO_3$	**458** (~ 50)	**512**	554	660	**692** (~ 30)	760	795	830	[3]
0.75 M $K_2C_2O_4$	442	494	550	657	680	712	810	865	[3]
45%-solution of K_2CO_3	**426**	472	**487**	**640**	764	794			[167]

Note: The main absorption bands are indicated in heavy type. Molar extinction coefficients are shown in parentheses.

The spectrophotometric method, as applied to solutions of plutonium (IV), is mainly utilized for the identification of this valency, determination of Pu (IV) in the presence of other valencies and for the study of processes of complex formation.

Table 15 gives the absorption bands of tetravalent plutonium in various media.

Determination of plutonium (VI) by self-absorption

The main absorption bands of nitric acid solutions of Pu(VI) are at 830–835, 958 and 990 mμ. The main absorption maximum of Pu(VI) at 830 mμ begins to disappear in solutions with nitric acid concentration higher than 4–5 M [125, 126]. At the same time a new absorption band appears at 812 mμ [125, 126], or, according to other authors [3, Chap. 9] at 819 mμ, and its intensity increases with the increase in the concentration of nitric acid. This is due to the formation of the anionic complex $PuO_2(NO_3)_3$.

It will be seen from Figure 58 that Pu(VI) may be determined by photometry of its nitric acid solutions at 830 mμ, if the concentration of nitric acid does not exceed 1.5–2.0 *M* (Tarakanov et al., 1953).

The optical density at 830 mμ is affected by the temperature; the variation is 2–3% per degree [348].

A serious drawback of the absorption band at 830 mμ is that Beer's law is not obeyed. One of the reasons for this fact is that the resolving power of a quartz spectrophotometer is not sufficient for the absorption to be determined at the maximum of the band [560, Tarakanov, 1951]. An undoubted advantage of this band is the high value of the molar extinction coefficient ($\sim$ 300), which makes it possible to determine as little as 0.02 mg/ml of plutonium. Moreover, in this part of the spectrum there is least interference due to the absorption of other elements. A method for the determination of plutonium by measuring the absorption of Pu(VI) has been proposed by Tarakanov, Grosheva and Zhelugova (1953).

Into a dry, tared 10 ml pycnometer is placed an aliquot of the sample solution. The volume of the aliquot should be large enough to result in a final concentration of plutonium of 0.02–0.8 mg/ml after the solution has been made up to 10 ml. The pycnometer is then reweighed. Saturated $K_2Cr_2O_7$ in 1 *M* HNO_3 (0.5 ml) is then added, and the bulb of the pycnometer is filled with 1 *M* HNO_3. The pycnometer with the solution is then placed on a water bath at 95 $\pm$ 5° for 10–15 minutes. After the plutonium has been oxidized to Pu(VI), the solution is cooled to 20°, made up to the mark with 1 *M* HNO_3 and the mixture is stirred by passing a stream of air through the capillary in the stopper. Two or three ml of the solution are withdrawn, placed in a 10 mm cell and the optical density of the solution is measured. The concentration of plutonium is read off the calibration curve (Figure 59).

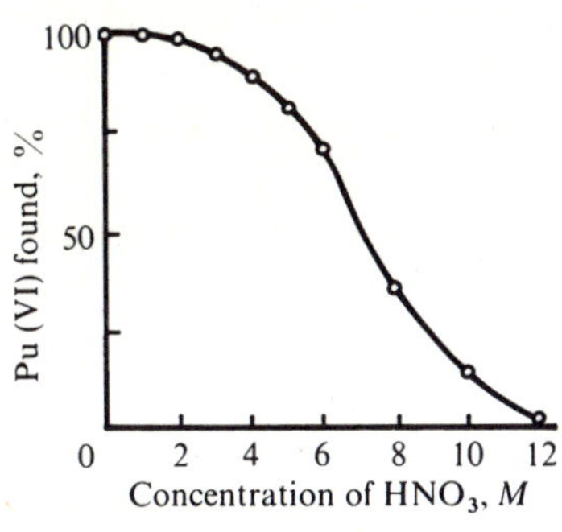

Figure 58

Effect of concentration of HNO_3 on the spectrophotometric determination of Pu(VI) at 800 mμ

The relative error in the determination of 0.02–08 mg of plutonium per ml is $\pm$0.5%.

The photometric determination of 0.02–0.8 mg/ml of Pu(VI) may be carried out in the presence of a 50-fold excess of U(VI) and sodium nitrate, 20- to 30-fold excess of Fe(III), Cr(III) and Pb. Under these conditions the relative error of the determina-

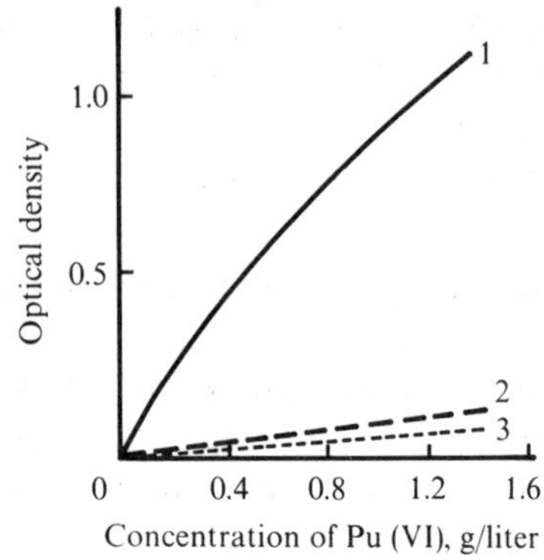

Figure 59

Dependence of the optical density of a solution of Pu(VI) in 1 *M* HNO_3 on the concentration of plutonium. Wavelengths: 1) 835 mμ; 2) 958 mμ; 3) 990 mμ.

tion is $\pm$ 1%. According to Allison (quoted from [547]), hexavalent plutonium at concentrations of 0.5 mg/ml and above may be determined using the 835 mμ absorption band in 10 mm cells.

Determination of a mixture of plutonium in several valency states

Tarakanov (1951) developed a method for the determination of Pu(III) and Pu(IV) and also of Pu(IV) and Pu(VI) in the presence of each other in solution.

Optical densities are measured at wavelengths corresponding to the absorption maxima of Pu(III) (565,* 605, 800 and 915 mμ), of Pu(IV) (476,* 660, 805 and 1085 mμ) and of Pu(VI) (835, 958 and 990 mμ).

The first step in the determination is to construct optical density vs. concentration curves for Pu(III), Pu(IV) and Pu(VI), when each is present alone. These graphs are constructed for a wavelength which is typical for this valency state and also for a wavelength which is typical of another valency state which may be present in solution at the same time. For example, in the analysis of mixtures of Pu(III) with Pu (IV), not containing any extraneous ions, the calibration graphs are plotted for the wavelengths of 565 and 476 mμ (Figure 60). The unknown concentrations of Pu(III) and Pu(IV) are found as follows.

The optical densities *D* of the solution containing Pu(III) and Pu(IV) are measured at 476 and 565 mμ (in Figure 60 they are 1.1 and 0.3 respectively). Since the value of *D*

The bands at 565 and 476 mμ are utilized only in the analysis of pure solutions of plutonium, since in this spectral region the absorption by U(IV) and U(VI), Fe(III), Cr(III) and Cr(VI) and by other elements is large.

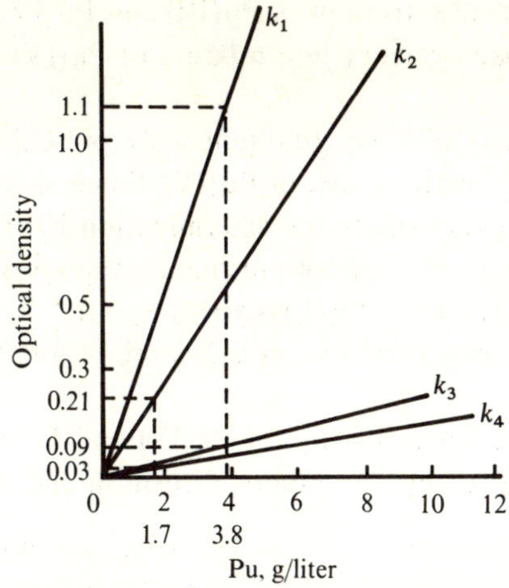

Figure 60

Calibration curves for the determination of Pu(III) and Pu(IV) by differential spectrophotometry:

k_1 Pu(IV) at 476 mμ; k_2 Pu(III) at 565 mμ; k_3 Pu(IV) at 565 mμ; k_4 Pu(III) at 476 mμ.

at 476 mμ is large, it is assumed to represent the absorption of Pu(IV) alone, and the concentration of Pu(IV) is read off the graph k_1 as 3.8 g/liter. According to the graph k_3, this amount of Pu(IV) corresponds to the optical density of about 0.09 at 565 mμ. This value must be deducted from the initial value $D = 0.3$, since the determination of Pu(III) from the graph k_2 (at 565 mμ) will not be accurate unless corrected for the absorption of Pu(IV) at this wavelength. The value $D = 0.21$ yields the concentration of Pu(III) as 1.7 g/liter. The optical density at 476 mμ, produced by this amount of Pu(III) is now read off graph k_4 as about 0.03; this is now deducted from the initial value 1.1. The concentration of Pu(IV) is now more accurately determined on k_1 as corresponding to $D = 1.07$, and the entire procedure is repeated. In this way, by introducing successive corrections into the initially found values of D, the true values of Pu(III) and Pu(IV) can be determined.

If the optical density at 476 mμ is low, the density at 565 mμ is attributed to Pu(III) alone and the mixture is analyzed as above.

The contents of individual valency forms of plutonium may also be obtained by calculation, by solving the following set of equations:

$$\begin{aligned} \varepsilon^{\mathrm{III}}_{\lambda_1} \cdot C_1 + \varepsilon^{\mathrm{IV}}_{\lambda_1} \cdot C_2 &= D_{\lambda_1}, \\ \varepsilon^{\mathrm{III}}_{\lambda_2} \cdot C_1 + \varepsilon^{\mathrm{IV}}_{\lambda_2} \cdot C_2 &= D_{\lambda_2}, \end{aligned} \tag{4}$$

where $\varepsilon^{\mathrm{IV}}_{\lambda_1}$, $\varepsilon^{\mathrm{III}}_{\lambda_1}$ are the molar extinction coefficients of Pu(IV) and Pu(III) at wavelength λ_1; $\varepsilon^{\mathrm{IV}}_{\lambda_2}$, $\varepsilon^{\mathrm{III}}_{\lambda_2}$ are the molar extinction coefficients of Pu(IV) and Pu(III) at the wavelength λ_2; D_{λ_1}, D_{λ_2} are the optical densities at wavelengths λ_1 and λ_2; and C_1 and C_2 are the unknown concentrations of Pu(III) and Pu(IV) in g/liter.

The determination of each valency in a mixture of Pu(IV) and Pu(VI) is carried out in a similar manner.

Three valency forms may also be simultaneously present in solution; this will be the result of disproportionation, say, of Pu(IV). Since in this case the content of Pu(VI) in solution will be very small, the determination by this method is difficult.

In solutions containing 1–10 g/liter of plutonium, the error of the method when applied to mixtures of Pu(III) with Pu(IV) is about ± 5 rel. %. In mixtures of Pu(IV) with Pu(VI), the accuracy of determination is 2–3 rel. % for Pu(IV), and 3–5 rel. % for Pu(VI).

Another method for the determination of Pu(IV) and Pu(VI) in the presence of each other has been proposed by Tarakanov, Grosheva and Zhegulova (1953).

A portion of the sample solution containing 2–8 mg Pu is placed in a 10 ml pycnometer, the solution is made up to the mark with 1 *M* HNO_3 and mixed. The solution is placed in a 5–30 mm cell and the content of Pu(VI) is determined by measuring the absorption at 835 mμ. The solution is then returned from the cell back into the pycnometer, a few crystals of potassium bichromate are added and the solution heated on a water bath at 95 ± 5° for 10–20 minutes. The solution is cooled to 20° and the concentration of total plutonium is determined as Pu(VI). From these data, the percentage content of Pu(VI) in the sample is calculated. The experimental error should not exceed ± 2 rel. %.

METHODS BASED ON REACTIONS WITH COLORLESS AND WEAKLY COLORED ORGANIC REAGENTS IN ORGANIC SOLVENTS

Plutonium is often isolated from irradiated preparations and complex technological solutions by extraction with organic solvents, including tributyl phosphate, thenoyltrifluoroacetone, methyl isobutyl ketone and other solvents. The spectrometric determination of plutonium directly in organic solvents is thus of interest. Figures 61–65 show absorption spectra of Pu(IV) in organic solvents (Milyukova, 1961).

The absorption spectrum of the complex of Pu(IV) with tributyl phosphate $Pu(NO_3)_4 \cdot 2TBP$ has typical absorption maxima at 430, 490 and 660 mμ (Figure 63). At these wavelengths the value of the optical density is practically independent of the concentration of nitric acid in the organic phase up to 0.4 *M*. The absorption remains unchanged for 28 hrs. The concentration of TBP in solution has considerable effect on the optical density. Molar extinction coefficients have their maximum values in 5–15% solutions of TBP in synthine (Figure 64).

Solutions of the complex of Pu(IV) with TBP obey Beer's law within the Pu(IV) concentration range studied, viz., 0.05–1.5 mg/ml.

If the concentration of TBP in solution is accurately known or else is known to be within the 5–15% range, the content of Pu(IV) is found from its absorption at

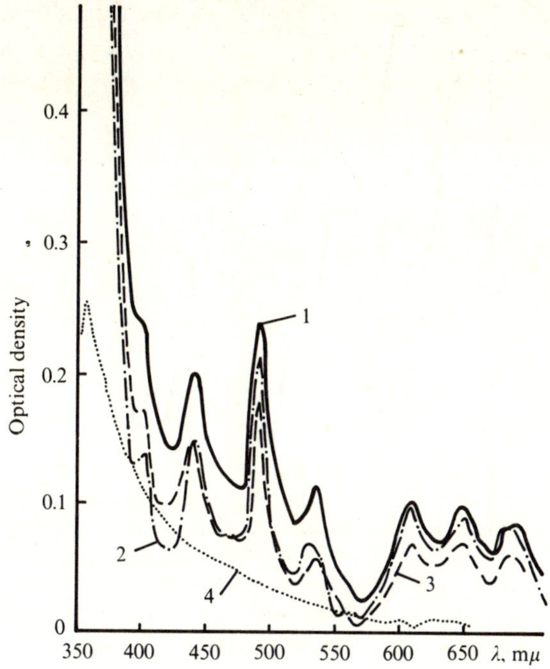

Figure 61

Absorption spectra of $Pu(NO_3)_4$ in organic solvents:

1) methyl ethyl ketone; 2) acetone; 3) diethyl ketone; 4) acetylacetone. The concentration of plutonium is 0.9 mg/ml for 1–3 and 0.015 mg/ml for 4.

490 mμ ($\varepsilon_{490} = 93$). In other cases it is preferable to use the band at 430 mμ, since the molar extinction coefficient does not show considerable variation when the TBP concentration is increased from 15 to 100%.

Milyukova and Palei (1961) proposed a spectrophotometric determination of plutonium in multicomponent solutions, in which the element is preliminarily separated from the impurities by extraction. The optical density of the complex of Pu(IV) in organic phase is measured at 490 mμ, after Pu(IV) has been extracted from 4 *M* HNO_3 with 20% TBP in synthine.

U(VI), Th, Fe(III), Cr(III), Mn(II), Ag, Cu and Ni do not interfere if present in the initial aqueous solution at a concentration of up to 10 g/liter each. Tartaric and citric acids do not interfere at concentrations up to 10 g/liter. The error in determining 0.3–1.0 g/liter of plutonium does not exceed ± 1%.

It is known that tetravalent plutonium is readily extracted into solutions of thenoyltrifluoroacetone (TTA) in benzene (see p. 313). Solutions of plutonium in TTA are strongly colored and can be used for spectrophotometric determination.

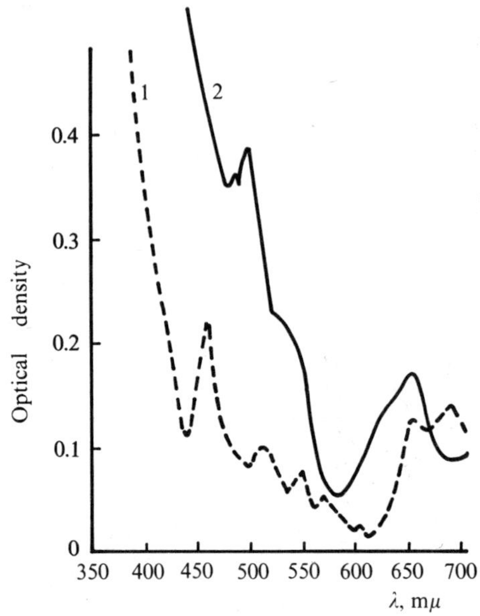

Figure 62

Absorption spectra of $Pu(NO_3)_4$ in organic solvents:

1) formamide; 2) dioxane. Concentration of plutonium is 0.9 mg/ml.

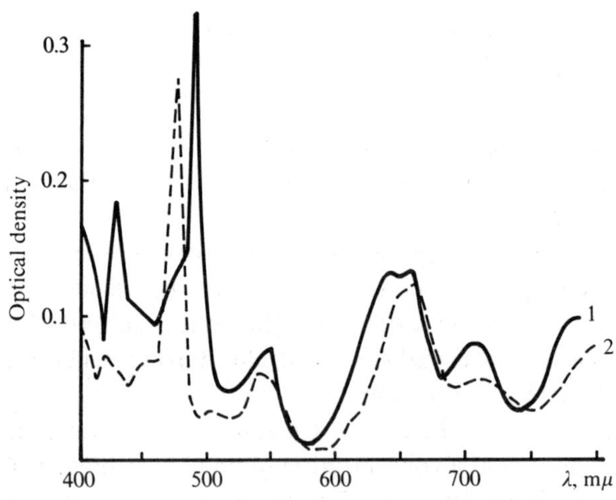

Figure 63

Absorption spectra of Pu(IV) in 20% solution of TBP in synthine:

1) solution of Pu(IV) in TBP containing 0.5 *M* HNO_3; 2) solution of Pu(IV) in 0.5 *M* HNO_3. Concentration of plutonium 0.94 mg/ml.

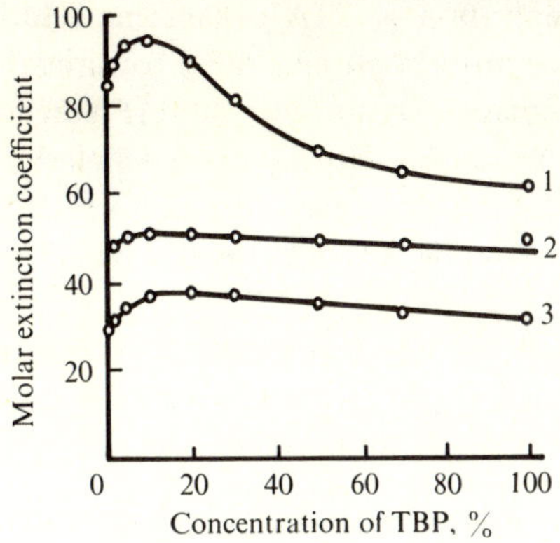

Figure 64

Effect of the concentration of TBP on molar extinction coefficients of Pu(IV) solutions in synthine.

Wavelengths: 1) 490 mμ; 2) 430 mμ; 3) 660 mμ.

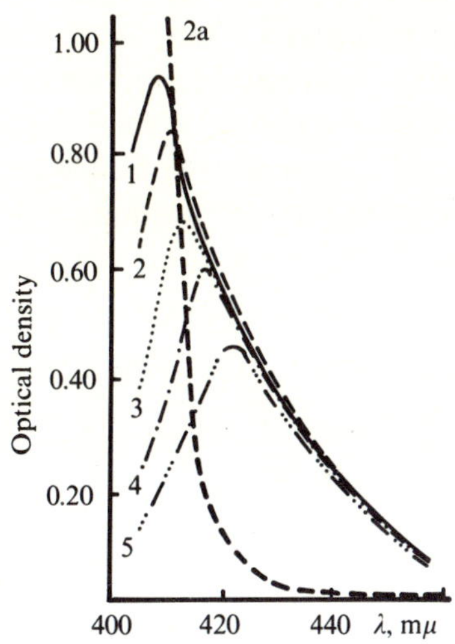

Figure 65

Absorption spectra of benzene solutions of the complex between Pu(IV) and TTA.

Concentration of Pu(IV) 0.024 mg/ml; concentration of TTA: 1) 0.05 M; 2) 0.1 M; 3) 0.2 M; 4) 0.5 M; 5) 1.0 M. Curve 2*a*—0.1 M solution of TTA not containing Pu(IV).

The absorption spectrum of the complex of plutonium(IV) with TTA in 0.1 M solution of TTA in benzene, studied by Milyukova and Palei (1961), has an absorption maximum at 410 mμ (Figure 65). The molar extinction coefficient is 10,000.

An increase in the concentration of TTA in benzene solution from 0.05 *M* and above results in a shift of the absorption maximum towards the long waves accompanied by a considerable decrease in its absolute value (Figure 65). The corresponding difference in the optical densities at 420–430 mμ is smaller as compared with the difference at 410 mμ. For this reason, unless the TTA concentration is maintained rigorously constant, it is preferable to work at 420 or 430 mμ, even though the sensitivity is thereby reduced.

Solutions of the complex of plutonium (IV) with TTA obey Beer's law at concentrations 0–35 μg/ml (410 mμ) and up to approximately 50 μg/ml at 420 and 430 mμ. The color of such solutions is stable for 6 hours.

An aliquot part of the sample solution ($\sim$5 ml), containing 5–30 μg Pu(IV) per ml and 1 mole HNO_3 per liter, is introduced into a 20 ml separating funnel. An equal volume of 0.1 *M* solution of TTA in benzene is added and the sample is extracted during 10 minutes. A 2.5–3 ml volume of the extract is transferred to a 10 mm cell and its optical density at 410 mμ is measured with respect to a 0.1 *M* solution of TTA in benzene. The amount of Pu(IV) is read off a calibration curve.

The extraction-photometric determination of 5.0–30 mg Pu per liter is not interfered with by U(VI) in concentrations up to 2.5 g/liter, or by Th, Cr(III), Al, Ni, Ag, Mn and Cu in concentrations up to 5 g/liter. Ferric iron and Zr are extracted in considerable amounts together with plutonium; the presence of iron increases while that of zirconium decreases the optical density. The accuracy of the determination is not less than $\pm$ 0.5%. The sensitivity of this method is some 100 times higher than that of the determination of plutonium in solutions of tributyl phosphate.

METHODS BASED ON REACTIONS WITH INTENSELY COLORED ORGANIC COMPOUNDS

The reagents utilized in these methods belong to the class of azo dyes and have atomic groupings of the type $R_1—N=N—R_2$:

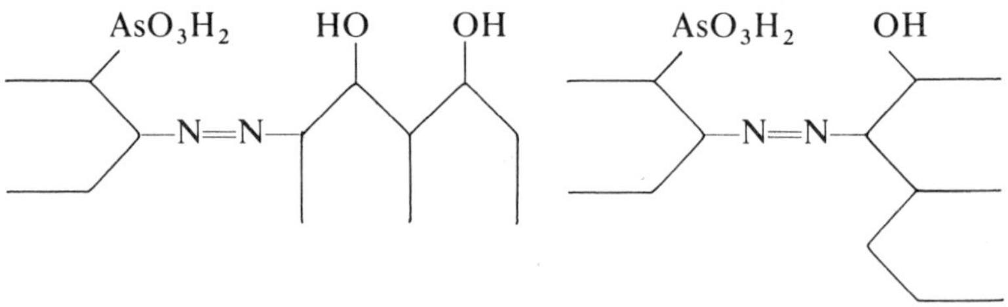

They form fairly stable inner complexes with plutonium ions, especially with Pu(IV), giving conspicuous and highly sensitive color reactions. These reagents, however, are not specific to plutonium, since they also yield colored complexes with other elements.

The most widely used reagents of this type are thoron and its analogs: Arsenazo I, Arsenazo III, etc.

Determination of plutonium (IV) by Thoron I

Thoron I, or benzene-2-arsonic acid-<1-azo-1>-2-hydroxy-naphthalene-3,6-disulfonic acid, was introduced by Kuznetsov [129]; it is a bright red, crystalline powder, soluble in water, dilute acids and alkalis and sparingly soluble in alcohol:

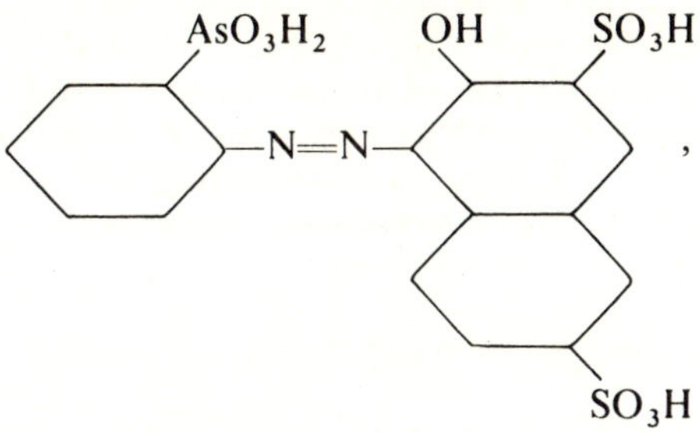

, MW 532.31.

Solutions of thoron are orange-colored. Adamovich and Didenko [1] determined a few physical constants and the molar extinction coefficients of thoron.

In acid media, thoron forms a raspberry-colored complex compound with tetravalent plutonium.

Lapshina and Mikhailov et al. (1956) carried out a spectrophotometric determination of the composition of the complex of Thoron I with plutonium. They found that plutonium (IV) reacts with Thoron I in the ratio 1:1. The reaction constant in 0.1–0.8 M HNO_3 is $(7.1 \pm 3.0) \cdot 10^3$. The compound is most intensely colored at pH 1.5–2.0. At higher acid concentrations the color is less stable on standing. Thus, for instance, in 0.1–0.3 M HNO_3 it remains unchanged during 2–3 hours, while in 0.5–1.0 M HNO_3 this time becomes much shorter. The reagent is probably decomposed by nitrite in more acid solutions.

In hydrochloric and perchloric acid solutions Thoron I gives a color reaction with Pu(IV) in a wider acidity range than in nitric acid.

It will be seen that the formation of the plutonium:Thoron I complex in various media is in accordance with the sequence $HNO_3 > HCl > HClO_4$. In 0.1–1.0 M sulfuric acid a colored Thoron I plutonium compound is not formed, apparently owing to the strongly competing reaction of complex formation with sulfate ions.

Lapshina, Mikhailov et al. (1956) proposed two variants of the photometric determination of Pu(IV) by Thoron I.

Colorimetric method

The reagent is colorimetrically titrated against a standard solution of Pu(IV), until the color matches a sample solution containing the same amount of reagent.

The determination may also be effected in the presence of extraneous elements. At the weight ratios Pu:U(VI) = 1:10; Pu:Fe(III) = 1:2; Pu:Cr(III) = 1:25; Pu:Pb = 1:40; Pu:Mn = 1:50, these elements do not interfere with the determination of plutonium.

Thorium, uranium (IV), lanthanum and nickel also do not interfere with the determination.

Oxidizing and reducing agents such as potassium permanganate, potassium dichromate, hydrogen peroxide, potassium bisulfite and sodium nitrite decompose the complex compound.

The sample solution containing 50–1150 μg of plutonium is placed in one of two colorimetric tubes; 10 ml of 0.2 *M* HNO_3 and 0.3 ml solution of Thoron I (2 g of reagent in 1 liter of water) are added and the solution is mixed. Ten ml of 0.2 *M* HNO_3, and 0.3 ml of Thoron I reagent are placed in the second colorimetric tube and the solution is titrated against a standard solution of plutonium containing 100 μg/ml, until the color of the standard matches that of the sample solution.

The plutonium content C (μg/liter) is calculated from the formula:

$$C_{Pu} = \frac{v_2 \cdot T \cdot 100}{v_1}, \tag{5}$$

where v_1 is the volume of sample, ml; v_2 is the volume of the standard solution consumed in the titration, ml; and T is the titer of the standard solution, μg/ml.

The sensitivity of the reaction is 0.2 μg/ml. The accuracy of the determination is ± 3–5%.

In the presence of major amounts of uranium, uranium is first separated by coprecipitating plutonium with nickel hydroxide in alkaline medium (p. 216). If the uranium content in the sample is not more than 60 mg, uranium is coprecipitated to the extent of 5% under these conditions.

Plutonium may also be determined by Thoron I by the standard scale method. A number of solutions are prepared (standard scale) which contain a fixed amount of the reagent and variable amounts of plutonium, and the colors are matched against a sample solution containing the same amount of reagent. Since the complex is not stable on standing, the standard scale must be freshly prepared every 2–3 hrs.

It is more convenient to use a standard scale prepared with thorium, which forms a similar, but more stable complex with Thoron I. The color intensity given by 0.8 μg Th corresponds to 1 μg Pu, by 2.4 μg Th to 3.0 μg Pu, etc. This scale remains constant for one month. The amounts of plutonium which can be determined by this method are 1–6 μg Pu in 5 ml solution.

Spectrophotometric method

The compound between plutonium (IV) and Thoron I gives an absorption spectrum with a maximum at 510 mμ (Kuznetsov and Mikhailov, 1955). If the comparison solution consists of a solution of Thoron I of the same concentration as that in the sample solution, the absorption curve (Figure 66) will have a maximum at 540–550 mμ. The molar extinction coefficient at 545 mμ is 9000 $\pm$ 1000, and according to other authors (quoted from [547]) 11,000.

Equilibrium in solution is established within 5 minutes, the color intensity remains constant for one hour.

The determination of Pu(IV) with Thoron I consists in the determination of the optical density of a solution of plutonium containing 0.2 M HNO_3 and $1.2 \cdot 10^{-4}$ M Thoron I, at 545 mμ. The comparison solution is identical except that Pu is absent.

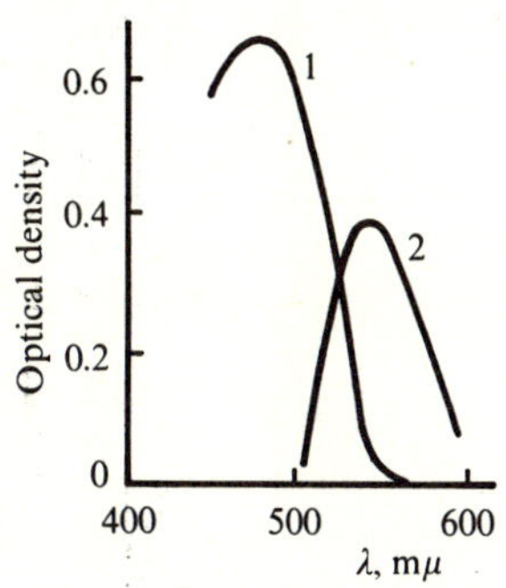

Figure 66

Absorption spectra of nitric acid solutions of Thoron I and its complex with Pu(IV):

1) $8 \cdot 10^{-5}$ M solution of Thoron I in 0.2 M HNO_3 (water used as comparison solution); 2) $8 \cdot 10^{-5}$ M solution of Thoron I, containing $5.4 \cdot 10^{-5}$ M Pu(IV) (comparison solution: $8 \cdot 10^{-5}$ M Thoron I in 0.2 M HNO_3).

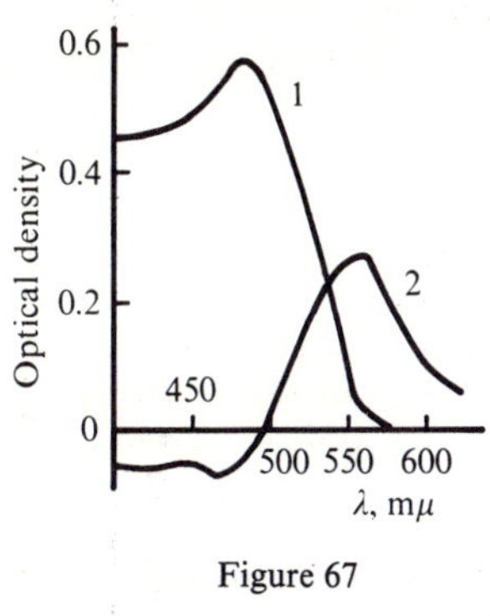

Figure 67

Optical density of hydrochloric acid solutions of Thoron II and its complex with Pu(IV):

1) $3.8 \cdot 10^{-5}$ M solution of Thoron II in 2 M HCl (comparison solution: water); 2) $3.8 \cdot 10^{-5}$ M solution of Thoron II in 2 M HCl, containing $4.1 \cdot 10^{-5}$ M Pu(IV) (comparison solution: $3.8 \cdot 10^{-5}$ M Thoron II in 2 M HCl).

The concentrations of plutonium which can be determined by this method are of the order of a fraction of a milligram per liter.

Determination of plutonium (IV) by Thoron II

As compared with Thoron I, a more selective reaction with Pu(IV) is given by Thoron II, synthesized by Kuznetsov and Savvin. Thoron II reacts with tetravalent cations in more acid media than Thoron I, and is therefore more selective.

Thoron II (a dimer-like analog of Thoron I) is a red powder, readily soluble in water. Its structural formula is

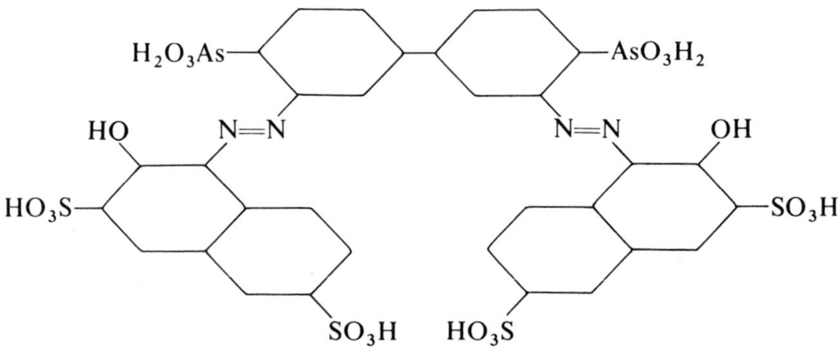

The reaction between Thoron II and tetravalent plutonium in acid solutions is accompanied by a color change from orange to raspberry red. Pu(IV) and Thoron II react in the ratio 1:1 (Kuznetsov and Mikhailov, 1959).

The absorption spectrum of the complex, given in Figure 67, is distinguished by a maximum at 555 mμ. The molar extinction coefficient of Pu(IV)–Thoron II complex is about 10,000. Solutions in 2 M HCl obey Beer's law between 1–10 mg/liter. The color intensity of the complex becomes stable within one minute. The sensitivity of the reaction is 0.6 μg/ml.

To a portion of the sample solution containing 50–100 μg Pu is added one drop of Rongalite solution (500 g/liter) to convert all the plutonium into the trivalent state. After one minute 2–3 drops concentrated HNO_3 are added and the solution is heated for 2–3 minutes on a water bath to decompose Rongalite and to oxidize Pu(III) to Pu(IV). The solution is diluted to 10 ml 2 M HCl, 0.2 ml solution of Thoron II (2 g/liter) is added and the mixture is held for 2 minutes. The optical density is then measured at 555 mμ in a 20 mm cell against a 2 g/liter solution of the reagent in 2 M HCl. The concentration of plutonium is read off a calibration curve.

Uranium (VI) (120-fold amount), chromium (III) (7-fold amount), lanthanum (20-fold amount), lead (II) (2500-fold amount) and Fe(III) (2-fold amount) do not interfere with the determination.

Oxalate, sulfate and nitrate ions do not interfere when present in weight ratios of 1:25, 1:25 and 1:5000 respectively, with respect to plutonium.

Determination of plutonium (IV) by chlorophosthoron

Chlorophosthoron is an analog of thoron; it has a phosphonic instead of an arsonic group and one atom of chlorine in the benzenic ring in the para-position:

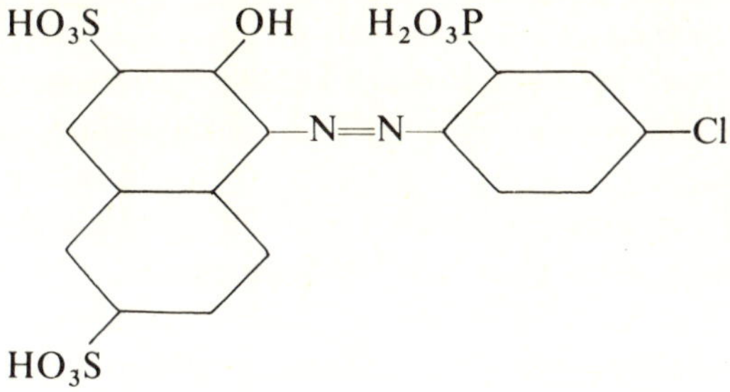

Neutral solutions of the reagent are orange-colored, and the color remains unaltered on acidification. An increase in the concentration of HNO_3, HCl, $HClO_4$, H_2SO_4 to 6 *M* results in the appearance of a pink tinge.

Palei, Kochetkova and Nemodruk (1961) noted that the solution turns red when chlorophosthoron reacts with tetravalent plutonium. From the point of view of color development, the optimum pH range is 1.5–4.0.

The absorption spectrum of a solution of the complex formed between plutonium (IV) and chlorophosthoron has a maximum at 515 mμ; the corresponding molar extinction coefficient is 17,300.

Because of the considerable overlapping of the absorption curves of the complex and the reagent, the determination is effected not at 515 mμ, but at 550 mμ, where the optical absorption by the reagent is insignificant.

The reaction between Pu(IV) and chlorophosthoron is more sensitive than the reactions with Thoron I and Thoron II, but has not yet been thoroughly studied.

Determination of plutonium (IV) by Arsenazo I

As compared with the reagents of the thoron type, a more sensitive reaction with Pu(IV) is given by Arsenazo I (benzene-2-arsonic acid-< 1-azo-2 >-1,8-dihydroxynaphthalene-3,6-disulfonic acid), proposed by Kuznetsov [129]:

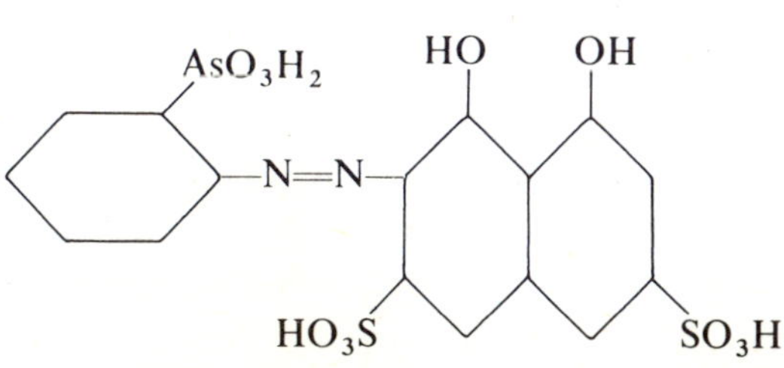

Aqueous solutions of Arsenazo I are pink-colored.

It was shown by Klygin and Pavlova in 1955 [109] that Arsenazo I reacts with tetravalent plutonium at pH 0.8–1.6 to yield a blue complex of the composition 1:1. Equilibrium is established some 5–10 minutes after the solutions have been mixed, and the color remains unchanged for 30–40 minutes. The complex has an absorption maximum at 590–600 mμ, with molar extinction coefficients $\varepsilon_{590} = 22{,}000$ and $\varepsilon_{600} = 21{,}700$. If the excess of Arsenazo I is sufficiently large, the optical density remains constant within a wide pH range (2–6), so that plutonium may be spectrophotometrically determined [109]. An aliquot of the solution of plutonium (IV) chloride or nitrate, containing between 1 and 100 μg of plutonium, is placed in a 50 ml volumetric flask. Five ml of 10^{-3} *M* Arsenazo I are then added, the pH is adjusted to 2–5 with the aid of ammonia, the solution is diluted to the mark, and the optical density is measured after 5–40 minutes at 590 mμ using a 50 mm cell. The sensitivity of the method is 0.02 μg Pu/ml.

Determination of plutonium (IV) by Arsenazo II

The reagent Arsenazo II, or biphenyl-4,4′-diarsonic acid-3,3′-bis-(<-azo-2> 1,8-dihydroxynaphthalene-3,6-disulfonic acid) was proposed by Kuznetsov and Savvin [130].

Arsenazo II and tetravalent plutonium forms a complex compound with plutonium: Arsenazo II = 2:1. The compound is violet-colored (Pavlova, 1958). The maximum development of the color intensity of the complex is noted at pH 2–4. At pH around 1.72 the equilibrium is established instantaneously and the color intensity of the complex remains unchanged for 1–2 hours; at pH 1.02 the equilibrium is only attained after 15–20 minutes and the color intensity remains constant only during 30–40 minutes.

The absorption spectrum of the complex of plutonium (IV) with Arsenazo II has a typical absorption maximum at 605–610 mμ with a molar extinction coefficient of 11,800.

A comparison of the molar absorption coefficients of the complexes of plutonium with Arsenazo I and with Arsenazo II shows that the sensitivity of the determination by Arsenazo I is almost twice that of the determination by Arsenazo II.

Determination of plutonium (IV) by Arsenazo III

Arsenazo III, or 1,8-dihydroxynaphthalene-3,6-disulfonic acid-2,7-bis-(<-azo-2>-phenylarsonic acid) was proposed by Savvin in 1959 [197]. It is a crystalline, dark-red powder. The solution of the reagent in a weakly acid medium is pink, in concentrated sulfuric acid it is green and in alkalis it is blue. The absorption curve has a maximum at 540 mμ.

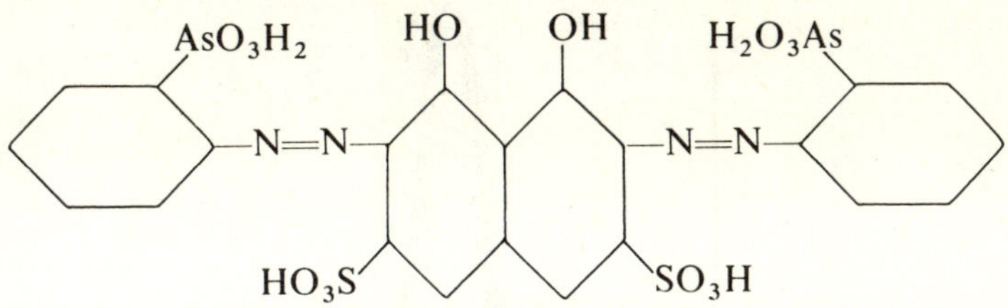

When Arsenazo III is added to an acid solution of tetravalent plutonium, a bluish-green complex compound is formed (Kuznetsov, Savvin, Mikhailov, Lapshina et al., 1960).

The maximum color development of the complex is attained in the presence of a threefold (molar) excess of the reagent. The composition of the complex formed under these conditions corresponds to the molar ratio plutonium:Arsenazo III = 1:1. The absorption spectrum of the solution has two absorption peaks, at 620 mμ and 665–670 mμ (Figure 68).

Since the absorption by the reagent at 620 mμ is large, the absorption band at 665–670 mμ (ε_{665} = 67,000) is employed in photometric work.

The optical density of plutonium–Arsenazo III complex largely depends on the acidity of the solution and on the nature of the acid. According to Palei, Kochetkova and Nemodruk (1961), in nitric acid solutions the maximum optical density is noted at high acid concentrations: as the nitric acid concentration is increased from 1 to 4 *M*, the optical density increases sharply, between 4 and 5 *M* the increase is slight, and remains constant between 5 and 7 *M* HNO_3 (Figure 68). The molar extinction coefficient in 4–7 *M* solutions is 136,000.

Since the color reaction is most intense in strongly acid media, the determination of Pu(IV) is highly selective. It is not interfered with by Mo(VI), U(IV), W(VI), Al, Cu, Ni or by other elements (Palei, Kochetkova and Nemodruk, 1962).

According to Kuznetsov, Savvin, Mikhailov, and Lapshina et al. (1960), the determination of plutonium may be effected in the presence of 10-fold amounts of U(IV), 20-fold amounts of Fe(III) and 1000-fold amounts of Cr(III) at a concentration of HNO_3 of 2.7 moles/liter. In 5 *M* HNO_3 5-fold amounts of U(VI) and Fe(III) and 1000-fold amounts of Cr(III) are permissible. The effect of Th, U(IV), and Zr, which, like plutonium, react with Arsenazo III in strongly acid media, is particularly strong.

The determination of plutonium may be effected in the presence of major amounts of oxalic acid.

Palei, Kochetkova and Nemodruk (1962) propose the following procedure for the photometric determination of Pu(IV) by Arsenazo III.

An aliquot of the sample solution, containing 0.1–30 μg of plutonium (IV), is placed in a 10 ml volumetric flask, and enough 10–12 *M* HNO_3 (treated with urea to decompose nitrogen oxides and nitrous acid, which oxidize the reagent) is added to ensure an acidity of 5–7 *M* in the final volume. Arsenazo III (1.2 ml of 0.001 *M* solution) is then added, the solution made up to the mark with water and the optical density

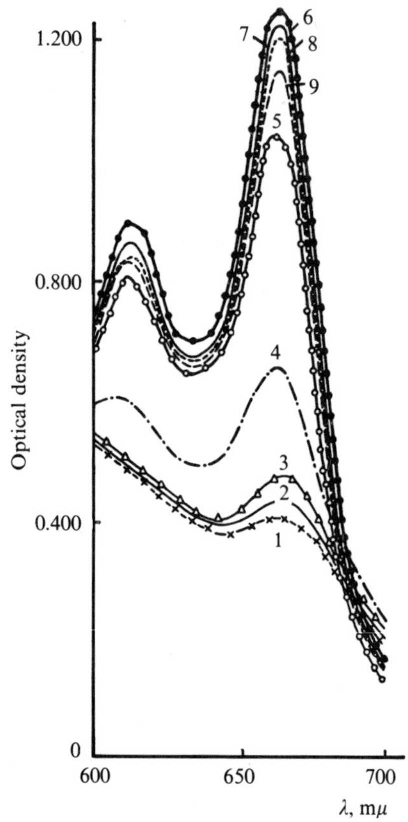

Figure 68

Absorption spectra of nitric acid solutions of Pu(IV) with Arsenazo III complex.

Concentration of Pu(IV) $0.92 \cdot 10^{-5}$ *M*; concentration of Arsenazo III $2 \cdot 10^{-4}$ *M*; concentration of HNO_3: 1) 0.1 *M*; 2) 0.5 *M*; 3) 1.0 *M*; 4) 2.0 *M*; 5) 3.0 *M*; 6) 4.0–7.0 *M*; 7) 8.0 *M*; 8) 9.0 *M*; 9) 10.0 *M*.

measured at 665 mμ in a 10 mm cell. The comparison cell is filled with Arsenazo III in the same concentration and with the same acid content as in the sample solution. The concentration of plutonium is read off a calibration curve. For convenience in working, two calibration curves are constructed: one for solutions with 0.01–0.25 μg/ml of plutonium in the final volume, the other for concentrations of plutonium between 0.1 and 3 μg/ml.

Under the experimental conditions used for the determination of plutonium, most elements, except thorium, zirconium, hafnium, U(IV) and partly U(VI), give no color reaction with Arsenazo III. The interference of small amounts of U(IV) is readily eliminated by oxidizing the element to U(VI).

Chmutova (1960) developed a photometric method for the determination of plutonium, in which plutonium is reextracted from tributyl phosphate solution with aqueous solution of Arsenazo III and the optical density is then determined.

The sample solution, which should be 4–6 M in HNO_3, is shaken with an equal volume of 23% solution of TBP in synthine. After phase separation, 10 ml of the organic solution containing 8–10 μg Pu is placed in a separating funnel, 25 ml of 0.05% solution of Arsenazo III in 3 M HNO_3 are added, and the solution is shaken for 2 minutes. The aqueous solution is placed in a 20 mm cell and the optical density is determined with a FEK-M instrument with No. 4 filter. The comparison cells are filled with a solution of Arsenazo III in 3 M HNO_3, previously brought into contact with a solution of TBP in synthine. The content of plutonium is read off a calibration curve.

The determination of plutonium may be effected in the presence of 60-fold amounts of U(VI) and Fe(III), 23-fold amounts of Cr(III) and 1000-fold amounts of nickel in the aqueous solution prior to extraction.

Arsenazo III and its analogs are readily soluble not only in water, but also in a number of organic solvents with a high dielectric constant, such as formamide, dimethylformamide and diethylformamide (Mikhailov, 1961).

Mikhailov (1961) found that complex formation between tetravalent plutonium and Arsenazo III also takes place in the organic solvents just listed. The color reaction also readily develops in their mixture with the extractants of plutonium.

Table 16 gives data on the miscibility of a number of extractants with dimethylformamide (DMFA), formamide (FA) and diethylformamide (DEFA). The table may be used to select the pair of solvents to be employed for photometric determination of Pu as complex with Arsenazo III or its analogs in nonaqueous media.

The absorption spectrum of the complex of plutonium with Arsenazo III in mixtures such as TBP and DMFA is not very different from the absorption spectrum in aqueous solutions, with only a very small blue shift.

The presence in organic solvents of nitric acid in concentrations above 0.6–0.8 M strongly interferes with photometric determinations. The concentration of HNO_3 should be maintained at 0.3 M, since at this acidity the color intensity of the plutonium–Arsenazo III complex will remain constant.

On the strength of his results, Mikhailov proposed a photometric method of determination of Pu(IV) by Arsenazo III in TBP solutions. The optical density of a 20% solution of TBP in CCl_4, which contains plutonium and to which a solution of Arsenazo III in dimethylformamide has been added, is determined.

Mikhailov performed his measurements on a FEK-M instrument with a red filter, using a 30 mm cell, and an empty cell as reference.

Determination of plutonium (IV) by reagents of the chlorophosphonazo group

Reagents of the chlorophosphonazo group, first prepared by Nemodruk, Novikov, Lukin, Kalinina [172, 173], contain a phosphonic instead of an arsonic group, as well as chlorine atoms in the benzenic ring in the para-position.

Table 16

MISCIBILITY OF ORGANIC SOLVENTS WITH FORMAMIDE, DIMETHYLFORMAMIDE AND DIETHYLFORMAMIDE
(Mikhailov, 1961)

Solvent	FA	DMFA	DEFA
Water	+	+	+
Hexone	×	+	+
TBP	−	+	+
Chlorex	−	+	+
Chloroform	−	+	+
Synthine	−	−	+
Benzene	−	+	+
Toluene	−	+	+
o-, m- and p-Xylenes	−	+	+
Ethyl ether	−	+	+
Ethanol	+	+	+
Methanol	+	+	+
Acetone	+	+	+

Note: The sign + means that the solvents are miscible in all proportions; the sign − means that the solvents are not miscible; the sign × means that the solvents are partly miscible.

The substitution of the phosphonic for the arsonic group results in a shift of the optimum intensity of the color reactions towards lower pH values, which indicates that the complexes formed are stronger as compared to those formed with Arsenazo type reagents.

Palei, Kochetkova and Nemodruk (1961) studied the reactions between tetravalent plutonium and the reagents of the chlorophosphonazo group.

Chlorophosphonazo I or 2-(4-chloro-2-phosphonobenzeneazo)-1,8-dihydroxy-naphthalene-3,6-disulfonic acid

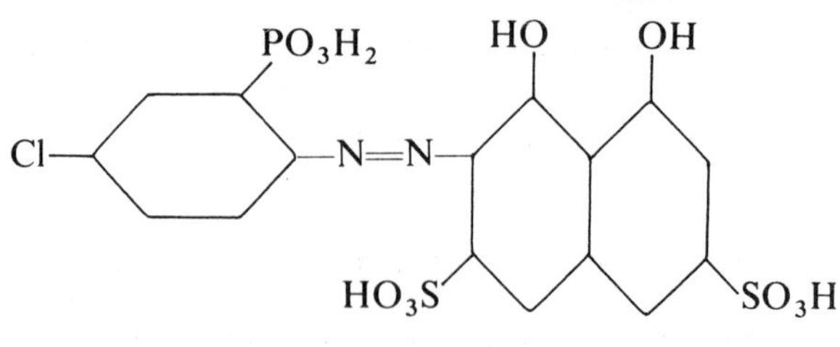

forms a complex compound with tetravalent plutonium; the reaction is accompanied by a color change from the red of the reagent to the blue of the complex.

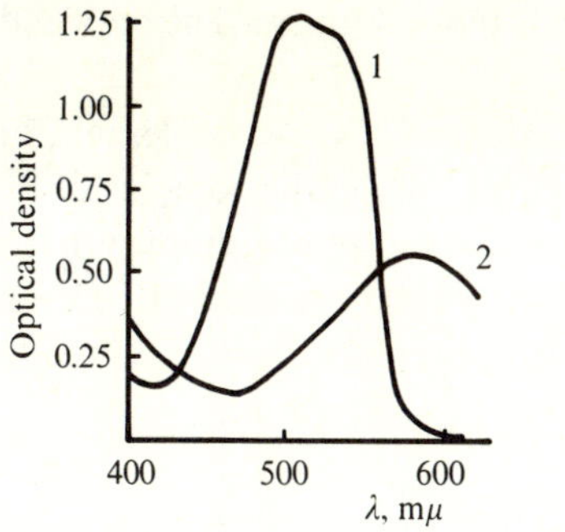

Figure 69

Absorption spectra of solutions of Chlorophosphonazo I and its complex with Pu(IV):

1) $4 \cdot 10^{-5}$ *M* solution of Chlorophosphonazo I; 2) $4 \cdot 10^{-5}$ *M* solution of Chlorophosphonazo I, containing $6 \cdot 10^{-5}$ *M* Pu(IV), pH 3.

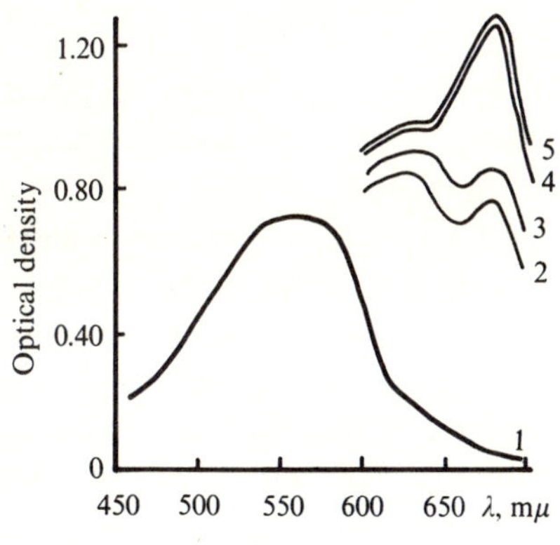

Figure 70

Optical density of nitric acid solutions of Chlorophosphonazo III and its complex with Pu(IV):

1) $1 \cdot 10^{-5}$ *M* solution of Chlorophosphonazo III; 2, 3, 4, 5) $1 \cdot 10^{-5}$ *M* solutions of Chlorophosphonazo III containing $2 \cdot 10^{-5}$ *M* Pu(IV) and 0.5, 1.0, 2.0 and 3.0 *M* HNO_3 respectively.

The absorption spectrum of the complex compound of plutonium with Chlorophosphonazo I has an absorption maximum at 580 mμ (Figure 69). The authors found that a variation in pH between 1.5 and 4.0 does not affect the height of the maximum or the shape of the absorption curve. A comparison of the color reactions of Pu(IV) with Chlorophosphonazo I and with Arsenazo I shows that in the former case the optimum pH range for the determination is somewhat wider (by about one pH unit) and that the width increment corresponds to lower pH values. This is an indication that the reaction is more selective. The optical density remains unchanged for more than two hours.

The molar extinction coefficient of the complex of plutonium (IV) with Chlorophosphonazo I is slightly higher than of the complex of Pu (IV) with Arsenazo I, and is 24,360, i.e., the reaction with Chlorophosphonazo I is more sensitive.

As regards its structure, Chlorophosphonazo III is similar to Arsenazo III, except that its molecule contains two phosphonic groups instead of two arsonic groups, and also two chlorine atoms in the para-position.

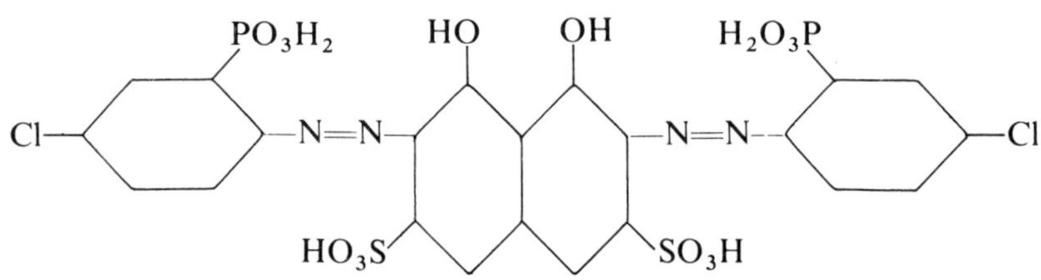

Aqueous solutions of Chlorophosphonazo III are reddish violet. When the pH of the reagent solution is increased above 3 the color turns violet, while in very acid solutions the color becomes bright green.

When Chlorophosphonazo III is made to react with Pu (IV), the reddish-violet color of the reagent changes to the green color of the complex. The absorption curve of the complex between plutonium (IV) and Chlorophosphonazo III displays two maxima, at 630 and 680 mμ. When the acidity of the solution is raised from 0.5 to 2.0 M HNO_3, the light absorption around these two maxima increases rapidly, the rise in the absorption at the second maximum being somewhat more rapid. Further increase in acidity from 2 to 5 M (Figure 70) has no significant effect on the absorption spectrum. It will be seen that the effect of the acid concentration on the complex formation of Pu (IV) with Chlorophosphonazo III is similar to that with Arsenazo III. The main difference is that with Arsenazo III constant values of optical densities are attained at acidities of 3 M or more, while in the case of Chlorophosphonazo III stability is attained at an acidity as low as 2 M.

The authors proposed to use the 680 mμ band for the determination of plutonium by Chlorophosphonazo III, since the absorption of the reagent alone is insignificant in this range. The molar extinction coefficient of the complex in 2–4 M solutions of HNO_3 at 680 mμ is 125,000, i.e., the reaction of plutonium with Chlorophosphonazo III is one of the sensitive color reactions. In addition, the reaction is highly specific. In 3–5 M nitric acid solutions Chlorophosphonazo III reacts only with tetravalent ions: Ce^{4+}, Ti^{4+}, Zr^{4+}, Hf^{4+}, Th^{4+}, U^{4+}.

Determination of plutonium (IV) by Alizarin S

The reagent Alizarin S (1,2-dihydroxyanthraquinone-3-sulfonic acid sodium salt) forms a complex compound with tetravalent plutonium (quoted from [547]). The absorption spectrum of the complex has a typical maximum at 530 mμ, which may be utilized in the determination of plutonium. The value of the molar extinction coefficient of the complex ($\varepsilon_{530} = 9800$) makes it possible to determine up to 2 μg/ml of plutonium. The experimental error at 2.5 μg Pu per ml is 1.84%, while at 15 μg/ml it is 0.46%.

The disadvantage of the method is its low specificity.

METHODS BASED ON COLOR REACTIONS IN THE SOLID PHASE

Kuznetsov and co-workers were the first to propose the use of solid phase color reactions with organic reagents for a colorimetric determination of plutonium.

When the reagent *Rhodamine 3B* is made to act on Pu(IV), a suspension of a sparingly soluble salt is formed (Kuznetsov and Pal'shin, 1957). The precipitation is accompanied by a color change from the pink color of the reagent to blue or violet. The color intensity remains constant at nitric acid concentrations in solution from 1.5 to 3.0%, and then rapidly decreases as the concentration of HNO_3 is increased.

It was found that the reaction with Rhodamine 3B is most complete in a solution containing 2% HNO_3, 30% NH_4NO_3, 0.5% urea, 0.03% gelatine and 0.090% Rhodamine 3B.

Pal'shin suggested that the colorimetric determination of plutonium by Rhodamine 3B could be effected by matching against an imitation scale consisting of Rhodamine 3B, methyl orange and methyl violet.

Since the reference solutions are unstable and have a reddish tinge, it is more convenient to use the standard scale proposed by Mikhailov and Lapshina (1958).

The standard scale is freshly prepared for each determination. To a mixture containing 2% nitric acid, 30% ammonium nitrate and 0.03% gelatine, varying amounts of plutonium, 0, 1, 3, 5, 7, and 10 μg, are added. The scale is stable for 30–40 minutes.

Amounts of 1–10 μg Pu may be determined in the presence of less than 0.5 mg uranium(VI), less than 1.0 mg Fe(III) and Cr(III), less than 4.0 mg Ni, less than 5.0 mg aluminum and lanthanum. Phosphates, sulfates and organic complexing agents interfere considerably.

The absorption spectrum of the compound of Pu(IV) with Rhodamine 3B has an absorption maximum at 605 mμ and a minimum at 560 mμ. However, solutions of the complex do not obey Beer's law. Spectrophotometric determination of Pu(IV) by Rhodamine 3B has yet another inconvenience: the reproducibility of the results is low; variations between single results may attain $\pm$ 30–50%.

A similar solid phase reaction with Pu(IV) is given by the reagent *butylrhodamine* proposed by Kuznetsov and Bol'shakova [137]. The resulting suspension has a blue-violet coloration, whose maximum development takes place 1–2 minutes after the addition of the reagent. The experimental procedure is the same as with Rhodamine 3B. The sensitivity of the reaction is 0.5 μg Pu per milliliter.

TITRIMETRIC METHODS

Despite the relative simplicity of titrimetric methods, which are highly important in the analytical chemistry as a whole, their relative importance in the analytical chemistry of plutonium is as yet only secondary. One reason for this is undoubtedly the existence of the radiometric determinations, which are very simple and selective. The limitations of the radiometric methods due to the dependence of the results on the isotopic composition of plutonium make it necessary to have recourse to other methods, including titrimetry.

As is the case for other elements, titrimetric methods of determining plutonium are very suitable for the analysis of relatively pure plutonium-containing products. They are chiefly used in control of the final stages of processing nuclear fuel, and for the standardization of solutions of plutonium prepared in the laboratory.

OXIDATION-REDUCTION TITRATIONS

Since plutonium has three stable oxidation states, two kinds of redox titrations are possible, viz., Pu(III)/Pu(IV) and Pu(IV)/Pu(VI). In a one-molar solution of an acid, they may be described by the following equations:

$$Pu^{3+} = Pu^{4+} + e^- \ (E_0 = +0.94\text{V}), \tag{1}$$

$$Pu^{4+} + 2H_2O = PuO_2^{2+} + 4H^+ + e^- \ (E_0 = +1.04\text{V}). \tag{2}$$

It has been pointed out (p. 50) that reaction (1) is readily reversible and is therefore the more convenient of the two in titrimetric work. Reaction (2) is slow and may proceed at an acceptable rate only at higher temperatures; it consists of several stages and is strongly affected by the nature of the oxidant or the reductant. On the other hand, reaction (2) may serve for the determination of plutonium in the presence of the most frequently encountered impurities—uranium and iron.

Since plutonium is present in solution, as a rule, in several coexisting valency states, titrimetric determinations usually comprise a preliminary treatment of the sample solution to convert all valency forms to one single degree of oxidation.

The main factor determining the course of a redox titration is the ratio of the formal potentials of the pairs of ions participating in the reaction. The titration of plutonium may be generally represented as

$$Pu_{red} + A_{ox} \rightleftharpoons Pu_{ox} + A_{red} \tag{3}$$

Since at equilibrium the potentials E of the interacting ion pairs are equal, i.e., the reaction potential is zero, we may write

$$E = E_0(Pu) + \frac{0.059}{n} \log \frac{[Pu_{ox}]}{[Pu_{red}]} = E_0(A) + \frac{0.059}{m} \log \frac{[A_{o}]}{[A_{red}]} \tag{4}$$

or

$$E_0(A) - E_0(Pu) = 0.059 \log \frac{[Pu_{ox}]^{1/n}[A_{red}]^{1/m}}{[Pu_{red}]^{1/n}[A_{ox}]^{1/m}}, \tag{5}$$

where E_0(A) and E_0(Pu) are the formal potentials of the ion pairs of the compound A and plutonium respectively, and m and n are the numbers of electrons required to oxidize or reduce the ions of A and plutonium respectively.

At the end point of the titration the number of equivalents of plutonium should be equivalent to the number of equivalents of the titrant:

$$\frac{[Pu_{ox}] + [Pu_{red}]}{n} = \frac{[A_{ox}] + [A_{red}]}{m}. \tag{6}$$

In view of the stoichiometry of reaction (3), we have:

$$\frac{[Pu_{ox}]}{n} = \frac{[A_{red}]}{m}$$

and at the equivalence point the concentration ratios are equal:

$$\frac{[Pu_{ox}]}{[Pu_{red}]} = \frac{[A_{red}]}{[A_{ox}]} = k. \tag{7}$$

Even though the equality (6) may be attained for any values of the formal potentials of the two pairs of ions at equilibrium, for purposes of titration the degree of completion of the reaction must be sufficiently high. If this is not the case, the potential jump at the end point will be small, and the point will not be identifiable with sufficient accuracy by potentiometric or any other kind of titration. The larger the difference between the concentration ratio k and unity, the higher—in accordance with Nernst's equation—the change in potential per unit amount of titrant added. It would appear that the potential change is sufficiently sharp only if the value of k is not less than 100.

At the equivalence point, equation (5) may be written as:

$$E_0(\mathrm{A}) - E_0(\mathrm{Pu}) = 0.059\left(\frac{1}{n} + \frac{1}{m}\right)\log k.$$

It follows from the equation that the formal potentials of the two pairs of ions in a given medium, at $k = 100$, for, say, 1–1 electron reactions must differ by at least 0.24 V. Most titrants which have been successfully employed in the determination of plutonium satisfy this condition, i.e., have potentials which differ to a much greater extent from the potential of the pair of plutonium ions in question. In this way the equivalence point may be more accurately found.

The position of the equivalence point on the potential coordinate will depend on the formal potentials of the reacting compounds and may be calculated from the expression:

$$E = \frac{mE_0(\mathrm{A}) + nE_0(\mathrm{Pu})}{m + n},$$

which follows from equations (4) and (7).

The above considerations will govern the choice of the titrant, provided the reaction is sufficiently fast for the titration to be possible. If the reaction is slow, it may be carried out at elevated temperatures, or else back-titration may be resorted to.

Another important factor which determines the accuracy of the titrimetric determination is the way in which the end point is determined. In the methods described below the techniques used include mainly potentiometric zero-current determinations of the equivalence point, and the use of redox indicators. Potentiometric titrations at set current intensity with two polarized electrodes are rarely used, while potentiometric titrations at set current intensity with one polarized electrode, and amperometric titrations at constant voltage with two polarized electrodes are not used at all [82]. The apparatus required for the last three methods is simple, and the methods may be convenient for end point detection in low reversibility systems.

Determination of plutonium (III)

The titration of the trivalent to the tetravalent plutonium is the most popular technique for titrating plutonium. Cerium (IV) sulfate is most often used as oxidant. Methods using potassium bichromate, potassium permanganate and ammonium vanadate are also known. With the exception of the special case of analysis of metallic plutonium, the dissolution of which in hydrochloric, perchloric or sulfuric acid yields a solution of trivalent plutonium directly, the sample solutions must be preliminarily treated with reducing agents to convert all the plutonium to the trivalent state. Such reducing agent was at first sulfur dioxide, but its excess may be removed from solution only by prolonged bubbling of an inert gas through the solution. It is preferable to use metals, including zinc and its amalgams, bismuth, cadmium and silver. In micro-determinations sample solutions may be reduced with bivalent chromium, the excess of which can be readily oxidized by atmospheric oxygen. The use of solutions of trivalent titanium involves the determination of two equivalence points. Most determinations are carried out in sulfuric acid media.

Titrations with cerium (IV) *sulfate*

The reaction between Pu^{3+} and Ce^{4+} is rapid and may be described by the equation

$$Pu^{3+} + Ce^{4+} \rightleftharpoons Pu^{4+} + Ce^{3+}.$$

One of the first workers to utilize the reaction was Koch [501]. Plutonium (about 2.5 mg) is reduced in a zinc reductor in 1 N H_2SO_4, and 10 ml of the solution thus obtained are then titrated against 0.0045 M cerium (IV) sulfate. The indicator is a 0.025 M solution of ferroin (complex of Fe(II) with o-phenanthroline). The standardization of plutonium solutions was carried out by radiometric and gravimetric methods. The results agreed to within 0.1%. The average deviation in eight experiments was ± 0.25%.

The author also checked the possibility of cerium (IV) sulfate producing an oxidation of cerium beyond the tetravalent stage. In the presence of a threefold excess of cerium in 1 N and 5 N H_2SO_4, after two hours standing, only 8 and 7% more plutonium than the true content was found. Thus, over-oxidation of plutonium is rather slow and does not interfere with the end point determination. The author mentions the difficulty of observing the color change of the indicator at the end point and recommends that the titration be carried out in daylight.

Spectrophotometric titration. The difficulties involved in the identification of the end point have been overcome with the aid of spectrophotometric titration, adapted for the analysis of samples of metallic plutonium (up to 500 mg) [333]. At large concentrations of Pu(IV), the brown color of the solution masks the color change of ferroin from red to light blue. This color change may, however, be observed in the spectrophotometer at 510 mμ, the point of maximum absorption by ferroin. Figure 71 shows the typical spectrophotometric titration curve of Pu(III) by Ce(IV).

The titer of the solution of cerium (IV) sulfate was established using very pure plutonium with a known impurity content, or with As_2O_3. The standard deviation obtained on a large number of samples was ± 0.07%. The average deviation of the analytical results from the plutonium concentration calculated from the impurity content was about 0.03%.

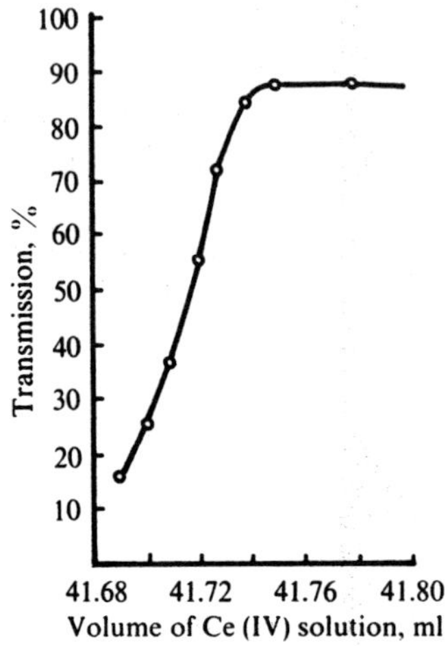

Figure 71

Spectrophotometric titration of Pu(III) against 0.05 *N* solution of Ce(IV).

Experimental conditions: 500 mg Pu(III), 5 ml of 6 *M* HCl, 100 ml of 1 *N* H_2SO_4 and 0.250 ml of 0.025 *M* ferroin solution.

The measurements were conducted with a modified type spectrophotometer described by Rost [619]. The instrument consisted of a light source, a 510 mμ interference filter with a line half-width of 10 mμ, and compartments for the beaker with the electrolyte (250 ml) and cadmium sulfide photoelement, connected with a 100 μA ammeter. The readings on the ammeter were proportional to the transmission of the solution.

Titrations were performed using an automatic zero-setting 50-ml buret, filled from the bottom, calibrated in 0.1 ml units. The efflux rate at the narrow end of the buret was 50 ml in $2\frac{1}{2}$ minutes. In order to eliminate corrections for the volume of the titrant, all titrations were carried out in a thermostatized compartment or from a thermostated buret.

The metal is cleaned from oxides by filing until a bright surface is obtained. If the metal is stored under oil, it is first rinsed in dichloromethane. About 500 mg of the metal is cut off and weighed with an accuracy to within 0.1 mg. The metal is placed in the 250 ml beaker and dissolved in 5 ml of 6 *N* HCl. During the dissolution the beaker is kept covered. When the dissolution is complete, the solution in the beaker is diluted to 100 ml with 6 *N* sulfuric acid. A solution of ferroin (250 microliters, 0.025 *M*) is added, the beaker placed in the spectrophotometer and the solution titrated against 0.05 *M* Ce(IV) sulfate, rapidly until the end point is approached, and then by adding very small portions. The jump in the transmission at the end point constitutes about 60% of the entire range of the instrument. The end point has been reached when further addition of cerium sulfate is no longer accompanied by a significant variation in the instrument reading (Figure 71).

To prevent oxidation, the samples are titrated immediately after dissolution. If not all the plutonium in the sample solution is in the trivalent state prior to titration, the solution is reduced in a zinc or lead reductor.

The result obtained is corrected for the volume of the titrant consumed in blank titration. The correction is only 0.1 ml and corresponds to the amount of cerium (IV) sulfate consumed in the titration of the ferroin indicator itself.

The authors showed that the dissolved samples contain only trivalent plutonium. With this end in view, the absorption spectra of the solutions were checked, and the titrations were conducted with and without a preliminary reduction in the reductor. No difference was noted. They also investigated the effect of the time elapsed between the dissolution and the titration on the accuracy of the results. Their results are shown below.

Time elapsed between dissolution and nitration, minutes	0	5	10	20	40	60
Pu found, %	100.00	100.02	99.91	99.84	99.75	99.69

Thus, unless the sample is titrated within 5 minutes after dissolution, the plutonium must be reduced in a zinc or a lead reductor.

The analysis of samples of metallic plutonium was subject to interference by iron, the content of which was 0.02–0.08 %. Since iron is titrated together with plutonium, it must be separately determined by another suitable method. In the work under discussion the iron was determined photometrically. Chromium, titanium, molybdenum, tungsten, uranium and vanadium interfere. Nitrates interfere, since they are reduced in the reductor. If plutonium has been separated from the impurities, the degree of completeness of the separation must be allowed for.

Waterbury [717] reported the results of a preliminary investigation into the possibility of effecting a direct spectrophotometric titration of Pu(III) against Ce(IV), using the absorption of Ce(IV) itself. A weighed portion of a standard solution of plutonium sulfate was treated with zinc amalgam, and about 98 % of Pu(III) were rapidly oxidized with a solution of cerium (IV) added from a weight buret. The titration was completed with the aid of a volumetric 2 ml buret, using a 0.01 *M* standard solution of cerium (IV). To prevent the oxidation of Pu(III), argon was bubbled through the cell. The absorption was measured at 380 mμ with the aid of a modified Beckman DU spectrophotometer after the addition of each 0.05–0.10 ml portion of the titrant near the end point. The absorption of Ce(IV) is at maximum at 320 mμ, but the measurements were conducted at 380 mμ, since at the former wavelength the absorption by Pu(IV) is considerable. The equivalence point was taken to correspond to the point of inflection on the volume of titrant vs. light absorption curve. For 200–400 mg samples of plutonium the accuracy obtained was up to 0.03 rel. %.

The authors gave a short description of the titration setup. A tall 250 ml beaker served as the cell. The beaker was fitted with a lid equipped with openings for the tip of the buret and for the argon inlet. The seals in the lid and blackened surface protected the photoelement from parasite light. The mixing was effected by a magnetic stirrer. Ferric iron, which absorbs strongly below 400 mμ, interferes.

In the same study, a variant of the spectrophotometric titration of Pu(III) against Ce(IV) in the presence of iron was verified. Plutonium and iron were reduced by zinc amalgam, o-phenanthroline was added to bind the iron and thus raise its oxidation potential to more than 1 V, and the Pu(III) was titrated against a solution of Ce(IV). Diphenylaminosulfonic acid was used as indicator.

The method was found to be insufficiently accurate, due to the absorption by ferroin, nonlinear absorption by the indicator and incomplete binding of Fe(II) by o-phenanthroline in acid solutions, leading to partial oxidation of Fe(II) by Ce(IV).

Potentiometric titration. The potentiometric variant of the determination of trivalent plutonium by titration against Ce(IV) sulfate is much more frequently employed. In sulfuric acid medium iron is also titrated because of the close potential values.

Anderson et al., as quoted by Metz [547], used a method in which plutonium was reduced to the trivalent state by zinc amalgam, and the solution was then potentiometrically titrated against a solution of cerium (IV) sulfate. The average deviation for samples containing 8–10 mg Pu was 0.15 rel.%. The correction for iron was introduced on the strength of an independent determination.

Boaz [303] obtained an accuracy of about 0.1% in a microdetermination of 8–10 mg Pu by this method. The reduction by zinc amalgam was conducted in an atmosphere of carbon dioxide.

In the reduction of plutonium by zinc amalgam uranium is also reduced. Since in sulfuric acid medium the potential UO_2^{2+}/U^{4+} is lower by about 0.35 V than the potential Pu^{4+}/Pu^{3+}, uranium and plutonium can be separately determined in the presence of each other. Bunce [329] carried out a microdetermination of plutonium and uranium by cerium (IV) sulfate after reduction by liquid zinc amalgam. The equivalence points were determined by the potentiometric method with the use of two polarized electrodes.

To 2 ml of the nitric acid solution of the sample in a porcelain dish some concentrated sulfuric acid is added, the solution is evaporated under an IR lamp until the appearance of SO_3 fumes, cooled, diluted with water to 1 ml and the evaporation is repeated twice again. The sample is then transferred to a reductor filled with nitrogen, a solution of 4 *N* H_2SO_4 is added and nitrogen is passed for 30 minutes. Three 0.5 ml portions of 4 *N* H_2SO_4 containing 33 mg $MnSO_4$ are added, and the uranium is titrated. The titration of uranium is terminated when a certain potential (slightly above 0.750 V) has been attained. The solution is then diluted until 1 *N* in H_2SO_4 and the plutonium is titrated. If uranium alone is titrated, 1 mg iron (as alum) is introduced towards the end of the titration. One ml of 1 *N* $Ce(SO_4)_2$ solution is equivalent to 119 mg U or 239 mg Pu.

The reproducibility of the method for pure solutions of plutonium is $\pm$ 0.3%. When uranium and plutonium are present together, the reproducibility is lower.

In a method [288] designed for industrial control of final nitric acid solutions of plutonium, plutonium was reduced to the trivalent state, after removal of HNO_3, in a silver reductor. Pu(III) was then titrated against a standard solution of $Ce(SO_4)_2$.

Plutonium may also be reduced by certain reagents, the excess of which after reduction may be titrated independently of plutonium. Metz [547] reported that one of the first methods was a potentiometric titration of Pu(III) to Pu(IV) after the sample has been reduced by excess of titanium (III) sulfate or chloride. The titration of the reduced solutions by cerium (IV) sulfate gives two equivalence points, corresponding to the oxidation of excess Ti(III) and to the oxidation of Pu(III) to Pu(IV). In the microdetermination of 4–6 mg Pu, the standard deviation was ~ 0.15%.

When automatic titrator was used with the method, an accuracy of ~ 0.12% was obtained [730]. The samples of plutonium were converted to sulfate prior to titration.

Reduction of plutonium by solutions of bivalent chromium gave satisfactory results [423]. 1–10 mg Pu in a sulfate solution were reduced to the trivalent state by excess chromium (II) sulfate in 1 *M* H_2SO_4. The excess of the reducing agent was oxidized by atmospheric oxygen; the oxidation was complete when a cell with platinum and calomel electrodes had attained a constant potential. Pu(III) was then titrated to Pu(IV) by a standard solution of Ce(IV) sulfate, added from a spring clip buret. The determination of 2–10 mg Pu could be performed with ± 0.2% accuracy, with a constant reproducibility.

The same method was utilized in the analysis of technological solutions of plutonium [705]. In an analysis of 250 samples containing 20–50 mg Pu, the accuracy was 0.7%, and the standard deviation was 0.41%.

A known amount of solution containing 20–50 mg plutonium is placed in a 5 ml beaker. Twenty drops of 9 *M* sulfuric acid are added, and the solution is evaporated to the appearance of fumes to expel HNO_3. If the sample consists of an oxide, a fluoride or an oxalate, the evaporation is repeated several times. The solution is cooled and about 2 ml of 0.05 *M* H_2SO_4 are added. If plutonium sulfate precipitates out, the solution is heated. The solution is transferred to the titration cell (Figure 72) with the aid of pipet with a drawn-out tip, and the beaker is washed with 0.05 *M* H_2SO_4. The electrodes are dipped into the solution, argon is bubbled through, and a freshly prepared 0.1 *M* $CrCl_2$ in 0.1 *M* HCl is slowly added (the $CrCl_2$ solution is prepared by reducing $CrCl_3$ in a column with zinc amalgam) until the indication of the potentiometer is zero. Air is then bubbled through the solution to oxidize the $CrCl_2$ in excess until the potential rises to 330 mV, and argon is again bubbled through until the potential is 358 mV. The solution is now titrated against 0.02 *N* solution of $Ce(SO_4)_2$ until a potential of 600 mV is attained, and the titration is then continued by small portions to a potential of 730 mV.

A blank titration in the absence of plutonium is carried out in exactly the same manner, except that the evaporation is omitted and 10 ml of $CrCl_2$ solution and 20 drops of 9 *M* H_2SO_4 are added. The consumption of ceric sulfate is recalculated to give the volume of the $CrCl_2$ solution used in the titration of plutonium, and a correction is introduced into the result of sample titration.

Two parallel determinations of plutonium in 0.1 ml standard solution containing about 300 mg/ml of plutonium are systematically effected and the results employed to calculate the deviation and the reproducibility of the method.

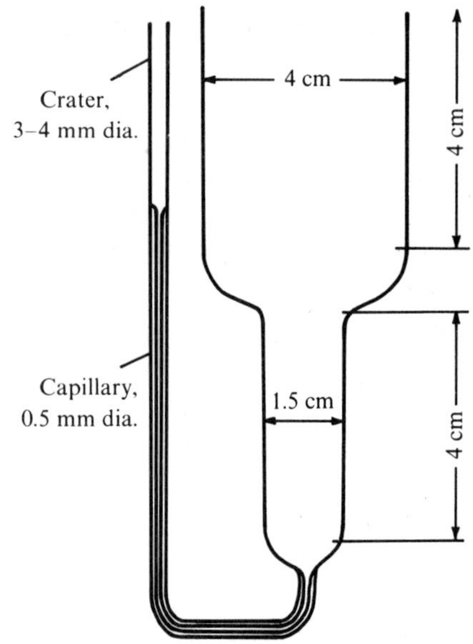

Figure 72
Titration cell

Since iron is titrated together with plutonium, it is separately determined by spectrophotometry, and a suitable correction introduced.

Helbig [458] used solutions of bivalent chromium as the reducing agent in his determinations of microgram amounts of plutonium. The ultramicroanalytical working procedure is described in [455–457]. In the determination of 0.7–0.8 μg Pu the average error of 20 determinations was $\pm$ 4.4%; in the determination of 0.10–0.12 μg Pu in 12 determinations it was $\pm$ 12%.

The reducing agent was chromium (II) sulfate, prepared in a zinc amalgam reductor. The solution of Cr(II) was transferred to a micropipet. The reduction and titration were effected in a microcone inside a humid chamber on a microscope stage.

The perchlorate, sulfate or chloride solution of plutonium in any valency state is placed in the microcone, containing 80–300 nl* H_2SO_4, and 90–180 nl of 0.05 *N* solution of Cr(II) in 0.1 *N* H_2SO_4 are added. A platinum and a calomel electrode is then introduced into the solution and the potential is measured. The solution is stirred by the vibration of the electrodes, actuated by a magnetic relay. In the presence of excess chromium the potential initially remains unchanged and then rapidly increases by 500–600 mV, after which it remains constant (Figure 73). The solution is titrated against 0.025–0.00625 *N* solution of cerium (IV). The consumption of the cerium solution in the titration is 0.4–0.7 μl. At the equivalence point there is a large (approximately 300 mV) potential jump (Figure 74). Working in an inert atmosphere is not necessary.

* 1 nl = 10^{-9} liter; 1 ng = 10^{-9} g.

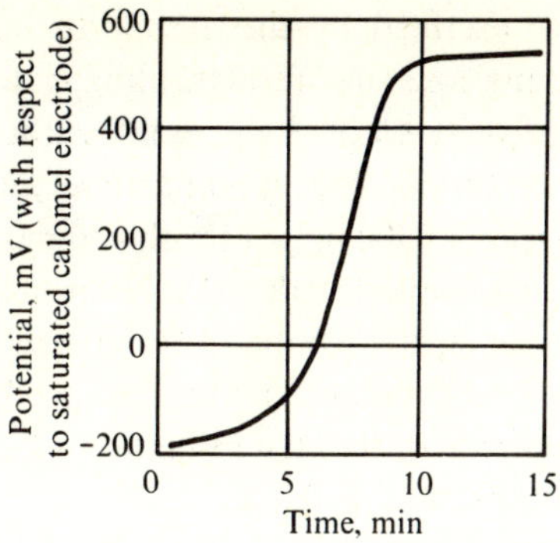

Figure 73

Oxidation of excess Cr(II) by atmospheric oxygen, represented as the variation of the platinum electrode potential as a function of time

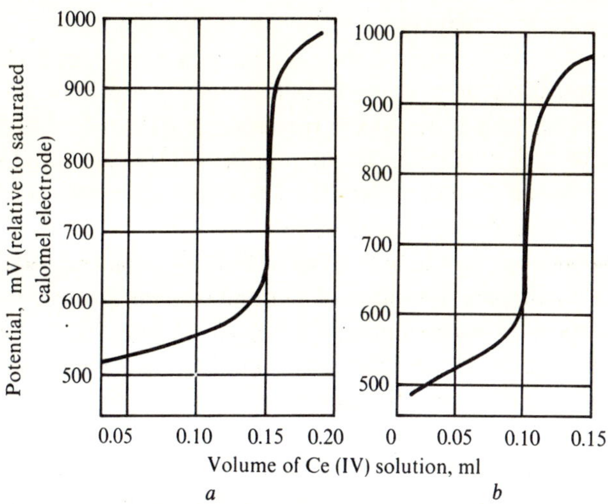

Figure 74

Potentiometric curves of Pu(III) titrated against a solution of Ce(IV):

a 800 ng Pu(III); 0.025 *N* Ce(IV); *b* 115 ng Pu(III); 0.00625 *N* Ce(IV).

In the development of the method, solutions containing 2.98 and 0.443 mg/ml of plutonium were used. The concentration of plutonium has been determined gravimetrically. Most errors are ascribed by the author to inaccuracies in measuring small volumes. The interfering ions are nitrates, phosphates and cations which are reduced by Cr(II) or oxidized by Ce(IV), e.g., gold, ruthenium and uranium. The error in the titration of a mixture of 999 ng manganese with 786 ng plutonium was 3%; in the titration of a mixture of 900 ng gold with 810 ng plutonium it was + 13%, and for a mixture of 450 ng ruthenium with 761 ng plutonium it was + 40%.

One variant of the potentiometric titration of Pu(III) against Ce(IV) was checked by Nebel [573]. Plutonium was preliminarily reduced to the trivalent state by electrolysis [574]. The average error in the determination of 1 mg plutonium in 2.5 ml 0.5 *N* H_2SO_4 (7 determinations) was $\pm$ 0.3%.

Titration with potassium bichromate

The main advantage of potassium bichromate as oxidant is its stability, because of which it may be employed as the primary standard in redox titrations. The reason why potassium bichromate has been relatively little used in determining plutonium is that the potential jump at the equivalence point is smaller. In addition, the equivalent weight of potassium bichromate relative to plutonium is low.

Palei and Zarinskii (1951) studied the suitability of potassium bichromate for a volumetric determination of plutonium. Satisfactory potentiometric curves were obtained, when about 2 mg Pu(III) were titrated to Pu(IV) with 0.01 *N* solution of $K_2Cr_2O_7$ in 3 *N* H_2SO_4. The potential jump at the equivalence point is about one-half that obtained when Ce(IV) sulfate is used as titrant (under comparable conditions). Over-oxidation of Pu(IV) is very slow.

In his very careful work on the hydrolysis of Pu(IV), Rabideau [604] used potassium bichromate to prepare solutions of plutonium with a given Pu(III)/Pu(IV) ratio. Oxide-free metallic plutonium was dissolved in 70% $HClO_4$. After dilution with water, aliquot parts of the solution of Pu(III) were mixed with calculated volumes of standard potassium bichromate solution to obtain the desired ratio Pu(III)/Pu(IV).

Most of the advantages of potassium bichromate as titrant for trivalent plutonium were utilized by Pietri and Baglio [592]. The method is intended for the analysis of high purity metallic plutonium (99.96%).

Metallic plutonium is dissolved in 4 *N* H_2SO_4, reduced to Pu (III) in a reductor filled with zinc amalgam, and titrated to Pu(IV) with a standard bichromate solution. In order to increase accuracy, weight burets are used to introduce about 99% of the titrant. In the final stage of the titration a very dilute solution of $K_2Cr_2O_7$ is introduced from a microburet. The equivalence point is determined potentiometrically with two polarized cold electrodes.

The average deviation in the determination of 70–140 mg Pu is 0.04%. The content of plutonium in the metal as determined by this method was 99.97% as against the value of 99.96% obtained from spectrographic data by subtraction of the impurity contents. When determining plutonium by this method, the results must be corrected for iron which is also titrated.

The authors studied the reduction of plutonium by zinc amalgams, the effect of the concentration of potassium bichromate, the effect of the kind of electrode system used and of the acidity on the end point determination and the rate of oxidation of the reduced solution by atmospheric oxygen.

It was found that the reduction may be effected both with solid and with liquid zinc amalgams, but in the case of liquid amalgams special apparatus and prolonged reduction time are required. In reductions with solid amalgams a simple reductor, 7.5 cm long and 2.5 cm in diameter, was employed. The only drawback of the reductor was that the solutions of plutonium had to be transferred. The possibility of reduction by a rotating spiral made of zinc amalgam was also investigated [675], and it was found that such reduction was not complete even after 6 hours at 100°.

In order to increase accuracy, the titration was effected in two stages. The first stage corresponded to the oxidation of 99% of plutonium and was carried out with the aid of a solution of potassium bichromate in 4 *N* H_2SO_4 (0.7 mg $K_2Cr_2O_7$ per gram of solution) from a weight buret of 30 ml volume. The second stage consisted in titrating to the equivalence point by a dilute solution of $K_2Cr_2O_7$ (0.25 mg/ml) in 4 *N* H_2SO_4 from a 5 ml microburet. One-stage titration from the weight buret resulted in an accuracy of only 0.2%.

A study of the electrode systems showed that platinum and calomel electrodes do not give a sufficiently large potential jump when plutonium is titrated against dilute solutions of bichromate. A considerable increase in the potential jump was attained using a platinum–platinum electrode system, which was polarized by a small direct current of $\sim 2\ \mu A$. The interelectrode distance was about 1 cm. The potential was measured by a vacuum-tube pH-meter. The solution was agitated by a magnetic stirrer with an adjustable revolution rate. The diagram of the apparatus employed is shown in Figure 75.

It was later found that platinum electrodes are frequently "poisoned" and lose their sensitivity to the variation in potential. They were accordingly replaced by polished gold electrodes, which proved sufficiently stable in this respect. The acidity of the sample solutions greatly affects the sharpness of the end point. In 2 *N* H_2SO_4 the end point is not sufficiently sharp (Figure 76a). Solutions of 4 *N* H_2SO_4 gave good results and were accordingly used in analytical work (Figure 76b).

A study of the rate of oxidation by atmospheric oxygen showed that such oxidation is insignificant if the titration is effected immediately after reduction. A solution left exposed to the atmosphere without stirring became oxidized to the extent of only 0.3% in two hours, while the extent of oxidation of a stirred solution was 4% within the same time. There is no need to work in an inert gas atmosphere, unless the reduced solution is left to stand prior to titration.

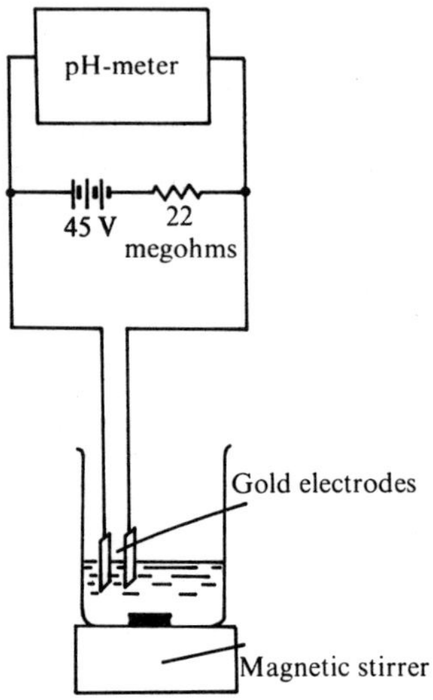

Figure 75
Apparatus for potentiometric titration with two polarized electrodes.

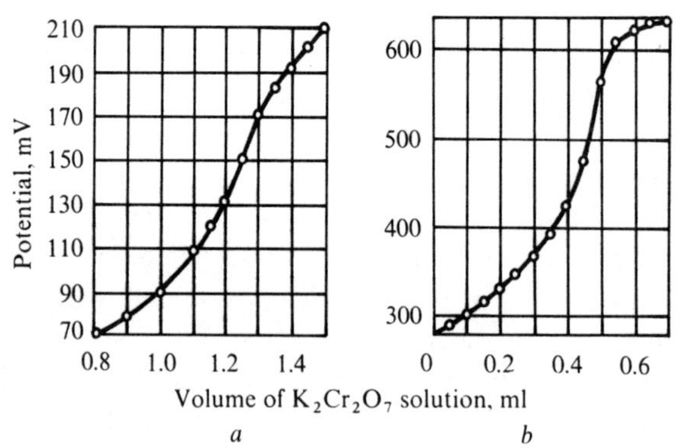

Figure 76
Potentiometric titration curve of Pu (III) against potassium bichromate:

a in 2 *N* H_2SO_4; *b* in 4 *N* H_2SO_4.

The authors suggest that the accuracy of the method be increased by using larger samples and increasing the sensitivity of the electrodes.

The solution of potassium bichromate was prepared from the U.S. Bureau of Standards primary standard, dried for 2 hours at 110° and cooled in a desiccator.

Into a clean, dry weight buret are poured 30–32 ml of potassium bichromate solution with a gravimetric titer, without filling the tip. The tip and the stopcock are closed and the buret weighed to within 1 mg. The microburet is filled with a more dilute solution of $K_2Cr_2O_7$.

The sample of plutonium was the α-phase of metallic plutonium, with the following impurity content, in %: C $2 \cdot 10^{-2}$; Si $1 \cdot 10^{-2}$; Fe $2.5 \cdot 10^{-3}$; Mn and B $2 \cdot 10^{-3}$; Mg $< 1 \cdot 10^{-3}$; Pb $5 \cdot 10^{-4}$; Sn and Ni $2 \cdot 10^{-4}$; Cu $1 \cdot 10^{-4}$; Cr $< 1 \cdot 10^{-4}$; Cd, Ag, Na, Li, K, Zn, P, Mo, V, Tl, Ca, Al, La, As, Be were not detected. The atomic weight of plutonium was 239.115 ± 0.005.

One hundred mg bright metallic plutonium are weighed out to within $\pm$ 0.02 mg and placed in a 150 ml beaker. If the plutonium is slightly oxidized, the oxide is removed by light scraping. If the oxide film is thick, the metal is electropolished or filed. Ten ml of 4 *N* sulfuric acid are introduced into the beaker and the beaker is covered with a watch glass. The dissolution is over within 15–20 minutes. In order to prevent spattering, the metal must be covered with a sufficiently high layer of acid. When all the metal has dissolved, the walls of the beaker and the watch glass are rinsed with the minimum amount of 4 *N* H_2SO_4. The solution is passed through the reductor and collected in a 150 ml beaker. The reductor is covered with a watch glass. The solution from the reductor should flow along the wall of the beaker. The rate of efflux from the reductor should not exceed 3 ml/min. When the liquid level in the reductor has descended to a height of 5 mm above the zinc layer, the walls of the reductor and the watch glass are washed down with 4 *N* H_2SO_4. The beaker with the sample solution is rinsed twice with 4 *N* H_2SO_4. The wash waters are added to the reductor. The reductor is washed with two 15 ml portions of 4 *N* H_2SO_4 and one 15 ml portion of water.

The electrodes are lowered into the beaker as shown in Figure 75. The titrant is rapidly added from the weight buret until the blue solution turns a pale orange color. The stirrer is switched on. The titrant is now rapidly added in 5 drop portions, until the potential begins to rise. When the potential variation becomes as large as 10 mV for each portion of titrant, the size of the portions is reduced, first to two drops per portion, then to one drop. The titration is then continued from the microburet in 0.05 ml portions. The potential change is recorded after each portion has been added. The titration is terminated two portions past the equivalence point which usually corresponds to 450 mV. The weight buret is reweighed and the calculation is carried out.

Titration with ammonium vanadate

Of all the oxidizing agents which are suitable for the titration of trivalent plutonium, ammonium vanadate is the mildest.

The formal potential of the $V(OH)_4^+$ and VO^{2+} ions in solution in 1 *N* H_2SO_4 is + 1.00 V. With the increase in the concentration of sulfuric acid the value of this potential increases rapidly and is + 1.45 V in the concentrated acid. In 5–6 *N* H_2SO_4 the rate of oxidation by vanadate increases, and the ion can be used to titrate Pu(III). Typically, ammonium vanadate is very stable when stored in 6 *N* solutions of H_2SO_4.

Palei and Zarinskii (1951) showed that a potentiometric titration of solutions of Pu(III) with ammonium vanadate was feasible. An aliquot part of a solution containing 2 mg Pu(III) was placed in a potentiometric cell, and an inert gas was bubbled through to remove excess SO_2 after reduction. The solution was diluted with several volumes of 5 N H_2SO_4 and Pu(III) was titrated to Pu(IV) with 0.02 N solution of NH_4VO_3 in 5 N H_2SO_4, the solution being stirred with the inert gas. The potential jump near the equivalence point was about 200 mV, which is similar to the potential jump during the titration of Pu(III) against 0.01 N solution of $K_2Cr_2O_7$, under comparable conditions, and one-half the potential jump in the titration of Pu(III) against 0.01 N solution of $Ce(SO_4)_2$. The potassium bichromate and ceric sulfate titrations of plutonium were carried out in 4 N H_2SO_4. Errors in single determinations were about 1% for all oxidants.

Redox titration of Pu(III) by ammonium vanadate was used by Palei and Kochetkova (1955) for the determination of the content of plutonium in plutonium dioxide. These authors studied the techniques of dissolution of plutonium dioxide, reduction of plutonium to the trivalent state in a bismuth reductor, experimental conditions of titration, effect of impurities and their separation.

It was found that the dissolution of plutonium dioxide may be carried out in a 1:10 mixture of sulfuric with hydrobromic acids. The dissolution rate is 5–6 times faster than in HBr or H_2SO_4 alone. The time required to dissolve 100 mg PuO_2 is about two hours. The dioxide is wetted with sulfuric acid, HBr is cautiously added, the mixture heated under a watch glass and evaporated to dryness. The blue precipitate formed is again treated with a small amount of H_2SO_4 and evaporated. The pink precipitate of Pu(IV) sulfate is dissolved in dilute acid and transferred to a 50 ml flask.

The dissolution can be accelerated somewhat by the use of the method of Rubinshtein, Zaitsev and Shubochkin, who work with a mixture of HCl and KI (see p. 98). The liberated free iodine interferes with the subsequent determination and is removed by evaporation with sulfuric acid. It was shown by centrifugation and α-count of the residue that plutonium dioxide was more than 99.99% dissolved.

Both dissolution techniques were employed. The reduction of plutonium to the trivalent state was carried out in a bismuth reductor. The optimum dimensions of the reductor are: inner diameter 10 mm, height of bismuth layer 11 cm, diameter of bismuth grains 0.7–1.0 mm. Solutions of up to 10 mg plutonium in 3 N H_2SO_4 were reduced by passing through the reductor at a rate lower than 2 ml/min. The reductor was washed with five 5 ml portions of 5 N H_2SO_4. The reduced solution and the wash waters were collected in a beaker, in which the titration was performed.

It was found that the use of a smaller size reductor and bismuth with a larger grain size results in an incomplete reduction of plutonium. Sample titration was carried out with 0.01 N or 0.005 N solutions of ammonium vanadate in 4 N H_2SO_4; phenylanthranilic acid was used as indicator (1–2 drops 0.03% solution in 0.3% Na_2CO_3) until the solution became pale pink. The concentration of vanadate solutions were determined by titration against solutions of known amounts of very pure metallic iron or metallic plutonium. Titrations of samples of standard solutions of plutonium, containing 4.5–7.5 mg Pu, had an average deviation of $\pm$ 0.12% in ten experiments. The average error in the determination of plutonium in the pure dioxide also did not exceed 0.12%.

Iron and uranium interfere with the determination, but since their content, including other possible impurities, in the dioxide did not exceed 0.6%, they had not been separated. If these elements are present in larger amounts, they may be separated, but the losses of plutonium must be followed by means of α-counts. The volatile admixtures in the sample of plutonium dioxide were taken to be equivalent to the loss in weight noted when PuO_2 was calcined at 1000–1100° during 3 hours.

Further oxidation of plutonium beyond the tetravalent state does not take place owing to the low value of the redox potential of the vanadate. Ions, such as Cr^{3+}, Mn^{2+}, Pb^{2+}, Ce^{3+} and other ions not oxidized by vanadate, probably do not interfere, but this has not been experimentally verified.

Titration by potassium permanganate

In titrating Pu(III), potassium permanganate has no special advantages over other oxidizing agents, except that its own color may serve to determine the equivalence point. It was found that the titration with permanganate in acid medium may be carried out in two stages [451]: the first at room temperature, resulting in the oxidation of Pu(III) to Pu(IV), the second at 60°, when Pu(IV) is further oxidized to Pu(VI). This feature makes it possible to determine plutonium in the presence of iron and uranium.

Determination of plutonium (IV)

Tetravalent plutonium may be determined both by oxidative and reductive titration. In the former case permanganate titrations are mainly employed; reductive titrations are carried out by iodometry.

Titration by potassium permanganate

It was found by Harvey et al. [451] that when plutonium (IV) sulfate in 1 N H_2SO_4 is titrated against 0.005 N $KMnO_4$, the reaction is very slow at room temperature and proceeds rapidly at 60°. A solution of about 0.8 mg $Pu(SO_4)_2$ was reduced with a solution of sulfurous acid in the heat. The excess SO_2 was removed by evaporation to a very small volume in a stream of purified carbon dioxide. The trivalent plutonium obtained was titrated to the tetravalent stage with $KMnO_4$. The end point was both sharp and constant. When the solution was heated to 60°, the pink coloration disappeared, and the titration could be continued to the hexavalent stage. In these first trials, the error was apparently as high as 2–3%.

Subsequently attempts were made to utilize the two-stage titration of plutonium in the determination of plutonium in the presence of iron and uranium.

The conditions of the titration were studied in detail by Palei and Zarinskii (1951) for amounts of plutonium of about 2 mg. It was confirmed that the optimum temperature for the realization of the second stage of the reaction is 60–70°. The oxidation of Pu(IV) to Pu(VI) proceeds slowly below 60°, and the stabilization of the potentials is accordingly slow; above 90° excessive permanganate consumption is noted, due to decomposition.

It was found when investigating the effect of the acidity that in solutions in 3–4 *N* H_2SO_4 the reaction becomes slower, apparently owing to the increase in the potential Pu(IV) → Pu(VI). The optimum concentration range of sulfuric acid is 0.25–1 *N*. The presence of Mn^{2+} ions accelerates the oxidation of Pu(III) to Pu(IV), but does not affect the rate of the reaction Pu(IV) → Pu(VI), since at this stage the solution already contains a sufficient amount of reduced manganese.

The concentrations of potassium permanganate solutions used in the titrations were 0.05 *N* and 0.01 *N*; better results were obtained with the former concentration. The average error was 1% in this case.

Solutions of trivalent plutonium were preliminarily prepared by reducing solutions of plutonium (IV) with sulfur dioxide. Prior to titration sulfur dioxide was displaced by a stream of purified nitrogen, which also served to stir the solutions during the titration.

The effect of iron and chromium, present as impurities, was investigated. Since ferrous iron is titrated together with Pu(III), plutonium was determined by the difference in volumes between the first and the second equivalence points. When the Fe:Pu ratio was 0.3–4, the average error in titrating 2 mg Pu in 0.8–2.2 *N* H_2SO_4 was + 2%. In the presence of chromium high results were obtained. At the ratio Cr:Pu = 0.5 the error was + 12%, and increased with increasing amounts of chromium.

Fomin, Vorob'ev et al. (1951) determined plutonium in the presence of uranium and iron by permanganate titration. A solution of plutonium in sulfuric acid, containing iron and uranium, was preliminarily reduced by SO_2 with heating; SO_2 was removed by passing through a stream of inert gas and the titration was effected first at room temperature and then at 70°. During the first stage, plutonium is oxidized to the tetravalent, iron to the trivalent and uranium to the hexavalent state. In the second stage plutonium alone is titrated to the hexavalent state. In the titration of 13–14 mg Pu against 0.04 *N* $KMnO_4$ in the presence of double the amount of uranium or half the amount of iron, the average deviation was −15%. When the amounts of uranium and iron were increased to correspond to U:Pu = 3 and Fe:Pu = 1, the error was −5%.

The authors attempted to titrate plutonium (IV), which had been prepared by evaporation with concentrated sulfuric acid to the elimination of SO_3 vapors. The evaporated solution, containing 20 mg Pu, was diluted with 2 *N* H_2SO_4 and titrated against 0.02 *N* $KMnO_4$ at 70°. It was found that under these experimental conditions, iron does not interfere if present in an amount (by weight) equal to that of plutonium. The presence of uranium causes nonreproducible results to be obtained, apparently because the Pu(VI) used as the initial material is not fully reduced to Pu(IV) during the evaporation with sulfuric acid if uranium is present.

Iodometric titration

Iodide ions reduce tetravalent to trivalent plutonium. It was pointed out by Harvey et al. [451] that the titration of iodine liberated in sulfuric acid solution as a result of the oxidation of I^- by Pu(IV) ions is not a convenient method for determining plutonium, since the reaction is slow. In chloride solutions the reaction is faster, but the titration is still inconvenient. The authors titrated the iodine liberated by ~ 2 mg $Pu(SO_4)_2$ against 0.01 *N* thiosulfate and obtained results which were low by ~ 5%.

Kuznetsov and Mikhailov (1956) showed that Pu(IV) may be iodometrically determined by titrating the excess thiosulfate introduced into the solution obtained by dissolving water-washed precipitate of plutonium selenite $Pu(SeO_3)_2$ in 1 *M* H_2SO_4. The selenite precipitation is not selective, since iron, uranium (VI) and chromium are also precipitated.

Titration with mercurous mercury

Moiseev, Borodina and Tsvetkova (1962) proposed a direct titration of tetravalent plutonium to the trivalent state by mercurous mercury in a thiocyanate solution.

In 1 *N* H_2SO_4 the potential of $2Hg^{2+}/Hg_2^{2+}$ is close to that of Pu^{4+}/Pu^{3+}, which is about + 0.75 V, but in thiocyanate solutions the mercury potential becomes much lower owing to the complexing of Hg(II) ions, and is + 0.24 V in 0.1 *M* KCNS and + 0.12 V in 1 *M* KCNS [215]. The authors showed that in solutions which are 0.25–1 *M* in H_2SO_4 and 0.1–1 *M* in KCNS tetravalent plutonium is readily reduced by mercurous nitrate or by mercurous perchlorate. Ferric iron is also readily reduced under these conditions, but only after plutonium (IV).

The end point is determined by potentiometry, or visually with the aid of an indicator. The indicator may be ferrous ions, which under the conditions of the titration reduce Pu(IV) to Pu(III) and the ferric ions formed give a colored complex with the thiocyanate. At the equivalence point all plutonium and iron are in a lower valency state and the solution is decolorized.

The method was applied to the analysis of plutonium dioxide containing up to 1.5% impurities (less than 0.05% Fe). When the content of plutonium was determined on 50–80 mg samples of PuO_2, the mean square deviation was 0.25% in 45 determinations carried out on six products. The difference between the results and those obtained gravimetrically by the peroxide–cupferron method was less than 0.25%. Ferric iron, molybdenum and copper interfere with the determination; Cr(III),

Mn(II), V, U(VI), La, Al, Ni, etc. do not interfere. Since Fe(III) is titrated together with plutonium(IV), a correction for the content of ferric iron must be introduced following a separate determination.

Reagents with a low iron content are used, or else are specially purified. A standard 0.02 *N* solution of $K_2Cr_2O_7$ is prepared. To prepare 0.02 *N* solution of $Hg_2(NO_3)_2$ in 0.02 *M* nitric acid, the salt is dissolved in half the amount of water and the solution is boiled for 3 hours to expel nitrogen oxides, filtered, and diluted with an equal volume of water. The titer of the solution is determined with potassium bichromate. To do this, a measured volume (10–20 ml) of 0.02 *N* standard bichromate is introduced into a 300 ml flask, 50 ml of 1 *N* sulfuric acid, 5–10 ml of 0.04 *N* $FeSO_4$ and 10 ml of 40% KCNS are added and the solution is titrated against $Hg_2(NO_3)_2$ until decolorized. Towards the end of the titration the last portions of the titrant are introduced at one minute intervals. A blank titration must be carried out.

To carry out the analysis, 50–80 mg PuO_2 are placed in a weighed 25 ml quartz flask and weighed to within at least 0.1 mg. Two grams of $NaHSO_4$ are added, a glass funnel introduced into the flask and the contents of the flask are fused at 500° on a hot plate until a homogeneous transparent melt is obtained. After cooling the melt is dissolved in 15–20 ml of 1 *N* H_2SO_4 and transferred to a 300 ml conical flask, and diluted to 80–100 ml with 1 *N* H_2SO_4. Oxygen is eliminated by blowing through pure CO_2, 8–10 ml of 40% KCNS and 1 ml 0.01 *N* $FeSO_4$ are added, and the solution is titrated against 0.02 *N* standard $Hg_2(NO_3)_2$ until decolorized, the last portions of the titrant being added at 1 minute intervals. Blank titration is carried out with 2 g $NaHSO_4$ and 100 ml 1 *N* H_2SO_4 as described above.

Determination of plutonium (VI)

Titrimetric determination of Pu(VI) is very important, since it may be carried out in the presence of iron and uranium.

Palei and Zarinskii (1951) proposed a potentiometric determination of Pu(VI) with ferrous iron. Plutonium was oxidized by ammonium persulfate in the presence of silver ions in the heat. It was found that the excess of the persulfate is decomposed on further heating with sulfuric acid. In preliminary experiments, 0.5 ml 12% $(NH_4)_2S_2O_8$, 0.1 ml 0.4% solution of $AgNO_3$ and 0.5 ml 1 *N* H_2SO_4 were added to 0.5 ml of a solution containing 2 mg plutonium, and the mixture was heated on a water bath for 40 minutes. The cooled solution was diluted with several volumes of 1 *N* H_2SO_4 and was titrated potentiometrically against 0.035 *N* solution of Mohr's salt. The method was not further investigated.

Helbig [457] described an amperometric determination of microgram amounts of plutonium by direct titration of Pu(VI) against ferrous iron in a solution of 2 *N* H_2SO_4.

On the strength of the current vs. voltage curves established for the ions PuO_2^{2+} (10^{-3} *M*), Pu^{4+} (10^{-3} *M*) and Fe^{2+} in 2 *N* H_2SO_4 (Figure 77), it was deduced that PuO_2^{2+} can be amperometrically titrated against Fe^{2+} ions at the anodic polarization of the platinum electrode of + 1.0 V (relative to saturated calomel electrode), since PuO_2^{2+} in sulfuric acid solutions is readily reduced to Pu^{4+}. The titration curves are of the inverted *L*-form (Figure 78), since considerable polarization of the platinum electrode at + 1.0 V is produced only by Fe^{2+} ions. The plutonium sample was

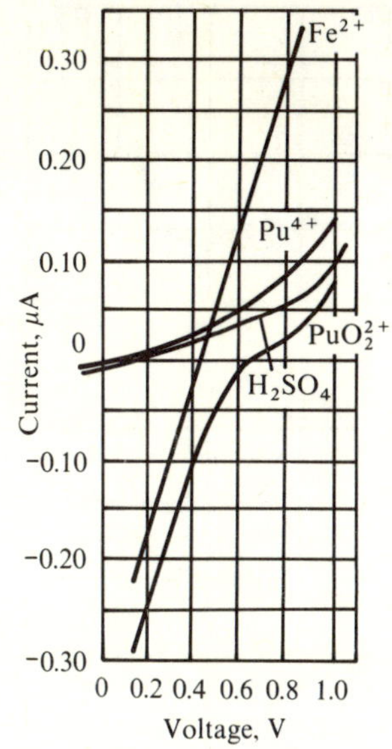

Figure 77

Current-voltage curves for Fe^{2+}, Pu^{4+} and PuO_2^{2+} in 2 *N* H_2SO_4

preliminarily oxidized to the hexavalent state by hot perchloric acid. There is no need to remove the excess oxidant; the oxidizing action of $HClO_4$ is suppressed by diluting the solution with water.

Inasmuch as the amount of plutonium in the sample was less than 1 μg, ultramicro methods were employed [455, 456]. Volumes of plutonium solutions were determined with the aid of volumetric capillaries, and titrations were effected using microburets reading to within ± 0.8 nl. The indicator electrodes consisted of a platinum wire 150 μ in diameter and a saturated calomel electrode which was in contact with the solution through a salt bridge in a capillary. The indicator electrode was made to vibrate and thus to stir the solution with the aid of an electric relay. The current was measured with a mirror galvanometer with a sensitivity of $5 \cdot 10^{-9}$ A.

Evaporations and titrations were conducted in a microcone with a 2 mm dia. hole placed in a humid chamber under a microscope. A platinum microheater was used for evaporation. Perchloric acid vapors were aspired with a glass tube connected to a water pump. The author recommends the following procedure.

The solution of plutonium, which should not contain sulfates or phosphates, is first strongly heated to remove volatile acids. About one microliter of concentrated $HClO_4$ containing 10% HNO_3 is then added, and the solution evaporated as rapidly as possible to a residual volume of 100–150 nl. If microamounts of plutonium are evaporated slowly, they are oxidized incompletely or not at all. The evaporation should be completed within 20 seconds, and the solution must not boil. The samples are evaporated 2–3 times and

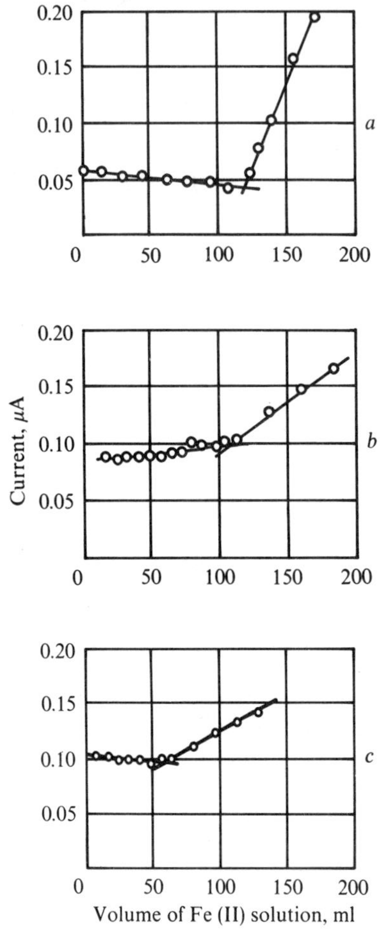

Figure 78

Amperometric titration curves of PuO_2^{2+} in 2 *N* H_2SO_4 against a solution of Fe (II):

a 807 ng Pu(VI), 0.05 *N* Fe(II); *b* 122.7 ng Pu(VI), 0.01 *N* Fe(II); *c* 57.3 ng Pu(VI), 0.01 *N* Fe(II).

finally cooled and mixed with 0.7–1.0 microliters of 2 *N* H_2SO_4. The solution of PuO_2^{2+} thus obtained is titrated with 0.05–0.01 *N* solution of Mohr's salt. To avoid liquid being captured by the tips of the buret and the electrode care must be taken to see that these do not touch the walls of the microcone.

The working voltage is fed from an accumulator through a variable resistance and checked with a valve voltmeter or potentiometer. The current intensities obtained are corrected using the equation $i = i_e(v + x)/v$, where i_e is the experimental intensity value, x is the volume of titrant added and v is the sample volume.

In the titration of 700–800 ng Pu against 0.05 *N* solution of ferrous iron, the average deviation was 3.5% in a series of 22 experiments. The average deviation during the titration of 100–200 ng Pu against 0.01 *N* solution of Fe(II) was 5.8 %; in the titration of 40–60 mg Pu against the same titrant it was 15.8%. Typical titration curves are shown in Figure 78.

The concentrations of standard solutions containing 2.98 or 0.433 mg/ml of plutonium were determined by a gravimetric method.

In the determination of 700–800 ng plutonium, equal amounts of Ca, Na, K, Ag, Al, Bi, Fe, Zr, Ce, Sn, Pt, or double the amount of uranium do not interfere; Mn and Ru interfere. All elements which are oxidized by $HClO_4$ and are then reduced by Fe^{2+}, including neptunium, gold and chromium, obviously interfere. Of the anions, sulfates and phosphates interfere. Sulfate ions prevent the oxidation of plutonium by perchloric acid from going to completion because of complex formation with Pu^{4+}; phosphates also form sparingly soluble and complex compounds with Pu.

According to the author, the accuracy could be considerably increased by using more sensitive burets.

Moiseev, Ognishcheva and Gudkova (1951) proposed a volumetric determination of plutonium in the presence of iron and uranium, in which plutonium is oxidized to the hexavalent state by sodium bismuthate or lead dioxide. The oxidation by sodium bismuthate in nitric acid solution is quantitative in the cold, while in sulfuric acid solution it is quantitative only if the solution is heated. Lead dioxide oxidizes plutonium in a boiling sulfuric acid medium. The solution of hexavalent plutonium thus obtained is separated by filtration from the excess oxidant and reduced by ferrous iron; the excess of the iron is then titrated against potassium permanganate or potassium bichromate.

Moiseev and Borodina (1955) used a modification of this method to determine plutonium in the presence of chromium and manganese. Plutonium was preliminarily separated from these two metals by precipitation as cupferronate. As distinct from the procedure just given, the excess ferrous ion was titrated after reduction of Pu(VI) by cerium (IV) sulfate, with ferroin as indicator.

The mean square error in the determination of 12–25 mg plutonium is 0.4%.

The sample solution of plutonium is placed in a quartz beaker, 100 ml in volume, 10–15 ml of 10% tartaric acid is added and the solution is neutralized with ammonia. The solution is diluted with water to 50 ml, 5 ml concentrated sulfuric acid are added and the solution cooled to 10–15°. Plutonium and iron are precipitated by 3% solution of cupferron, 5–10 ml SO_2-saturated water are added, and after 20–30 minutes the precipitate is filtered off and washed with 4% H_2SO_4 containing 0.2% cupferron. The precipitate is dried, calcined at 600–800° and fused with $NaHSO_4$ at the same temperature. After the melt has been dissolved in 20–40 ml of 1 *N* H_2SO_4, 200 mg sodium bismuthate are added, the beaker is covered and the solution boiled for 15–20 minutes. The solution is cooled to 50° and filtered through a filter paper which has been preliminarily washed with 2 *N* H_2SO_4 boiled with sodium bismuthate. The beaker and the precipitate are washed with cold 2 *N* H_2SO_4, bringing the volume of the solution to approximately 100 ml. To the resulting solution 0.02 *N* Mohr's salt is added from a buret until the color changes from yellow to greenish-blue. A drop of ferroin solution is added and the solution titrated against 0.02 *N* cerium (IV) sulfate solution. At the equivalence point the color changes from orange-red to pale yellow-green.

The duration of a determination of plutonium in the presence of chromium and manganese is about 6 hours, and without chromium and manganese it is 2–3 hours.

The most promising redox titration method for the determination of plutonium as regards accuracy and small degree of interference by many foreign admixtures is the method of Waterbury and Metz [717], mentioned somewhat earlier by Metz [547]. The method involves a quantitative oxidation of plutonium to the hexavalent state by evaporation with perchloric acid; Pu(VI) is then reduced to Pu(IV) by a small excess of a standard solution of ferrous iron, which is then back-titrated by automatic titration against a solution of cerium (IV). For samples of very pure metal the average content of plutonium experimentally obtained was 99.98% with a standard deviation of 0.02% in 11 determinations. The samples taken consisted of 3–5 g solution of plutonium, with a concentration of about 60 mg per gram of solution. The value obtained was in agreement with that arrived at by spectroscopic determination of impurities and by high-vacuum fusion of the metal.

To obtain highly accurate results, large sample weights were taken, the sampling was effected gravimetrically, weight burets were used, and the back-titration was effected with a dilute solution of cerium (IV). The evaporation of the sample solutions to remove undesirable anions and the oxidation of plutonium to the hexavalent state were effected simultaneously for the same purpose. After a sample of the solution has been weighed, all operations were carried out in the same beaker, without any transfers. To eliminate the subjective error in estimating the equivalence point and to accelerate the titration, an automatic titrator was used [712], even though ordinary potentiometric titrations can be used.

The weight of the samples of metallic plutonium used in the method must be at least 2 grams. If the samples are weighed on a semimicrobalance, their weight can be determined to within 10^{-5} g. Prior to weighing the metal is carefully cleaned of oxide films. Large lumps of metal may be cleaned by filing the oxide away from the surface. If this proves insufficient, electropolishing is applied.

The electropolishing apparatus consisted of a glass crystallizer, containing a little 20% K_2CO_3, a tantalum cathode, a tantalum wire anode and a dry 22.5 V battery. To clean the metal, the sample was placed in the crystallizer and the anode was kept in contact with it until the surface began to shine. The sample was then transferred onto a glass filter, and washed with water, acetone and ether; the solvents were removed by suction. The cleaned samples were deep-frozen to prevent reoxidation or else were dissolved immediately after electropolishing.

To check the necessity for cleaning the surface of the metal, some plutonium which had not been cleaned was dissolved in hydrochloric acid and analyzed by the same method. The oxide film on the surface was not thick, but the color of the surface was altered. The average experimental value of the plutonium content was 99.903% with a standard deviation of 0.013% in five titrations, instead of 99.98% for the cleaned sample.

Metallic plutonium and most of its alloys readily dissolve in excess dilute hydrochloric acid (1–3 *M*). To prevent losses by spattering, the dissolution was effected in a long-necked flask of a large volume (a 250 or 300 ml volumetric flask held at a 45° angle is suitable). When the dissolution was completed, the solution was diluted with 3 *M* HCl to 100–200 ml and weighed. Portions of this solution, weighing at least 2 grams and containing more than 200 mg Pu, were weighed out on a semimicrobalance. If the sample is not fully soluble in HCl or in a HCl–HNO_3–$HClO_4$ mixture, the authors recommend using the sealed flask technique [432] with HCl and $HClO_4$ solutions. The sample solutions may consist of plutonium dissolved in hydrochloric, nitric and perchloric acids, but not in sulfuric or phosphoric acid.

The solution, containing more than 200 mg Pu, is placed in a tall 250 ml beaker covered with a ribbed protective glass. The solution is evaporated almost to dryness in a special box, using mild electric heating from below and heating with an IR lamp from above. The solution is then cooled, the walls of the beaker are washed down with 6 *M* HNO_3 and 5–10 ml 70% $HClO_4$ are added. The solution is evaporated until vigorous evolution of perchloric acid fumes, increasing the heat without letting the solution boil. The solution is cooled and 5–10 ml 20–30% $HClO_4$ are added. The solution is again evaporated to 2–3 ml residual volume; the solution should not be evaporated to dryness (low results may be obtained). If this was inadvertently allowed to happen and the residue is not fully soluble in perchloric acid, a few ml of 16 *M* solution of the acid are added together with 1 *M* hydrofluoric acid and the solution is heated until the residue is fully dissolved. The evaporation is then repeated.

After the portion has been evaporated to 2–3 ml, the solution is rapidly cooled by placing the beaker on a cold metallic surface and is then diluted with 50 ml water while avoiding spattering. The beaker is then placed on the magnetic stirrer of the automatic titrator and the electrodes, the stirrer and the tip of the buret are immersed in the solution. (An ordinary potentiometric titration with platinum and calomel electrodes can also be effected.) The stirring is begun, the titrator is set at the maximum sensitivity and 10 ml of 50% sulfuric acid solution are added. The weight buret containing standard 0.5–0.7 *M* solution of Fe(II) in 0.5 *M* H_2SO_4 is weighed; the reagent is added to the beaker until a large potential jump is recorded. Since the fluctuations of the instrument serve as a warning that the end point is being approached, a small excess (1 drop) of ferrous iron may be employed. The buret is again weighed. The sensitivity scale of the instrument is now set at a position corresponding to the maximum deflection during the titration of Fe(II) against Ce(IV) and the null point of the measuring instrument is checked. The initial reading of the 1 ml buret of the titrator, filled with 0.05 *M* solution of cerium(IV) in 1 *M* H_2SO_4, is taken and the titrator is switched on. When the equivalence point has been reached, the titrator automatically switches off. The buret is again read, and the amount of Fe (II) solution required to reduce the plutonium sample taken is calculated.

In manual titration the end point is found from the inflection point on the titration curve, since the potential of the end point may vary from one determination to another. The differential-potentiometric type titrator used automatically switches off at the inflection point, irrespective of the initial or the final potential of the solution.

0.05 *N solution of cerium* (IV) is prepared by dissolving 30 g of $Ce(SO_4)_2 \cdot 2(NH_4)_2SO_4 \cdot 2H_2O$ in 500 ml of 2 *M* H_2SO_4, and diluting with water to one liter; the solution is then thoroughly mixed and left to stand overnight. The titer of the solution is determined with an alkaline solution of arsenite and a solution of sodium oxalate.

0.5–0.7 *M solution of ferrous iron* is prepared by dissolving 200 grams of Mohr's salt ($FeSO_4 \cdot (NH_4)_2SO_4 \cdot 6H_2O$) in 500 ml 1 *M* H_2SO_4 and diluting with water to one liter. Nitrogen is passed through the solution to expel oxygen and the solution is transferred to a ground glass stoppered bottle. The titer of the solution is determined by titration against potassium bichromate of primary standard grade against metallic plutonium of high purity or against a solution of ceric sulfate of known titer; this must be done every day or immediately prior to use. If potassium bichromate is used, not less than 2 grams of the reagent dried at 240° are weighed out, dissolved, and aliquots of the solution containing ~ 200 mg $K_2Cr_2O_7$ are placed into tall 250 ml beakers with the aid of a weight buret. The titration is effected as in the case of plutonium solutions. It is desirable to determine the titer of the solution against metallic plutonium of high purity; if this is done, correction must be introduced for the extraneous matter in the metal, as determined by spectrography. The content of oxygen in plutonium is determined by melting in a platinum crucible. The plutonium used by the authors was 99.98% pure. The titer of the ceric sulfate solution used for the standardization of the ferrous solution must be known to within 0.01%. A portion of cerium (IV) sulfate containing about 4 milli-equivalents of Ce (IV) is withdrawn gravimetrically and titrated against the solution of ferrous iron.

The authors investigated the effect of more than 25 different ions on the titration of plutonium. It was found that the following ions interfere little or not at all: 1–3-fold amounts of K; 0.2–1-fold amounts of U; 0.12–1-fold amounts of Ag; 0.1–1-fold amounts of Ce(III), Co, Cu, Pb(II), Mo, Nb, Rh, Ru, Tl, Sn and Ti; 0.08–1-fold amounts of Zr; 0.075–1-fold amounts of Bi; 0.05–1-fold amounts of Sr and Mg and 0.036–1-fold amounts of Cd, Nd and Ni. Cr, Mn, V, Au and Pt interfere strongly. To eliminate the interfering effect of Ce(IV), it was reduced by adding a few drops of hydrogen peroxide prior to evaporation. In the absence of H_2SO_4 or H_3PO_4 perchloric acid does not oxidize trivalent cerium [676]. Repeated evaporations with HCl and $HClO_4$, carried out to volatilize chromyl chloride, diminish but do not eliminate the effect of chromium. In the presence of sulfate ions the oxidation of plutonium by perchloric acid is not quantitative, while phosphates bring about the precipitation of plutonium. Iron does not interfere with the determination; iron in amounts up to 40% of that of plutonium was added to the portions of a standard solution of plutonium prior to evaporation without any undesirable results. If the solution contains major amounts of salts (threefold amount of potassium), the residual volume of the evaporated solution cannot be reduced to 2–3 ml, but this is without effect on the results of quantitative determinations.

The authors also studied a number of separations of plutonium from extraneous elements and showed that in such cases the accuracy of the determination of plutonium is limited by the efficiency of the separation.

The authors also verified the possibility of spectrophotometric determination of the end point in the method by the absorption of Ce(IV) at wavelengths below 400 mμ. The considerable absorption by Fe(III) was found to interfere seriously. The use of complexing agents (fluoride or phosphate) to bind Fe(III) failed to give satisfactory results, since precipitates of plutonium fluoride and phosphate were obtained. Under conditions of controlled acidity and phosphate content, the accuracy of the determination of plutonium was 0.2%.

COMPLEXONOMETRIC TITRATIONS

Up till now, complexonometric determinations of plutonium involved the use of only ethylenediaminetetraacetic acid (EDTA, Complexone II), or its disodium salt (Complexone III, Trilon B). EDTA forms stable and readily soluble complexes with many metals, the various uses of which have been described in detail by Přibil [185]. The EDTA titration of plutonium makes it possible to determine the metal in the presence of a large number of cations and anions in hydrochloric, sulfuric, nitric and acetic acid media.

Foreman and Smith [407, 408] studied the complex formation of Pu(III), Pu(IV) and Pu(VI) with EDTA. The complexes with Pu(III) and Pu(IV) were found to have the composition 1:1, and their stability constants were determined. For the hexavalent plutonium these authors also established complex formation, with the composition of the complex 1:1; its stability constant was also determined. The authors noted that the spectrum is similar to the absorption spectrum of Pu(V), and the increase in the intensity of the spectrum as a result of the increase in pH from 1 to 6.5 was ascribed to complex formation.

The data of Foreman and Smith on hexavalent plutonium would seem to require reexamination in view of the results obtained by Kabanova, Danushenkovy and Paley [483] on the study of EDTA complexes of Pu(V) and Pu(VI). It was found by the latter authors that Pu(VI) in 0.5 *N* acid forms a very weak complex with EDTA or else no complex at all is formed, while at pH 3–5 hexavalent plutonium is rapidly reduced by the reagent to the pentavalent state (1 gram-molecule of EDTA reduces 6 gram-molecules of Pu(VI)). Plutonium (V) forms a 1:1 complex with EDTA, whose absorption spectrum is identical with that given by Pu(V).

Klygin, Smirnova and Nikol'skaya [110] studied the behavior of ethylenediaminetetraacetic acid in aqueous solutions and obtained more accurate data on the dissociation constant of EDTA itself, and determined again, more accurately, the stability constants of the complexes of EDTA with U(IV) and Pu(IV).

Determination of plutonium (III)

Milner and Woodhead [550] studied the complex formation between Pu(III) and EDTA in order to develop a volumetric determination of milligram amounts of plutonium.

The method involves back-titration of the excess EDTA not bound to Pu(III) as a 1:1 complex against a standard solution of thorium nitrate at pH 2.5, in the presence of a mixed indicator of Alizarin S and methylene blue. In the determination of about 10 mg Pu the error is less than 1%. Elements which form stable complexes with EDTA at pH 2.5, viz., iron, titanium, thorium and probably gallium and vanadium, interfere with the determination.

In preliminary experiments the composition of the complex of Pu(III) with EDTA was established by means of acid-base titration. The authors concluded that the presence of Pu(IV) should not affect the stoichiometry of formation of the complex, while Pu(VI) probably interferes.

Attempts to effect a direct titration of Pu(III) against EDTA were unsuccessful. Indicators used during direct titration of other elements failed to give positive results in the titration of Pu(III), and back-titration was accordingly resorted to. For back-titration solutions of ferric iron with salicylic acid as indicator are suitable. A sharper equivalence point is given by solutions of thorium nitrate with Alizarin S as indicator at pH below 3.5. Titration was usually conducted at pH $\sim$ 3.0, and the pH value was maintained constant throughout the titration. Some difficulty was caused by the blue coloration of Pu^{3+} ions. When methylene blue was added, the color change at the equivalence point becomes sharper—from green to purple—and the pH should be somewhat below 2.5.

The authors also noted that the end point is more or less sharp, depending on the age of solutions of plutonium. The use of freshly prepared samples eliminates this difficulty.

Standard 0.01 *M* solution of thorium nitrate is prepared by dissolving the salt in water. The titrimetric factor of the solution is determined gravimetrically, by precipitating the thorium as oxalate salt and calcining to ThO_2. With the aid of the standard solution of thorium, the titer of 0.01 *M* solution of Complexone III is established.

Metallic plutonium is dissolved in HCl so as to obtain a final concentration of plutonium of 1 mg/ml and a final HCl content of about 2%. An aliquot of the solution (about 10 mg Pu) is transferred to a 150 ml beaker and is diluted with water, if necessary, to 40 ml. An excess of standard 0.01 *M* solution of Complexone III is added from a buret, and then some 4 drops of 0.05% solution of Alizarin S and one drop of 0.1% solution of methylene blue. The solution is mechanically stirred and a dilute solution of ammonia (1:10) is added until the pH is 2.5. The pH value is controlled with a glass electrode. The sample is then titrated against a standard solution of thorium nitrate until the green color turns purple. The pH of the solution should be kept steady at about 2.5 with the aid of the ammonia solution.

If amounts of plutonium larger or smaller than 10 mg are determined, the accuracy of the determination is impaired.

Determination of plutonium (IV)

Palei and Chubukova (1956, 1957) developed a titrimetric determination of Pu(IV) in nitric, hydrochloric and sulfuric acid solutions, in which the excess Complexone III is titrated either against a standard solution of Fe(III) at pH 4–5, with pyrocatechol as indicator, or potentiometrically. Uranium does not interfere. When the indicator was employed, the average deviation in the titration of 10–40 mg amounts of plutonium and 4–13 mg plutonium in the presence of 1–10-fold amounts of uranium

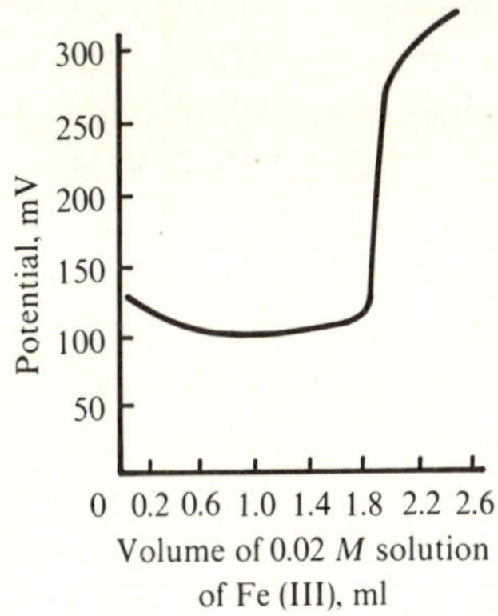

Figure 79

Potentiometric titration curve of excess Complexone III against a solution of Fe (III) in the determination of Pu (IV)

was $\pm$ 1%. When uranium is present in larger amounts, its typical coloration interferes. With potentiometric titration between platinum and calomel electrodes under the same conditions, the average deviation for 7–15 mg plutonium was $\pm$ 0.8%; it was $\pm$ 1.8% for solutions containing 1–50-fold amounts of uranium relative to plutonium.

The titration was carried out in acetate buffer solution at pH 4–5. At pH below 3 both the complex formation and potential stabilization are slow. In the presence of uranium the titration is effected at 50–60° in order to displace uranium rapidly from its complex with the added iron. Unless this is done, the establishment of the equivalence point is slow and the results will be high. A typical potentiometric curve is shown in Figure 79.

A 0.02 *M* solution of Complexone III is prepared from the recrystallized reagent, and its titer determined by titration against a standard solution of iron (III), using pyrocatechol or sulfosalicylic acid as indicator. The 0.02 *M* solution of iron (III) is prepared by dissolving the appropriate amount of iron chloride or sulfate in 0.5 *N* acid. The titer of the iron solution is established against potassium bichromate after reducing the former in a zinc reductor.

A weighed portion of the solution of plutonium (IV) is placed in a cell with a jacket, through which hot water is circulated. The solution is diluted to 10–15 ml with a 0.1 *N* solution of acid excess (not more than 2 ml) of standard solution of Complexone III is added, and then 2–3 ml 50% ammonium acetate until the pH is 4–5 (indicator paper). In the visual titration a few crystals of pyrocatechol are added and the solution is titrated against standard solution of iron (III) until the color changes from pale yellow to dark-green. In the potentiometric titration the electrodes are immersed in the cell, and the iron solution is introduced by portions, first 0.2 ml at a time, then drop by drop, and the volumes and corresponding potentials are recorded. The equivalent point is determined from the potential jump on the titration curve.

One indicator titration takes 8–10 minutes.

Iron, and apparently also a number of other cations, which at pH 4–5 form complexes stronger than iron with Complexone III, interfere with the determination. The authors did not study the interference by cations other than uranium.

The authors studied other substances as possible indicators. Thiocyanate gives a very slow color change at pH 1–2 from yellow through orange to red because of the formation of colored complex with Pu(IV) in the presence of Complexone III. Sulfosalicylic acid is also unsuitable as indicator between pH 1 and pH 5, since the color development at the equivalence point is slow. A more rapid color change from orange to raspberry red is observed when the sample at pH 1–2 is heated to 40–50°, but nevertheless the average error in the titration of 10 mg of Pu is around 7%.

An attempt was made to titrate Pu(IV) directly against Complexone III. Negative results were obtained with Alizarin S at pH 3–4, and with Eriochrome Black T and murexide in alkaline media. When thoron was used as indicator, both direct and back-titration of Pu(IV) by thorium nitrate in 0.1–0.5 *M* solutions of nitric and sulfuric acid could be effected.

The 0.05 *M* solution of Complexone III used in the determination was prepared by dissolving the salt after recrystallization from alcohol. The titer of 0.05 *M* solution of thorium nitrate was determined with the standard solution of Complexone III [438]. A 0.1% solution of thoron was used as indicator.

An aliquot volume of the solution of plutonium (IV) in nitric or sulfuric acid is diluted to a plutonium content of 0.1–1.0 mg/ml and acid concentration of 0.5 *N*. Thoron (5–10 drops) is added and the solution is slowly titrated against the standard solution of Complexone III until the color changes from raspberry red to orange.

In the back-titration, an excess of Complexone III is first added, then 5–10 drops solution of thoron, and the solution is titrated against the standard solution of thorium until the color changes from orange to raspberry red.

The average error in the direct titration of 7–20 mg Pu(IV) was ± 2.9%, while in the back-titration it was ± 2.5%. The equivalence point is not sufficiently sharp, and its sharpness is not improved when the solution is heated to 50–80°.

Palei and Chang Wen Ch'ing [179] described a complexonometric determination of tetravalent plutonium using Arsenazo I as indicator. With Arsenazo I in acid medium, Pu^{4+} ions form compounds which have a bright blue-violet coloration; on the addition of excess Complexone III the color turns pink. The sharpest color change and the minimum error in the titration are obtained in 0.1–0.2 *N* HNO_3. The average deviation for amounts of 2.5–25 mg Pu was ± 0.5%, in addition to the systematic deviation of −0.45%. Uranium in 140-fold amounts, chromium in 4-fold amounts, lanthanum and lead in 3-fold amounts and nickel in 2-fold amounts did not interfere. The average deviation in the determination of 1.4–5.3 mg Pu in the presence of the above listed extraneous elements was ± 1.2%. Iron in the amount of 3–4% of the weight of plutonium does not interfere. Elements giving unstable complexes in acid medium, including Ca, Mg, Zn, Al, Cd and Mn, probably do not interfere; Th and Zr probably interfere.

The method is sensitive to the acidity changes outside the range shown. An alteration of the acidity to 0.04–0.08 *N* HNO_3 gives low results; at acidities of 0.3–0.6 *N* HNO_3 high results are obtained. At HNO_3 concentrations above 1 *N* the titration cannot be effected.

The initial solution of plutonium (IV) nitrate is prepared by evaporating the plutonium sample with sulfuric acid almost to dryness. The evaporation is repeated twice and the residue dissolved in 0.5 *N* HNO_3. A solution of Complexone III about 0.005 *M* is prepared by dissolving 1.86 g recrystallized salt in one liter of water. The titer is established against standard calcium chloride [214]. The titration is carried out in the presence of 0.4% solution of Arsenazo I.

A portion of the solution of plutonium (IV) is diluted to a concentration of 0.1–0.2 mg/ml Pu, the acidity being adjusted at the same time to 0.1–0.2 *N* HNO_3, 10–20 drops of Arsenazo I solution are added for each 40–100 ml, and the solution is then titrated against a standard solution of Complexone III until the color changes from blue to violet to pink. The blue-violet color returns after the titrated solution has stood for some time; this should be ignored. For a better determination of the end point a comparison solution may be employed, consisting of an aqueous solution of the indicator in the same concentration and at the same acidity.

Chang Wen Ch'ing [241] studied the titration of Pu(IV) against Complexone III in acid media in the presence of a number of indicators other than Arsenazo I. Of the compounds with a structure related to Arsenazo I, i.e., which have an azo group with an arsonic and a hydroxylic group in an ortho-position to it, Thoron I gave the best results. Even though the color change from pink to orange at the equivalence point was not sufficiently sharp, an average deviation of $\pm 0.7\%$ was obtained in the determination of 3–5 mg Pu in 0.1 *N* HNO_3.

Less sharp color changes from pink to orange at the end point was obtained with Thoron II in 0.1 *N* acid and with o-aminophenylarsonic acid-azo-phenyl-J-acid at pH 0.5–1.5 as indicators.

Arsenazo II, Arsenazo III and benzene-2-arsonic acid-4-nitro-<1-azo-7'>-1-amino-8'-hydroxynaphthalene-3',6'-disulfonic acid form in acid medium, respectively, a blue-violet, a bright green and a raspberry-red complex, which are more stable than the complex of plutonium with Complexone III, and are not suitable for use as indicators.

Of other type indicators, Alizarin S may be employed. The experimental conditions for titrating plutonium with this indicator have been determined. The acidity of the solution being titrated must be comprised within a very narrow range, viz., 0.1–0.2 *N* HNO_3. It has been mentioned above that at pH 3–4 the complex of plutonium(IV) with Alizarin S is quite stable, while in solutions more acid than 0.5 *N* HNO_3 it is not formed at all. The color change from raspberry red to orange is not sufficiently sharp, while the reproducibility is not better than 1.8% for 1.5–3 mg Pu in 20 ml.

For various reasons, the use of the following indicators (pH 0.5–1.5) gave negative results: Pyrocatechol Violet, 2,4-disulfobenzaurin-3,3-dicarboxylic acid; nitroso-R-salt; Eriochrome Black T; 1,8-dihydroxy-2-(2'-carboxybenzeneazo)-naphthalene-3,6-disulfonic acid; Acid Chrome Dark Blue; 2,4-dihydroxybenzene-<1-azo-1>-2-

hydroxynaphthalene-4-sulfonic acid and diphenylamine-2-sulfonic acid-<4-azo-2>-dihydroxynaphthalene-3,6-disulfonic acid. The last two indicators were simply decomposed by nitric acid.

Complexonometric titration of Pu(IV) with Arsenazo I was employed by Palei and Kochetkova (1959) to determine the content of plutonium in plutonium dioxide. The dissolution of the dioxide was conducted as in the redox titration of Pu(III) by ammonium vanadate, described above (p. 180).

Determinations of plutonium in 12–40 mg pure dioxide gave an average deviation of $\pm$ 0.15%. In the presence of 10% U, 10% La, 2.5% Cr and 2.5% Fe the average error of 3–4 determinations was about $\pm$ 0.25%.

A 0.005 *M* solution of Complexone III, purified by recrystallization from methanol [299], is prepared. A 50–60 mg sample of the dioxide is placed in a quartz beaker, 500–600 mg KI and 4–5 ml concentrated HCl are added, then the beaker is covered with a watch glass and heated at 100°. After dissolution, sulfuric acid is added to eliminate iodine and the solution is evaporated twice. The residue is dissolved in 2.5 ml of 1.5 *N* HNO_3, transferred to a 25 ml volumetric flask and diluted to the mark with water. An aliquot containing about 10 mg Pu is withdrawn, diluted with 0.1 *N* HNO_3 to 25 ml, 2–3 drops of 0.5% solution of Arsenazo I are added and the solution titrated until pink.

Boase, Foreman and Drummond [302] recently used complexonometric titration for the determination of plutonium in nitric acid solutions of irradiated uranium.

Plutonium as Pu(IV) is extracted by quaternary amine, i.e., p-diisobutylphenoxyethoxyethylbenzyldimethylammonium chloride (Hyamine 1622) dissolved in benzene. Benzene is expelled by evaporation and, after the residue has been dissolved in acetone, an excess of EDTA is added. The excess of EDTA is titrated photometrically against a standard solution of zinc chloride, using dithizone as indicator. Under carefully controlled conditions the method gives a standard deviation of 0.4%; the accuracy in the determination of 100 μg Pu in the sample is $\pm$ 0.3%.

The samples analyzed in this study consisted of solutions in which the uranium content was several thousand times higher than that of plutonium, also containing fission products with β- and γ-activity of about 15 curie per liter of solution. The method was tried out on an artifact solution containing 300 mg/ml of U, 50–150 μg/ml Pu, 3 *M* nitric acid and specially added products of fission.

Prior to extraction all plutonium must be converted into the tetravalent state, since only Pu(IV) is extracted by the amine. This is done by adding a solution of Fe(II) to the sample solution, when Pu(VI) is readily reduced to Pu(IV). In the presence of nitrites further reduction to Pu(III) does not take place. The optimum acidity in the extraction of plutonium is 1 *M* HNO_3 at a concentration of uranium of 300 mg/ml, 2 *M* HNO_3 at a concentration of uranium of 100 mg/ml and 4 *M* HNO_3 if the concentration of uranium is less than 10 mg/ml. The purification factor during the extraction is 100 with respect to uranium and about 10^5 with respect to the fission products. The behavior of the amine during the extraction was studied in detail in [599].

Small amounts of uranium in the titrated solution (5–6 mg U) and the iron employed in the reduction of Pu(VI) do not interfere with the determination. Chromium and nickel do not interfere in concentrations up to 0.5 mg/ml. Fission products in amounts corresponding to the content of plutonium in the sample to be analyzed (Sr—5, Zr—15, Mo—12, Ru—7, Cs—15, La—25, Ba—5 and Ce—10 μg/ml) do not affect the results of the determination of plutonium. When the concentration of the fragments in the solutions was increased to 1 mg/ml, high results were obtained ($\sim$ 4 μg Pu in 2 ml samples). Neptunium interferes, being reduced by Fe(II) to Np(IV), which is then extracted together with plutonium (IV). However, the element is present in the solutions only in small amounts. Oxidants which decompose dithizone also interfere. Ce(IV) decomposes dithizone, but is reduced by Fe(II) to Ce(III).

A sample, containing 100–200 μg Pu, is withdrawn with the aid of a remote manipulation pipet and is transferred in a 3 ml polythene dish into the extractor. Dilute HNO_3 is added, until the acidity is 2 *M* and the concentration of uranium is 100–150 mg/ml. The contents of the extractor are mixed and 0.3 ml 10% solution of Mohr's salt in 0.004 *M* H_2SO_4, 0.15 ml of 1 *M* $NaNO_2$ and 2 ml of 1% solution of the amine (nitrate salt) in double-distilled benzene are added with the aid of the polythene dish. The phases are mixed for 90 seconds, the aqueous phase is separated out and transferred with the aid of another 3 ml polythene dish to a second extractor. The dish is washed with 2 ml solution of the amine and the washings added to the extractor. The phases are mixed, the aqueous phase is separated out and run to radioactive waste. The organic phase is transferred from the second to the first extractor. The second extractor is washed with 2 ml of 4 *M* HNO_3; this solution is used to reextract for 90 seconds in the first reactor. The aqueous phase is separated out and the organic phase is transferred to a 5 ml polythene dish. The extractor is washed with 1 ml solution of the amine in benzene and the washings are introduced into a 3 ml polythene dish.

One ml of the basic organic phase is transferred by means of a pipet into a clean, dry titration cell, equipped with an electric heating coil, and the evaporation of the solution is begun, while stirring with a weak air stream. As the solution evaporates, the remainder of the organic phase is added by portions. The wash liquor is used to wash out the 5 ml dish and then transferred into the cell. The solution is evaporated to a residual volume of about 0.7 ml, the electric current is interrupted and the last traces of benzene are expelled by bubbling air through. After cooling, 50 μl of 4 *M* HNO_3 are added to prevent Pu (IV) from polymerizing. One ml of acetone is added and then an accurately measured (usually 0.12 ml) volume of 0.01 *M* EDTA from a microburet. The liquid is cautiousty stirred, 1 ml of acetate buffer solution is added (pH 4.75) and acetone, until the solution clears up. A solution of dithizone in acetone (0.15 ml, 0.04% w/v) is now added and the excess EDTA is titrated against 0.004 *M* standard solution of zinc chloride from another 0.5 ml microburet. The end point is determined photometrically, using a filter transmitting at 510 mμ. The tip of the buret should dip into the solution. Between the additions of the titrant the solution is stirred by a weak current of air. Near the end point the volume of each added portion should be 0.002 ml, and the galvanometer readings should be recorded after each addition. The end point is determined from the inflection point on the titration curve. The titer of EDTA solution is determined, as in the titration of plutonium, by zinc chloride.

To determine the consumption of the titrant in the blank titration (usually about 2% of the total volume) a portion of a nitric acid solution of uranium not containing plutonium is taken and all the analytical operations are repeated.

The working technique and the apparatus required for this type of analysis are described in detail in [302].

PRECIPITATION TITRATIONS

A few studies deal with methods involving precipitation titrations.

Gokhshtein and Zaritskii (1951) developed a method in which tetravalent plutonium is amperometrically titrated against potassium iodate; the sparingly soluble $Pu(IO_3)_4$ precipitates out in aqueous acetone solution containing 0.6 *N* HNO_3. The course of the titration is followed amperometrically on the reduction wave of iodate ions on a rotating platinum electrode. The equivalence point is located near the minimum of the curve. The average deviation in titrating 1.5–6 mg Pu was about $\pm$ 1.5%. Fourfold amounts of Cr(III) and up to 25% Fe(III) (average deviation in the presence of these ions was $\pm$ 1.6%) and uranium do not interfere. Lanthanum and lead interfere with the determination.

The authors investigated the composition of the precipitate as a function of the composition of the solution, of the solubility of plutonium iodate, the polarographic behavior of the iodate ions and other factors. It was found that the solubility of plutonium(IV) becomes much smaller in water–acetone solutions. The volume ratio between acetone and the aqueous solution at the equivalence point should be about two.

The authors found that when iodate ions are reduced on platinum electrode in acid aqueous solutions, two waves result, at + 0.3 V and −0.02 V with respect to mercury sulfate electrode in 1 *N* H_2SO_4. When the concentration of iodate ions in aqueous solution is less than 0.01 *N*, and also in water–acetone solutions, the second wave shows a strong shift towards negative potentials and may be used to follow the course of titration at −0.4 to −0.6 V.

The initial 0.2 *N* aqueous solution of KIO_3 is prepared and its titer determined iodometrically with sodium thiosulfate. The water–acetone solution of KIO_3 for the titration is freshly prepared prior to each determination by adding to the initial solution an equal volume of acetone. The titer of the water–acetone solution is also checked iodometrically against sodium thiosulfate.

Electrodes: platinum electrode, 8 mm long, 0.5 mm in diameter, rotating at 960 rpm, and a mercury sulfate electrode Hg/Hg_2SO_4, 1 *N* H_2SO_4.

The measurements were carried out with an ordinary polarograph; the voltage applied to the electrodes was −0.4 to −0.6 V.

Three ml of the solution of Pu(IV) to be analyzed in 1.5 *N* HNO_3 are placed in a 25 ml electrolytic cell, 6–7 ml acetone are added and the electrodes are dipped into the solution. The applied voltage is set at −0.4 to −0.6 V on the polarograph and the galvanometer is set at a sensitivity which will ensure a suitable deflection by the current. The solution is then titrated against a water–acetone solution of 0.1 *N* KIO_3, added from a 10 ml buret in 0.1 ml portions, and the values of the current are recorded. The solution is stirred by bubbling through an inert gas during the titration.

When constructing the titration curves in the coordinates volume of titrant vs. current intensity, a correction must be introduced for the dilution by the titrant. This is done by using the formula $i = i_e(V + v)/V$, where i_e is the experimental value of the current intensity, V is the initial volume of the solution and v is the volume of the titrant at each point. The equivalence point is obtained by continuing the two branches of the curve as straight lines to their intersection point. The curves have an inverted *L*-form.

The method was subsequently further improved by Gokhshtein (1952). It was found that traces of chlorides increase the sensitivity of the indicator electrode by a factor of more than two. This effect can be produced simply by employing a calomel half-cell as the second electrode. The increase in sensitivity is apparently due to a decrease in the overvoltage of the reduction of IO_3^- ions, accompanied by merging of the two iodate reduction waves into one. The titration potential is then + 0.10–+ 0.08 V relative to a saturated calomel electrode.

While studying the dependence of the solubility of plutonium iodate on the composition of the solution it was found that the presence of iron, chromium and uranium cations increases the solubility of $Pu(IO_3)_4$. The same effect is produced by increasing the acidity of the solution; at the equivalence point this should not exceed 0.5 *N* HNO_3.

The graphic determination of the end point was made more precise and was conducted according to the method of Langer and Stevenson [517], in which the equivalence point is taken to be the intersection of asymptotes of a hyperbola. (Figure 80 shows an example of this method. The first asymptote is the abscissa, the second is the line DD′ found by marking off equal segments on any two parallel lines so that C′D′ = A′B′ and CD = AB.) It should be borne in mind that if 0.2 *N* or more concentrated KIO_3 solutions are employed, excess KIO_3 may precipitate out with consequent distortion of the ascending branch of the titration curve.

To preserve the sensitivity, the platinum electrode was periodically treated with aqua regia for a few seconds.

On introducing the improvements just described in the titration technique, the average deviation, obtained in titrating samples containing about 20 mg Pu in the presence of up to 10 mg U, 6 mg Cr and 0.6 mg Fe, was ± 0.75%. An example of a titration curve is shown in Figure 80.

The amperometric titration of Pu(IV) against iodate in water–alcohol media is carried out in a similar manner (Gokhshtein, 1953). In the titration of 10–20 mg Pu in the presence of up to 300% U, up to 100% Cr, up to 30% Mn, up to 10% Fe and up to 5% La relative to the weight of plutonium, the average deviation was 0.9% and the reproducibility was ± 0.6%.

The volume ratio of ethanol to water prior to titration should be 2.2–2.4, and the acidity should not be higher than 0.4 *N* in HNO_3. At the equivalence point the volume ratio is not less than 1.6. The titrant is standard 0.1 *N* KIO_3 in 50 vol. % ethanol.

The applied emf should be + 0.1 V relative to a saturated calomel electrode. At this potential value the diffusion current is produced by the reduction of plutonium, iron and other admixtures, but mainly by the reduction of iodate ions which have a much higher diffusion coefficient and accept four electrons during the reduction.

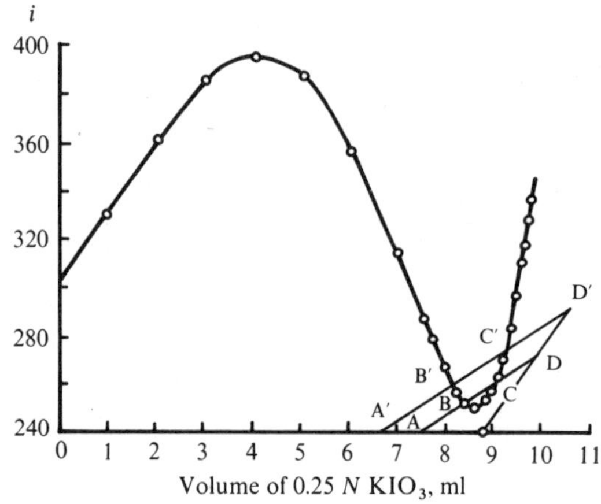

Figure 80
Amperometric titration curve of Pu(IV) against potassium iodate in the presence of Fe(III), Cr(III) and U(VI)

The effect of lanthanum, which forms an insoluble iodate, is limited by its slow precipitation rate. The rate of separation of lanthanum iodate becomes much slower in the presence of citric acid (~ 100 mg per sample), owing to the complexing of lanthanum. The addition of citric acid is also recommended to bind iron and other foreign elements. The acidity of the solution prior to titration should not exceed 0.2 *N*.

Dokuchaeva (1953) studied the amperometric titration of Pu(IV) against iodate with the use of a mercury dropping electrode. The titrations were conducted in 50 vol. % ethanol containing 0.1 *N* HNO_3, by standard 0.5 *N* HIO_3 in 50 vol. % ethanol. The voltage varied from −0.6 V at the beginning to −1.2 V at the end of the titration. A calomel half-cell was used as the other electrode.

An average deviation of about 1% was obtained for plutonium concentrations between $3 \cdot 10^{-4}$ *M* to $2 \cdot 10^{-3}$ *M*. The lower limit of determination is about $1 \cdot 10^{-4}$ *M*. In the presence of fivefold amounts of uranium and equal amounts of iron and chromium the average error was 2.5%.

Lead greatly interferes with the determination. Satisfactory results were obtained after separation of lead by three successive extractions as diethyldithiocarbamate complex from 2.5–3 *N* nitric acid into chloroform. The effect of the excess sodium diethyldithiocarbamate on the titration was eliminated by introducing a few milligrams of cobalt nitrate into the solution. In the presence of lead one determination takes 30 minutes.

Ions (other than iodate) giving sparingly soluble precipitates with plutonium may probably also be used in precipitation titrations of plutonium. This is the case, for example, of potassium hexacyanoferroate, the interaction of which with plutonium ions was studied by Anderson [273]. According to Nebel [573], when Pu(III) and Pu(IV) are titrated against ferrocyanide ions, a diffusion current is produced at + 1 V when the precipitation is terminated.

ELECTROCHEMICAL METHODS

Electrochemical methods of analysis are particularly conveniently adapted to automatic control and remote control, which is particularly necessary in the industrial manufacture of plutonium. Intensive research is now in progress to adapt the highly selective, sensitive and accurate coulometric and polarographic methods to the determination of plutonium.

COULOMETRIC DETERMINATION OF PLUTONIUM

Coulometry at controlled potential

Coulometric analysis at controlled potential is based on the well-known laws of electrolysis and is distinguished by the wide range of concentrations of plutonium which can be determined (from a few dozen micrograms to several hundred milligrams), by its high accuracy (better than 0.1%) and sensitivity (up to a few micrograms). The selectivity of the method is based on the differences in the redox potentials of the different elements, which may be adjusted by the addition of suitable complexing agents.

In coulometric determinations the main requirement is that the electrode systems used ensure a 100% efficiency of the passing current. In other words, side reactions at the electrodes must be absent. Then, in accordance with Faraday's law, the amount of the reacted substance in grams will be

$$W = \frac{M \cdot Q}{96{,}484\, n}, \tag{1}$$

where M is the atomic or molecular weight of the oxidized or reduced substance in the sample; Q is the measured quantity of electricity and n is the number of electrons participating in the reaction. The number 96,484 is the number of coulombs required to oxidize or reduce one gram-equivalent of a substance (one faraday). Thus, the accuracy of the determination will depend on the accuracy of the determination of the quantity of electricity. Coulometry at controlled potential does not involve an end point determination, since the process of electrolysis ends by the establishment of an equilibrium, corresponding to the potential applied to the working electrode.

The potential of the working electrode in the absence of externally applied emf is determined by the concentration ratio of the two ions forming the reversible redox system, in accordance with Nernst's law:

$$E = E_0 + \frac{RT}{nF} \ln \frac{C_{ox}}{C_{red}} \tag{2}$$

or, at 25°:

$$E = E_0 + \frac{0.05916}{n} \log \frac{C_{ox}}{C_{red}},$$

where E_0 is the formal potential of the system, and C_{ox} and C_{red} are the respective concentrations of the oxidized and the reduced ionic forms.

When an external emf is applied to the working electrode, a new concentration ratio between the ions participating in the electrochemical process is rapidly established near the electrode surface; this ratio is a function of the applied potential. The electrolytic current is limited by the rate of diffusion of the ions in the layer near the electrode and will depend on the rate of stirring of the solution. When the concentration at the electrode has become equal to the concentration in the bulk of the solution, the current intensity becomes zero and the solution is at equilibrium with the applied potential. The amount of electricity Q, determined by a suitable instrument, is a measure of the amount of the electrolyzed substance in accordance with equation (1).

The potential of the working electrode is a fundamental parameter in this type of coulometry, and therefore the stability of the potential must meet certain requirements. Figure 81 shows the dependence between the degree of completion of the reaction and the electrode potential according to Nernst's law. It follows from the diagram that the effect of the applied potential on the final state of the system is not the same at various potential values, and is at a maximum at the degree of completion of 50%, i.e., at the formal potential E_0.

The differentiation of equation (2) yields the relationship between changes in concentration and changes in potential:

$$\frac{dC_{ox}}{dE} = \frac{nF}{RT} C_{ox} (1 - C_{ox}) = \frac{nF}{RT} C_{ox} C_{red}, \tag{3}$$

which is convenient for the calculation of errors, and in which the concentrations are expressed in relative units, i.e., $C_{ox} + C_{red} = 1$. At 25° we have

$$\frac{dC_{ox}}{dE} = 38.92\, n\, C_{ox} (1 - C_{ox}). \tag{3a}$$

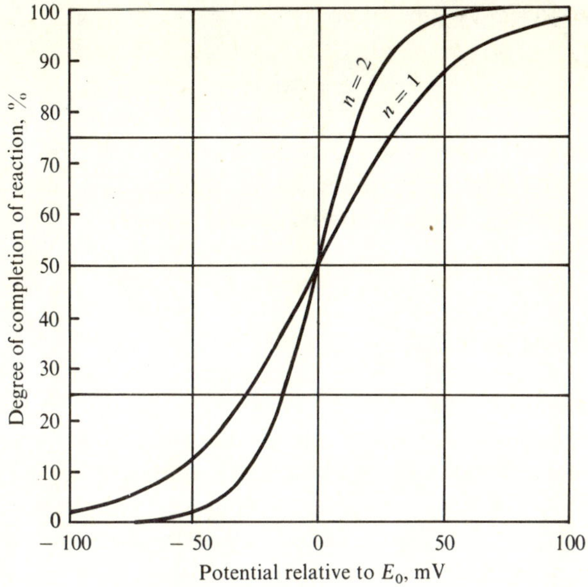

Figure 81

Dependence of the degree of completion of the electrode reaction on the potential and on the number of electrons participating in the reaction

Figure 82 plots the errors associated with the deviation of the electrode potential from the applied value, for one-electrode reactions. The curves have been calculated for various degrees of completion of the reaction from Nernst's equation for finite increments:

$$\Delta E = \frac{RT}{nF} \ln \frac{C_{\text{red}}(C_{\text{ox}} - \Delta C)}{C_{\text{ox}}(C_{\text{red}} + \Delta C)}. \tag{4}$$

It will be seen from these calculations that, for an experimentally attainable degree of stability of the applied potential, which is about 1 mV, and $n = 1$, the electrolysis may be performed with an error of 0.004%, 0.04% and 0.35% for the degrees of completion of 99.9%, 99% and 90% respectively. The corresponding potential values differ from the value of the formal potential by 180, 120 and 60 mV. Since the process usually involves both oxidation and reduction, the error must be approximately doubled.

In spite of the assumption that the electrolysis has proceeded to the end, i.e., to the state of equilibrium at the given applied potential, the reaction may be interrupted earlier, which may be advisable if it is desired to save time. The unreacted material may be corrected for with the aid of the equation proposed by Lingane [522, 523] for the current vs. time curve:

$$I = nFAD\,\frac{C_0}{\delta_0}\,\exp\left(-\frac{AD}{V\delta_0}\,t\right), \tag{5}$$

where A is the area of the electrode, D is the diffusion coefficient, C_0 is the initial concentration of the substance, δ_0 is the thickness of the diffusional layer and V is the volume of the solution. Since the initial current intensity

$$I_0 = nFAD\,C_0/\delta_0,$$

we may write

$$I = I_0 e^{-t/T}, \tag{6}$$

where $T = V\delta_0/AD$ is the time constant of the electrolyzer, defined as the time in which the electrolysis current drops by a factor of $e = 2.73$. The exponential nature of the decrease in current intensity is often noted experimentally. Figure 83*a* represents the current recorded in our experiments during the reduction of Pu(IV) and oxidation of Pu(III) in 1 *N* HNO_3, while Figure 83*b* represents the same curves in $\log I$ vs. t coordinates; their linear character demonstrates that the current intensity decreases exponentially.

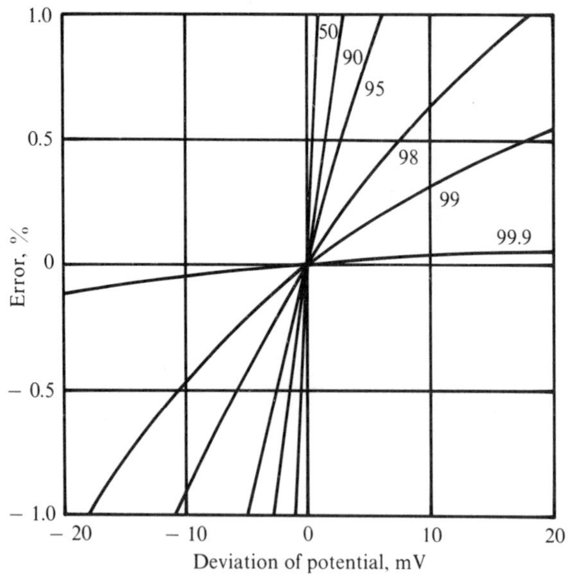

Figure 82

Variation of the degree of completion of a one-electrode reaction as a function of the deviation of the potential from the applied value. (If the signs of both coordinates are changed, the curves will correspond to 50, 10, 5, 2, 1 and 0.1% completion respectively.)

The amount of electricity consumed during electrolysis is

$$Q = \int_0^t I(t)\,dt$$

or, in view of equation (6)

$$Q = \int_0^t I_0 e^{-t/T}\,dt = I_0 T\,(1 - e^{-t/T}) = Q_0\,(1 - e^{-t/T}), \tag{7}$$

where Q_0 is the full amount of electricity required for all the substance to be electrolyzed (at $t \to \infty$). Eliminating t from equations (6) and (7), we obtain the following important relationship:

$$\frac{I}{I_0} = 1 - \frac{Q}{Q_0}, \tag{8}$$

in which the current and the quantity of electricity in relative units are directly interrelated. This equation may serve to follow the course of electrolysis. Thus for instance, if the current intensity is 1% of the initial value, the amount of unreacted substance is also 1%. Unless the determination is to be highly accurate, the electrolysis may be interrupted at this point, and the proportion of the unreacted substance corrected for.

In view of the above, equation (6) may be written in a form suitable for estimating the time required for electrolysis:

$$t = 2.3\,T \log \frac{I_0}{I}. \tag{9}$$

The time constant T, which will depend on the cell volume, area of the electrode, rate of stirring, etc., is a fundamental constant of the electrolyzer and may be determined from equation (6) or (9), or else from the relationship

$$T = \frac{Q_0}{I_0}$$

For practical purposes, a more useful measure of the efficiency of the apparatus is the time of electrolysis $T_{0.01}$, defined as the time at the end of which the current intensity drops to one-hundredth of its initial value: $T_{0.01} = 4.6\,T$. If the decrease in current is exponential, the electrolysis will be 99% complete at the end of the time $T_{0.01}$.

The current relationships given above are a faithful representation of the course of electrolysis for high degrees of completion and are therefore independent of the

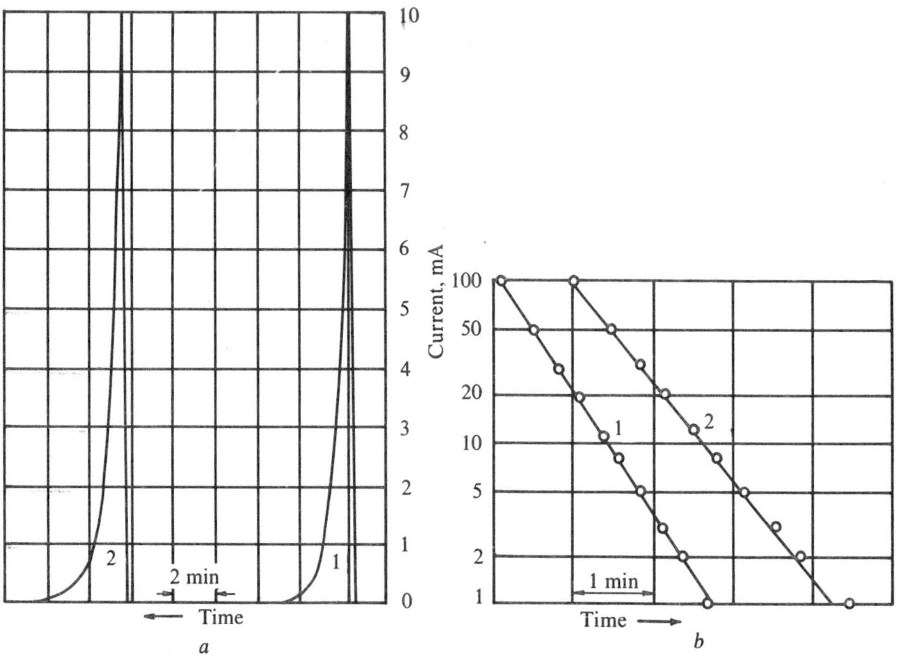

Figure 83

Current curves in ordinary (*a*) and semilogarithmic (*b*) coordinates during the coulometric oxidation of Pu(III) and reduction of Pu(IV) at controlled potential.

Amount of plutonium: 0.005 mole. 1) oxidation at + 830 mV; 2) reduction at + 470 mV (with respect to saturated calomel electrode). The time of electrolysis $T_{0.01}$ for curve 1 was 2.6 minutes; for curve 2 it was 3.3 minutes.

potential of the working electrode. If the degree of completion of the reaction at equilibrium is lower, the currents, and consequently, the efficiency of the electrolyzer decrease and become a function of the applied potential.

The measurement of small amounts of the substance to be analyzed introduces special factors, which are mostly connected with the decrease in the efficiency of the current. Two main processes may be distinguished. Firstly, the electrolytic current becomes of the same order of magnitude as the residual current due to the background electrolyte. Even though the redox potentials of the background ions are usually very different from those of the ions being analyzed, there is still a possibility of their participating in the electrolysis. Secondly, the capacitance current of the double electric layer on the working electrode, which passes through the cell at the initial moment of the electrolysis, becomes of major importance. Both these components of the current are sufficiently reproducible to be determinable in a blank experiment in the presence of the background electrolyte; suitable correction can then be introduced.

The cells used in coulometry at controlled potential have an anodic and a cathodic compartment divided by a porous partition. One compartment is the working compartment and is filled with the solution to be analyzed; the background electrolyte is placed in the other compartment. The potential of the working electrode is controlled by a comparison electrode placed in a special compartment in the electrolyzer, in electrolytic contact with the working compartment of the cell. The working electrode may be constituted by the mercury surface, or by a solid platinum electrode. To reduce the time of the determination, the surface area of the electrode should be sufficiently large in relation to the cell volume under conditions of efficient stirring. The electrolyzer can be considered as satisfactory if its time constant does not exceed 100 seconds.

It should be pointed out that the recent development of this technique became possible due to advances in electronics. Coulometers work on the following principle. The potential of the working electrode, which is measured by the auxiliary electrode, is compared in the instrument with the desired potential. The difference between the two values is amplified and picked up by the current source which feeds the electrolyzer so as to maintain the potential of the working electrode at the desired level. The electrolytic current is integrated in a special instrument; the sum total of the current is shown by the integrator as a potential difference which may be measured and recorded. Instruments with full electronic control have a rapid response to any alterations in the parameters of the electrolysis and cell and ensure a highly accurate measurement.

Determination of plutonium in citrate, nitrate and hydrochloric acid solutions

The first paper on coulometric determination of plutonium at controlled potential was read by Scott and Peekema [641] at the Second International Conference on the Peaceful Uses of Atomic Energy. The results of this work were confirmed in [642].

The following readily reversible electrode reaction was used in the determination:

$$Pu^{3+} \rightleftharpoons Pu^{4+} + e^- . \tag{10}$$

It was found that acetates, citrates, oxalates and tartrates produce a change in the formal potential which is sufficient to oxidize plutonium to the tetravalent state without dissolution of mercury at the anode. Citrates, which had been previously used [306] in coulometric determination of uranium, were investigated particularly thoroughly.

Citrate background electrolyte was prepared by dissolving 1 mole of citric acid in a 0.1 molar solution of aluminum sulfate in water, adjusting the pH to 4.5 with a solution of KOH and diluting to one liter. The Pu(IV)/Pu(III) redox potential in this medium is −0.19 V relative to saturated calomel electrode.

When working with a platinum cell, nitric, sulfuric, hydrochloric and perchloric acid solutions may be used. Solutions of 1 *N* HNO_3 and 1 *N* HCl were studied, in which the potentials E_0 are approximately + 0.67 and + 0.71 V relative to saturated calomel electrode. Solutions of HNO_3 were made 0.01 *M* in sulfamic acid to eliminate the interference by nitrite ions.

As a rule, the reduction was effected at a potential 0.120 V lower, and the oxidation at a potential 0.120 V higher than the formal potential in the given medium.

Accuracy. The method is suitable for the determination of concentrations of plutonium between 0.05 and 50 grams per liter of solution. In citrate medium, at a concentration of plutonium of 20 g/liter, the standard deviation was ± 0.05% and the accuracy was ± 0.1%. Comparable results were obtained with a platinum electrode. At a concentration of 50 mg/liter the standard deviation was ± 0.3% and the accuracy ± 0.5%. The determination of plutonium in solutions of irradiated uranium is possible in citrate media after the separation of plutonium by lanthanum fluoride precipitation, or in an acid medium without any separations. It was pointed out [642] that for solutions of irradiated uranium a standard deviation of 0.5% and an accuracy better than 1.4% were obtained in the determination of 100–200 μg Pu in the sample.

Interfering elements. Certain elements may interfere with the determination of plutonium to the extent to which they participate in redox reactions at the potentials employed. The effect of As, Cd, Cr, Co, Cu, Au, I, Fe, Pb, Mn, Hg, Mo, Ni, Pd, Ru, Ag, Sn, U and Zn was studied. Data were collected on the values of the formal potentials of reversible reactions from which the values of the deviations can be calculated. Table 17 gives the calculated values for the elements which may react during the determination of plutonium. An experimental verification of the effect of iron and mercury in 1 *N* HNO_3 showed that it is in agreement with the calculated values (Table 17). The effect of gold in this medium is less than the calculated value, probably due to the irreversibility of the electrode reaction on the platinum surface.

The effect of iron in all these media is considerable, but owing to the high reversibility of the electrode reaction of iron, an accurate correction may be introduced in the results by determining the concentration of iron. Solutions with iron concentrations 50 times higher than the concentration of plutonium can be analyzed, but the error in the determination of plutonium will clearly increase as a result.

Of the irreversible species which are titrated together with plutonium and for which a correction cannot be introduced, oxygen and nitrites are the most important. Oxygen is removed prior to electrolysis by bubbling an inert gas through the solution. The concentration of nitrite ions is reduced to a permissible value by adding a small excess of sulfamic acid to the background electrolyte.

Substances which catalyze competing electrode reactions, decomposition of water or disproportionation of Pu(IV) to Pu(III) and Pu(VI) may also interfere. Thus, for instance, ruthenium removes the overvoltage of hydrogen evolution on the mercury surface. The interference by disproportionation of Pu(IV) was not noted in the study.

Table 17

PERMISSIBLE AMOUNTS OF INTERFERING ELEMENTS [641]

Element	Weight ratio element : Pu corresponding to an effect of 1% ($E = E_0$ (Pu) $\pm$ 0.120 V)		
	citrate ion	1 N HNO_3*	1 N HCl
Cu	0.006	> 100	Not detd.
Au	> 100	~ 0.02	" "
Fe	0.014	0.17	1.0
Pb	4.6	> 100	Not detd.
Hg	> 1000	0.025	" "
Mo**	—	> 100	" "
Pd	100	0.04	" "
Ru†	—	> 10	" "
Ag	> 100	1.7	" "
Sn**	—	> 100	" "
U(VI)	~ 20††	> 1000	> 1000

* The solution contained 0.01 M sulfamic acid.

** Tin and molybdenum in citrate medium give irreversible reactions. Samples of plutonium which contain molybdenum and tin may be analyzed in an acid medium.

† Ruthenium in citrate medium interferes by decreasing the overvoltage of hydrogen on mercury. The permissible amount of ruthenium is not a function of the concentration of plutonium and was about $2 \cdot 10^{-7}$ g-molecule in the cell used.

†† The reduction of U(VI) is irreversible.

Cells. Electrolytic cells are electrolyzers with a well-developed surface of the working electrode and separate anodic and cathodic compartments, designed for work with highly radioactive solutions. Figure 84a shows the diagram of a cell with a mercury electrode and Figure 84b shows a cell with a platinum electrode. The three tubes with glass filters of the mercury cell are shown in the same plane, even though they are actually grouped around the stirrer planchet in order to save space. Both cells have a capacity of 5–10 ml of solution. In mercury electrode cell the agitation is effected with a magnetic stirrer, while in the platinum electrode cell it is effected with the gas serving for deaeration of the solution. The platinum electrode is made of platinum gauze with a star-shaped cross section. The porous filters are saturated with silica gel to reduce solution flow and to increase their electrical conductivity. The auxiliary electrodes in both cells (not shown in the diagram) are platinum spirals, while the reversible electrodes consist of saturated calomel half-cells. The compartments of these electrodes in the mercury electrode cell are filled with 1 N H_2SO_4 and in the platinum electrode cell with 1 *N* HNO_3. The filters consist of a layer of fine glass wool 125 mm thick in a 19 mm tube. The solutions are drained by vacuum. In the study under discussion the deaeration of the solutions was effected by helium, since this gas need not be purified from oxygen.

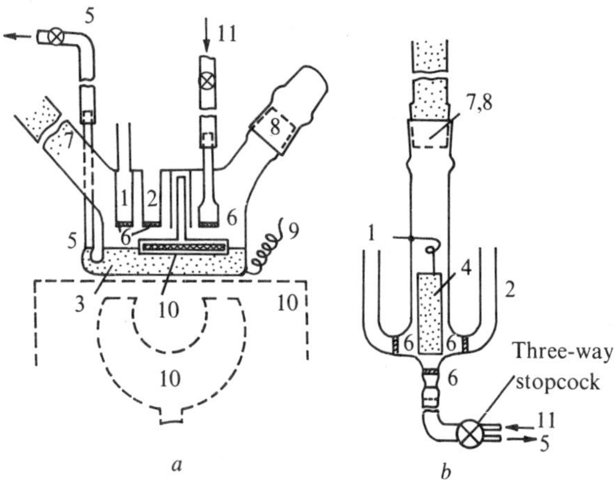

Figure 84

Diagrams of electrolyzers with mercury (*a*) and platinum (*b*) electrodes:

1) anodic compartment; 2) reference electrode compartment; 3) mercury electrode; 4) platinum electrode; 5) tube in evacuated vessel to remove the solution; 6) porous glass filters; 7) vent filter; 8) feed inlet; 9) mercury contact; 10) parts of magnetic stirrer; 11) inlet of helium.

Instrument. The coulometer used in the study was similar to the one designed by Boomen [305]. The block diagram of the instrument and the diagram without amplification have been given in the paper. The main constituent parts of the instrument are typical difference amplifiers, with a very large amplification, used in model electronic instruments. Three such amplifiers are used, two of which form part of the potentiostat circuit, while the third acts as current integrator. The working principle of coulometers has been discussed above.

The instrument has a provision for continuous adjustment of the applied potential between 0 and 1–2 V, of either polarity, five working current ranges (0.05, 0.5, 5, 50 and 500 mA) and six ranges of amounts of electricity (0.1, 0.25, 0.5, 1.0, 2.5 and 5 coulombs). The instrument also has a switch which can be set at *reduction*, *oxidation* or *initial state*. A connection to a 50 mV autopotentiometer and another to an accurate one-volt potentiometer serve to record and to measure the amount of the current which has passed through the cell. In Boomen's instrument [305] the potential control was accurate to within 3 mV on a stirred mercury electrode; the accuracy of the instrument under discussion was probably the same.

Experimental procedure. The background electrolyte and the sample are placed in the appropriate cell, and the solution is stirred and deaerated for 5 minutes. The potential required to reduce plutonium to Pu(III) is set, and the current is switched on. The time required for complete reduction may vary from 5 to 30 minutes, depending on the initial valency state of the sample. When the reduction current has decreased to the value of the residual current, the electrolysis is interrupted, the potential is set at a new value, which is required to oxidize plutonium to Pu(IV) and the current is again switched on. With the cells described above the time required for the oxidation is about 10 minutes. At the end of the oxidation the coulometer reading is taken and corrected, if necessary, by the amount corresponding to the blank determination. The residual currents are particularly small for the platinum cell (up to 10 μA).

It will be seen that the experimental technique is very simple and gives very accurate results.

Determination of plutonium in perchloric acid solution

A similar work on the coulometric determination of plutonium at controlled potential was carried out by Schults. Some information on the work may be found in [491, 664].

Plutonium was determined in 1 *M* $HClO_4$. The formal potential of Pu(IV)/Pu(III) in this medium under the conditions of the titration was + 720 mV, while that of Fe(III)/Fe(II) was + 470 mV with respect to a saturated calomel electrode [665], which is sufficient for a selective determination of plutonium in the presence of moderate amounts of iron.

The coulometer which was employed in the work is described in [490, 491] and is simpler than Boomen's instrument [305]. It contains only two difference amplifiers, a transistor-type current amplifier in the potentiostat circuit, which is fed exclusively from AC mains and may be manually or automatically adjusted. In the work the quantities of electricity measured were up to 4 coulombs (about 5 mg U), with a standard deviation of 0.1 %, and up to 0.3 coulomb (about 0.6 mg Pu), also with a standard deviation of 0.1 %. The cell current may be as high as 300 mA, but the current intensity is reduced to 25 mA with the aid of a resistance when using small sized cells. When the automatic control is employed, the electrolysis is interrupted as soon as a preset residual current has been attained.

The electrolytic cell had a platinum gauze working electrode, a saturated calomel electrode, or any other suitable reference electrode and a mechanical stirrer.

The experimental procedure is similar to the one just described.

Eight to ten ml of a solution of plutonium in 1 *M* $HClO_4$ are placed in the cell and oxygen is removed by bubbling through helium, while stirring. An automatic reduction of plutonium to trivalent plutonium is carried out; according to [491], this should be done at a potential of + 0.135 V relative to an Hg/(Hg_2SO_4)Li_2SO_4 electrode to the residual current value of 5 μA. The integrator is then set at zero and the solution is oxidized at + 0.435 V with respect to this electrode until the process is automatically arrested. The integrator reading is taken and the amount of plutonium calculated. These conditions correspond to an electrolysis potential of ± 150 mV relative to the formal potential of plutonium, i.e., for a reaction 0.3% short of completion; this must be allowed for in the calculations.

For amounts of plutonium of the order of 1000 μg the relative standard deviation was 0.1%, for 250 μg it was 0.3% and for 50 μg it was 1.5%. The deviations are mostly due to the drift of the current integrator.

The method is applicable to the analysis of samples containing 40–60 mg/ml Pu in 4.5 *M* HNO_3 in the presence of uranium and iron.

A portion of solution containing about 5 mg Pu was heated with two drops of concentrated HNO_3 to remove reducing impurities. Fifteen ml of 1 *M* $HClO_4$ were added as the electrolyte and 5 drops of a saturated solution of sulfamic acid. The reduction was effected at + 0.560 V relative to a saturated calomel electrode until the current intensity has decreased to the residual value. The Pu(III) thus obtained was oxidized to Pu(IV) at + 0.880 V. The quantity of electricity required for the oxidation corresponds to the amount of plutonium in the sample.

To prevent interference by sulfates, the concentration of sulfate ions should be less than 0.007 *M*.

The working potentials of the reaction differ by + 0.160 V from the formal potential of Pu(IV)/Pu(III), which is 0.720 V with respect to saturated calomel electrode. This means that the reaction is 0.2% short of completion in either direction. In order to be able to neglect the effect of iron (to within 0.1%), its content must not exceed 3% of the content of plutonium.

According to Cooper [360], the method is experimentally used in the analysis of industrial samples of plutonium, along with potentiometric titration with cerium (IV) sulfate.

Handshuh [447] determined plutonium in highly radioactive nuclear fuel reactors by coulometric titration at controlled potential, after quantitative separation of microgram amounts of plutonium from a large excess of uranium, iron and fission products, by means of anion exchange. The average yield of plutonium was 100.0% with a standard deviation of 0.92%.

Determination of plutonium and iron in the presence of each other

We have carried out a detailed study of the coulometric behavior of plutonium and iron at controlled potential in nitric acid media for purposes of its determination in solutions of irradiated uranium (Palei, Sentyurin and Udal'tsova, 1962). Our results were confirmed in [641, 642]. In addition to the problems studied in [641], we also studied the coulometric behavior of the background electrolyte, and the effect of various factors (concentration of background electrolyte, temperature, etc.) on the determination.

Coulometer. The measurements were carried out on TsLA universal-type coulometer. The instrument has two high stability difference amplifiers, especially designed as part of a coulometer. One of them acts as a potentiostat, the other as current integrator.

The controlled potential unit has two independent circuits for oxidation and reduction respectively. The applied potential may vary between 0 and $\pm$ 2.4 V with an accuracy better than 1 mV. Change of potentials, and at the same time the reversal of polarity of the measured and integrated currents is effected by means of an *oxidation-reduction switch.*

The ranges of current, and consequently also of the amounts of electricity can be altered in two steps from 0.125 to 320 mA. The connection between the current and amount of electricity ranges is given by the formula

$$Q = D \cdot \tau \cdot 10^{-2}, \tag{11}$$

where Q is the amount of electricity, coulombs; D is the current range, mA; and τ is the time constant of the integrator, seconds. The time constant of the integrator may assume the values of 2, 4, 8 and 16 seconds. Thus, the lower range of the amount of electricity is $2.5 \cdot 10^{-3}$ coulombs, while the upper range is 51.2 coulombs.

If the initial electrolysis current and the final quantity of electricity cover the full scale of the instrument, the electrolysis time $T_{0.01}$ of the electrolyzer should be between 1.5 and 12 minutes. If less efficient electrolyzers are employed, the initial current will correspond only to part of the scale.

The coulometer has control and measuring instruments for the electrolytic current, amount of electricity and output voltage of the potentiostat. The amount of electricity and the cell current are recorded as a function of time by means of two autopotentiometers, with a paper band 272 mm wide. Such an arrangement is very convenient for following the course of electrolysis and evaluating the results, especially when working with small amounts. For accurate measurements of the amount of electricity the integrator has a special connection to a high accuracy one-volt potentiometer.

The inherent error in the determination of the quantity of electricity in the coulometer, without special calibration, does not exceed 1%. With an electrical calibration of the integrator and with the use of a PPTV potentiometer with it, the experimental error may be less than 0.1%. The checking and adjustment of the zero points of the amplifiers may be carried out with the aid of the control devices in the front panel of the instrument, set at maximum sensitivity. The amplifiers are highly stable, and periodical check of zero points is required only for very accurate measurements.

The coulometer is fully fed from AC mains and the autopotentiometers are the only part which requires systematic maintenance.

Experimental procedure. The instrument is switched on 15 minutes prior to the beginning of the determination, in order to heat up. The cell is inserted, and while the inert gas is bubbled through the cell to remove oxygen, the current range, the time constant of the integrator, the oxidation and the reduction potentials and the operation to be performed in the first electrolysis are all set. The automatic recorder is switched on. The oxidation-reduction switch is set at *electrolysis* and the electrolysis is begun by switching the tumbler of the potentiostat to *electrolysis.* If reduction was the first operation, it is discontinued after the residual current value has been attained by switching the potentiostat tumbler to *control.* If no recording of the quantity of electricity has been performed, the grip of the integrator is set at *measure* and the reading taken. The reading of the integrator is cancelled by returning the grip to *control.* The oxidation-reduction switch is now set at *oxidation*, and the oxidation is performed, the order of the operations being the same as during the reduction.

It will be seen that the operational procedure with a coulometer of this type is very simple and requires minimum attention on the part of the operator. Since the results are recorded by the instrument, they may be evaluated all together at the end of a series of determinations.

Cell. The cell consists of an electrolyzer with separate anodic and cathodic compartments; the working electrode is made of platinum gauze. It is diagrammatically represented in Figure 85 and its design is similar to that described in [641]. The working volume of the electrolyzer is 20 ml. The ratio of the area of the platinum gauze electrode to the volume is 5 cm^2/ml. The platinum gauze is bellows-shaped with its ribs turned towards the diaphragms. The auxiliary electrode consists of 8 coils of 1 mm platinum wire. The comparison electrode is a saturated calomel electrode. The compartments are separated by porous No. 3 glass filters with the diameter of the working surface of 10 mm. The size of diaphragms is sufficient to transmit a 300 mA current. The diaphragms are sealed practically in contact with the central vessel in order to be able to stir the entire volume of the sample solution. A No. 2 glass filter is sealed into the bottom of the working compartment; it serves to drain the spent solution, and to deaerate and stir the solution with the inert gas.

The advantage of the convective stirring with a stream of gas is that moving solid particles are absent. In addition, electrolytic currents reach saturation values at slow bubbling rates of the inert gas, and the parameters of the electrolyzer do not depend on the stirring rate. In the electrolyzer used the required bubbling rate was 25–35 cm^3/minute.

The time of a single electrolysis $T_{0.01}$ in a cell with a nonactivated working electrode is about 5 minutes and depends only slightly on the amount of the substance in the sample (about 20%). This time may be reduced by half if the electrode is activated by applying to it, successively, the potentials of evolution of oxygen and hydrogen.

The inner resistance of the cell when working in 1 N HNO_3 is about 70 ohms.

In order to avoid spattering of minute droplets when the gas is bubbled through the solution, a filter, consisting of a tube filled with glass wool (not shown in the diagram) is placed on top of the cell.

All parts of the instrument are interconnected by means of standard ground glass joints NSh-14.5 and NSh-24.

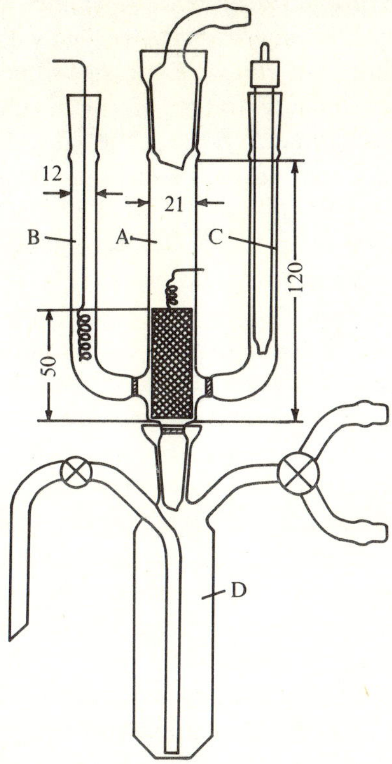

Figure 85

Diagram of coulometric cell with platinum working electrode.

Compartments: A working electrode; B auxiliary electrode; C comparison electrode; D vessel for spent solution.

The working procedure is very simple. The bubbling is begun. The side compartments of the cell are filled with the electrolyte. The filter is removed and a few ml of the background electrolyte, and then the sample solution are introduced into the working compartment. The background electrolyte is added until the working volume is 20 ml and the filter is replaced. The solution is left for 5 minutes until all oxygen has been removed. The measurements are then taken. The cell is disconnected from the inert gas source by means of a three-way stopcock and is connected to a vacuum line. The cell is rinsed and again connected to the source of inert gas. The spent solution is evacuated from the lower vessel by expression with the inert gas, with the three-way stopcock open. All these operations can be effected by remote control.

Solutions. In order to test the method for the determination of plutonium in the presence of uranium and iron, a solution of pure plutonium (IV) in 2 N HNO_3 was prepared and its concentration determined coulometrically as 1.070 mg/ml, with an accuracy of $\pm$ 0.17%. This value was in agreement with the result of a radiometric determination. The solutions of iron were prepared from iron wire and diluted with 2 *N* HNO_3 to the required concentration. Uranium solutions were prepared from recrystallized uranyl nitrate in 2 *N* HNO_3. The sample solutions contained up to 25 mole % of iron, calculated on the plutonium, and up to 2000 times as much uranium.

The background electrolyte was 1–2 *N* HNO_3 containing 1 g/liter of sulfamic acid. The auxiliary cell compartments were filled with 1 *N* HNO_3 (AR) to a level somewhat above that in the main vessel.

Deaeration and stirring of solutions were carried out by a stream of argon from a tank; the gas contained less than 0.005% oxygen. The argon was additionally purified by passing through a solution of pyrogallol in 10% KOH.

Potentials. We studied the dependence of the potentials both on the composition of the background electrolyte and on the temperature.

The potentials were measured coulometrically, by setting the controlled potential at successive values differing by 30 mV and carrying out a complete electrolysis each time. The quantity of electricity vs. potential curves obtained obeyed Nernst's law within the accuracy range of the coulometer. The potential of the middle point on the curve is thus the formal potential of the system.

The determination of the temperature coefficients of the potentials were carried out potentiometrically in a special thermostated cell with platinum and calomel electrodes. To do this, a portion of the solution of plutonium was electrolyzed at the formal potential of the system at a given temperature, or else the solution was simply electrolyzed until the current fell to half its initial value after which the solution was transferred to a potentiometric cell in an atmosphere of argon. By varying the temperature of the cell, the variation of the formal potential as a function of the temperature could be measured.

The calomel half-element was connected to the cell bridge and kept at a constant temperature. The potential Pu(IV)/Pu(III) in 1–2 *N* HNO_3 was 663 mV relative to a saturated calomel electrode at 20° with a temperature coefficient of 1.7 mV/degree (average) between 20 and 80°.

Various working potentials were employed. The reduction potential of plutonium was usually chosen 120 mV lower than the formal potential, while the oxidation potential was 180 mV higher than the formal. In the determination of iron, the reduction potential was set at 120 mV below $E_{Fe(III)/Fe(II)}$; during the oxidation the potential chosen for the oxidation of plutonium was employed.

Behavior of Pu(VI). Since plutonium may be present in the sample solution as a mixture of various valency forms, including hexavalent plutonium, it was necessary to examine its behavior under the conditions of the determination. Solutions of Pu (VI) were prepared by boiling the initial solution in dilute nitric acid, or by evaporation with concentrated perchloric acid followed by the dissolution of the residue in nitric acid.

It was found that Pu(VI) is rapidly reduced at the potential values chosen for the reduction of Pu(IV) to Pu(III). The electrode reduction process involves the transfer of one electron and apparently corresponds to the reversible reaction Pu(VI) + $e^- \rightarrow$ Pu(V). Further irreversible reduction of Pu(V) to Pu(III) probably takes place with the participation of the ions in the nitric acid present as background electrolyte. The reduction processes were controlled by the spectrophotometric method at all stages of the electrolysis. Thus, the presence of Pu(VI) in the sample solution does not introduce any additional complications.

Residual currents. Residual currents in the working potential range (+500–+800 mV) for the determination of plutonium are 20–30 μA. A study of the residual currents in a wide range of potentials showed that at low potentials they are identical with the ascending parts of cathodic reduction currents of hydrogen ions, while at high potentials they correspond to the ascending parts of the anodic oxygen separation current. For various reasons, at the formal potential of Pu(IV)/Pu(III) the residual currents have a minimum value; near the formal potential they change sign and pass through zero. The value of the residual current is a function of time and gradually decreases with the duration of the electrolysis. The contribution of the residual current to the results of the determination increases with the duration of the electrolysis, as the current is integrated. In addition to the residual currents due to the participation of the background ions in the electrode process, the capacitance currents of the double electric layer on the surface of the working electrode also play a major part. These currents appear at the initial moment, when the current is switched on, and contribute to the amount of the electricity measured, depending on the difference between the oxidation and reduction potentials. The value of the capacitance current is determined by the differential capacitance of the double electric layer, which also has a minimum value at the particular potentials used in the determination of plutonium.

The residual and capacity currents are readily reproducible and may be corrected for with satisfactory accuracy if a blank determination is made once a day on the background electrolyte at the same potential values and duration of electrolysis as in the actual determinations. Figure 86 shows a recording of the amount of electricity as a function of time in the determination of about 100 μg Pu and the curve of the integrated capacity and residual currents of the background under the same conditions, for the same sensitivity of the coulometer. The slope of the plateaus of the curves is the result of integration of the residual current, while the initial jump of the background curve corresponds to the total capacity current.

The experimental procedure to be followed when working with a coulometer and a coulometric cell has been described in detail above. The general procedure is as follows.

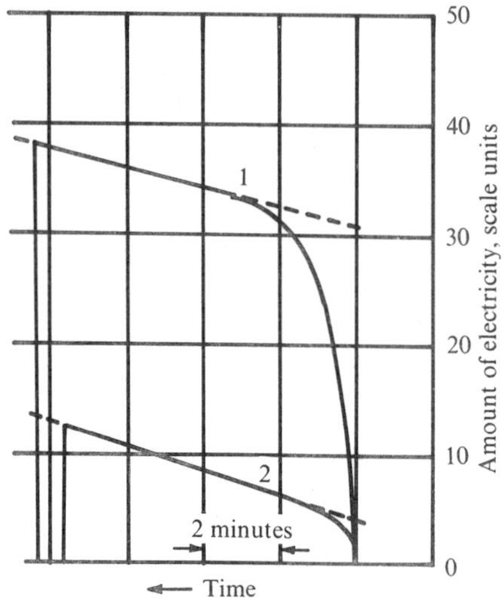

Figure 86

Recording of the amount of electricity in a coulometric determination of 100 μg Pu: 1) curve obtained during oxidation of sample; 2) curve of supporting electrolyte.

A portion of the solution of plutonium, containing 100–200 μg Pu, is introduced into the working compartment of the cell and the supporting electrolyte is added until the working volume is 20 ml. Argon is bubbled through the solution during 5 minutes. A preliminary reduction is now carried out at a potential of + 545 mV relative to saturated calomel electrode, until a residual current of about 30 μA is attained. This operation converts plutonium in all states of valency into trivalent plutonium, including Pu(IV) and polymerized Pu(IV). The time needed for the reduction is about 10–12 minutes. The trivalent plutonium is now oxidized to Pu(IV) at 845 mV until the residual current value is obtained. The experimental conditions, the number of sample and the coulometer range may be recorded directly on the coulometer tape which registers the amount of electricity as a function of the time of electrolysis. The content of iron is now determined; this is necessary in order to be able to calculate the amount of plutonium. The preliminary reduction of iron is carried out at + 330 mV. Since this is accompanied by the reduction of plutonium, the current range of the coulometer must be increased. The oxidation of iron is conducted at 545 mV.

The sequence of the determinations may be reversed, i.e., iron may be determined first, and plutonium next. Moreover, the oxidation and reduction of plutonium may be repeated several times on the same sample solution in the cell; if equilibrium is attained each time, agreeing results will be obtained.

If the potentiostat is working properly there is no need to adhere to the exact potential values as given above and the potentials may be varied according to circumstances; they must not, however, differ from E_0 by less than 90–100 mV.

In order to allow for the residual and the capacitance current the cycle is repeated on the supporting electrolyte alone.

The calculation of the results takes place as follows. The first step is to calculate—from the values of formal potentials, using Nernst's equation (2)—the equilibrium composition of plutonium and iron ions, in molar percent relative to their concentration, at each one of the three working potentials, allowance being made, if necessary, for the temperature. The amounts of plutonium and iron participating in the reaction are calculated in the range of potentials for the determination of plutonium and iron.

Tangents are now drawn to the recorded curve, as shown in Figure 86 by a broken line. The value of the ordinate of the tangent at the initial moment of electrolysis corresponds to the amount of electricity not including residual current (if equilibrium has been attained). By subtracting from it the value obtained in the blank determination, we have corrected for the value of the capacitance current. The results of the oxidation only are used in the calculation, since the reduction is only a preliminary operation which is carried out to convert plutonium from several valency forms into one.

Generally, the amount of plutonium may be calculated by solving two simultaneous equations with two unknowns, the solution of which yields the formulas

$$q_{\mathrm{Pu}} = \frac{Q_1 b_2 - Q_2 b_1}{a_1 b_2 - a_2 b_1} \cdot 100 \quad \text{and} \quad q_{\mathrm{Fe}} = \frac{Q_2 a_1 - Q_1 a_2}{a_1 b_2 - a_2 b_1} \cdot 100, \tag{12}$$

where q_{Pu} and q_{Fe} are the number of coulombs corresponding to plutonium and iron respectively for a quantitative reaction; Q is the number of coulombs established experimentally; a is the amount of plutonium participating in the reaction, %; b is the amount of iron participating in the reaction, %; indexes 1 and 2 refer, respectively, to the potential range of the determination of plutonium and iron. The calculation may be carried out in scale divisions, which are then recalculated to coulombs from the coulometer range as given by (11), and then to micrograms of plutonium in the sample according to (1).

Results. In order to check the reproducibility of the method, about thirty determinations were carried out of plutonium contents in the sample ranging between 100 and 200 μg (a few determinations were made on amounts as high as 1 mg Pu). The results, recalculated to give the content of plutonium in the initial solution, gave an average value of 1.070 mg/ml plutonium, with a mean square deviation of a single determination of $\pm$ 0.7%. A special graduation of the coulometer was not effected.

Determination of plutonium by generation of auxiliary reagent

The scope of application of coulometry at controlled potential may be substantially extended when the method is made to include generation of an auxiliary reagent and a coulometric back-titration of its excess.

Schults [665] determined plutonium coulometrically at a controlled potential by using the reaction between Pu(VI) and Fe(II):

$$PuO_2^{2+} + 2Fe^{2+} + 4H^+ \rightarrow Pu^{4+} + 2Fe^{3+} + 2H_2O. \tag{13}$$

The reaction proceeds at a satisfactory rate in the presence of excess Fe(II). Ferrous iron is electrolytically regenerated at a controlled potential, in accordance with the readily reversible reaction

$$Fe^{3+} + e^{-} \rightleftharpoons Fe^{2+}, \tag{14}$$

and the amount of the reagent thus produced may be accurately determined by coulometry. The excess of the reagent, and also the amount of Pu(III) obtained by its further reduction is obtained by coulometric back-oxidation to Fe(III) and Pu(IV) at controlled potential, since the reactions are reversible.

The method of Schults is similar to the most accurate columetric determination of plutonium known, reported by Waterbury and Metz [717]. The method was applied to the analysis of plutonium samples of the order of 25 mg and gave a standard deviation of 0.1% when one-tenth of the sample had been electrolyzed. Tenfold amounts of uranium and iron do not interfere with the determination. The effect of other impurities was not studied as this had already been done [717].

Apparatus. The electronic coulometer and the electric cell have been briefly described above in connection with an earlier paper by Schults on the coulometric determination of plutonium. The work was carried out using the coulometer range of 8 coulombs per volt. Air was expelled from the cell by bubbling through CO_2 from a tank. The gas was preliminarily purified by passing through a single washbottle with distilled water.

Reagents. Metallic plutonium contained 99.93% Pu (atomic weight 239.07). The initial solution was prepared by dissolving the weighed metal and diluting the solution obtained to an accurately known volume with 6 *M* HCl. The concentration of the initial solution was 25.593 mg Pu per ml.

The solution of iron in 1 *M* HCl was prepared from iron wire. The concentration of the iron solution was 5.00 mg/ml. The solution of uranium in 1 *M* HNO_3 was prepared from U_3O_8. Its concentration was 199.1 mg U per ml. All other reagents were of analytical grade.

Preparation of a solution of Pu(VI). Prior to electrolysis, plutonium was oxidized to the hexavalent state with the aid of hot concentrated $HClO_4$; this reagent had been previously employed for the purpose [717]. Satisfactory results will be obtained only if the experimental conditions of the oxidation are rigorously adhered to in order to prevent interference by impurities. The oxidation is carried out in the presence of a small amount of HNO_3, which apparently facilitates the process and reduces the extent of the decomposition of $HClO_4$, as well as ensuring more reproducible titration results. Under these conditions the oxidation takes 3–5 minutes. Prolonged evaporation for one hour does not affect the result of the determination. The solution of Pu(VI) obtained by evaporation is diluted to 10 ml with 1 *M* HCl in a volumetric flask; 1 *M* H_2SO_4 may also be used as diluent. One ml portions (about 2.5 mg Pu) were taken for each titration.

The author studied the effect of aging on solutions of plutonium diluted in this manner. It was shown that the reduction of Pu(VI) is slow, and in 1 *M* H_2SO_4 is only 1% per day at a content of 25 mg Pu in the sample. The aging is less in 1 *M* HCl and still less in 1 *M* $HClO_4$. Nevertheless, in order to obtain an accuracy of 0.1%, the titration is carried out immediately after the solution is diluted.

Conditions of electrolysis. The electrolysis may be carried out in 1 *M* $HClO_4$ or 1 *M* H_2SO_4; these media have been studied in more detail.

Figure 87 shows the ratio between the amount of electricity *Q* consumed in the electrolysis of 5.119 mg Pu(IV) and 1.00 mg Fe(III) and the applied potential *E* (with respect to a saturated calomel electrode) in these media. It will be seen from the figure that the redox potentials Fe(III)/Fe(II) and Pu(IV)/Pu(III) differ by some 250 mV in 1 *M* $HClO_4$, but are almost equal in 1 *M* H_2SO_4. From these, and other data, the following experimental values of formal potentials E_0 for the two elements (with respect to saturated calomel electrode) have been obtained:

	Pu(IV) / Pu(III)	Fe(III) / Fe(II)
1 *M* $HClO_4$	+ 0.72	+ 0.47
1 *M* H_2SO_4	+ 0.48	+ 0.45

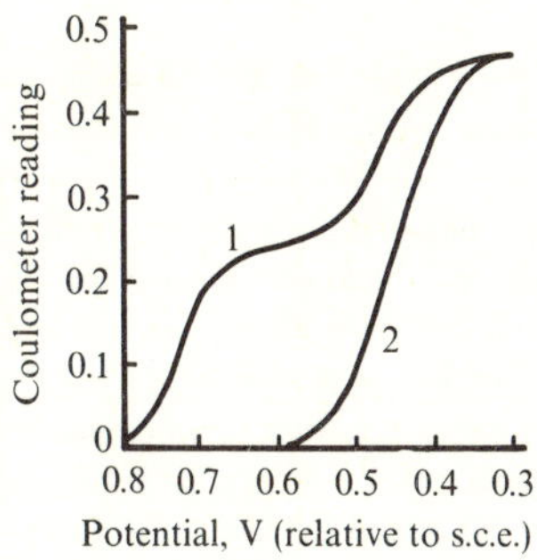

Figure 87

Coulometer readings as function of the given potential at the reduction of 5.119 mg Pu(IV) and 1.00 mg Fe(III):

1) in 1 *M* $HClO_4$; 2) in 1 *M* H_2SO_4.

In the electrolytic regeneration of ferrous iron the potentials chosen were 180 mV below the formal $E_{0\ Fe}$: + 290 mV in $HClO_4$ and + 270 mV in H_2SO_4. At these potential values Pu(IV) itself may be reversibly reduced to Pu(III), and, in the presence of ferrous iron in excess of that required for reaction (13), the solution will contain a mixture of Fe(II), Fe(III), Pu(III) and Pu(IV) ions. It was found that in order to obtain quantitative results the excess of Fe(II) must be of the order of 10–20 μg-equiv per 10 μg-molecule of Pu in the sample to be titrated.

In order that, as a result of the reversible oxidation, all iron and plutonium be obtained as Fe(III) and Pu(IV) respectively, the potentials should be higher than the formal potential of Pu(IV)/Pu(III). The author used potentials higher by 180 mV than $E_{0\ Pu}$ = 900 mV relative to saturated calomel electrode in 1 *M* $HClO_4$ and higher than + 660 mV in 1 *M* H_2SO_4.

The accuracy of the method was evaluated both in perchloric and in sulfuric acid medium. The results were similar. For plutonium samples of the order of 13 mg the standard deviation was 0.2%, while for samples of 25 mg Pu it was 0.1%, in more than five determinations.

Larger amounts of plutonium (up to 50 mg) in the sample do not improve the accuracy of the determination.

Sulfuric acid was chosen as the supporting electrolyte, since the full reduction-oxidation cycle in this medium requires a substantially narrower range of working potentials.

When studying the interfering effects of uranium and iron it was found that the elements do not interfere when each is present in a tenfold amount relative to plutonium, or when the ratio is 100:1 for uranium and 1:1 for iron. Above these values the results of electrolysis are unreliable owing to the formation of the solid phase during the evaporation.

Procedure. To prepare a solution of Pu(VI), an aliquot of the sample solution containing 10–50 mg Pu is withdrawn by means of a pipet and placed in a borosilicate glass beaker of 30 ml capacity and 1 ml conc. $HClO_4$ and 2 drops of conc. HNO_3 are added. The solution is evaporated under an IR lamp until white fumes appear. The beaker is transferred onto a hotplate, 2 ml conc. $HClO_4$ are added and the evaporation is continued for another 10–15 minutes. The beaker is taken off the hotplate, cooled for 1 minute and is then quantitatively transferred into a borosilicate glass 10 ml volumetric flask, while washing out the beaker and diluting to the mark with 1 *M* HCl or 1 *M* H_2SO_4.

The preliminary titration of the electrolyte is carried out at the same time. To do this, 13–14 ml of 1 *M* H_2SO_4 and about 0.5 ml of a standard solution of iron are placed in the cell. Gaseous CO_2 is passed through the cell for deaeration. The electrolyte solution is oxidized at + 660 mV (relative to a saturated calomel electrode) until a residual current of 50 μA is obtained. The electrolysis is discontinued and the integrator is zeroed. If 1 *M* $HClO_4$ was used as the background electrolyte, the preliminary titration should be carried out at + 900 mV relative to a saturated calomel electrode.

To the electrolyte in the cell thus prepared 1.00 ml of the solution of Pu(VI) is introduced with a pipet. The solution is reduced at + 270 mV relative to a saturated calomel electrode (+ 290 mV if 1 *M* $HClO_4$ is used) until the excess of Fe(II) is 10–20 μg-equiv. The electrolysis is discontinued and the reading Q_{red} of the integrator is taken. The integrator is zeroed. The solution is now oxidized at + 660 mV relative to a saturated calomel electrode (+ 900 mV in 1 *M* $HClO_4$) until the current decreases to 50 μA. The electrolysis is discontinued and the integrator reading Q_{ox} is recorded.

In the calculations the difference between the integrator readings ($Q_{red}-Q_{ox}$) is multiplied by the equivalence (mg Pu per volt) and the result is multiplied by 10 to obtain the amount of plutonium in the initial sample. In the investigation, the equivalence of the instrument range chosen was determined by electrical calibration and was 8.1875 coul/V or 10.14 mg Pu per volt for the two-electron reaction under study.

Coulometry at controlled current intensity

Coulometry at controlled current intensity would seem to be less complicated than the controlled potential technique. The main simplification is that during the electrolysis a constant current intensity is maintained. The number of coulombs consumed in the titration is obtained by simply multiplying the current intensity by its duration. Both these values may be determined with an error not exceeding 0.1 %, so that the number of coulombs is obtained with satisfactory accuracy. In order to ensure a constant current intensity, a high-voltage DC source is usually employed, in conjunction with a high resistance, connected in series with the cell and with a battery. This compensates for any changes in the cell resistance during the electrolysis.

Since the current intensity and the potential of the working electrode cannot both be controlled at the same time, the current efficiency decreases towards the end of the electrolysis owing to potential shift. To prevent this effect, an intermediate redox system is introduced into the sample solution. The result is that at the end of the electrolysis, or even throughout its duration, the element being determined is oxidized or reduced by an electrolytically generated reagent. If the amount of the reagent is large, as compared with the amount of the element to be determined, this is not accompanied by a decrease in current efficiency.

Coulometry at controlled current intensity is the electronic analog of the volumetric redox methods, and displays all the characteristics of these methods including low selectivity. The equivalence point may be determined in several ways. It is usually determined potentiometrically, when the current is interrupted during the determination.

In the coulometric determination of plutonium at controlled current intensity both the readily reversible reaction (10) and the largely irreversible reaction (13) have been employed.

Kharlamov (1954) described the coulometric titration of trivalent plutonium by the electrolytically generated Ce^{4+} ions in sulfuric acid solution. The preliminary reduction of plutonium was effected in a reductor filled with zinc amalgam. A bismuth reductor was used for small amounts of plutonium. The optimum acidity of the sample solution is 1–3 N H_2SO_4. The potential jump at the equivalence point was 30–40 mV per 0.04 coulomb. The increase in the acidity to 4 N H_2SO_4 decreased the potential jump. The error for 10 mg Pu in an approximate volume of 50 ml was $\pm 1\%$.

Kharlamov studied the possibility of a coulometric titration of hexavalent plutonium against generated Fe^{2+} ions. The preliminary oxidation of plutonium was carried out with lead dioxide:

$$Pu^{4+} + PbO_2 \rightleftharpoons PuO_2^{2+} + Pb^{2+}.$$

The reaction is independent of the hydrogen ion concentration, but depends to a certain extent on the identity of the acid. An increase in the concentration of sulfuric acid to 4 N impairs the oxidizing power of lead dioxide (apparently by screening the PbO_2 particles by the lead sulfate formed). Since the reaction between Pu(VI) and the generated Fe(II) ions

$$PuO_2^{2+} + 2Fe^{2+} + 4H^+ \rightleftharpoons Pu^{4+} + 2Fe^{3+} + 2H_2O$$

is accelerated when the acidity is increased, the titration was effected in 6 *N* sulfuric acid, in which the potentials became stabilized more rapidly. The error for 3.5–9.0 mg Pu in ~ 40 ml was ± 1%, while the potential jump near the equivalence point was about 12–15 mV per 0.05 coulomb. When the amount of plutonium was decreased to 1 mg, the error increased to ± 1.5%.

Uranium, molybdenum, aluminum, beryllium, gallium and (obviously) iron do not interfere with the determination of plutonium. Manganese and chromium interfere, since the permanganate and the bichromate obtained as a result of the oxidation by lead dioxide are very readily titrated by Fe^{2+} ions. Permanganate may be preliminarily reduced with oxalic acid until the solution becomes colorless. The bichromate may be reduced to trivalent chromium by arsenious acid, which does not reduce plutonium.

Coulometric determination of plutonium by titration against Fe(II) was described in detail by Carson, Vanderwater and Gile [338]. Plutonium was first oxidized to the hexavalent state by excess permanganate in dilute perchloric acid at 75°. The excess permanganate and the MnO_2 formed were removed by a small excess of formaldehyde, which was then decomposed by the nitric acid in solution when the temperature was increased. The oxidized solution of plutonium was cooled, transferred to the titration cell which contained 0.6 mole/liter ferrous ammonium alum in a mixture of 25% sulfuric acid and 100% H_3PO_4 as the electrolyte. The sample was then reduced electrolytically by the generated Fe(II) ions. The end point was determined potentiometrically. The amount of plutonium varied between 0.003 and 1 mg in 0.5 ml. The error (in terms of the standard deviation of a single result) was about 5% for 3 μg Pu and 1% for 1 mg. The study contains a detailed discussion of the possible interferences and the corresponding modifications of the procedure. Automatic titrator may be used.

Up to 10 mg Pu may be determined by this method with an error of ± 0.5% [324].

Carson et al. [338] mentions his attempt to titrate Pu(III) coulometrically against electrolytically generated Ce(IV) ions. Better results were obtained when the solutions were evaporated with concentrated sulfuric acid, diluted with water and the plutonium reduced with an excess of titanium (III) sulfate. The solutions were then titrated with the generated Ce(IV) ions, the end point being determined potentiometrically. Fairly sharp end points were obtained both for plutonium and for titanium, but the consumption of the reagent in the blank experiment was high because of the presence of iron in the reagents. The authors admitted their method to be unsuitable for the titration of microgram amounts of plutonium.

Tentative experiments on the coulometric titration of Pu(III) by electrolytically generated Ce(IV) ions are also found in [385].

Singer (1957) investigated the possibility of coulometric titration of plutonium (IV) by the electrolytically generated Ti(III) ions. The determination was carried out in a solution of plutonium (IV) sulfate, obtained by the evaporation of the sample solution with H_2SO_4 and dissolving the residue in 6 *N* H_2SO_4. The amount of plutonium was about 1 mg/ml. To the sample 1 ml of 10% solution of $TiCl_4$ was added and the reduction was performed on a mercury cathode. The end point was determined potentiometrically. The accuracy of the determination was about 0.3% with a potential jump of 100 mV at the end point. Threefold amounts of U(VI) and small amounts of Cr(III), Ni(III) and Mn(II) do not interfere. As in other cases, in which the system Pu(IV)/Pu(III) is present in solutions of H_2SO_4, iron interferes with the determination.

POLAROGRAPHIC DETERMINATION OF PLUTONIUM

The most convenient systems for the polarographic determination of plutonium are the readily reversible Pu(IV)/Pu(III) and Pu(VI)/Pu(V). Their redox potentials in the most commonly employed media—solutions of perchloric, hydrochloric, nitric and sulfuric acids—are positive, above +0.7 V (see p. 43). By choosing media containing strong complexing agents, the redox potentials can be considerably altered. Since most complexing agents form stronger complexes with more highly charged ions, their addition usually results in the potentials becoming more negative. Polarographic determinations may accordingly be carried out in two ways, with two different kinds of electrodes: in the positive range of potentials, polarographic waves of plutonium may be obtained on solid, usually platinum electrodes; in the negative potential range, mercury electrodes are most convenient.

Both methods have their advantages and disadvantages. Solid electrodes, as has already been mentioned, make it possible to work directly in ordinary mineral acid media, at potentials corresponding to simple hydrated plutonium ions. On the other hand, when used for a prolonged time, such electrodes give poorly reproducible results, because of the difficulties involved in the renewal of the electrode surface. The classical mercury dropping electrode gives more reproducible results if certain precautions are taken, but special complexing media must be chosen to ensure a shift of the plutonium potential into the working range of the electrode.

The selectivity of the polarographic method is based on the difference between the redox potentials of plutonium and those of the accompanying elements. The sensitivity will depend on the polarographic technique employed and on the apparatus; it may be said that, as a rule, the minimum determinable concentration is about 10^{-6} *M*. The accuracy is 1–2%.

Polarography on solid electrodes

In the studies which will now be described only platinum electrodes were used, but other metals, e.g., gold, may also be employed.

Harvey et al. [451] were the first to describe plutonium waves obtained on a stationary platinum microelectrode. The polarographic work was effected on solutions of trivalent and tetravalent plutonium in 1 *N* H_2SO_4, 1 *N* HCl and 1 *N* $HClO_4$. The half-wave potentials in these solutions ranged from + 0.78 to + 1.0 V. In all cases the diffusion current was lower for Pu(IV) than for Pu(III).

Palei and Gokhshtein (1951) employed platinum electrodes in a polarographic determination of plutonium in nitric and sulfuric acid media.

Preliminary work showed that Pu(III) and Pu(IV) are oxidized and reduced on a stationary platinum microelectrode, but the reproducibility of the waves is poor, the dependence between the height of the wave and the concentration of plutonium is not linear and the results are strongly affected by the presence of oxygen.

Satisfactory results were obtained on a rotating platinum electrode. The electrodes were cylindrical in shape, 0.2 mm in diameter and 1.5–2 mm long. When the rate of rotation of the electrode was above 800 rpm, the current intensity was independent of the rotation rate.

The half-wave potentials are situated near the formal potentials of Pu(IV)/Pu(III) in the corresponding media. The linear dependence of the limiting current and the concentration of Pu(III) or Pu(IV) has been verified up to a concentration of 250 mg Pu per liter. The accuracy of the determination is 5–7%, depending on the concentration of plutonium.

The preparation of the electrode is of major importance in obtaining reproducible results and distinct waves. A few techniques of electrochemical and chemical activation of the surface of platinum electrodes have been tried out. The simplest method is the following: the electrode is treated for 2–3 minutes with hot aqua regia and is then washed with distilled water. It is then immersed in 1 *N* H_2SO_4 and an anodic voltage of 1.2–1.5 V applied for 20 minutes. The voltage is then altered three times from + 1 V to zero. The electrode thus prepared retains its properties during 12 hours.

If 1 *N* HNO_3 is used as the background electrolyte, plutonium may be determined in the presence of uranium, lanthanum, large amounts of chromium and equal amounts of iron. Plutonium in 1 *N* HNO_3 solutions containing 50–100 μg Pu per ml, in which the above mentioned elements are present, is reduced to Pu(III) by SO_2 during 10–15 minutes. (In 1 *N* H_2SO_4 the quantitative reduction takes 2–3 hours.) The excess SO_2 is removed by passing hydrogen through the solution for 40 minutes, and the measurement is carried out. The amount of plutonium is determined from a graduated graph.

Fomin, Vorob'ev, Andreev and Kiryushina (1951) studied the electrochemical behavior of Pu(IV), Pu(III), Fe(III) and Fe(II) by the polarographic method on a stationary platinum electrode in a solution of sulfuric acid.

The polarograph used in the study was of the V-301 type and the surface of the platinum electrode was 4 mm^2.

The half-wave potential of Pu(IV) shows a linear dependence on the logarithm of the concentration of sulfuric acid and varies from + 525 mV in 0.05 M H_2SO_4 to + 480 mV in 1 M H_2SO_4, relative to a saturated calomel electrode. The half-wave potential of Fe(III)/Fe(II) in 1 M H_2SO_4 is about + 470 mV relative to a saturated calomel electrode. The reduction of Pu(IV) and Fe(III) is a one-electron process ($n \cong 1$).

Bakh, Medvedovskii and Gorokhova (1961) studied the polarographic behavior of tri-, tetra-, penta- and hexavalent plutonium on a rotating disc platinum electrode. A special feature of the rotating disc electrode is that it is one of the few cases for which the full set of hydrodynamic equations has been solved and the theory is fully confirmed by experiment [147].

The measurements were conducted on a TsLA recording polarograph. The rotating disc platinum electrodes had diameters varying between 2 and 5 mm and were electrochemically treated before use. The stirrer, in the form of side paddles on the rotating electrode, ensured a vertical agitation of the solution. The second electrode consisted of a normal calomel half-cell. The measurements were carried out in an atmosphere of nitrogen.

The polarographic determination of plutonium was carried out in 1 M $HClO_4$; at low acidities $NaClO_4$ was added to have an ionic strength of unity. Solutions of Pu(III), Pu(IV) and Pu(VI) were obtained electrolytically; solutions of Pu(VI) were also obtained by evaporation with $HClO_4$. Solutions of Pu(V) were prepared from Pu(VI) by treatment with hydrogen peroxide at pH 3–4; the excess hydrogen peroxide was expelled by gentle heating.

The control of the total concentration and of the concentrations of individual valency forms in the sample solutions was effected radiometrically and spectrophotometrically at 603, 470 and 833 mμ. It was found that, just as in the case of rotating wire electrode, trivalent and tetravalent plutonium in 1 N $HClO_4$ give one wave with a half-wave potential of + 0.60 V with respect to a saturated calomel electrode. The current is proportional to the concentration between $2 \cdot 10^{-4}$ and $3 \cdot 10^{-3}$ M plutonium. The calibration curves of Pu(III) and Pu(IV) coincide to within 3–5%, which indicates that the diffusion coefficients are similar. The dependence of the diffusion currents of Pu(III) and Pu(IV) on the rotation rate is in agreement with the theoretical relationship $i_d = k\omega^{\frac{1}{2}}$, where ω is the angular velocity between 0 and 3000 rpm [147].

Hexavalent plutonium in 1 M $HClO_4$ gives one wave with the half-wave potential of about 0.60 V relative to a saturated calomel electrode, as does Pu(IV). The wave corresponds to the transition Pu(VI) $\rightarrow$ Pu(V). The rectilinear calibration curve giving the dependence of the limiting current on the concentration coincides to within a few percent with that of Pu(III) → Pu(IV).

Pentavalent plutonium at pH 0.55 gives an anodic wave, somewhat extended along the potential axis, with a half-wave potential of about 0.70 V. At pH 2.5 the wave is even more extended and the half-wave potential is shifted to the value of + 1.1 V. The values of the diffusion currents coincide with those for Pu(VI).

The result of this behavior is that when plutonium is present as a mixture of different valency forms, the individual waves merge into one, the height of which is proportional to the overall concentration of plutonium. If the predominant species in the mixture are Pu(IV) and Pu(VI), the half-wave potential is close to + 0.60 V, while in the presence of large amounts of Pu(III) and Pu(V) the potential shifts to + 0.70 V with respect to a saturated calomel electrode. In the overall polarographic wave the cathodic and the anodic branches may be distinguished, which correspond, respectively, to the concentrations of Pu(VI) + Pu(IV) and Pu(III) + Pu(V). In weakly acid solutions Pu(V) is unstable and its concentration may be neglected; the anodic branch of the wave is then exclusively due to Pu(III).

In order to determine each valency state of plutonium separately, the authors supplemented the polarographic determinations with potentiometric work. From the potential of the system and the concentration of Pu(III), the concentration of Pu(IV) may be found from Nernst's equation; the concentration of Pu(VI) may then be found from the sum Pu(IV) + Pu(VI).

When the overall concentration of plutonium, present as a mixture of Pu(III) with Pu(IV) and Pu(VI), was determined in the concentration range $1.5\text{–}2 \cdot 10^{-3}$ *M*, the systematic deviation from the results of the radiometric determinations was less than 1% in ten determinations, while the standard deviation of a single determination was about ± 3.5%. The deviations from the results obtained by spectrophotometry were larger; the systematic deviation was about 3% and the standard deviation was ± 8%. The accuracy in the determination of individual valency forms depends on their relative contents.

The last few years have witnessed an intense development of certain polarographic methods which may be described as the application of nonstationary diffusion effect to analysis, and also of AC polarography. Of the methods which have been applied to the determination of plutonium we may mention square-wave polarography [506] and voltammetry at constant current intensity [160].

Square-wave polarography was first introduced by Barker and Jenkins [289]. In this method square-wave alternating voltage of low amplitude is superposed on the electrode potential. The parameter measured is the alternating component of the electrolysis current as a function of the potential, which is varied, as in conventional polarography, linearly with time. In order to eliminate the capacity component of the alternating current, the measurements are carried out at the end of each half-period, when the double electric layer on the electrode surface has been charged anew. According to Barker and Jenkins [289], concentrations of reversibly reducible substances of the order of $2 \cdot 10^{-7}$ *M* can be determined. The analytical applications of the method have been described in [398–401], and the theory of the method is given in [484].

Koyama [506] described the determination of small amounts of plutonium in hydrochloric and nitric acid solutions with the aid of square-wave polarography, using a stationary platinum electrode.

This worker used the improved instrument of Hamm [445], in which the frequency of the square wave was reduced from 200 to 85 cycles/sec and its amplitude to 10 mV. The reduction in frequency was caused by the larger surface area of the solid electrode as compared with the dropping mercury electrode which had been formerly employed. The polarograms were obtained by superposing a square wave on the voltage of a direct current voltage scan of 50 mV per minute.

A glass polarographic cell, about 70 mm tall and with an external diameter of 32 mm was used. The cell had entry ports for the introduction and draining of solutions and three top inlets for the saturated calomel electrode and two platinum electrodes. The electrodes consisted of a platinum wire, 1 mm in diameter, protruding to the extent of 1 cm from a tube made of lead glass, 6 mm in diameter. The ends of the electrodes were fire-polished to eliminate any roughness produced by the preparation treatment. A piece of porous Vycor acting as a salt bridge separated the calomel electrode from the electrolyte.

One platinum electrode served as the working electrode. The other was connected through a 300 μF capacitor to the calomel electrode and transmitted the alternating component of the electrolysis current, by-passing the impedance of the calomel electrode. This technique reduces the time constant of the cell circuit and thus suppresses the effect of the capacitance current.

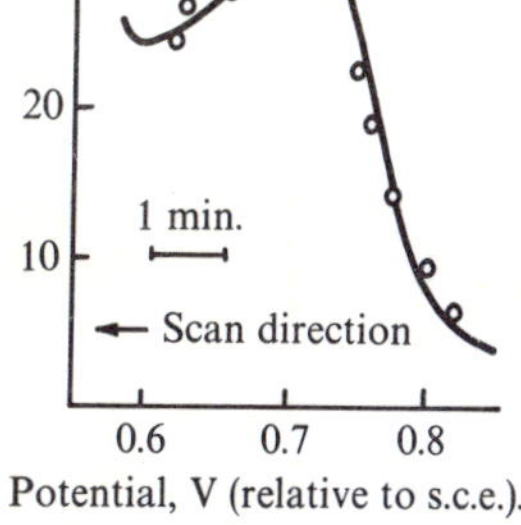

Figure 88

Square-wave polarogram of a solution of Pu(IV) in 1 *M* HCl.

Concentration of Pu(IV): $7.5 \cdot 10^{-6}$ *M*.

The fundamental equation of square-wave polarography is the equation proposed by Kambara [484]:

$$S = n^2 k E C_0 D^{\frac{1}{2}} f^{\frac{1}{2}} \frac{P}{(1+P)^2}, \tag{15}$$

where S is the square-wave current, n is the number of electrons transferred in the electrode reaction, k is a constant which depends on the shape and size of the electrode, the amplification employed and the temperature, E is the square-wave amplitude, C_0 is the concentration of the substance, D is the coefficient of diffusion, f is the frequency of the current and

$$P = \exp\left[\frac{nF}{RT}(E - E_{\frac{1}{2}})\right],$$

where $E_{\frac{1}{2}}$ is the half-wave potential and E is the actual potential applied.

Figure 88 shows a typical polarogram obtained during the reduction of Pu(IV) at a concentration of $7.5 \cdot 10^{-6}$ M in 1 M HCl. The circles on the curve correspond to the current values calculated from equation (15), which are in good agreement with the experimental curve.

It follows from the table given in the original paper, which illustrates the linear dependence of the current on the concentration, that the proportionality coefficient in this relationship varies by only 4% between $3.75 \cdot 10^{-6}$ M and $2.0 \cdot 10^{-5}$ M Pu(IV).

The half-wave potential of the reduction of Pu(IV) to Pu(III), which corresponds to a maximum on the differential square-wave polarograms was found to be + 0.71 V relative to saturated calomel electrode in 1 M HCl and + 0.66 V in 2 M HNO_3.

Square-wave polarography is not affected by the stirring of the sample solution and is only little affected by the presence of irreversibly reducible substances such as oxygen in the determination of plutonium.

The presence of nitrite ions in nitric acid media results in a very high residual current, which interferes with the determination of small amounts of plutonium. This indicates that the reactions given by nitrite ions are highly reversible. The nitrite interference is eliminated by the addition of sulfamic acid. The residual current in solutions of HNO_3 nevertheless remains higher than in HCl solutions.

The effect of Fe(III), the half-wave potential of which is + 0.45 V relative to a saturated calomel electrode, is stronger in nitric than in hydrochloric acid. The permissible amount of iron in nitric acid media is five times that of plutonium.

Just as in other cases in which solid electrodes are employed, a systematic regeneration of the electrodes is essential if reproducible results are to be obtained. In the study under discussion, the electrode surface was regenerated by periodic application of a small negative potential to the electrode, to evolve hydrogen. Chemical methods, such as treatment with aqua regia, failed to yield positive results.

Thus, Koyama's work [506] has confirmed the high sensitivity of square-wave polarography and its suitability for the determination of plutonium. It should also be pointed out that the method is more selective than classical polarography. This is because the polarograms are of the differential type and the wave height is measured at the half-wave potential and not from the current intensity values on both sides of it. The half-wave potential is located on the potential axis farthest away from the potentials of interfering elements, both to the right and to the left of the element being determined. The gain is thus approximately 120 mV, and the effect of the interfering elements is reduced by one or two orders of magnitude, provided their waves do not merge with that of the element being determined. Thus, for instance, the effect of iron on the determination of plutonium in this method is about one order of magnitude less than in the coulometric method of Scott and Peekema [641], which has the same selectivity as ordinary polarography for the same nitric acid media.

Also, as has already been mentioned, square-wave polarography is not very sensitive to irreversibly reduced ions.

A method which is similar to square-wave polarography is vector polarography. This method is now being developed in the Soviet Union, owing to the appearance on the market of the TsLA-type vector polarograph. The difference between this method and the square-wave polarography is that a low sinusoidal voltage rather than a square-wave alternating voltage is superposed on the electrode potential. To separate the capacity component of the alternating current, with a phase shifted by 90° with respect to the Faraday current, a phase-sensitive amplifier is employed. This design, which is technically simpler, retains all the advantages of square-wave polarography and is suitable for the determination of plutonium.

Voltammetry at constant current intensity was described by Gierst and Juliard [428] and Delahay [82], while its analytical applications were described in [429, 377, 616]. Through electrodes immersed in the sample solution, one of which is polarized, a given constant intensity current is passed and the variation in the potential of the polarized electrode is recorded as a function of time with the aid of a reversible auxiliary electrode. The current intensity is so chosen that it is higher than the diffusion current; the result is that the concentration of the substance being oxidized or reduced in the layer next to the electrode varies from the initial to zero concentration during a time called "transition time". Since the electrolysis is conducted at a constant rate, the electrode potential gives two jumps, at the beginning and at the end of the transition time, which can be recorded and the transition time measured.

For a reversible two-ion system the bulk concentration C_0 is connected with the transition time τ and the current intensity i_0 by the equation [82]:

$$C_0 = \frac{2\tau^{\frac{1}{2}} i_0}{\pi^{\frac{1}{2}} n F D^{\frac{1}{2}}}, \tag{16}$$

where π, n, F and D have the usual significance. The time $\tau/4$ corresponds to the half-wave potential in ordinary polarography, while the dependence of the electrode potential E on the time t has the form

$$E = E_{\frac{1}{2}} + \frac{RT}{nF} \ln \frac{\tau^{\frac{1}{2}} - t^{\frac{1}{2}}}{t^{\frac{1}{2}}}. \tag{17}$$

The polarized electrode may be a dropping mercury electrode with a controlled surface, or a stationary platinum electrode.

Mechelnik [160] applied voltammetry at constant current to the determination of uranium and plutonium, using platinum electrodes. The measurements were effected on an automatic instrument—a transitometer of the type described by Gierst [427]. The accuracy is better than 1%. The determination of uranium is described in more detail.

The selectivity of voltammetry at constant current intensity is not different from that of ordinary polarography, since the form of the polarograms is described in the two cases by similar equations (17) and (19), which become identical if i_d is substituted for $\tau^{\frac{1}{2}}$ and i for $t^{\frac{1}{2}}$. The effect of interferences on the reversible ion couples may be calculated from equations (16) and (17) from the known values of half-wave potentials.

Polarography on mercury electrodes

We have already pointed out that the determination of plutonium on a mercury electrode is only possible in those media in which plutonium ions are bound into stable complexes, e.g., in oxalate, citrate, tartrate and acetate solutions. In all studies which will now be discussed, a dropping mercury electrode and a polarograph of the conventional type were employed.

The fundamental principles of the method have been adequately exposed by Kolthoff and Lingane [113], while numerous analytical applications have been described by Kryukova, Sinyakova and Aref'eva [128].

The most important relationship for the mercury dropping electrode is the Ilkovic equation, which gives the dependence of the diffusion current i_d on the concentration C_0:

$$i_d = 607\, nC_0 D^{1/2} m^{2/3} t^{1/6}, \tag{18}$$

where i_d is the current, μA; C_0 is the concentration, millimoles per liter; n is the number of electrons; m is the weight of mercury issuing from the capillary in one second, mg; D is the coefficient of diffusion, $cm^2 \cdot sec^{-1}$; and t is the lifetime of a drop, seconds. The shape of the polarographic wave for reversible processes may be described in this case by the equation which connects the potential E and the current i at any point of the curve:

$$E = E_{\frac{1}{2}} + \frac{RT}{nF} \ln \frac{i_d - i}{i}. \tag{19}$$

The information on the composition of the complexes of plutonium and their stability constants, which are used in the polarography of plutonium, will be found in Chapter II.

Behavior of plutonium in oxalate solutions

Fomin, Vorob'ev and Andreeva [227] investigated the polarographic behavior of plutonium (IV). The authors paid special attention to the technique of measuring the potentials of plutonium, and obtained the value -0.205 V for the half-wave potential of the reduction of Pu(IV) in 1 N $K_2C_2O_4$ at pH 5.5.

The diffusion current varies linearly with the concentration of plutonium in the approximate range $1 \cdot 10^{-5}$–$1 \cdot 10^{-3}$ M.

The same authors (1953, 1955) determined the half-wave potentials of iron and uranium under the same conditions and obtained the values of -0.245 and -0.432 V respectively, relative to a saturated calomel electrode. Thus, iron interferes strongly with the determination of plutonium, while uranium can be independently determined. The effect of the strong oxidizing agents Cr(VI) and Mn(VII) is eliminated by adding oxalic acid prior to the evaporation of the samples with sulfuric acid to obtain Pu(IV). Small amounts of copper and bismuth interfere with the determination.

The picture of the polarographic behavior of plutonium in oxalate media has been completed by the study of Cook, Foreman and Kemp [359] who studied the reduction of hexavalent plutonium on dropping mercury electrode. It was found that in solutions containing $Na_2C_2O_4$ and $H_2C_2O_4$ in a total concentration of 0.18 M, Pu(VI) gives a reversible reduction wave to Pu(V) ($n \cong 1$) with a half-wave potential of -0.207 V relative to a saturated calomel electrode. To ensure stability of Pu(V), the solutions should have pH 3.5–4.5. The measurements were conducted at pH 4.2. It was noted, nevertheless, that the limiting diffusion current increases in a linear dependence on $[\mathrm{Pu(VI)}]^{1.38}$ between 4.5 and 900 mg/liter of Pu(VI), which indicates that secondary reactions take place. According to the authors, such a process may be the disproportionation of Pu(V), which is accelerated by the complexing of Pu(IV) and Pu(VI), and also direct reduction of Pu(V) by oxalate ions.

Behavior of plutonium in citrate solutions

The reduction of Pu(IV) in citrate solutions was studied by Gryzin (1951). In 1 N sodium citrate, tetravalent plutonium present at a concentration of 220 mg/liter is reversibly reduced at pH 3–6.5, with the participation of one electron in the reaction. In this acidity range the diffusion current is constant, and the half-wave potential may be expressed as $E_{\frac{1}{2}} = 0.170 - 0.072 \cdot \mathrm{pH}$ volts relative to s.c.e.

At pH 7.25–10.0, tetravalent plutonium is irreversibly reduced and $E_{\frac{1}{2}} = -0.318 - 0.065 \cdot (\mathrm{pH} - 6)$. The diffusion currents are not reproducible to the same extent and vary with time. The slope of the lines E vs. $\log \frac{i_d - i}{i}$ corresponds to $n \cong \frac{1}{3}$.

The diffusion current is proportional to the concentration of plutonium both at pH 5.5–6 and at pH 8.5–9, the sensitivities being equal. The diffusion coefficient D calculated from Ilkovic's equation (18) was $0.18 \cdot 10^{-5}$ cm^2/sec. The diffusion current and the half-wave potential depend on the concentration of citrate. At pH 5 the half-wave potential is $E_{\frac{1}{2}} = -0.185 + 0.060 \log C$, and at pH 6 $E_{\frac{1}{2}} = -0.308 + 0.058 \log C$, where C is the concentration of citrate.

In the study under discussion, plutonium was determined at concentrations above 50 mg/liter. Uranium, iron and chromium interfere with the determination. Nebel [573] pointed out that plutonium may be determined in citrate solutions at pH 4.

Behavior of plutonium in tartrate solutions

Gryzin (1952) studied the polarographic behavior of Pu(IV) in solutions of tartaric acid and showed that in 1 N tartaric acid tetravalent plutonium gives reversible reduction waves at pH 3.5–5.0. The half-wave potential depends on the acidity, in accordance with the expression $E_{\frac{1}{2}} = 0.126 - 0.060 \cdot \text{pH}$. At pH 2–3 waves with a maximum are formed.

The decrease in the acidity of solutions impairs the reversibility of the reduction of Pu(IV). At pH 5–6.5 the diffusion current decreases, the slope of the E vs. $\log \frac{i_d - i}{i}$ curves decreases and the half-wave potential swiftly becomes more negative $E_{\frac{1}{2}} = -0.160 - 0.106 \cdot (\text{pH} - 5)$. Between pH 6.5 and pH 10 another wave develops, at more negative potentials, for which $E_{\frac{1}{2}} = -1.08 - 0.069 \cdot (\text{pH} - 7)$. The slope of the polarographic curves corresponds to $n \cong 0.25$. The diffusion current increases up to pH 10 and then remains constant.

The most suitable for the determination of plutonium are solutions with pH 5. The half-wave potential in these solutions does not depend on the concentration of the tartrate between 0.1 and 1.2 M and is -0.182 V relative to saturated calomel electrode. The diffusion current is proportional to the concentration of plutonium; the diffusion coefficient in Ilkovic's equation (18) is $0.17 \cdot 10^{-5}$ cm^2/sec. Plutonium was determined at concentrations above 50 mg/liter. A large excess of uranium, chromium and iron interfere. At pH 5 uranium gives an independent wave, and at pH 1.25 the uranium wave may be obtained without the plutonium wave.

Cook, Foreman and Kemp [359] reported that hexavalent plutonium in 0.1 N sodium potassium tartrate is irreversibly reduced with a half-wave potential of about -0.5 V relative to saturated calomel electrode.

Behavior of plutonium in acetate solutions

Nebel [573, 171] studied the polarographic reduction of Pu(IV) in acetate solutions and used sample solutions of an ionic strength $\mu = 0.5$ and pH 4.2. The diffusion current constant $I_d = \frac{i_d}{C_0 m^{2/3} t^{1/6}}$ was 0.97 which corresponds to the formal diffusion coefficient $D \cong 0.25 \cdot 10^{-5}$. Trivalent plutonium is easily oxidized to Pu(IV) by atmospheric oxygen under these conditions.

In the analytical determination, plutonium was first electrolytically reduced to Pu(III), the acetate was then added, plutonium was electrolytically oxidized to the tetravalent state and Pu(IV) was polarographically determined.

Cook, Foreman and Kemp [359] failed to notice the waves of the reduction of Pu(VI) in 1 *M* acetate solution at pH 4 on a dropping mercury electrode.

Behavior of plutonium in other media

Information on the polarographic determination of plutonium in other media is very scanty, though other complexing agents could undoubtedly be employed.

Supplementing the studies discussed above, Gryzin (1953) reported that he had obtained reduction waves of Pu(IV) in malic, lactic and butyric acid media.

In 1 *M* solution of malic acid at pH>4 a plutonium wave was obtained, the half-wave potential of which undergoes a negative shift when the pH is increased.

In 1 *M* lactic acid Pu(IV) gives a reduction wave at pH 2–4 with a half-wave potential of about + 0.06 to + 0.08 V relative to a saturated calomel electrode.

In solutions of butyric acid at pH 2 a wave with a maximum is obtained, the beginning of which is situated at + 0.06 V relative to saturated calomel electrode. These solutions have not been studied in detail.

Cook, Foreman and Kemp [359] studied the reduction of Pu(VI) in 0.1–1.0 *M* solutions of NaOH. At a constant concentration of plutonium the height of the wave decreases nonlinearly with the decrease in NaOH concentration. In 0.7–1.0 *M* NaOH there is one wave with $E_{\frac{1}{2}} = -0.90$ V relative to a mercury oxide electrode.

When the alkalinity further decreased, the reduction is evidenced by two imperfectly separated waves with $E_{\frac{1}{2}}$ of -0.86 and -1.13 V. At the concentration of 0.1 *M* NaOH, the appearance is still that of a double wave, but the resolution is poorer.

According to the authors, these results indicate an irreversible reduction of one or more hydrolyzed forms of PuO_2^{2+}, while the wave height characterizes the concentration of the forms in solution. The content of these forms shows a nonlinear dependence on the concentration of hydroxyl ions.

GRAVIMETRIC METHODS

Gravimetric methods were employed in the very first stage of investigation of the properties of plutonium and its compounds [48, 367]. Since the amounts of the element at the disposal of the workers were very small, special ultramicrogravimetric methods of determination were developed [367]. At present, gravimetric methods are mainly used for the determination of milligram or gram amounts of plutonium, whenever a high accuracy (0.1–0.2 rel. %) is required.

In most gravimetric methods compounds of tetravalent plutonium are precipitated out of solutions. In this valency state plutonium forms a large number of sparingly soluble compounds. The precipitation methods involving tri- or hexavalent plutonium [3, p. 351; 54, 56] are mainly employed to determine the content of one valency form of plutonium in the presence of another.

The gravimetric methods of determination of plutonium usually consist of two operations: precipitation of plutonium ion as a sparingly soluble precipitate, and conversion of the precipitate into a gravimetric form. Even though the number of sparingly soluble compounds of tetravalent plutonium is large, the *gravimetric form* is usually *plutonium dioxide*, which fulfils the requirements of a gravimetric form better than other compounds. Detailed investigations of plutonium dioxide were carried out in [48, 189, 237, 388, 554, 726, 732]. Calcination and weighing as PuO_2 is the final stage in the determination of plutonium following the precipitation of plutonium hydroxide and peroxide, oxalates of tri- and tetravalent plutonium and of many organic compounds of plutonium. Drummond and Welch [388] showed that, depending on the method of preparation, the composition of plutonium dioxide may vary from $PuO_{2.00}$ to $PuO_{2.09}$. This small change in the composition is caused by additional absorption of oxygen at calcination temperatures below 1000°. Palei and Milyukova (1952) showed that for sample weights of about 20 mg, weighed to within 0.1–0.2 mg, the weight of the dioxide becomes stoichiometric at 1050–1100° after 3 hours calcination. When the calcination temperature is raised to 1200°, the dioxide of the stoichiometric composition is invariably formed. Roberts et al. [189] studied the oxidation by oxygen of a number of samples of plutonium dioxide at various temperatures and pressures. The results of these workers agreed with those of Drummond and Welch. The mechanism of the oxidation to PuO_2 and the formation of nonstoichiometric compounds have not been clarified.

Plutonium dioxide with least amount of impurities, and with a composition which is almost stoichiometric is obtained by calcination of plutonium peroxide and of tri- and tetravalent plutonium oxalates [3, p. 169, 237]. The calcination of dried crystalline Pu(III) and Pu(IV) oxalates and of Pu(IV) peroxide is carried out while slowly raising the temperature from 25 to 700°. In order to prevent the atomization of the precipitate, owing to the simultaneous dehydration and decomposition processes, the heating between 90 and 200° should be particularly cautious. The sulfate ion, which is a structural constituent of plutonium(IV) peroxide, is volatilized as SO_3 at 600–700°. In the absence of sulfates, plutonium peroxide is precipitated as fine crystalline precipitate, containing other anions, which are removed at higher temperatures (about 1000°). Plutonium oxalates should be calcined not below 900–950°. As has been mentioned above, the calcination must be carried out at 1050–1200° if plutonium dioxide precipitates of a stoichiometric composition are to be obtained.

Plutonium(IV) iodate [48, 206, 367] has also been proposed as a gravimetric form. Accurate results are obtained only if the precipitate has been carefully washed free of the excess precipitant. The calcination to constant weight should be effected not above 250°. The gravimetric factor is 0.254.

Of the organic compounds of plutonium, only the 8-hydroxyquinolate is used as a gravimetric form. It is dried to constant weight at 120–130° (Moiseev, 1953). The compound is nonhygroscopic. The gravimetric factor is 0.293 [206].

Precipitation as plutonium (IV) hydroxide

One of the principal and most convenient methods of determination of plutonium is precipitation as the hydroxide [3, p. 329; 48, 60, p. 168]. Plutonium (IV) hydroxide is precipitated from acid solutions of plutonium by solutions of alkalis. To ensure quantitative precipitation of plutonium, carbonates, which form soluble complex compounds with plutonium, must be absent during the precipitation [3, p. 331; 60, p. 140]. For this reason, the ammonium hydroxide or alkali hydroxide used in the precipitation must not contain carbonate ions.

Aluminum, lead, zinc and potassium and ammonium salts do not interfere with the quantitative precipitation of plutonium. In solutions containing calcium, magnesium, manganese, colbalt, copper, chromium and other elements, the precipitation of plutonium hydroxide is effected with 20% ammonium hydroxide. Some of the extraneous elements (Cu, Ni, Co) form soluble complexes with the excess of ammonia. The precipitation of plutonium hydroxide is prevented by anions which form complex compounds with it. These include in the first place carbonates, oxalates, fluorides, phosphates, tartrates and other ions. Tri- and tetravalent cations, which form insoluble precipitates with hydroxyl ions, also interfere with the determination of plutonium. The ammonium hydroxide precipitation is carried out as follows.

The nitric or hydrochloric acid solution of tetravalent plutonium (optimum concentration of Pu about 1 mg/ml), not containing interfering cations or complexing anions, is heated to boiling to remove CO_2. To the solution 20% ammonium hydroxide (free of carbonate) is added drop by drop, with continuous stirring, until the odor of ammonia becomes noticeable. The solution with the precipitate is heated for 3–5 minutes to convert the precipitate into an easily filterable, coarse-grained form. A small excess of NH_4OH solution is then added, the solution is left to stand for 5–10 minutes to ensure better coagulation, and is then filtered through a White Ribbon filter paper. The precipitate is washed a few times with water to which a few drops of 20% ammonia have been added. The filter with the precipitate is transferred to a platinum crucible and dried on a hotplate. It is then calcined at 1000° during 2–3 hours to constant weight, cooled in a desiccator and weighed.

The accuracy of the determination of plutonium by the above method essentially depends on the accuracy of weighing and the sample weight. The error may vary between 0.01 and 0.1 rel. %.

Plutonium may be determined by precipitation with 20% alkali hydroxide in exactly the same manner.

Precipitation as plutonium (IV) peroxide

Plutonium peroxide may be precipitated both from acid solution (nitric, hydrochloric and sulfuric acid solutions) and from acetate-oxalate and carbonate solutions. The completeness of precipitation depends in each case on several factors: the acidity of the solution, the valency of plutonium, the excess of the peroxide, time of standing, and the presence of impurities which decompose H_2O_2 [504, 519].

Precipitation from acid solutions

If plutonium (IV) is precipitated by a large excess of hydrogen peroxide, the effect of acidity (between 0.4 and 4 *M*) on the completeness of the precipitation will be negligible. Hexavalent plutonium is not precipitated by hydrogen peroxide, but is reduced by it to the tetravalent state; the reduction process is slow. For this reason, if plutonium is present in part as Pu(VI), the precipitation will be incomplete. In the precipitation of plutonium from solutions containing major amounts of Pu(VI), it is more convenient first to reduce Pu(VI) in the usual way (sulfurous acid, hydroxylamine, rongalite, etc.) and only then precipitate by peroxide.

If large amounts of Fe(III) (equal amounts of plutonium and iron) are present, the precipitation of plutonium may be incomplete, owing to the partial oxidation by the iron to the hexavalent state [519].

The determination of plutonium in acid solutions as peroxide may be effected in the presence of a number of elements (Cr, Mn, Ni, La) and small amounts of iron. Uranium, hafnium and zirconium interfere.

The precipitation of plutonium peroxide is carried out from $\sim 1.7\,N$ solutions of nitric acid, at plutonium concentrations varying from 1 to 20 g/liter. To a nitric acid solution of plutonium 30% H_2O_2 is added ($\frac{1}{5}$ of the volume of the initial solution) and the solution is left to stand for about one hour at room temperature. During this operation small amounts of Pu(VI) are reduced to Pu(IV). The solution is then cooled to 15° and a cooled solution of 30% H_2O_2 is slowly introduced, with vigorous stirring, until a precipitate is formed. The hydrogen peroxide is then added more rapidly, until the total added volume of the reagent is one-half the volume of the initial solution. The concentration of H_2O_2 in the solution after precipitation should be 9%. The precipitate of the peroxide together with the solution are stirred for 30 minutes at 15–20° and filtered through a Blue Ribbon filter paper. Plutonium peroxide is washed with 1.5 *N* HNO_3, containing 6% H_2O_2. The filter with the precipitate is placed in a platinum crucible, carefully ashed on an electric hotplate and calcined at 1000° for 3 hours. Plutonium is weighed as PuO_2.

Precipitation from carbonate solutions

In order to separate plutonium from uranium it was proposed (Kondrashova, 1953; Deberdeeva, 1960) to precipitate Pu(IV) peroxide from carbonate solutions. Soluble carbonate complexes of plutonium and uranium are obtained by adding a saturated solution of $NaHCO_3$ and 20% ammonium carbonate to the initial solution in 1 *N* HNO_3. From the solution thus obtained plutonium is precipitated by hydrogen peroxide. A more complete precipitation of plutonium is obtained when the concentration of hydrogen peroxide in solution is 7–8%. The optimum concentration of ammonium carbonate is 3–4% (at higher concentrations the peroxide precipitate is more soluble and requires more time to precipitate). The precipitate should be left to stand for about 20 hours. Under these conditions the residual concentration of plutonium is low enough to satisfy the requirements of gravimetric analysis, but true equilibrium in solution does not appear to be reached even after 24 hours.

When plutonium is precipitated from solutions containing uranium as the only other element, the beginning of the reaction may be accelerated by heating the solution to 60–80° prior to the addition of H_2O_2 and, on the contrary, solutions containing major amounts of impurities catalyzing the decomposition of hydrogen peroxide should be cooled. In the latter case complete precipitation is attained at the end of 3–4 hours.

Plutonium is determined (Deberdeeva, 1960) in the presence of less than threefold amounts of uranium as follows.

The aliquot sample solution is introduced into 10% solution of ammonium carbonate and the solution is diluted with water so that the final carbonate concentration is 5–6%. Hydrogen peroxide is then added in an amount sufficient to ensure a concentration of 5–7 vol. % in solution. To ensure complete precipitation, the solution with the precipitate is left to stand for 20 hours. If the content of uranium is large (of the same order as that of plutonium), the peroxide must be reprecipitated. After the first precipitation, plutonium peroxide is filtered through a No. 4 fritted glass filter, and washed twice with 5% solution of $(NH_4)_2CO_3$ containing 5% H_2O_2. The washed precipitate is dissolved on the filter with the aid of 2–3 ml hot concentrated HNO_3. The filter is rinsed with 1.5 *N* solution of HNO_3. The nitric acid solution is evaporated to dryness in a quartz dish. The dry precipitate is dissolved in 1–2 ml 1 *N* HNO_3 and this solution is introduced drop by drop into a solution of ammonium carbonate, and the precipitation of plutonium peroxide is repeated as described above. After the second precipitation, the precipitate is filtered through a Blue Ribbon filter paper, washed with the washing solution, dried and calcined to constant weight at 1000°. The losses in the filtrates are determined radiometrically.

The error does not exceed $\pm$ 0.5 rel. %.

Precipitation from ammonium carbonate solutions in the presence of Complexone III

In order to improve the selectivity of the precipitation by hydrogen peroxide, Deberdeeva (1960) used Complexone III. It was found that in the presence of Complexone III plutonium is precipitated from acid solutions (0.2 *N* HNO_3) only to the extent of 50%, from neutral solutions to the extent of 98.7%, and from ammonium carbonate solutions (pH 10) it is precipitated quantitatively (99.9%) in 24 hours. Longer standing of the solution increases the amount of plutonium in the mother liquor, due to a partial oxidation of Complexone III by hydrogen peroxide. If the solution is heated, complete precipitation may be attained within a shorter period of time. With this method plutonium may be determined in multicomponent solutions even in the presence of zirconium, hafnium and vanadium. The experimental procedure is as follows.

The nitric acid solution containing 20–60 mg Pu is introduced into ammonium carbonate solution with Complexone III. To accelerate the precipitation rate, the solution is heated to boiling and left to cool slowly. To the cooled solution H_2O_2 is added. The composition of the solution during precipitation is as follows: 5% in ammonium carbonate, 2.5–2.8% in Complexone III, 7–8% in H_2O_2 and 2% in NH_4OH (pH > 10). The precipitate is left to stand for 5 hours, is then filtered through No. 4 fritted glass filter and washed with 5% solution of ammonium carbonate containing 0.7% Complexone III, 5–7% H_2O_2 and made alkaline with NH_4OH. The precipitate is dissolved in 2–3 ml conc. HNO_3, and plutonium peroxide is reprecipitated as described above. The precipitate is filtered through an ashless filter paper (Blue Ribbon) and calcined to PuO_2.

With the method it is possible to determine plutonium in the presence of threefold amounts of uranium. If the concentration of uranium is lower than that of plutonium, only one precipitation is carried out. In this case one determination takes 5–6 hours. In the presence of large amounts of uranium a second precipitation must be effected, and the time of one determination is 17–18 hours. The experimental error does not exceed $\pm$ 0.5%.

Precipitation as plutonium (IV) iodate

When iodic acid or an iodate is added to an acid solution of trivalent or tetravalent plutonium, plutonium is quantitatively precipitated [3, p. 324; 48, p. 15]. Plutonium (VI) iodate has a relatively high solubility even in weakly acid solutions, so that the precipitation of plutonium as iodate can be carried out only in solutions of trivalent or tetravalent plutonium [277]. The usual procedure is to precipitate plutonium tetraiodate from nitric acid media. The teraiodate is quantitatively precipitated within a wide acidity range (0.1–6 *M* HNO_3).

The composition of the precipitate of plutonium iodate depends on the acidity of the medium and the concentration of the precipitant. When $Pu(IO_3)_4$ is used as the gravimetric form, the precipitate must be very carefully washed free from the excess precipitant; for this reason it is more usual to calcine to PuO_2 rather than to $Pu(IO_3)_4$ [206]. Plutonium may be determined by this method in the presence of many elements, including uranium and lanthanum [203, p. 320]. To separate tenfold amounts of uranium, a single precipitation of the tetraiodate is enough (Konarev et al., 1951). At higher concentrations of uranium a reprecipitation must be carried out.

Varying the temperature between 5 and 80° during the precipitation had no significant effect on the completeness of precipitation or on the rate of formation of the precipitate.

The iodate method is used to separate plutonium in amounts of 5–100 mg in the presence of uranium and rare-earth elements.

The precipitation is conducted from 1 *M* solution of HNO_3 containing 5 mg/ml of plutonium.

To the sample solution an equal volume of 0.25 *M* solution of KIO_3 in 1 *M* HNO_3 is added, with constant stirring. The precipitate is filtered off after 2 hours and is washed with a 0.125 *M* solution of KIO_3 in 1.5 *M* HNO_3. Plutonium iodate is reprecipitated if necessary. The precipitate is dissolved in a 6% solution of sulfurous acid in 1.5 *M* HNO_3. The solution obtained is heated to 60–80° to remove SO_2 and plutonium iodate is precipitated as described above. After washing, the precipitate with the filter paper is dried, ashed and calcined to PuO_2 at 1000–1050°, and is then weighed.

The capture of a small amount of the excess precipitant does not cause high results to be obtained, since at this temperature potassium iodate is volatilized [206]. In the determination of 5–100 mg Pu the error does not exceed ±0.2–0.3 rel.%. If plutonium is to be gravimetrically determined as $Pu(IO_3)_4$, it is calcined to constant weight at 250°.

Precipitation of plutonium (III) and plutonium (IV) as oxalates

Determination of plutonium (III)

Oxalic acid and oxalates quantitatively precipitate trivalent plutonium from weakly acid solutions as $Pu_2(C_2O_4)_3 \cdot 9H_2O$ [3, p. 347; 587]. The solubility of this compound in water is 1 mg/liter, and in weakly acid solutions (0.2–0.8 *M* HNO_3 or HCl) it is 5–20 mg Pu per liter [54, 56]. When the acidity is increased, the residual concentration of plutonium in solution increases. Sulfates, phosphates, fluorides and tri- and tetravalent cations interfere with the determination. The precipitation is carried out as follows. The acidity of the nitric acid solution is adjusted to 0.5 *M*. If the initial solution contains Pu(IV), it is reduced by adding potassium iodide and occasionally stirring during 30 minutes. One-molar sodium oxalate is then added in an amount which is sufficient to precipitate all the plutonium and produce an 0.2 *M* oxalate concentration in the solution. The precipitate is filtered off after 30 minutes, washed with water, dried and calcined in the air to PuO_2 at 800–1000° during 2–3 hours. The error is 0.3–0.5 rel. %.

Determination of plutonium (IV)

Tetravalent plutonium is quantitatively precipitated as the oxalate hexahydrate [48, p. 348]. The solubility of the oxalate in water is $10.3 \cdot 10^{-5}$ moles/liter [57, 168]. As distinct from the trivalent oxalate, the residual concentration of plutonium in solution during the precipitation of Pu(IV) oxalate decreases with increasing acidity and the optimum concentration of HNO_3 or HCl is 3–4 *M*. The solubility of plutonium (IV) oxalate is considerably reduced in the presence of ethanol. The method makes it possible to determine plutonium in the presence of most elements except thorium, zirconium and rare-earth elements. Oxalates of certain elements (Ba, Mn, Co, Ni, Pb, Sn, Sr), which are precipitated in neutral solutions, remain in solution if the solution is sufficiently acid (>3 *N*). If present in small amounts, these elements can be quantitatively separated by one precipitation; if present in large amounts, a reprecipitation must be carried out.

Fluorides and sulfates interfere with the determination of Pu(IV) as oxalate. Acetic acid does not materially affect the precipitation of plutonium (IV) oxalate if its concentration in solution does not exceed 5%. Organic compounds which precipitate plutonium or form complex compounds with it must be absent. The oxalate method is particularly valuable because it affords a way of separating Pu(IV) from U(VI). If the concentration of uranium in the sample solution is large, the precipitant is added in an amount sufficient not merely to precipitate all the plutonium, but also to bind UO_2^{2+} ions.

Precipitation of plutonium (IV) cupferronate

Cupferron quantitatively precipitates plutonium from sulfuric acid solutions, irrespective of its valency state. The composition of the precipitate is $Pu(C_6H_5N_2O_2)_4$ [161]. The oxidation of Pu(III) is probably effected by hydrogen ions, while the cupferron in solution has only an auxiliary role, in that it binds the Pu(IV) ions formed into a stable complex compound [161]. The authors of this work consider that the oxidation of plutonium takes place according to the equation

$$Pu_2(SO_4)_3 + 8C_6H_5N_2O_2NH_4 + H_2SO_4 = 2Pu(C_6H_5N_2O_2)_4 + 4(NH_4)_2SO_4 + H_2 .$$

The oxidation takes place in 2–3 minutes in the absence of reducing agents (H_2SO_3 or hydrazine hydrate). In the presence of reducing agents full reduction takes 30 min.

Hexavalent plutonium in acid solutions is first reduced by cupferron to Pu(IV), and is then precipitated as $Pu(C_6H_5N_2O_2)_4$. The required completeness of precipitation of plutonium is attained 50–60 minutes after the addition of cupferron. However, if a small amount of H_2SO_3 or hydrazine hydrate salt is added to the acid solution of Pu(VI) immediately after the addition of cupferron, the plutonium is quantitatively precipitated within 5–10 minutes.

Equimolar amounts of uranium (VI), chromium (III), manganese (II), aluminum, silver, nickel, lanthanum and americium do not interfere with the precipitation of plutonium cupferronate. The losses of plutonium in the filtrate do not exceed ~0.3 mg/liter. The determination is carried out as follows [161].

To 30–50 ml sample solution in 1.5–3 *M* H_2SO_4, containing 10–30 mg Pu(IV), is added, with stirring, 5% aqueous solution of cupferron in an amount corresponding to 6 g-equiv. cupferron per 1 g-atom Pu (which produces a 2.5–3.5 mg/ml excess of cupferron). After 3–5 minutes the flocculent, orange-yellow precipitate is filtered off, washed with 1% solution of H_2SO_4 containing 0.1% cupferron and then with water. The filter with the precipitate is dried on an electric hotplate and calcined at 1000–1050° for 2 hours. The precipitate is weighed as PuO_2.

A quantitative precipitation of plutonium as cupferronate is also possible from 1–1.5 M HNO_3, in the presence of excess cupferron (2–3 mg/ml), according to the above procedure.

The precipitation of plutonium by cupferron from nitric acid solutions (>2 M), in which Pu(VI) is present as well as Pu(IV), has certain special features [161]. To the solution tartaric acid is first added and only then a solution of ammonium hydroxide to pH 2–5. The solution is then acidified by nitric acid to 1–1.5 M, the required amount of cupferron solution is added, then a little hydrazine nitrate, and plutonium (IV) cupferronate is filtered off after 15–20 minutes. The error of the method is ± 0.5%.

Precipitation as plutonium (IV) 8-hydroxyquinolate

8-Hydroxyquinoline precipitates tetravalent plutonium from alkaline and weakly acid solutions. The precipitation of plutonium from sulfate solutions by 8-hydroxyquinoline does not yield a precipitate of constant composition.

Moiseev (1953) found that tetravalent plutonium is quantitatively precipitated by 8-hydroxyquinoline from ammoniacal or bicarbonate solutions (pH 4.5–12) in the presence of tartrate ions. Under these conditions trivalent plutonium is oxidized, and the precipitate consists only of the hydroxyquinolate of tetravalent plutonium. This fact is of great practical importance, since the precipitation by 8-hydroxyquinoline may be conducted from solutions containing both trivalent and tetravalent plutonium. Hexavalent plutonium must not be present, since it forms a precipitate of a different composition with 8-hydroxyquinoline. Moiseev (1953) proposed the use of plutonium (IV) 8-hydroxyquinolate as the gravimetric form in the determination of small (less than 30 mg) amounts of plutonium. The precipitate is carefully washed and dried at 120–130°, when its composition becomes perfectly constant at $Pu(C_9H_6NO)_4$. The content of plutonium in the precipitate is 29.3%. Plutonium may be determined by this method in the presence of U(VI) and of alkali metals; small amounts of phosphates, fluorides, oxalates and hydroxylamine do not interfere [10, p. 39].

The procedure is as follows (Moiseev and Ognishcheva, 1954).

To the acid sample solution, containing 10–50 mg Pu and 20–50 mg U, 10–20 ml of 10% solution of ammonium tartrate and 5 ml of water, saturated with SO_2, are added. The solution is neutralized with a solution of ammonia until a faint odor of ammonia is noticeable, heated to 50–70° on a water bath and 20–30 ml of 10% solution of $NaHCO_3$ (heated to 50–70°) are added. A 3% solution of 8-hydroxyquinoline is added drop by drop, with stirring, when a fine crystalline, red-brown precipitate of plutonium 8-hydroxyquinolate is formed. The precipitate is then heated on a water bath for 1–2 hours and is filtered through a tared No. 4 fritted glass filter. The precipitate is washed 3–4 times with warm 1% solution of sodium bicarbonate, containing 0.05% of 8-hydroxyquinoline, then with hot water to which a few drops of NH_4OH have been added (pH 7–8), until the filtrate becomes colorless. The crucible with the precipitate is dried in the oven at 120–130° to constant weight. The gravimetric form has the formula $Pu(C_9H_6NO)_4$.

Precipitation as plutonium (IV) benzenesulfinate

Benzene sulfinic acid and its alkali metal salts quantitatively precipitate tetravalent plutonium from weakly acid solutions as an orange-colored amorphous precipitate [48]. Its composition corresponds to the formula $Pu(C_6H_5SO_2)_4$. Hexavalent plutonium is reduced by this reagent to the tetravalent state. Trivalent plutonium is not precipitated by the reagent (Milyukova, 1953). The solubility of plutonium (IV) benzenesulfinate under conditions corresponding to a quantitative precipitation (0.15 M HNO_3 and 2.5% excess precipitant) is $1.2 \cdot 10^{-5}$ mole/liter (Milyukova, 1959). The determination of plutonium as benzenesulfinate is interfered with by tartaric and citric acids at concentrations above 10%, and also by U^{4+}, Th^{4+}, Zr^{4+} and Fe^{3+}. The determination may be carried out in the presence of nickel, cobalt, chromium, lanthanum, iron (II), manganese, copper, and other elements. Milyukova (1953) proposed the following method for the determination of Pu(IV) as benzenesulfinate.

To 10–30 ml nitric acid solution (0.2 N HNO_3), containing 10–50 mg Pu, a 15–30% solution of ammonium benzenesulfinate is added with stirring, in an amount sufficient to produce a 2.5–3.0% concentration of the precipitant in the solution. The solution with the precipitate is left to stand for 15–17 hours, after which the precipitate is filtered through a paper filter and washed with a 2.5% solution of the precipitant in 0.15 N HNO_3. The precipitate with the filter is ashed and calcined at 1000–1100° and weighed as PuO_2. The accuracy of the determination is ± 0.2%.

Precipitation as plutonium (IV) phenylarsonate

Tetravalent plutonium is quantitatively precipitated from acid solutions by phenylarsonic acid. Trivalent plutonium does not form a precipitate with phenylarsonic acid. Hexavalent plutonium forms plutonyl phenylarsonate, the solubility of which is much higher than that of plutonium (IV) phenylarsonate (Mikhailov, 1956–1959). The optimum concentration of nitric and phenylarsonic acid for a selective precipitation of plutonium (IV) phenylarsonate is 0.4–0.5 and 0.04–0.06 M respectively. Mikhailov determined the limits of interference by certain elements with the precipitation of Pu(IV) as phenylarsonate under these conditions. According to his data, the following do not interfere with the determination of 30–100 mg Pu: up to 10 mg Fe; up to 100 mg Cr, Pb, Mn; up to 50 mg U, La, Zn, Cu, Ni, Ca, and major amounts of sodium nitrate. The method makes it possible to determine plutonium as the complexing ions at the following molar ratios to plutonium: SO_4^{2-} (5:1); F^- (0.02:1); $C_2O_4^{2-}$ (0.02:1) and Complexone III (0.005:1). Hydrogen peroxide interferes with the determination. The gravimetric form for the determination of plutonium by this method is PuO_2.

Mikhailov (1957) recommends the following procedure for the determination of Pu(IV).

To the nitric acid solution (0.5–0.7 M HNO_3) containing 30–100 mg Pu(IV) is added a 4% aqueous solution of phenylarsonic acid, in amount sufficient to render its concentration in a solution 0.04–0.06 M. The solution with the precipitate is heated on a water bath at 60–90° for 15–30 minutes. After cooling, the precipitate is filtered through an ashless filter paper (Blue Ribbon) and is washed with 0.02% solution of phenylarsonic acid in 0.004 M HNO_3. The filter with the precipitate is dried, ashed on a hotplate and calcined at 1050–1100° for 3–4 hours. The precipitate is weighed as PuO_2. The error of the determination does not exceed ± 0.5%.

Precipitation as plutonium (III) salicylate

When an excess of salicylic acid is made to act on solutions of trivalent plutonium (the final concentration of salicylate ion in solution is 100 g/kg), plutonium is quantitatively precipitated as $Pu(Sal)_3 \cdot 1.5\ H_2O$ [100]. It is calcined and weighed as PuO_2.

Precipitation as plutonium (IV) salicylate

It was found by Zvyagintsev and Sudarikov [100] that a quantitative precipitation of Pu(IV) from solution as salicylate takes place at pH 4–5 at minimum excess concentration of the precipitant. The composition of the precipitated salicylate is $PuO(Sal)_2$. The solubility of plutonium (IV) salicylate in water at 25° is 0.3 mg Pu/kg of solution. Major amounts of U, Fe, Cr, Th, Zr and other elements interfere with the determination. The determination is terminated by calcining the precipitate to PuO_2 at 1000°.

CHAPTER V

Separation of Plutonium from the Accompanying Elements

The determination of plutonium is carried out in various products of technological processing of nuclear fuel, in which the content of plutonium, fission products and various impurities may vary within very wide ranges. The impurities present in the samples render quantitative determination of Pu difficult. Thus Pu is usually separated from accompanying elements by a chemical method before determination.

SEPARATION BY PRECIPITATION WITH INORGANIC AND ORGANIC REAGENTS

The behavior of a substance present in trace concentrations is very different from the behavior of the same substance when present in weighable amounts. The main criterion in the separation of macroamounts of plutonium is the low solubility of the compound being isolated. When present in trace concentrations, plutonium is usually incapable of forming a solid phase, and a carrier is introduced into the solution, with which plutonium is coprecipitated. The precipitation and coprecipitation methods are still being used in analytical practice. The precipitation of weighable amounts of plutonium is mainly employed in umpire analysis and in high precision methods, while coprecipitation is widely employed in the concentration of plutonium and purification of its solutions.

COPRECIPITATION OF TRACE AMOUNTS

These methods were used in the very first studies on plutonium [368]. Coprecipitation is understood to mean the transfer of microamounts of a substance from a solution into a solid phase. The microcomponent (in our case plutonium) may be coprecipitated by one of two mechanisms [232, 236]. In one case crystalline precipitates separate out, in which the microcomponent forms part of the crystal lattice and is continuously (though not necessarily homogeneously) distributed throughout the precipitate (isomorphous and isodimorphous crystallization, formation of anomalously mixed crystals, etc.). In the other case, the distribution takes place from the surface into the bulk of the precipitate and is governed by adsorption laws (primary, secondary and internal adsorption). Co-crystallization takes place when the solid phase separates out as well-formed crystals, in which the surface is not very well developed, while adsorption is noted when the microcomponent is captured by finely crystalline or amorphous precipitates with a strongly developed surface.

The fundamental laws governing the co-crystallization process were established by Khlopin and by Hahn et al. The laws governing the absorption capture of small amounts by macroscopic precipitates were formulated by Paneth, Hahn, Starik, Ratner and others.

Co-crystallization of a system consisting of a macro- and a microcomponent is characterized by a physicochemical constant—the crystallization coefficient.

The microcomponent in the crystal may be distributed homogeneously, or inhomogeneously with a logarithmic distribution.

When a true thermodynamic equilibrium has been attained between all the precipitate and the solution, the distribution of the microcomponent is homogeneous and obeys the distribution law. Khlopin [236] proposed the following equation for the distribution law, referred to unit weight:

$$D = \frac{x(1-y)}{y(1-x)},$$

where D is the crystallization coefficient, x and $(1-x)$ are the fractions of the microcomponent in the precipitate and in the solution; y and $(1-y)$ are the fractions of the macrocomponent in the precipitate and in the solution respectively.

If the distribution of the microcomponent in the precipitate is logarithmic, and there is free crystal growth from solution with the formation of mixed crystals, the concentration of the microcomponent varies smoothly from the center to the periphery, and only the surface layer of the crystal is in equilibrium with the solution. In this case the crystallization coefficient is calculated by the formula of Doerner and Hoskins [174] and is denoted by λ:

$$\lambda = \ln\frac{a}{a-x} : \ln\frac{b}{b-y},$$

where a and b are the amounts of the micro- and macrocomponents in solution prior to the crystallization of the solid phase, and $(a-x)$ and $(b-y)$ are the amounts of the micro- and macrocomponents in solution after the crystallization of the solid phase.

If, during the formation of the precipitate (for logarithmic distribution) a true equilibrium is established between the surface layer of the crystal and the solution at any moment of time, and the precipitated solid phase does not recrystallize, the crystallization coefficient λ will have its maximum value of D, which is the crystallization coefficient for a homogeneous distribution of the microcomponent in the solid phase. The crystallization coefficient λ is, in this case as well, a physicochemical constant, characteristic of the given system.

To clarify the mechanism of the capture and in order to study the relationships governing the coprecipitation process, the crystallization coefficient of a given pair of salts is determined under various conditions as a function of several factors. The equilibrium between the precipitate and the solution is attained by methods developed by the school of Khlopin [236]; by isothermal crystallization of the precipitate ("from above" and "from below") and by the method of partial recrystallization of the solid phase. Depending on the experimental conditions, the crystallization coefficient is calculated by the Khlopin formula, or by the formula of Doerner-Hoskins.

Numerous technological and analytical methods of separation and purification of plutonium are based on its coprecipitation. In order to be able to develop a practical method for the separation of plutonium, it is necessary to consider not only the theoretical principles, but also the available information on the stability of the various valency forms under the experimental conditions, the solubility of various compounds of plutonium, the kinetics of redox reactions, behavior of plutonium on being coprecipitated with various carriers and the behavior of the impurities under these conditions. All these problems except for the last two have been discussed in detail in preceding chapters. Literature data on the behavior of impurities accompanying plutonium, and on the coprecipitation of plutonium on various carriers are given below.

Table 18 lists the reactions of uranium, thorium, plutonium and fission products with the chemical reagents usually employed in the separation and purification of plutonium from irradiated uranium. Since the analyst will do most of his work with nitric acid solutions, the data in the table refer to such solutions. It is assumed, moreover, that the uranium is present in the initial solution as $UO_2(NO_3)_2$, cerium is present as a mixture of trivalent and tetravalent compounds, cesium, strontium, barium, all rare-earth elements, yttrium, rhodium are present as nitrates, zirconium is present as zirconium nitrate, niobium is present as nitrate or as a suspension of niobium pentoxide; ruthenium is present as the soluble ruthenium nitrosonitrate (or possibly also as $HRuO_4$ or H_2RuO_4), tellurium is present as $Te_2O_3 \cdot OH \cdot NO_3$, antimony is present as trivalent and pentavalent antimony and iodine is present as a mixture of iodide with iodate.

Table 18

ACTION OF VARIOUS REAGENTS ON NITRIC ACID SOLUTIONS OF IRRADIATED URANIUM [101]

Reagent	Pu(VI)	Pu(IV)	Pu(III)	U(VI)	Th	Cs	Ba	Sr	La	Ce(IV)	Y	Nb	Ru	Rh	Sn(II)	Sn(IV)	Zr	Sb(III)
Potassium sulfate	NP	P	P	NP	P	NP	P	P	P	P	NP	NP	NP	NP	NP	NP	NP	P
Phosphoric acid and phosphates	NP	P	P	SA	P	NP	SA	SA	P	SA	P	P	P	NP	P	P	P	P
Hydrofluoric acid	NP	P	P	NP	P	NP	P	P	P	P	P	SA	NP	P	NP	NP	SA	NP
Sodium acetate	P	NP	NP	P	P	NP	NP	NP	NP	P	NP	ND	NP	ND	NP	ND	P	ND
Oxalic acid	P	P	P	P	P	NP	SA	SA	P	P	P	NP	NP	ND	NP	ND	SA	ND
																ND	P	ND
Potassium iodate	P	P	P	P	P	NP	P	P	P	P	P	ND	ND	ND	ND		ND	ND
Hydrogen peroxide	$\overrightarrow{\text{Pu}}$(IV) P	P	$\overleftarrow{\text{Pu}}$(IV) P	P	P	NP	NP	NP	P	P	P	P	P	P	$\overrightarrow{\text{Sn}}$(IV)			
									in the presence of alkali (La–Ru)									
Alkali hydroxides	P	P	P	P	P	NP	NP	NP	P	P	P	NP	NP	P	SP	SP	P	SP
Carbonates	P	SP	SP	SP	SP	NP	P	P	P	SP	SP	ND	NP	P	P	SP	P	P

Note: P—cation forms a precipitate with reagent; NP—no precipitate formed on addition of reagent; SA—precipitate formed is soluble in mineral acids; SP—precipitate formed is soluble in excess precipitant; ND—no data available.

The extraction of plutonium from the solutions of irradiated uranium and its purification is a complex task, which is usually carried out with the aid of precipitation, extraction or ion-exchange methods. The purification by precipitation is based on the capacity of plutonium to display various properties, depending on the degree of oxidation, which in turn may be altered at will throughout the purification process. In the separation of plutonium from uranium the differences in the redox reaction rates of these elements are taken advantage of (thus, in the hexavalent state uranium is the more stable element, whereas in trivalent and tetravalent states plutonium is the more stable of the two).

The separation of plutonium from the fission products is based on the fact that if a fission product element is similar to plutonium in one valency state, it will differ from it in other valency states. By performing a sufficient number of identical purification cycles, plutonium may be purified to the desired degree. It will be seen from the data in Table 18 and Table 19 that if it is desired to extract and purify plutonium with potassium sulfate, phosphoric acid, hydrofluoric acid, phytic acid and phenylarsonic acid, a carrier must be introduced (salts of zirconium, lanthanum, bismuth, etc.). When precipitating carbonates or acetates from solutions containing uranium, uranium itself serves as carrier.

It is occasionally more expedient to carry out a number of precipitations with various reagents, rather than repeat the same precipitation several times.

In such cases, after preliminary purification (when plutonium has been separated from the bulk of uranium and fission products by one of the reagents listed above), other reagents are used to effect the separation of plutonium from the residual impurities, such as potassium iodate, hydrogen peroxide and oxalic acid. Of the carriers listed in Table 19 we must mention zirconium phenylarsonate, zirconium phytate and zirconium benzenesulfinate as the most specific for the isolation of tetravalent plutonium, and sodium uranyl acetate and potassium uranyl carbonate for the isolation of hexavalent plutonium.

In what follows we shall discuss first the fundamental methods for the separation of plutonium from irradiated uranium, and then the methods of concentration of plutonium and its separation from transuranic and certain other elements.

Lanthanum sulfate method

The lanthanum sulfate method for the separation of small amounts of plutonium, proposed by Kurchatov (1947), is extensively used in analytical practice in the Soviet Union. A detailed study of this method was carried out by a number of authors [72–74, 143]. The method [143] is based on the capacity of the double sulfate of potassium and lanthanum to coprecipitate Pu(III) and Pu(IV) from aqueous solutions. Plutonium (VI) is not coprecipitated under these conditions. Plutonium is precipitated with potassium lanthanum double sulfate in a reducing medium, after

Table 19

COPRECIPITATION OF TRACE AMOUNTS OF PLUTONIUM IN MAIN VALENCY STATES WITH VARIOUS CARRIERS

Carrier	Pu(III)	Pu(IV)	Pu(VI)	References
Hydroxides	C	C	C	[394, 503]
Lanthanum fluoride	C	C	NC	[368, 521, 547, 637, 655]
Potassium sulfate	C	C	NC	[68]
Potassium lanthanum sulfate	C	C	NC	[23, 72–74, 143]
Potassium lanthanum selenate	C	C	NC	[244]
Phosphates and arsonates				
Bismuth phosphate	C	C	NC	[145, 628]
Zirconium phosphate	NC	C	NC	[145, 532]
Thorium pyrophosphate	NC	C	NC	[3]
Thorium hypophosphate	—	C	NC	[3]
Uranium (IV) hypophosphate	—	C	NC	[3]
Zirconium phytate	NC	C	NC	[205]
Bismuth arsonate		C	NC	[691]
Zirconium phenylarsonate	NC	C	NC	[211, 707, 708]
Oxalates				
Thorium oxalate	C	C	NC	[3]
Uranium (IV) oxalate	C	C	NC	[3]
Bismuth oxalate	C	C	NC	[3]
Lanthanum oxalate	C	C	NC	[203]
Iodates				
Zirconium iodate		C	NC	[3]
Thorium iodate		C	NC	[212]
Cerium (IV) iodate		C	NC	[3]
Lead iodate		C	NC	Chaikovskii, 1954
Zirconium benzenesulfinate	NC	C	NC	Milyukova, 1953
Thorium peroxide		C		[3]
Sodium uranyl acetate	NC	NC	C	[210, 442]
Potassium uranyl carbonate	NC	NC	C	[256]

Note: C—quantitative coprecipitation; NC—coprecipitation less than 1–2%.

which lanthanum is separated as the double sulfate in an oxidizing medium. The second precipitation of plutonium in a reducing medium is carried out with a smaller amount of lanthanum. In one cycle plutonium can be extracted fairly completely from the initial solution and the amount of the carrier can be reduced by a factor of

10–15. At the same time a substantial purification of plutonium from the fission products takes place.

The *reduction stage* is as follows.

Through the initial nitric acid solution, sulfur dioxide is passed to reduce plutonium, after which its solution is saturated with solid potassium sulfate. Lanthanum nitrate is then added, the solution is left to stand, and the precipitate is separated out. The composition of the solution during precipitation from the reducing medium is: 5–7% HNO_3, about 1% SO_2, 0.2–0.3 mg La per ml. The reduction is carried out for 1 hour at room temperature, after which the solution is saturated with potassium sulfate (up to ~ 1.5 *M* K_2SO_4). The precipitate is separated by centrifuging for 10–12 hours, washed with 1.5 *N* HNO_3 saturated with K_2SO_4 and dissolved in 0.5 *M* HNO_3. Under these conditions 95–99% Pu are coprecipitated with the precipitate of potassium lanthanum double sulfate. It is desired to separate plutonium quantitatively in the presence of large amounts of uranium, it is preferable to carry out a double precipitation, with the introduction of a new portion of lanthanum into the solution.

The *oxidation stage* is conducted in order to discard lanthanum and the fission elements entrained by the precipitate in the reduction stage. Plutonium is oxidized to the hexavalent stage either by potassium bichromate or by ammonium persulfate.

To the solution from the reduction stage solid bichromate is added in the amount of 10 grams per liter of solution. The oxidation is carried out for one hour at 95–98°. When using ammonium persulfate, the oxidation is conducted with gentle heating (35–40°) for one hour. The salt should be added in the amount of 10 grams per liter of solution, together with one gram of silver nitrate.

After the oxidation lanthanum is precipitated by saturating the solution with potassium sulfate, and the silver by adding potassium chloride. During this operation, the double sulfate precipitate occludes about 2% of plutonium, irrespective of the amount of plutonium and the manner of oxidation. In order to extract plutonium more fully, the precipitate of the double sulfate is dissolved in 0.5 *M* HNO_3, and the precipitation in the oxidizing medium is repeated. In order to concentrate lanthanum and plutonium, they are precipitated as hydroxides by an aqueous solution of ammonium hydroxide. The determination of plutonium is terminated radiometrically.

In this method, plutonium is sufficiently fully separated from uranium, iron and chromium and a number of other elements, and to the extent of 90–95% from the fission products. Neptunium accompanies plutonium in all these operations and must be specially separated.

In the presence of zirconium, sodium sulfate rather than potassium sulfate should be employed, in order to prevent occlusion of zirconium.

From the analytical point of view the sulfate method has certain advantages over the well-known fluoride method, since it may be carried out in glass vessels; random errors due to the presence of reducing substances originating from the plastic material of the test tubes are thus obviated.

The authors [143] showed that the percentage of the precipitated plutonium remains the same, both for trace and for weighable amounts of plutonium. The method may accordingly be used not only for analytical purposes, but also to extract plutonium from dilute technological solutions.

A number of authors [23, 71–74, 143] studied the effect of various factors on the completeness of the precipitation, as well as the mechanism of separation of plutonium with the crystals of the double sulfate.

The separation of plutonium with the double sulfate of potassium and lanthanum in 1.5 *N* HNO_3 was studied [143] as a function of the time of precipitation of the solid phase, concentration of K_2SO_4 in solution and the amount of lanthanum added. The results are given in Table 20.

Table 20

AMOUNT OF COPRECIPITATED PLUTONIUM AS A FUNCTION OF THE TIME OF STANDING AND THE CONCENTRATION OF CARRIER [143]

Amount of lanthanum, mg/ml	Amount of coprecipitated plutonium, %		
	10 minutes	2 hours	12 hours
0.2	83	—	98
0.4	88	93	99
0.6	88	99	—
1.2	98	—	100

The same work describes experiments on the effect of a more intensive reduction of plutonium by SO_2 (at 90–95° for 30 minutes). The amount of plutonium passing into the precipitate was 98%, i.e., approximately the same amount as at room temperature.

Further detailed investigation of the coprecipitation of Pu(IV), Np(IV) and Am(III) with the double sulfate of lanthanum and potassium, undertaken by Grebenshchikova and Chernyavskaya to clarify the mechanism of the coprecipitation [72–74], was carried out in solutions unsaturated with respect to potassium sulfate.

The behavior of Pu(IV) and Am(III) during separation with potassium lanthanum double sulfate was shown [72] to be different. While during the coprecipitation of Pu(IV) with potassium lanthanum double sulfate the distribution coefficient of plutonium decreases sharply when the concentration of potassium sulfate in solution is decreased from 0.19 *M* ($D = 27$) to 0.38 *M* ($D = 6$), it remains constant at $D = 0.45$ when Am(III) is coprecipitated under like conditions. It was shown [151] that in 0.19 *M* potassium sulfate Pu(IV) is present as the complex $Pu(SO_4)_3^{2-}$, similar in composition to the complex ion $La(SO_4)_3^{3-}$, but differing in its lower charge and smaller size; it follows that the complex ion of plutonium may form part of the crystal lattice of the macrocomponent. The formation of the complex ion of plutonium explains the different behavior during the coprecipitation of Am(III) and Pu(IV) with the double sulfate. The available data on the respective distribution coefficients of americium and plutonium make it possible to select the optimum conditions not only for their simultaneous separation from solution, but also for their separation from each other by the sulfate method.

The separation of the two elements from each other by the sulfate method is facilitated by two factors: the large difference in the values of the distribution coefficients of Am(III) and Pu(IV), and the fact that when the supersaturation is isothermally removed, the microcomponents are distributed between the liquid and the solid phases logarithmically, and not in accordance with Khlopin's law. Consequently, a more complete separation of plutonium may be effected in one operation. Table 21 gives the results of a study on the distribution of Pu(IV) and Am(III).

Table 21

DISTRIBUTION OF Pu(IV) AND Am(III) WHEN PRESENT TOGETHER IN SOLUTION [72]

Initial amount of $K_3La(SO_4)_3$, grams	Relative content in liquid phase, %				
	$K_3La(SO_4)_3$	Pu	Am	λ Pu	λ Am
0.1363	95.6	31.0	98.2	26 ± 3	0.4 ± 0.2
0.1433	87.0	3.1	94.0	25 ± 2	0.4 ± 0.1
0.1380	86.3	2.5	94.1	25 ± 2	0.4 ± 0.1
0.1378	86.3	1.9	92.5	27 ± 2	0.5 ± 0.1

Note: Initial amount of Pu $66 \cdot 10^3$ cpm; initial amount of Am $25.5 \cdot 10^3$ cpm; concentration of K_2SO_4 0.19 *M*.

The value which was chosen for the molarity of the sulfate (0.19 *M*) corresponds to the maximum difference between the distribution coefficients of Pu(IV) and Am(III). It will be seen from the results in Table 21 that when 97–98% of plutonium separates out, the precipitate is contaminated with 6–8% of the initial amount of americium. A reprecipitation will result in a complete purification of plutonium from americium (the precipitate will contain a few tenths of one percent of americium).

In view of the fact that the distribution coefficient of plutonium decreases in saturated sulfate solutions, it is sometimes more convenient to precipitate from unsaturated solutions [72]. In more recent studies [23, 71, 74] the coefficients of crystallization of Np(IV), Pu(IV) and Am(III) were determined in solutions unsaturated with potassium sulfate.

It is more expedient to separate Pu(IV) from 0.7 *M* K_2SO_4, since the crystallization coefficient of Pu(IV) in solutions of this concentration is much higher than in 1.3 *M* K_2SO_4. In order to decrease the duration of precipitation, the precipitate should be left to stand for 30 minutes, the concentration of potassium sulfate then increased to 1.1 *M* and the precipitate again left to stand for 30 minutes; the total standing time is one hour.

Coprecipitation with lanthanum potassium double selenate

Shvedov et al. [244] described the preparation and properties of lanthanum potassium double selenate $KLa(SeO_4)_2 \cdot xH_2O$. They found the solubility of the compound to be 1.524 g/liter. Plutonium (IV) (98.6%) and Ce(IV) (97.4%) are coprecipitated with the precipitate of $KLa(SeO_4)_2 \cdot xH_2O$. The method was not noted to have any advantages over the precipitation with potassium lanthanum double sulfate [244].

Coprecipitation with potassium sulfate

Grebenshchikova and Bobrova [68] studied the coprecipitation of Pu(IV) and Am(III) with potassium sulfate and showed that these elements are coprecipitated by way of formation of anomalous mixed crystals. In view of the fact that Pu(IV) and Am(III), when coprecipitated with potassium sulfate, display a similar behavior to La(III) and Ce(III), these authors consider that the formation of mixed crystals of Pu(IV) and Am(III) with potassium sulfate in the case under consideration is due to the cocrystallization of the double sulfates of potassium and plutonium or potassium and americium, formed on the surface of the precipitate, with the crystals of potassium sulfate.

The high value of the crystallization coefficient found for plutonium ($\lambda = 30$), the logarithmic character of the distribution and the absence of a lower miscibility limit indicate that potassium sulfate may be used as carrier for plutonium. Obviously, rare-earth elements are fully coprecipitated with plutonium.

Bismuth phosphate method

This method has been discussed in detail in American literature [145; 203, p. 278; 628, 658]. Bismuth phosphate $BiPO_4$ is used as carrier. It follows from Table 19 that bismuth phosphate coprecipitates Pu(III) and Pu(IV), and tetravalent plutonium is most completely coprecipitated. Pu(IV) is quantitatively coprecipitated with the precipitate of bismuth phosphate from 0.1–1.0 *N* solutions of nitric acid. Further increase in the acidity of the solution impairs the separation of plutonium. Bismuth phosphate gives a dense, crystalline, easily filterable precipitate. It is practically insoluble in moderately concentrated (less than 1 *M*) nitric and sulfuric acids. This makes it possible to prevent the precipitation of uranyl phosphate when bismuth phosphate is precipitated from nitric acid solution containing sulfuric acid, which forms a stable complex compound with uranyl ion. Bismuth phosphate is highly selective with respect to Pu(IV) and occludes fission products only in insignificant quantities. In order to attain a high degree of purification, the precipitation is carried out from an oxidizing and a reducing medium in turn. Since bismuth phosphate is soluble in concentrated nitric acid, this operation is readily performed; this is an advantage over the lanthanum fluoride method, in which lanthanum trifluoride dissolves only with difficulty.

When this method is employed, plutonium can be isolated in the presence of major amounts of iron; for this reason the reduction of hexavalent plutonium is carried out by Fe(II), which is easily separated during the precipitation of Pu(IV) [92] from dilute nitric acid. Large amounts of calcium do not interfere with the precipitation of plutonium on bismuth phosphate. This makes it possible to isolate ultramicro-amounts of plutonium from animal bones (Chaikhorskii and Lapshina, 1953).

In order to remove radioactive zirconium and niobium during the precipitation of bismuth phosphate under reducing conditions, fluoride ions are added to complex these elements. In the subsequent oxidative precipitation inactive cerium or zirconium is introduced for the sake of more complete coprecipitation of radioactive zirconium and niobium [395].

Rydberg [628] gave an analytical procedure for the separation of trace amounts of plutonium from the irradiated material.

The sample of irridiated uranium is dissolved in concentrated nitric acid and heated at 90° for 30 minutes, the solution is diluted with acidulated water to 5 *M* HNO_3 and solid $NaBiO_3$ is added to oxidize plutonium to the hexavalent state. The solution is then diluted so as to obtain a final solution which is 0.1 *M* in HNO_3 and 0.1 *M* in phosphate ion. The precipitate of $BiPO_4$ is centrifuged; it occludes most of the fission products.

The solution containing Pu(VI) is made 1 *M* in HNO_3, 0.05 *M* in hydrazine and 0.005 *M* in ferrous iron. Under these conditions plutonium is readily reduced to Pu(III). The solution is diluted to 0.1 *M* in HNO_3 and 0.1 *M* in phosphate ion. A solution of Be(III) is then added (1 drop of 0.1 *M* Be(III) solution per ml of sample solution) to precipitate $BiPO_4$ which occludes all the plutonium. The precipitate of $BiPO_4$, which usually also occludes some uranium, is centrifuged and washed with some acidulated water. The phosphate precipitate is then converted into hydroxides by treating it with hot 10 *M* KOH. The solution with the precipitate is centrifuged, washed and dissolved in hot concentrated HNO_3. However, the reduction of plutonium to Pu(IV) by sodium nitrite [203, p. 280] is more efficient.

According to Lavroskii [145] a high degree of purification of plutonium from the fission products may be attained by this method. The overall purification coefficient is 107. The insignificant content of uranium in the plutonium thus obtained is a proof that the separation of the two elements is satisfactory.

It is recommended [628] that plutonium be finally purified by extraction with thenoyltrifluoroacetone in benzene.

In order to facilitate the subsequent dissolution of the precipitate of $BiPO_4$-plus-plutonium in acid solutions, it has been recommended [591] that the precipitate be preliminarily treated with solutions of alkalis. As a result of such treatment, bismuth and plutonium phosphates are converted to hydroxides which are readily soluble in solutions of acids.

Faris and Strassel [396] accelerate and facilitate the purification of plutonium by the bismuth phosphate method by adding a solution of Ce(III) (1–5 g/liter) prior to the reductive precipitation, and by adding Hg(II) salts (1–5 g/liter) prior to the oxidative precipitation of the fission products.

It was shown in another investigation [395] that the purification from γ-activity is much improved if, after the precipitation of $BiPO_4$ from an oxidizing medium, the fission products are reprecipitated on cerium phosphate. To do this, 2 grams of cerium (IV) ammonium double nitrate are added to the nitric acid solution 30 minutes after the precipitation of $BiPO_4$, and the precipitate is separated by centrifugation after another 30 minutes.

Coprecipitation with bismuth arsenate

Thompson et al. [691] proposed the use of $BiAsO_4$ as carrier. The mode of separation is similar to the one used in the bismuth phosphate method. Pu(IV) is coprecipitated with $BiAsO_4$ from a solution of 0.5–1 *N* HNO_3. In order to form the carrier, a soluble bismuth salt is introduced into the solution and then arsenic acid to the concentration of 0.4 mole/liter of each reagent. The precipitate of $BiAsO_4$ which is formed occludes plutonium, while uranium and products of fission which are not isomorphous with plutonium, remain in solution. As distinct from the bismuth phosphate precipitation, this method makes it possible to achieve a fuller precipitation of plutonium. Under otherwise identical conditions, the bismuth arsenate precipitate includes 99.4% Pu, 10.9% of β-active and 20.7% γ-active products; the corresponding figures for the precipitation with bismuth phosphate are 97.2, 22.0 and 13%. The authors [691] suggest that bismuth arsenate precipitation be carried out prior to the bismuth phosphate precipitation, the latter being carried out from oxidizing media only. This combination increases the overall purification from β-active fission products by a factor of six, and from the γ-active products by a factor of two as compared with the conventional bismuth phosphate method.

Coprecipitation with zirconium phosphate

Like tetravalent thorium and cerium, Pu(IV) is quantitatively coprecipitated with zirconium phosphate, while trivalent and hexavalent plutonium remain in solution.

The zirconium phosphate method makes it possible not only to separate plutonium and neptunium from fission products, but also to separate these elements from each other [532].

The experimental procedure described by Magnusson, Thompson and Seaborg [cf. 145] is as follows.

The solution of the uranium target is brought to the following approximate composition: 3 *M* in HNO_3, 0.01 *M* in Zr(IV), 0.001 *M* in Ce(IV) and 0.02 *M* in $NaBrO_3$. The solution is then heated for one minute to oxidize plutonium and neptunium to the hexavalent state and H_3PO_4 added to a concentration of 0.5 *M* to precipitate zirconium phosphate. The solution is heated for one minute to coagulate the precipitate and centrifuged. The solution containing Pu(VI) and Np(VI) is reduced with excess hydrazine sulfate to decompose the bromate, the concentration of hydrazine sulfate is adjusted to 0.01 *M* and Fe(II) is added to a concentration of 0.005 *M* to reduce plutonium to Pu(III) and neptunium to Np(IV). This reduction takes place rapidly. A solution of Zr(IV) is added in the amount corresponding to 1 mg Zr per ml of solution and the solution is heated with stirring. The zirconium phosphate which precipitates out occludes neptinium. Uranium and Pu(III) remain in solution. Lanthanum is added to the solution as carried and lanthanum hydroxide is precipitated with ammonia; plutonium is occluded. Lanthanum hydroxide is dissolved in nitric acid and the solution is heated to 60–80° to oxidize plutonium to the tetravalent state. Zirconium is added as carried and zirconium phosphate is precipitated as above; all the Pu(IV) is coprecipitated. The precipitate is mixed with a solution of lanthanum nitrate (0.1 mg/ml on the metal) and dissolved in a mixture of 1 *M* HNO_3 and 1 *M* HF. The LaF_3 precipitate formed contains all the plutonium.

The lanthanum fluoride method

This is the procedure by which weighable amounts of plutonium were isolated for the first time from neutron-irradiated uranyl nitrate [368].

Neptunium (VI), plutonium (III) and plutonium (IV) are quantitatively precipitated from acid solutions with the precipitate of lanthanum fluoride. Plutonium (VI) is not coprecipitated with the LaF_3 precipitate. During the precipitation from a reducing medium uranium and the bulk of fission products are discarded, and there is some concentration of plutonium. The procedure given below has been recommended by Hyde [655].

The irradiated sample is dissolved in concentrated HNO_3, the solution is partly evaporated to remove excess acid and diluted so as to obtain a concentration of uranium not higher than 0.5 *M*. Lanthanum is then added in the amount of 0.1–0.5 mg/ml. Plutonium is reduced to Pu(III) by passing SO_2 through the solution for a few minutes. The solution is then placed in an HF-proof vessel and hydrofluoric acid is added to a concentration of 1–3 *M*. The precipitate of lanthanum fluoride is centrifuged and washed with a mixture of 1 *M* HNO_3 with 1 *M* HF containing SO_2. The precipitate is then dissolved in concentrated HNO_3 containing some H_3BO_3, Al(III) or Zr(IV) to complex the fluorides. The solution is diluted to a few ml with SO_2-containing water, and lanthanum fluoride is again precipitated. The washed fluoride precipitate is treated twice in the heat with a solution of carbonate-free KOH. The plutonium and lanthanum hydroxide obtained are washed and dissolved in 1 *M* HNO_3.

Plutonium is then oxidized to the hexavalent state by 0.15 *M* solution of $KBrO_3$ at 95° for 20 minutes (Ag^{2+} ion may also be employed) and hydrofluoric acid is added to the solution. Lanthanum trifluoride precipitates out, which carries almost the entire residual activity, while hexavalent plutonium remains in solution. The precipitate is separated by centrifugation and is discarded. Plutonium is reduced to Pu(III) by SO_2 (15–30 min), lanthanum is added and the fluoride is again precipitated. The hydrofluoric acid used in the operation must not contain any reducing admixtures and is accordingly pretreated with a solution of persulfate ion. The fluoride is converted into the hydroxide, as in the first hydroxide, as in the first stage. The hydroxide obtained is dissolved in acid.

By repeating this sequence of operations several times, a fairly pure plutonium (less than 1% impurities) may be obtained.

In order to separate plutonium from neptunium, the oxidation of the solution by bromate should be carried out without heating for 20 minutes. When this is done, Pu(III) is rapidly oxidized, whereas the oxidation of Pu(IV) and Np(V) to the hexavalent state in a cold bromate solution is very slow. It was shown that the oxidation of Np(IV) by bromate is strongly catalyzed by the fluoride ions, so that as soon as lanthanum fluoride is precipitated, a rapid and complete oxidation to Np(V) takes place. Pu(IV) does not become oxidized to a higher valency and is coprecipitated with lanthanum fluoride.

If a high degree of separation from uranium is required, the lanthanum fluoride cycle should be repeated, with 1 M H_2SO_4 instead of nitric acid, and with a smaller (not less than 0.2 mg/ml) amount of carrier. Under these conditions the amount of the coprecipitating uranium decreases. Zirconium and alkaline-earth elements are extracted with the carrier when present in trace amounts, but are not captured when present in milligram amounts. In order to remove the radioactive isotopes of these elements, inactive carriers—zirconium, barium and strontium—should be added to the sample solution prior to precipitation.

Magnusson, Thompson and Seaborg [532] recommend that, before the lanthanum fluoride cycle is performed, protactinium be separated by coprecipitation on manganese dioxide, to prevent it from coprecipitating with lanthanum fluoride. By employing the lanthanum fluoride scheme in conjunction with extraction, very small amounts of plutonium have been isolated from uranium ores [424, 521, 655]. The coprecipitation of plutonium on lanthanum fluoride was utilized by Peppard et al. [590] to isolate the former from the waste waters of a uranium plant. The completeness of the extraction of plutonium was determined from the yield of Pu^{238}, which is added prior to analysis to a parallel sample. About 20–30% of the plutonium originally present may be separated. The lower limit of the determination of the content of plutonium by this method is 1–10 μg Pu per liter [547].

According to [724], the coprecipitation of small amounts of plutonium on lanthanum fluoride is not applicable to its full separation from U(VI). The adsorption of hexavalent uranium is suppressed by the addition of zirconium. To decrease adsorption in these cases it has also been recommended to add small amounts of coprecipitant (not more than 50 μg La per ml).

A number of improved modifications of the lanthanum fluoride method, and its applications to various purposes have been reported [107, 152, 188, 245, 503, 505, 533, 618, 637, 639, 709]. Monk [553] proposes the acceleration of the lanthanum fluoride method, by reducing the time of standing of the fluoride precipitates from 17 to 1 or 2 hours, and increasing the centrifugation time to 5 minutes instead. The new procedure, subsequent to the addition of hydrofluoric acid, is as follows: the solution is stirred and centrifuged for 2 minutes, 0.15 ml of $La(NO_3)_3$ solution (1.6 mg La per ml) is added drop by drop and the solution is again stirred. The solution with the precipitate is left to stand for 1.5 hours and is then centrifuged. The solution over the precipitate is withdrawn to a residual volume of 0.1 ml and the residue is suspended in 1 ml of 0.5 M HF. The solution with the precipitate is again centrifuged for 5 min; the solution is decanted and the procedure continued according to Hyde [655].

Other workers [107] proposed to reduce the time of standing by introducing $PbSO_4$ as coprecipitant (lead is introduced into the solution as a soluble salt, after which H_2SO_4 is added). The time is thus reduced to 3 hours. Lead is extracted from the precipitate with a solution of KOH.

To dissolve the precipitate of lanthanum fluoride which contains plutonium, it is treated with 1 *M* HNO_3 in the presence of Fe^{3+} ions, which bind fluoride ions into stable FeF_6^{3-} complexes. In the solution thus obtained plutonium is oxidized to Pu(VI) and LaF_3 is precipitated by supplementary introduction of fluoride ions in order to occlude radioactive admixtures [709].

Other authors [505] dissolve the precipitate of LaF_3 by treating it with a solution of a zirconium salt. The presence of zirconium increases the solubility of LaF_3 and render Pu(IV) more readily oxidizable to hexavalent state.

For solutions containing only uranium and plutonium (in the valency state of four or less) Maddock and Smith [152] proposed to desorb plutonium from the precipitate by treatment with a half-saturated solution of ferric iron or aluminum nitrate. After desorption, the concentration of nitrate is raised to the saturation value and the extraction carried out.

Ritter [188] proposed the combination of the lanthanum fluoride process with the bismuth phosphate process. Dalton [371] used the lanthanum fluoride method as the first stage in the isolation of plutonium from urine. The method proposed by Aron [11] for the separation of lanthanides from plutonium is based on their coprecipitation with CaF_2 in an oxidizing medium; Pu(VI) remains in solution. Plutonium was oxidized with silver oxynitrate, which oxidizes trace amounts of plutonium within 1 minute at room temperature within a wide acidity range (1–11 *N* HNO_3).

Coprecipitation on hydroxides

This method is one of the most convenient and most widely used in the concentration of plutonium. Lanthanum or nickel [503] are usually employed as carriers. In addition, data are available (Chaikhorskii, 1953) to the effect that the hydroxides of the following elements may be employed as carriers of plutonium: Cd, Cr, Al, Mn, Fe, Co, Be, Mg, Ti, Sn, Pb. Plutonium may be precipitated by both alkali hydroxides and ammonia. In the presence of aluminum, lead, zinc, sodium salts, potassium salts and ammonium salts in solution, plutonium is readily precipitated as hydroxide by 20% alkali hydroxide solutions. When plutonium is to be determined in solutions containing Ca, Mg, Mn, Co, Cu, Cr, etc., it is precipitated by 20% ammonia. Some of these elements form soluble compounds in the excess of ammonia and thus do not interfere with the coprecipitation of plutonium.

It should be pointed out that in the presence of gold the coprecipitation of plutonium with lanthanum hydroxide is practically absent, since, as suggested by Chaikhorskii et al. (1953), the aurate ions adsorbed on the precipitate of lanthanum hydroxide alter the electric charge on the precipitate and thus prevent the coprecipitation of plutonium.

If platinum is present in the solution, the precipitation of lanthanum hydroxide should be carried out only with a solution of sodium hydroxide, since potassium and ammonium ions form the insoluble compounds $(NH_4)_2[PtCl_6]$ and $K_2[PtCl_6]$, with the chloroplatinate ion.

Derbenova, Khramtsova and D'yakova (1954) showed that small amounts of plutonium (less than 1 mg/liter) can be isolated from oxalate and lactate media by precipitation on nickel hydroxide. The utilization of lanthanum as carrier is impossible in this case, since an insoluble precipitate of lanthanum oxalate is formed. Nickel, on the other hand, forms with oxalate ions a soluble complex ion of the composition $[Ni(C_2O_4)_2]^{2-}$, which is decomposed in alkaline medium, while nickel precipitates out as the hydroxide.

When the sample solution contains major amounts of uranium (up to 250 g/liter) and chromium, together with elements which are not precipitated in alkaline and ammoniacal media, plutonium may be isolated by coprecipitation with nickel hydroxide in the presence of hydrogen peroxide (Chaikhorskii et al., 1953). The separation from uranium and chromium is based on the capacity of these elements to form soluble peruranates and chromates in alkaline medium (NaOH, KOH) in the presence of hydrogen peroxide. If, in addition to uranium and chromium, copper, zinc, cadmium, silver, calcium, etc., which are not precipitated in ammoniacal medium, are also present, plutonium is coprecipitated on lanthanum hydroxide with CO_2-free ammonia in the presence of hydrogen peroxide. The great advantage of the coprecipitation with nickel hydroxide is the speed of the method (the separation and the determination of plutonium takes about one hour), and also the fact that plutonium may be determined in the presence of large amounts of uranium (up to 250 g/liter) in solution. The accuracy of the determination is $\pm 6\%$.

About 0.3–0.5 ml of the sample solution (~ 1 μg Pu/ml) are placed in a centrifuge tube, the solution of nickel (5–6 mg Ni) is added, with 7–8 ml 40% solution of alkali hydroxide (carbonate-free) and 3–5 drops of 30% H_2O_2. The solution with the precipitate is heated on a water bath for 5–10 minutes to coagulate nickel hydroxide and partly to decompose the excess hydrogen peroxide. The precipitate of nickel hydroxide and plutonium is separated by centrifugation, washed with water, dissolved in 5–10 ml of 1.5 *N* HNO_3 and the plutonium content is determined by the radiometric method.

Beaufait, Stevenson and Rollefson [24] proposed a method for the separation of plutonium from uranium and the fission products by coprecipitating tetravalent plutonium with ferric hydroxide from acetate solutions at pH 5–6. Zirconium and niobium are also coprecipitated under these conditions. They are removed by precipitating ferric hydroxide from an oxidizing medium, when hexavalent plutonium remains in solution.

Sodium uranyl acetate method

The method takes advantage of the capacity of hexavalent plutonium to form the sparingly soluble sodium plutonyl acetate, which is isomorphous with sodium uranyl acetate. Tri- and tetravalent plutonium are not coprecipitated with sodium uranyl acetate. The acetate method was proposed by Hahn and Strassman [442] and had been originally employed in the separation of neptunium. The method was subsequently employed in the separation and purification of plutonium from large amounts of uranium.

Starik et al. [210] employed the coprecipitation of plutonium with uranyl diacetate to purify plutonium from naturally α-active elements (Pa, Th, Ra, Po) found in uranium pitchblende, and showed that plutonium may be fully separated from these elements in two precipitations. The isolation takes place from 0.1 N HNO_3. Plutonium is first oxidized to Pu(VI) by $KBrO_3$ in this solution, when the manganese present in the ore precipitates out as peroxide. This makes for a better purification of plutonium from radioelements, in particular protactinium. After the separation of the precipitate of manganese peroxide, Pu(VI) is precipitated with the precipitate of uranyl diacetate at 90° by twice the volume of 45% solution of sodium acetate from 2 N HNO_3. The dense, crystalline precipitate of uranyl diacetate is separated by decantation, washed and dissolved in 2 N HNO_3. The operation is then repeated. After the dissolution of the precipitate it is precipitated from a reducing medium, when plutonium is separated from uranium. A more complete separation of uranium may be obtained [210] by concentrating the plutonium (coprecipitation with hydroxide) and extracting with ether.

A similar method [67] has been reported in two variants: plutonium is oxidized and coprecipitated with uranyl acetate, or else it is first reduced and uranium eliminated. The second variant is only applicable if plutonium is present in larger amounts; it is precipitated without a carrier, since uranium is precipitated in the first stage.

Coprecipitation with zirconium phenylarsonate

Phenylarsonic acid $C_6H_5AsO(OH)_2$ is a highly selective reagent for tetravalent cations, mainly for zirconium; under certain conditions of precipitation zirconium may be separated from a number of elements, including uranium, thorium, aluminum, iron, rare-earth elements, etc. [243]. Voigt et al. [707, 708] showed that Np(IV) and Pu(IV) can be coprecipitated with zirconium phenylarsonate.

Starik et al. [211] determined the optimum conditions for the separation of neptunium with the precipitate of zirconium phenylarsonate and developed the following procedure for the separation of Np(IV) from Pu(III).

To a 1 N solution of HCl, containing neptunium and plutonium, with lanthanum as carrier, $ZrCl_4$ (3–4 mg Zr per 100 ml of solution) and an excess of hydroxylamine are added. The solution is heated for

$2^1/_2$ hours on a water bath, and a 4% solution of phenylarsonic acid is then added (in the volume ratio 1:2). The resulting precipitate of zirconium and neptunium phenylarsonates is left to stand for 30 minutes, centrifuged, and washed with 1% HCl. It is then treated, with gentle heating, with a 5% solution of alkali, centrifuged and dissolved in HCl. Neptunium is determined radiometrically. The supernatant and the washings, containing plutonium, are combined, and plutonium is isolated from the solution by coprecipitation with lanthanum hydroxide. After dissolution in acid, the α-activity is measured.

In another paper [180, p.51] phenylarsonic acid was employed to separate Pu(IV) and Am(III). The method is not suitable for solutions with a very high or a very low plutonium-to-americium ratio.

An aliquot of the nitric acid solution (total α-activity ~10 cpm) is placed in a 2 ml centrifuge tube. Pipets are washed twice with 1 *M* HNO_3. To the sample is added one drop of carrier (a solution of zirconium in 0.1 *N* HCl with an approximate concentration of 1 mg/ml), after which the volume is made up to 1 ml with 1 *M* HNO_3. The solution is thoroughly stirred and 2 drops of a saturated solution of phenylarsonic acid are added. The precipitate is left to stand for about 5 minutes, and the tube is then placed in a boiling water bath for 5 minutes in order to coagulate the precipitate. The mixture is cooled and centrifuged for a few minutes. The solution is placed in a volumetric vessel of 2 ml capacity, while the precipitate is washed with 10 drops of washing solution (10 ml of saturated solution of phenylarsonic acid, diluted to 100 ml with 1 *M* HNO_3). After centrifugation the washing solution is placed in the same volumetric vessel. Americium is determined radiometrically on the solution.

The precipitate is suspended in a few drops of water and transferred to the target. The tube is washed out several times with water and the washings are also placed on the target. The excess nitric acid is neutralized with concentrated NH_4OH. The liquid is evaporated to dryness, and NH_4NO_3 and the zirconium complex are decomposed by cautious heating. The target is calcined and plutonium determined radiometrically.

Coprecipitation with lanthanum oxalate

It was shown by Grebenshchikova and Bryzgalova [69] that Pu(IV) is coprecipitated with lanthanum oxalate with the formation of anomalous mixed crystals without a lower miscibility limit in the concentration range of the microcomponent 10^{-6} and 10^{-9} *M*. The distribution of Pu(IV) between lanthanum oxalate and the solution may obey Khlopin's law or the logarithmic law, depending on the experimental conditions of coprecipitation. It was also found that in solutions containing HNO_3 (0.5–1.5 *N*), with a concentration of oxalic acid or ammonium oxalate of not more than 0.1 *M*, the enrichment of lanthanum oxalate crystals with plutonium reaches its maximum value ($D = 21$). Further increase in the concentration of oxalate ions in solution results in a decrease in the crystallization coefficient because of the formation of soluble oxalate complexes of Pu(IV). Thus, Pu(IV), when coprecipitated with a trivalent carrier, behaves similarly to a trivalent element, the only difference being a more rapid variation in the crystallization coefficient with the variation of the oxalate ion concentration in solution. This is explained by its greater tendency to form complex compounds of a more involved structure.

Uranium when present in amounts of several hundreds of milligrams does not interfere [48].

Coprecipitation with lead iodate

Chaikhorskii et al. (1954) studied the coprecipitation of Pu(IV) on lead iodate, in order to separate it from uranium, iron, chromium and certain other elements. About 20% iron is coprecipitated on lead iodate together with plutonium, irrespective of its content in the sample. Uranium is coprecipitated to the approximate extent of 6% (at uranium contents between 0.1 and 200 mg). Only about 0.3% plutonium remains in the filtrate.

Coprecipitation with thorium iodate

The iodate method [212] of separation of plutonium from irradiated uranium is based on the different solubilities of Pu(IV) and Pu(VI) iodates.

The irradiated uranium is dissolved in HNO_3, and H_2O_2 is added to convert all the plutonium to Pu(IV). Thorium nitrate $Th(NO_3)_4$ (0.3%) is introduced into the solution, and thorium iodate, which serves as carrier for Pu(IV), is precipitated from 2–3 *N* HNO_3 by adding a solution of potassium iodate. The bulk of uranium and fission products remains in solution. The washed precipitate is dissolved in 12 *N* HCl, Pu(IV) is oxidized to Pu(VI) with sodium bichromate at 65°, the solution is cooled, aqueous KIO_3 or HIO_3 is introduced, and a 2–3-fold excess of a thorium or cerium salt is added. The fission products are coprecipitated with $Th(IO_3)_4$, while plutonium remains in solution as PuO_2^{2+}. If necessary, the cycle of operations is repeated several times.

After concentrating plutonium according to the lanthanum fluoride technique, it is separated from the carrier by the iodate technique [368].

Coprecipitation with alkali metal carbonates

It was shown by Yakovlev and Gorbenko-Germanov [256] that in concentrated solutions of potassium carbonate in an oxidizing medium plutonium forms a sparingly soluble complex salt $K_4[PuO_2(CO_3)_3]$. This compound is isomorphous with the analogous compound of uranium $K_4[UO_2(CO_3)_3]$, with which it is coprecipitated. When precipitated from an oxidizing medium, 99% of plutonium is extracted from solution by one single precipitation of uranyl potassium tricarbonate. Because of the relatively high solubility of the double carbonates of rare-earth elements, their occlusion is small (1–3%). Barium and strontium carbonates crystallize in another system, so that they probably will not be occluded. It may also be expected that the occlusion of cesium and ruthenium will be small. Zirconium will probably coprecipitate.

The carbonate precipitation from oxidizing medium ensures a relatively high purification of uranium and plutonium from the accompanying elements. A behavior similar to that of U(VI) and Pu(VI) is also displayed by Am(V) [256]. These elements must be separated by another method.

Coprecipitation with zirconium phytate

Tetravalent plutonium is quantitatively precipitated with phytic acid [205]. This is the principle of coprecipitation of small amounts of plutonium on zirconium phytate. If the concentration of plutonium is 0.5 μg per 100 ml and that of the carrier is not lower than 0.2 mg per 100 ml, the separation of plutonium is 95–98% of the quantitative. At Pu(IV) concentration of 0.1 μg per 100 ml and the same concentration of carrier, the coprecipitation is 80–90% complete. The precipitation is carried out from 3 *N* HNO_3. The method permits not merely the isolation of plutonium, but also its purification from Cr, La, Mn, Pb, Al, Cu, Zn, Ni, rare-earth elements and certain other elements.

Coprecipitation with zirconium mandelate

It was shown by Merz [546] that mandelic and bromomandelic acids separate trace amounts of Pu(IV) on a carrier from acid solutions (0.1–1.0 *N*). Hexavalent uranium remains in solution under these conditions, and its precipitation begins only at pH 4.0. Trivalent plutonium also coprecipitated on a carrier by the action of these acids, but only in weakly acid solutions (pH 3). Ytterbium is used as carrier for Pu(III), while zirconium is used as carrier for Pu(IV). Hexavalent plutonium remains in solution in both cases. The variation in the acidity of the initial solution affects the coprecipitation process of tetravalent plutonium. The data on the coprecipitation of plutonium (IV) with zirconium mandelate and bromomandelate are shown in Table 22.

The author [546] separated trace amounts of plutonium from major amounts of uranium with the aid of bromomandelic acid under the following conditions: concentration of plutonium 0.2 μg/liter, concentration of U 5 g/liter, acidity 0.5 *N* HCl. Under these conditions 84% of the plutonium were coprecipitated with the carrier (1 mg Zr per 10 ml). The coefficient of purification from uranium was 400. With mandelic acid, 51% of the plutonium were found in the precipitate and the purification coefficient was somewhat lower (380).

Coprecipitation of plutonium with potassium rhodizonate

It was found by Weiss and Shipman [723] that plutonium is quantitatively precipitated by potassium rhodizonate from aqueous solutions (pH 5); these workers used the reagent for the separation of plutonium from urine. The precipitation is carried

Table 22

COPRECIPITATION OF Pu(IV) WITH ZIRCONIUM IN THE PRESENCE OF MANDELIC AND BROMOMANDELIC ACIDS [546]

Concentration of acid, N	Coprecipitation of Pu(IV), %: bromomandelic acid, HCl	bromomandelic acid, HNO_3	mandelic acid, HCl	mandelic acid, HNO_3	Concentration of acid, N	Coprecipitation of Pu(IV), %: bromomandelic acid, HCl	bromomandelic acid, HNO_3	mandelic acid, HCl	mandelic acid, HNO_3
5.0	49	0	22	0	1.0	87	85	82	75
2.0	64	42	45	13	0.1	90	87	88	80
1.5	83	62	69	41	0.01	89	88	86	87

Note: Concentration of Pu in solution 0.5 μg/liter; concentration of carrier 2 mg Zr per 20 ml solution.

out as follows. Solid potassium rhodizonate (20 mg) is added to 20 ml urine (pH 6). When potassium rhodizonate has dissolved, 10 ml absolute ethanol are added. The precipitate of potassium rhodizonate formed coprecipitates plutonium quantitatively. The authors showed that below pH 4 plutonium is not coprecipitated with the reagent. The fullest (100%) separation of plutonium is noted at pH 9. The method is suitable for the separation of plutonium for concentration purposes. To separate plutonium from impurities the rhodizonate method must be combined with some other method (coprecipitation with lanthanum fluoride, ion-exchange purification, etc.).

Coprecipitation of plutonium with organic compounds

Kuznetsov and Akimova [135, 136] proposed a number of organic reagents for use in concentrating plutonium from very dilute solutions. The coprecipitation of plutonium takes place by two different mechanisms: coprecipitation by the nitrate mechanism and coprecipitation as cyclic salts.

Coprecipitation by nitrate mechanism

The coprecipitation of plutonium takes place from solutions containing nitric acid and ammonium nitrate. In such solutions Pu(IV) forms a typical weakly ionized complex anion $Pu(NO_3)_6^{2-}$, which may be coprecipitated with the precipitates of nitrates of heavy organic cations.

A number of such reagents were tried out [136]: methyl violet chloride, nitron acetate, diphenylquanidine hydroacetate, Rhodamine B chloride, diantipyrylphenylmethane hydrochloride, diantipyrylmethane hydrochloride, Rhodamine B amyl ether chloride, and Rhodamine B butyl ether chloride (butylrhodamine chloride). Of the reagents studied, butylrhodamine proved to be most suitable for coprecipitation.

Tetravalent plutonium is also coprecipitated almost quantitatively (90–95%) with amylrhodamine and diantipyrylmethane in 1.8 *N* HNO_3 containing 50% ammonium nitrate.

In concentrated nitrate solutions, only a small number of other elements—U(IV), Th(IV) and Ce(IV)—form anionic hexanitrate complexes, which have a tendency to be coprecipitated with organic reagents. Plutonium may be separated from Zr, V, Sb, Fe, Cr, Mn, Co, Cu, etc. by this method.

To 200 ml of 3 *N* HNO_3, containing ~1 μg/liter of Pu(IV), are added about 130 g NH_4NO_3, 1 gram of phenolphthalein powder (to facilitate the filtration of the precipitate which tends to adhere to the walls of the flask) and 40 ml slightly warmed 1% solution of butylrhodamine. The raspberry-colored precipitate is left to stand for one hour and is then filtered through a White Ribbon filter paper. The precipitate on the filter is washed twice with ~ 50% solution of ammonium nitrate in 1.8 *N* HNO_3, to which a solution of butylrhodamine has been added in an amount sufficient for a small, but distinctly visible precipitate to be formed. The precipitate together with the filter paper is placed in a crucible, 2 ml of a saturated solution of ammonium sulfate are added, dried somewhat, ashed in a muffle furnace by gradually raising the temperature from room temperature to 550–600° (if the nitrate-containing precipitate is ashed in the absence of ammonium sulfate, flashes may occur). After ashing, the inner walls of the crucible are washed with a few drops of HNO_3. In order to solubilize plutonium oxide, the precipitate is treated first with hydrogen iodide, then with concentrated HNO_3; the solution is evaporated, the residue dissolved in a few drops of 1.5 *N* HNO_3 and plutonium is determined radiometrically. The error of the determination is ± 5%.

The completeness of precipitation of plutonium as a function of the dilution is given in Table 23.

Table 23

COPRECIPITATION OF Pu(IV) WITH BUTYLRHODAMINE NITRATE [135]

Concentration of Pu, μg/liter	Coprecipitation of Pu(IV), %	Concentration of Pu, μg/liter	Coprecipitation of Pu(IV), %
5	100	0.1	89–90
1	100	0.05	76–82
0.5	97–100		

The results of coprecipitation of plutonium with simultaneous separation from other elements are given in Table 24.

All metals, except for the elements the salts of which are shown in brackets, were introduced as nitrates in the amount of 100 mg, calculated as the element; in each experiment one element, or a sum of the elements shown was taken.

It will be seen that the coprecipitation of Pu(IV) with butylrhodamine nitrate ensures a fairly complete separation of plutonium even at dilutions of $1:10^{10}$.

Table 24

COPRECIPITATION OF Pu(IV) WITH BUTYLRHODAMINE WITH SIMULTANEOUS SEPARATION FROM OTHER ELEMENTS [135]

Valency of metal	Elements accompanying Pu(IV) during coprecipitation	Weight of ashed precipitate, mg	Coprecipitation of Pu(IV), %
	No other elements	—	99; 100
I	Li + Na + K + Rb + Cs	1	95
II	Cu; Be ($BSO_4 \cdot 4H_2O$); Mg; Ca	In each case $\leqslant$ 1–2	93; 100; 100
	Sr + Ba; Zn + Cd; Pb; Mn; Co + Ni	In each case 1–2	95; 92; 90; 96; 97
III	Rare-earth elements Bi, Sb ($SbCl_3$)	$\leqslant$ 1; ~ 7; ~ 15	101; 103; 96
	B($Na_2B_4O_7$); Al; Ce; Fe	In each case $\leqslant$ 1–2	90; 85; 83; 95
IV	Ce; Th; U (UCl_4)	~ 20; ~ 43; ~ 43	87; 55; 54
	Zr; Sn ($SnCl_4$); V ($VOSO_4$)	In each case $\leqslant$ 1–2	92; 90; 100
V	P (Na_2HPO_4)	$\leqslant$ 1	86
VI	U; Mo[$(NH_4)_6Mo_7O_{24}$]	~ 7; 1	94; 90

Note: In each experiment 0.1 μg Pu were coprecipitated from 200 ml of a solution 50% in NH_4NO_3 and 1.8 *N* in HNO_3; other elements were introduced in amounts up to 100 mg, calculated on the element; in each experiment up to 40 ml of 1% solution of butylrhodamine were added.

It will be seen from Table 24 that Pu(IV) may be separated from practically all elements except thorium. The precipitation of Ce(IV) is easily prevented by effecting a preliminary reduction of Ce(IV) to Ce(III). The conversion of U(IV) to U(VI) considerably reduces the occlusion of uranium. A complete separation of plutonium from hexavalent uranium may be attained by carrying out repeated reprecipitations. Thus, after four successive coprecipitations 80% of plutonium may be separated. The final precipitate contains only 0.0025% of the amount of uranium originally present in a mixture of plutonium with uranium with a ratio Pu(IV):U(VI) = $1:10^7$.

Coprecipitation as cyclic salts

Tetravalent plutonium forms soluble cyclic salts with many organic reagents containing sulfonic groups. When thus bound, plutonium may be coprecipitated with the precipitates formed by the cations of basic dyes with organic reagents (anions), which bind plutonium into a stable complex compound.

Since Pu(IV) is hydrolyzed even in acid solution, the formation of plutonium complexes also takes place in acid solutions. The method [135] is as follows.

To 200 ml of an acid solution containing Pu(IV), 100 mg of complex-forming reagent is added as an aqueous solution, pH is adjusted to the value required for complex formation (see Table 25), the solution is left to stand 10–20 minutes; 5–7 ml of a 1% solution of methyl violet or methylene blue is introduced with stirring. The precipitate is filtered, partly dried and ashed by slowly raising the temperature to 500°

it is then dissolved in concentrated HNO_3, the solution is evaporated and taken up in 1.5 *N* HNO_3. Plutonium is determined by radiometry. At a content of Pu(IV) of 0.1 μg in 200 ml and at pH 1–4, plutonium is coprecipitated with a number of organic precipitants (Table 25).

It can be seen from Table 25 that 95–100% of plutonium coprecipitated at pH 1–4, if Chromotrope 2B, Phensulfazo, Arsenazo or Stilbazo are used as reagents. As compared with the nitrate mechanism, the precipitation as cyclic salts is more complete, but less selective; this means that plutonium may be precipitated from more dilute solutions. Tetravalent plutonium is separated only from elements the cations of which are hydrolyzed to a smaller extent than Pu(IV) cations: from alkali metals, doubly charged cations, U(VI) and certain trivalent elements including rare-earth elements.

Table 25

COPRECIPITATION OF Pu(IV) AS CYCLIC SALTS [136]

Components of coprecipitant	pH	Coprecipitation of Pu(IV), %
Calcone and methyl violet	3–4.2	35–51
Acid Chrome Yellow N and methyl violet	3.2–4.7	82–100
Benzene-4-sulfonic acid < 1-azo-5 > 8-hydroxyquinoline and methyl violet	1.3–4.7	89–100
Chromotrope 2 B and methyl violet	1.7–4.3	95–100
Phensulfazo and methylene blue	2.4–5	98–100
Arsenazo and methylene blue	2–6.3	100–101
Stilbazo and methyl violet	1–4.3	100–102

Trivalent plutonium may be coprecipitated from very dilute solutions, at pH 5–7, in a similar manner. Its precipitation is less specific than that of tetravalent plutonium.

Coprecipitation of plutonium (IV) *with alizarin*

Pasternak [586] studied the mechanism of coprecipitation of plutonium in various valency states. The coprecipitation of tetravalent plutonium was shown to be the most complete. The mechanism of the coprecipitation was interpreted in terms of a selective adsorption on alizarin.

PRECIPITATION OF WEIGHABLE AMOUNTS

It was shown in the preceding section that a quantitative coprecipitation of microamounts of plutonium may also take place when the solubility of the plutonium compound is larger than its concentration in solution. When precipitating macroamounts, on the contrary, the solubility of the compound separating out under the given experimental conditions is the fundamental criterion of the suitability of the given method.

Plutonium is present in weighable amounts in solutions which have been subjected to a preliminary concentration (coprecipitation with carriers, extraction), in the course of which plutonium has been separated from the bulk of uranium and fission products, and with them also from the bulk of β- and γ-activity [3, p. 469; 230, 404]. Depending on the specific technological process employed, solutions with macro concentrations of plutonium may also contain significant amounts of various other elements: K, Na, Ag, Mg, Cu, Fe, Ca, Zn, Pb, Mn, Cr, Ni, Ce, La, Th, U, Np, Am, etc., of which the most difficult to separate are U, Fe, Cr, Zr, Th, Ce and Np.

Depending on the nature of the sample and on the final method of determination of plutonium (radiometric, titrimetric, gravimetric, spectrophotometric), the methods of determination of plutonium must meet varying requirements. Thus, if the final determination of plutonium is done by radiometry, plutonium must be carefully separated from all other α-emitters and from most nonvolatile salts. For a volumetric determination, all extraneous substances reacting with the titrant must be carefully removed. In a gravimetric determination it is important to remove all extraneous substances which are not eliminated in the conversion of plutonium to its gravimetric form. It will be seen that the choice of the method to be used will be governed by the composition of the sample to be analyzed and by the degree to which it has been purified prior to the determination.

For solutions containing many components it is difficult to find a method of precipitation of plutonium by which it could be completely separated from all extraneous substances. In such cases several methods of separation must be combined—successive precipitation of several sparingly soluble compounds, combination of precipitation with extraction, etc. [230, 336, 633].

It is not possible to give an exhaustive description of all the available methods of separation by precipitation; all that can be done is to give a critical survey of individual operations. The most difficult task in these separations is to separate plutonium from uranium. This is, firstly, because uranium is an analog of plutonium in all its valency states, and secondly, because it is usually present in plutonium solutions in large amounts. For this reason, one of the main criteria in evaluating a given separation method is the extent of separation of these two elements. When working with macroamounts of plutonium, disproportionation reactions must be reckoned with [485]. If plutonium is present in solution in more than one valency state, its separation becomes much more difficult. Accordingly, many separation procedures are preceded by the conversion of all the plutonium to one valency form. The precipitation operations discussed below refer to separation methods most often encountered in analytical practice.

Precipitation of plutonium (IV) by ammonium hydroxide

The precipitation of Pu(IV) by ammonium hydroxide [3, p. 329; 48, 170] is extensively employed to separate it from alkali and alkaline-earth metals, from small amounts of ammoniate-forming metals (Cu^{2+}, Ni^{2+}, Zn^{2+}, Cd^{2+}, Ag^{+}), and also to concentrate plutonium in a small volume [66, 368]. Since the selectivity of the method is low, the reaction is employed to separate plutonium in the last stages of purification. Anions which complex plutonium, such as carbonates, oxalates, fluorides, phosphates, tartrates and some others, interfere with quantitative precipitation. Unsatisfactory results will also be obtained if the solution contains major amounts of zinc, chromium and boron. These three elements are occluded by the precipitate of plutonium hydroxide.

The effectiveness of the separation of plutonium from copper, nickel, silver and other ammoniate-forming elements will depend on their concentration. As a rule, a complete separation, even from very small amounts of these elements is attained only after 2 or 3 reprecipitations.

The solution of ammonium hydroxide must obviously be free from carbonates. The precipitation of plutonium by ammonium hydroxide has been described on p. 237.

Precipitation of plutonium by hydrogen peroxide

The precipitation of plutonium by hydrogen peroxide does not discriminate between the different valencies of plutonium. However, the advantage of this precipitation is that plutonium in all valency states is quantitatively precipitated, both Pu(VI) and Pu(III) being converted into Pu(IV). The precipitation of plutonium by hydrogen peroxide is extensively employed to purify concentrated solutions of plutonium [144]. The method is quite selective, but the selectivity can be further increased by varying the precipitation medium. A discussion of the precipitation of plutonium from various media is given below.

Precipitation of Pu(IV) *peroxide from acid solutions*

The precipitation of plutonium from acid solutions (1–2 N) is quantitative; it is accompanied by the separation of plutonium from many elements (Table 26) [519].

It can be seen from Table 26 that the separation of the elements of Groups I, II and III is satisfactory. Of Group IV elements, all are separated satisfactorily except zirconium, which gives only a partial separation. The behavior of Group V elements was not investigated. Of the elements of Groups VI and VII, chromium, molybdenum and manganese (in amounts shown in Table 26) are not coprecipitated with plutonium. A precipitation was conducted from solutions containing equal amounts by weight of chromium and plutonium; the plutonium peroxide obtained contained less than 1 mg Cr per 10 g Pu. All elements of Group VIII which were studied could be separated by peroxide precipitation, including cobalt and ruthenium, which were present in major amounts.

Table 26

PURIFICATION OF PLUTONIUM BY PEROXIDE PRECIPITATION [519]

Group of element in the periodic system	Element	Content of element, wt. % of Pu		Group of element in the periodic system	Element	Content of element, wt. % of Pu	
		before precipitation	after precipitation			before precipitation	after precipitation
I	Na	> 0.01	~ 0.001	VI	Cr	1.4	< 0.002
	Cu	~ 0.01	~ 0.0001		Mo	0.28	< 0.002
	Ag	~ 0.0005	~ 0.0002	VII	Mn	0.0015	< 0.0001
II	Be	~ 0.01	0.0003	VIII	Fe(III)	0.62	< 0.001
	Mg	0.1	< 0.0005			> 2.0	0.037
	Ca	~ 0.01	0.0025		Co	6.5	< 0.002
	Zn	0.001	0.001		Ni	> 0.13	< 0.002
	Pb	0.01	0.0002		Ru	~ 1.6	0.1
III	B	~ 0.01	~ 0.0001	Lanthanides	La	0.005	< 0.001
	Al	> 0.05	0.0005		Ce	1.0	~ 0.015
	In	~ 0.01	0.0005	Actinides	Th	3.4	3.2
IV	Si	0.12	0.002		U	277.0	253.0
	Ti	~ 0.04	< 0.02				
	Zr	0.34	0.1				
	Sn	0.001	0.0001				

The method also gives an efficient separation of lanthanides. Thorium and uranium are precipitated together with plutonium under these conditions. Hall and Herniman [443] reported that peroxide precipitation from acid media can be used to separate plutonium from americium.

Leary et al. [519] showed that if iron in amounts equal to that of plutonium is present in solution, Pu(IV) is incompletely precipitated due to its oxidation (which takes place in the presence of H_2O_2 only) to Pu(VI) by the ferric iron. Iron, copper, manganese, etc. catalyze the decomposition of hydrogen peroxide and therefore should be present in solution only in insignificant amounts. The addition of hydroxylamine or hydrazine to the solution prior to the precipitation of plutonium improves its quantitative separation [290].

The effect of various factors on the precipitation of plutonium (IV) peroxide, and the precipitation technique from acid media are discussed on p. 239.

Precipitation of plutonium (IV) *peroxide in carbonate media*

A quantitative separation from uranium may be achieved by precipitating plutonium peroxide from carbonate media (Deberdeeva, 1960). However, under these conditions most elements either precipitate out as well, or else interfere with the precipitation of plutonium. For this reason, in order to remove the bulk of the impurities from plutonium, the precipitation from carbonate medium is usually preceded by a precipitation from an acid medium.

If the amounts of uranium and plutonium in solution are equal, uranium will be completely separated from plutonium by one precipitation; if more uranium is present, another precipitation is required. The experimental procedure will be found on p. 239.

Precipitation of plutonium (IV) *peroxide in the presence of Complexone III*

Deberdeeva (1960) successfully employed the complexing action of Complexone III to eliminate the effect of bivalent and trivalent elements during the precipitation of plutonium (IV) peroxide from ammonium carbonate media. Complexone III binds these elements into a stable soluble complex. Ammonium carbonate is added to keep uranium in solution. The procedure developed by this worker for the precipitation of plutonium peroxide from ammonium carbonate medium makes it possible to isolate plutonium from multicomponent solutions in one single operation.

The effect of various factors and the procedure to be followed in precipitating plutonium from ammonium carbonate media will be found on p. 239.

Precipitation of plutonium (IV) iodate

The precipitation of Pu(IV) as iodate is employed to separate it from many elements, mainly from rare-earth elements and hexavalent uranium [368]. The method is extensively employed by analysts, due to the fact that the precipitate filters very fast and is readily dissolved. If the amounts of plutonium are large (more than 50 mg), the extraneous elements will be more fully eliminated if the precipitation is conducted in 6 *M* HNO_3; if the content of plutonium is lower, the acidity is preferably reduced to 0.5–1.0 *M* HNO_3 for the sake of a quantitative separation. Separation from thorium, zirconium and titanium is not achieved by this method. Tetravalent cerium and uranium are also precipitated by the iodate, but if the solution is preliminarily treated with hydrogen peroxide, both elements will remain in solution, since the former will be reduced and the latter oxidized. The peroxide treatment is also desirable for plutonium, since the element is thus converted to the tetravalent state. Trivalent rare-earth elements are, as a rule, readily separated during the iodate precipitation, but if they are present in major amounts, the precipitation must be repeated.

Vinogradov and Zakolupin (1947) employed iodate precipitation to separate plutonium from multicomponent solutions. Konarev et al. (1951) studied the isolation of plutonium from solutions, in the presence of varying amounts of uranium and in the presence of many other elements. The concentration of plutonium in solutions varied from 0.1 to 1 mg/ml. At uranium contents not less than 10 mg/ml, plutonium iodate must be reprecipitated for complete separation. In addition to uranium, the solutions under study contained Ba, Cd, Ce, Cr, Co, La, Fe, Mn, Mo, Ni, Sn, V, each element being present in a concentration not exceeding 10% of the content of plutonium. If the terminal stage is a gravimetric determination, the standard deviation is $\pm$ 0.4%, even with a double precipitation of plutonium iodate. The authors consider that the error is due to the imperfect separation from iron. A careful washing of the precipitate with a solution of the precipitant, diluted by a factor of two, reduces somewhat the occlusion of iron. In the presence of large amounts of iron, the authors propose that it be preliminarily removed by extraction with ether from a hydrochloric acid solution.

The detailed experimental procedure for the precipitation of plutonium (IV) iodate is described on p. 240.

Precipitation of plutonium sulfates

Separation of plutonium (IV)

Anderson [272] determined the solubility of plutonium sulfate in the presence of excess sulfuric acid and methanol.

The solubility of plutonium sulfate depends to a considerable extent on the concentration of sulfuric acid. In 4 *M* sulfuric acid the precipitation of plutonium sulfate is more complete, due to the common ion effect. Further increase in the concentration of the free acid is favorable to the formation of sulfate complexes of plutonium, up to the formation of a complex with eight sulfate groups [150]. The addition of alcohols, acetone and perchloric acid strongly decreases the solubility of plutonium sulfate. Palei and Sklyarenko (1952) proposed to precipitate plutonium by perchloric acid from sulfate solutions, with the purpose of separating it from certain elements, the solubility of the sulfates of which is much higher than that of plutonium sulfate. When the concentration of perchloric acid in the sulfate solution is 30%, plutonium is quantitatively precipitated as $Pu(SO_4)_2 \cdot 4H_2O$. The authors found that plutonium could be separated from lanthanum, chromium and nickel, if each element is present in an amount not more than 10% of that of plutonium. Ferric iron and U(VI) are partially separated. The extent of precipitation of plutonium in pure solutions is 99.5–99.9%; in the presence of extraneous substances it drops to 90–97%.

Separation as double plutonium (IV) *sulfates*

Anderson [270] pointed out that Pu(IV) could be precipitated from sulfate media in the presence of alkali metal sulfates and ethanol as the double sulfates $M_4Pu(SO_4)_4 \cdot \cdot (1\text{–}2)H_2O$ where M is the alkali metal ion or ammonium ion. Double salts of plutonium with ammonium, potassium and rubidium were obtained, which under the conditions employed by the authors were only sparingly soluble.

The data obtained by Palei and Sklyarenko (1956) on the precipitation of Pu(IV) from nitrate solutions by introducing excess sodium sulfate into the solution and adding ethanol, proved that the precipitation of plutonium was quantitative. The precipitation was conducted from 1.5 *N* HNO_3, by adding a saturated solution of sodium sulfate and a double volume of ethanol. Uranium, when present in amounts equal to that of plutonium, is eliminated to a large extent as a result of this operation. If plutonium is to be determined gravimetrically as a final stage, it must be additionally purified from uranium by another method (iodate precipitation, etc.). Elements forming sparingly soluble sulfates (Ag, Pb, Ca, etc.) are not separated by this precipitation. Iron, nickel, chromium and manganese can be satisfactorily separated by this method.

The precipitation is usually conducted as follows. To 0.5 ml solution of plutonium nitrate (1–10 mg/ml) in 1.5 *N* HNO_3 are added 50–100 mg sodium sulfate, and, after careful stirring, a double volume of ethanol. The resulting precipitate is centrifuged and washed with a solution of the precipitant, diluted to half the concentration. If further purification is required, the precipitate is dissolved in 2–3 *N* HNO_3. The separation is usually terminated by the precipitation of plutonium hydroxide to eliminate the alkali metals.

Separation of plutonium (III)

Anderson [274, 275] also described the formation of double sulfates of trivalent plutonium of two types, $MPu(SO_4)_2H_2O$ and $M_5Pu(SO_4)_4$ where M is the alkali metal ion. The compound of the first type is precipitated only in the presence of alcohol, while the second is precipitated only in the absence of alcohol, but with a higher excess of alkali sulfate. More recent investigations (Palei and Sklyarenko, 1956) deal with the possible utilization of the precipitation of plutonium (III) double sulfates to separate plutonium from a number of other elements. It was found that Cr, Ni, Co, Mn, partly U(VI) and other elements which form soluble sulfates can be successfully precipitated.

Quantitative precipitation of trivalent plutonium from nitrate media has the advantage that plutonium can be easily converted to Pu(III); the precipitation of sulfates is the technique which is often preferred in practice.

The procedure, according to Palei and Sklyarenko (1953) is as follows.

A 30–50 mg sample of plutonium is placed in a 50 ml beaker, the solution is diluted with three times the amount of 0.5 *N* HNO_3, and gaseous SO_2 is passed through for 3–4 minutes. The solution is left to stand for 15–20 minutes, to ensure a more complete precipitation of plutonium. Saturated solution of Na_2SO_4 in 1.5 *N* HNO_3 (0.8 ml) is then added and then a double volume of ethanol. The precipitate consists of sodium plutonium double sulfate and sulfates of lead, silver, with some U(VI), manganese and traces of iron. The precipitate of the sulfates is filtered off after half an hour through a porous glass filter No. 4, and is washed with three 5 ml portions of the washing (20 vol.% saturated solution of Na_2SO_4 in 1.5 N HNO_3, 40 vol. % of ethanol and 40 vol. % of 0.5 N HNO_3) preliminatily saturated with SO_2 for 5 minutes. The result of the washing is to reduce the content of uranium and manganese in the precipitate and to remove small amounts of iron completely. The precipitate is then dissolved in 15 ml of 1.5 N HNO_3, heated to 60–80° and plutonium is precipitated as the hydroxide.

In high accuracy gravimetric determinations, plutonium must be further purified by any suitable method (e.g., iodate precipitation).

Precipitation of plutonium (IV) fluorides

Hydrofluoric acid precipitates Pu(IV) from acid solutions as $PuF_4 \cdot 2.5H_2O$. By this method it is possible to separate plutonium from hexavalent uranium, iron, zirconium, tantalum and other elements which form soluble complex compounds with fluoride ions. The method is not very frequently employed because of the considerable solubility of plutonium fluoride [3, p. 314].

Deichman and Tananaev [81] proposed the use of the double salt $NaPuF_5$ for analytical purposes. This compound has a very low solubility ($1.25 \cdot 10^{-6}$ moles/liter) and may be employed for a quantitative separation of plutonium. It is formed when an excess of NaF is added to a nitric acid solution of plutonium (IV). It may be assumed, by analogy with U(IV) [158, 178], that precipitation from 1 N HNO_3 by this method will separate plutonium from Fe(III), V, Mo, Ti, Ni, Co, Mn, Cu, B, Zr, Ta and certain other elements forming soluble fluorides. Cerium is partly coprecipitated with plutonium [273, p. 229]. Plutonium may probably also be separated as the potassium salt $KPuF_5$, which is even less soluble [271].

Precipitation of plutonium (IV) selenite

Mikhailov (1956) showed that tetravalent plutonium is quantitatively precipitated as the selenite $Pu(SeO_3)_2$ from <0.2 N nitric acid solutions. Selenious acid in excess is added to a weakly acid solution of plutonium (IV) at 60°, the mixture is stirred, and the solution and the precipitate are left to stand for 30 minutes at 60–80°. The final determination of plutonium may be carried out by gravimetry, if interfering elements are absent, or radiometrically. The optimum conditions of quantitative precipitation are: 0.1 M HNO_3, 0.1 M final concentration of H_2SeO_3 and 0.01–0.001 M concentration of H_2SO_4. Sulfuric acid is added to prevent possible disproportionation of Pu(IV). Under these conditions, iron, chromium and uranium, if present in equimolecular amounts, are coprecipitated with Pu(IV) to the extent of 70, 5 and 7–9% respectively.

Precipitation of plutonium (III) by potassium ferrocyanide

The sparingly soluble $KPu[Fe(CN)_6] \cdot 7H_2O$, which precipitates out of 0.8 *M* hydrochloric acid solution of plutonium (III) on the addition of potassium ferrocyanide, may be employed in the separation of plutonium from phosphates and arsenates [273]. One of the few separations which can be carried out with ferrocyanide is the separation from beryllium. The reaction is not very selective and its uses are limited.

Precipitation of plutonium (IV) phosphate

The separation of plutonium as phosphate has not been described in the literature. The probably reason for this is that the amorphous precipitate which separates out may adsorb large amounts of impurities and that its further treatment is difficult. It would appear, nevertheless, that the method could be successfully used in separations of plutonium from acid media for the purpose of separation from U(VI) and many other elements. It is known that tetravalent plutonium reacts with orthophosphoric acid in acid media with the formation of the sparingly soluble $Pu(HPO_4)_2 \cdot xH_2O$ [8, 194, 205]. The solubility data available for the phosphate precipitates in various acid media (cf. Figure 30) show that plutonium may be quantitatively separated in this way. By analogy with zirconium, the precipitation of plutonium phosphate under these conditions will probably separate Al, Cu, Cd, Bi, Ni, Co, Mg, Mn, alkali and alkaline-earth metals, V, W, Mo and U(VI). Only tetravalent cations Ti^{4+}, Zr^{4+}, Hf^{4+}, Ce^{4+}, U^{4+} and Th^{4+} are precipitated together with plutonium. It will be seen from Figure 30 that the solubility of plutonium phosphate increases with increasing concentration of phosphoric acid, which indicates formation of soluble phosphate complexes [3, p. 325].

Precipitation of plutonium (IV) phytate

Organic derivatives of phosphoric acids, which are widely used in analytical work, have also been employed in the precipitation of plutonium and in its separation from the accompanying elements [194, 205]. It was found that phytic acid and phytin (the salt of inositolpyrophosphoric acid) precipitate tetravalent plutonium from strongly acid media (up to 10–12 *N* HNO_3) fairly completely. It was found that [205] the ignited precipitate has no constant composition and therefore cannot be recommended for use in accurate gravimetric determinations, but can be successfully employed in separations. The behavior of phytin and phytic acid in the precipitation is very similar. The selectivity of these reagents strongly depends on the acidity of the initial solution. Whereas in neutral solutions phytic acid and phytin precipitate practically all elements except alkali metals, in 1 *N* HNO_3 the selectivity is much higher: elements of the titanium subgroup and also Ce(IV), Th(IV), U(IV), Cr(III) and Fe(III) are coprecipitated with Pu(IV). At higher acidities of the initial solution (up to 6 *N*) iron and chromium also remain in solution [8]. The phytates of Ti, Zr, Hf, Sc, Ce(IV) and Th do not dissolve even in 12 *N* HNO_3. The completeness of precipitation of Pu(IV) varies with the acidity; in 0.1 *N* HNO_3 99.9% Pu is precipitated, whereas 98% are precipitated in 12 *N* HNO_3. The optimum conditions for the precipitation of Pu(IV) are as follows [205]: medium—2 *N* HNO_3; concentration of Pu—above 5 mg/ml; precipitant—$1\frac{1}{2}$ times in excess; temperature—40°. The precipitate is immediately filtered off and washed with 0.1% solution of phytic acid in 2.2 *N* HNO_3. Under these conditions plutonium is precipitated to an extent of not less than 99.6%. Experiments performed in order to study the precipitation of plutonium in the presence of many other elements showed that the precipitation from such solutions is not any less complete than from pure solutions. Quantitative precipitation of plutonium may be carried out in the presence of the following elements.

Element	Ca	Al	Cr	Fe	La	W	Mn	Pb	U	Ni
Element : plutonium ratio	0.2	2	0.5	0.07	1	0.5	0.15	0.1	0.1	1

When the acidity is increased to 6–7 *N* HNO_3, plutonium may be precipitated from solutions containing uranium in amounts equal to, or even double that of plutonium. To solubilize the plutonium, the precipitate is treated with a saturated solution of trisubstituted ammonium or sodium citrate.

Phytic acid has also been employed to separate the various valencies of plutonium. Figure 89 shows the dependence of the completeness of precipitation of plutonium phytate in various states of valency on the acidity of the initial solutions. It will be seen from the figure that neither trivalent nor hexavalent plutonium is precipitated by phytic acid beginning with concentrations of 0.1 *N* hydrochloric or nitric acid.

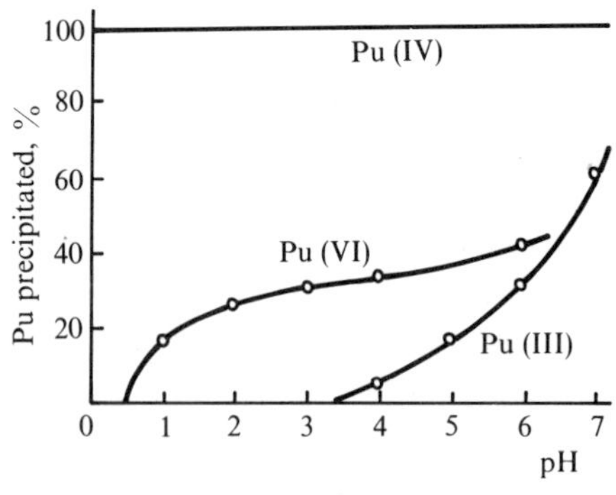

Figure 89
Effect of the pH of the medium on the completeness of precipitation of plutonium phytates

It has been experimentally demonstrated that even if the various valency forms are present together in equal amounts, it is only tetravalent plutonium that is quantitatively precipitated by phytic acid, while the trivalent and the hexavalent plutonium remain in solution. If it is desired to extract plutonium from the solution quantitatively, all of it must first be converted to tetravalent plutonium.

Precipitation of plutonium oxalates

Separation of plutonium (III)

When crystalline oxalic acid or an oxalate is added to a weakly acid (0.3–0.8 *N*) solution of trivalent plutonium, the sparingly soluble oxalate $Pu_2(C_2O_4)_3 \cdot 9H_2O$ precipitates out [3, p. 347; 587]. The excess oxalic acid after the precipitation of the oxalate is 0.2 moles/liter [3, p. 347]. When the mineral acid content increases, plutonium (III) oxalate becomes more soluble [3, p. 347]. Sulfates, fluorides, phosphates and trivalent and tetravalent cations interfere with the precipitation [9, p. 277]. The precipitate of plutonium (III) oxalate is stable only in an inert atmosphere. This fact, in conjunction with the low selectivity of the method makes it useless for analytical purposes. The procedure for the precipitation of $Pu_2(C_2O_4)_3 \cdot 9H_2O$ is described on p. 241.

Separation of plutonium (IV) *oxalate*

The method may be employed for the separation of plutonium from those elements which can be separated by precipitation of tetravalent uranium [9, p. 277]. The solubility of Pu(IV) oxalate decreases as the acidity is increased to 1.0 *M* HNO_3 [34, p. 310]. However, in order to achieve a more complete separation of Pu(IV) from other elements, the precipitation is preferably carried out from a 2 *M* solution of an acid (the increase in the solubility of $Pu(C_2O_4)_2$ at this acidity is negligible). Thorium, tetravalent uranium and rare-earth elements are quantitatively precipitated together with plutonium under these conditions. Niobium may also partly coprecipitate, depending on its content. Sulfates, phosphates, fluorides and certain organic complex-forming substances interfere [9, p. 277].

Precipitation of plutonium (IV) cupferronate

In sulfuric acid solutions, irrespective of the valency state of plutonium, cupferron precipitates the element as $Pu(C_6H_5N_2O_2)_4$. Any trivalent plutonium is oxidized by cupferron to tetravalent plutonium, and any hexavalent plutonium is reduced to the tetravalent state [161].

The oxidation is rapid (2–3 minutes), whereas the reduction takes about an hour. In order to accerlerate the process, a reducing agent is added (hydrazine hydrate, hydroxylamine). The precipitation of plutonium (IV) cupferronate from 3–4 *N* solutions of H_2SO_4 results in the separation of plutonium from equimolecular amounts of the following elements: U(VI), Cr(III), Mn(II), Al, Ni, Ag, La, Be, Zn, Am, and alkali and alkaline-earth metals. The elements Fe, Ti, Zr, Ta, V, Ga, Nb, Sn, Sb, Hf and U(IV) are precipitated together with Pu(IV). These elements must be separated prior to the cupferronate precipitation of plutonium. Major concentrations (higher than 2 *N*) of nitric acid decompose the precipitant [161]. Separation of different valency forms with cupferron is not possible, but the reagent has the advantage of quantitatively precipitating plutonium in all valency states. The experimental details of the precipitation of plutonium (IV) cupferronate are given in p. 242.

Precipitation of plutonium (IV) 8-hydroxyquinolate

It was shown by Moiseev (1953) that when plutonium 8-hydroxyquinolate is precipitated from ammoniacal or bicarbonate media (pH 4.5–12) in the presence of tartrate ions, the composition of the precipitate exactly corresponds to the formula $Pu(C_9H_6NO)_4$. Trivalent plutonium is oxidized to Pu(IV) under these conditions and precipitates out as well. Hexavalent plutonium precipitates out as 8-hydroxyquinolate of a different composition. Plutonium may be separated from phosphates and tartrates in this manner. 8-Hydroxyquinoline precipitates most elements including uranium, so that plutonium is only isolated by this method in the ultimate purification stage to obtain a gravimetric form (Moiseev, 1954). On account of its low selectivity, the method cannot be employed in separations to any significant extent.

Mikhailov (1957) attempted to increase the selectivity of the method by studying the precipitation of hexavalent plutonium by 8-hydroxyquinoline in the presence of Complexone III (pH 4–8). He succeeded in separating plutonium from Al, Cu, Co, Ni, Zn, Cd, Bi and certain other elements, but not from hexavalent uranium.

Precipitation of plutonium (IV) benzenesulfinate

The precipitation of plutonium (IV) benzenesulfinate from 0.2 *N* nitric acid solutions will separate Ni, Co, Cr, Fe(II), Mn and Cu if present in equimolecular amounts, and also Al, Be, La and other rare-earth elements in tenfold or higher amounts with respect to plutonium (Milyukova, 1958). Partial separation of U(VI), Fe(III) and Pb(II) is also achieved. Hexavalent uranium is most probably occluded by the precipitate, forming a double salt of benzenesulfinic acid. This is indicated by the fact that the content of U(VI) in the precipitate of plutonium (IV) benzenesulfinate is constant at 30%, irrespective of its initial amount, conditions of precipitation, and whether or not double reprecipitation has been effected. The occlusion of ferric iron by the precipitate may be minimized by increasing the concentration of nitric acid in solution. According to Alimarin and Alikberov [7], Re(III) is not precipitated as benzenesulfinate from solutions more acid than 0.3 *N*. The reagent is not sufficiently selective, and, which is most important, is not capable of separating Pu(IV) from U(VI).

Milyukova (1958) proposed to use the reagent to separate Pu(III) from U(IV). Trivalent plutonium is not precipitated by the benzenesulfinate ion, while U(IV) is precipitated. It was shown that during the precipitation of uranium (IV) benzensulfinate, the precipitate captures less than 0.1% of the initial amount of Pu(III), but U(IV) itself is not precipitated quantitatively (only to about 70%). Thus, the precipitation may be employed only as a preliminary purification of plutonium from the bulk of uranium. Milyukova (1958) proposed a double precipitation by benzenesulfinic acid: one from Pu(III)-containing solution, the second from Pu(IV)-containing solution.

In a beaker are placed 3–5 ml of nitric acid solution of plutonium (5–50 mg/ml) containing up to 50 mg U; the volume is made up to 20 ml so as to have the final acidity at 0.5 *N* HNO_3. The solution is heated to 60–80°, 250–300 mg Rongalite are added and the mixture is left to stand for 25–30 min to convert Plutonium to the trivalent and uranium to the tetravalent state. One gram of sodium benzenesulfinate is added and the solution left to stand for one hour. The precipitate, containing uranium benzenesulfinate and other elements, is filtered through a No. 4 glass filter, washed with 3–4 ml water, dissolved in 1.5–2.0 *N* HNO_3, and the content of the occluded plutonium is determined radiometrically. The filtrate containing Pu(III), a part of U(IV) and Fe(II), as well as La, Mn, Ni, Co and other elements which have not precipitated, is treated with 10% solution of NaOH; the precipitate formed is washed with 2–3 portions of water, 2–3 ml each portion, and is dissolved in 2–3 ml hot concentrated HNO_3. The solution is then evaporated to dryness, dissolved in 20 ml 0.15 *N* HNO_3 and plutonium (IV) benzenesulfinate is precipitated, so that the precipitant is present in a 5% excess, and the solution is left to stand for 8–10 hours. The precipitate is filtered off on No. 4 glass filter, washed with three 3–4 ml portions of the washing solution (2.5% solution of sodium benzenesulfinate in 0.15 *N* HNO_3) and dissolved in hot concentrated HNO_3. The solution is then diluted with water to about 50 ml and plutonium hydroxide is precipitated. As the result of all these operations many elements are eliminated, with the exception of U(VI), which is occluded by the precipitate of plutonium (IV) benzenesulfinate and cannot be separated by the hydroxide precipitation.

Various methods are available for the separation of the residual uranium, in particular iodate precipitation, extraction, etc. It was proposed (Milyukova, 1951) to use sodium benzenesulfinate to separate small amounts of thorium from large amounts of plutonium (see p. 377). The reagent was also used in the separation of plutonium from neptunium. A preliminary treatment of the acid solution with hydrazine hydrochloride converts neptunium into the tetravalent and plutonium into the trivalent state. Tetravalent neptunium is then precipitated by sodium benzensulfinate. The trivalent plutonium residual in the solution may then be precipitated with ammonium hydroxide (Borisova, 1954).

Precipitation of plutonium (IV) m-nitrobenzoate

m-Nitrobenzoic acid precipitates tetravalent plutonium both from acetic acid–acetate media with pH 3–5 [588], and from weakly acid solutions (0.1 *M* HNO_3) (Palei and Milyukova, 1956). Under these conditions trivalent and hexavalent plutonium are not precipitated. Tetravalent plutonium is more fully precipitated from acid solutions when acted upon by a 20% alcoholic solution of m-nitrobenzoic acid. Under these conditions Pu(IV) can be separated from lanthanum, cerium and most rare-earth elements and also from Al, Be, U(VI), Bi, Pb, Mg and Mn [584].

It was, however, shown by Milyukova (1956) that plutonium is not separated quantitatively in 0.1 *N* HNO_3 even on the addition of ethanol, acetone and other substances which reduce the solubility of plutonium m-nitrobenzoate; in the presence of extraneous elements plutonium is only precipitated to the extent of 90–92%.

Quantitative precipitation of plutonium may only be attained by precipitating from acetic acid–acetate media [588], but under these conditions the selectivity of the reagent becomes much lower. For this reason m-nitrobenzoic acid is not much used in analytical work.

Precipitation of plutonium (IV) by phenylarsonic acid and its derivatives

Phenylarsonic acid is a specific reagent for tetravalent cations; it precipitates Ti^{4+}, Zr^{4+}, Hf^{4+}, Th^{4+}, U^{4+}, and a few other elements such as Sn(IV), Nb and Ta. Titanium, zirconium and hafnium are precipitated from strongly acid solutions, U(IV) from weakly acid solutions, while thorium precipitates out in an acetate medium. Rare-earth elements present in amounts much higher than equimolecular with respect to plutonium must be reprecipitated.

According to Mikhailova (1959) the following are the optimum experimental conditions for the quantitative precipitation of tetravalent plutonium: the solution after the separation of the precipitate should be 0.4–0.5 *M* in HNO_3 and 0.04–0.06 *M* in phenylarsonic acid. Under these conditions, 30–100 mg Pu may be separated in one precipitation from 10 mg Fe, from 50 mg each of U(VI), Cr, La, Zn, Cu, Ni and Ca, from 1 mg Bi and from 2000 mg $NaNO_3$. The complexing anions do not interfere in the following molar ratios: 5:1 (SO_4^{2-}), 0.2:1 (F^-), 0.02:1 ($C_2O_4^{2-}$) and 0.005:1 (Complexonate III). If the extraneous elements are not present in solution in amounts larger than those indicated above, one phenylarsonate precipitation will separate the impurities and quantitatively isolate plutonium, which may then be determined gravimetrically by ignition of plutonium phenylarsonate at 1100° to plutonium dioxide.

Mikhailov (1958) studied the interaction of plutonium with 4-nitrophenylarsonic, 1-naphthylarsonic and anthraquinone-1-arsonic acids. As regards their specificity, all these acids are about equal. Some differences exist in regard to the completeness of precipitation of plutonium; the author recommends that 1-naphthylarsonic acid be employed for quantitative precipitation.

Precipitation of plutonium salicylates

Salicylate ions quantitatively precipitate plutonium (III) and plutonium (IV) at pH 4–5. Trivalent plutonium is precipitated as $Pu(Sal)_3 \cdot 1.5H_2O$, while tetravalent plutonium is precipitated as $PuO(Sal)_2$ [100]. The salicylate precipitation of plutonium results in the separation of the bulk of numerous elements, which form under these conditions fairly stable soluble complex salicylates: U(VI), Fe(III), Ti, Al, Cr(III), etc. Careful investigations conducted by Zvyagintsev and Sudarikov [100] showed that the plutonium precipitate occludes 2–3% uranium, chromium and 1% iron. The separation from Al and Ti is quantitative. Pu can be precipitated by salicylic acid, as part of gravimetric or other determination.

Precipitation of sodium plutonyl triacetate

Hexavalent plutonium is quantitatively precipitated from nitric acid solutions as $NaPuO_2(CH_3COO)_3$ on the addition of CH_3COONa and $NaNO_3$. The concentration of acetic acid in the initial solution should be 1 *M* [3, p. 351; 738].

The method is not selective and the permissible amount of impurities in solution is very low (~0.1%). The precipitation is employed in order to separate hexavalent plutonium in the presence of other valency forms. Hexavalent uranium is quantitatively precipitated by the reagent and cannot be separated.

Precipitation of plutonium (IV) acetylacetonate

Tetravalent plutonium is precipitated from weakly acid solutions by acetylacetone as $Pu(C_5H_7O_2)_4$ [381]. The volatile compound thus obtained may be employed in the volatilization of plutonium performed in order to apply plutonium to various supports as a homogeneous layer. The high volatility of plutonium acetylacetonate is the principle of its separation from neptunium. The acetylacetonates of plutonium and neptunium are precipitated and heated to 170–180°. At this temperature plutonium acetylacetonate sublimes, while neptunium acetylacetonate remains in the solid state. Plutonium acetylacetonate is condensed in a specially cooled vessel.

SEPARATION OF PLUTONIUM BY EXTRACTION

The extraction methods for the separation and isolation of various elements are frequently employed in analytical practice. Extraction has found particularly extensive application in nuclear technology and in the processing of nuclear fuel, and also in the laboratory in the separation of actinide elements from other elements and from each other. This is because extraction methods have a number of advantages over other purification and separation techniques, in particular over the precipitation methods. The small interface area between the immiscible phases renders the adsorption or mechanical occlusion of impurities practically impossible. Also, extraction methods are distinguished by their selectivity, rapid element separation, suitability for continuous, as opposed to batch, processing, and by the fact that remote control techniques can be readily applied, which means that highly radioactive solutions can be analyzed. Another advantage of the extraction methods is that the lowest extractable amounts of the element are very small, which means that extraction methods may be suitable in cases in which the concentration of the element lies below the limiting solubility of conventional precipitates.

For the theory, mechanism and classification of extraction processes, the reader is referred to [78, 134, 175, 191, 226, 565, 566]. It should be noted, however, that there is as yet no theory or classification which would conclusively interpret the physicochemical principles of extraction.

The fundamental parameter of extraction processes is the distribution coefficient, which gives the ratio between the analytical concentrations of the given element in the organic (C_0) and aqueous (C_{aq}) phases at equilibrium:

$$K_d = \frac{C_o}{C_{aq}}. \tag{1}$$

If the compound being extracted is undissociated in both the water and the organic phase, the distribution coefficient is independent of the ionic concentration of the metal being extracted. If, on the other hand, the compound dissociates in the organic phase, the distribution coefficient increases with the increase in the concentration of the metal being extracted [226]. The distribution coefficient is effected by the concentration of hydrogen anions, the nature of the extractant and its concentration. Numerous workers [2, 162, 253, 267, 383, 666] noted the effect of the temperature on the distribution coefficient during the extraction of metals by various extractants. Thus, for instance, in the extraction of microamounts of plutonium (IV) nitrate with tributyl phosphate from 0.5 N HNO_3, the distribution coefficient decreases with increasing temperature. However, when plutonium is extracted from more acid solutions, the distribution coefficient increases with increasing temperature and passes through a maximum at 40° (Figure 90) [253]. Fomin [226] remarks that the variation of the distribution coefficients with the temperature is connected with temperature dependence of the equilibrium constant of the extraction reaction, as well as of other reactions, in the aqueous and organic phase.

The presence of complexing ions strongly affects the value of the distribution coefficient [226]. As a rule, the presence of such ions decreases the value of the distribution coefficient and interferes with the extraction of the element. However, complex formation in the aqueous phase may be utilized to carry out an effective extractional separation of certain elements.

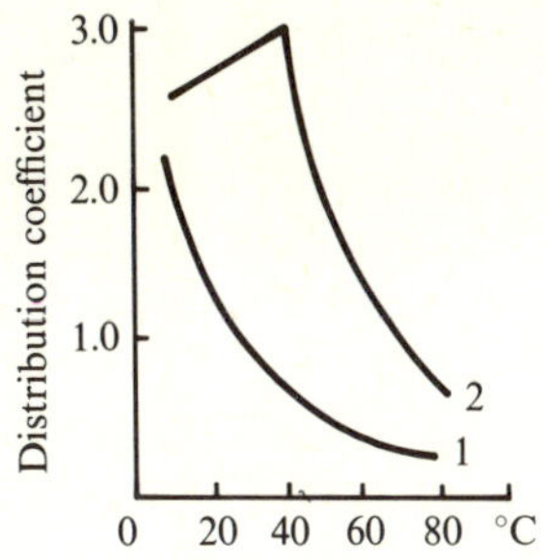

Figure 90

Effect of temperature on the extraction of indicator amounts of Pu(IV) with tributyl phosphate from nitric acid solutions.

Initial HNO_3 concentration: 1) 0.5 *M*; 2) 1.7 *M*. TBP concentration: 0.74 *M*.

In order to increase the value of the distribution coefficients, the so-called salting-out agents are added to the solution. Such agents are usually employed in the extraction of small amounts of the substance. Salting-out agents should be readily soluble in the aqueous phase and insoluble in the organic phase. As a rule, they have a common anion with the salt being extracted. In certain cases the acid itself may act as the salting-out agent. The mechanism of the salting-out process was studied by many workers [134, 304, 479, 564], most of whom came to the same conclusions. According to Rozen [191], the distribution coefficient increases, because the salting-out agent raises the activity of the compound being extracted in the aqueous phase. Salting-out agents are more effective, the higher their extent of hydration, and, consequently, the larger the decrease in the degree of the activity of water, and also the stronger the hydration of the cation of the substance being extracted [134, 190, 226]. The salting-out effect is sometimes interpreted as being due to the decrease in the water vapor pressure over the solution of the agent.

In separation processes the main parameter which determines the efficiency (selectivity) is the separation coefficient of the two elements, which is their distribution coefficient ratio:

$$E = \frac{K_{d_1}}{K_{d_2}}. \tag{2}$$

or

$$E = \frac{C'_o \cdot C''_{aq}}{C''_o \cdot C'_{aq}},$$

where C'_o and C''_o are the equilibrium concentrations in the organic phase, and C'_{aq}

and C''_{aq} are the equilibrium concentrations in the aqueous phase of the first and second element respectively.

The larger the difference between the value of this ratio and unity, the better will be the separation. If $K_{d_1} = K_{d_2}$, the separation is impossible. By varying the conditions of the extraction, it is possible to carry out a highly selective separation of any particular element.

In extraction processes plutonium behaves differently in its various valency states, and may be separated from a large number of elements by conversion from one valency state into another. In most cases the extractability of plutonium increases in the sequence Pu(V), Pu(III) < Pu(VI) < Pu(IV), which is parallel to the stability sequence of their complex compounds.

The choice of the organic extractant is determined not only by the values of the separation and the distribution coefficients, but also by a number of other factors. These include the resistance of the reagent to chemical and radiation-induced changes, the ease and rapidity of reextraction of the extracted products, the density of the organic as compared with the aqueous phase, the value of the mutual solubilities of the extractant and water, the value of the surface tension, viscosity, flammability and price.

The commonly used extractants are oxygenated compounds such as ethers and ketones, or phosphorus-organic compounds, or else amines. In certain cases strong complex-formers of the cupferron, thenoyltrifluoroacetone, acetylacetone, etc. types are employed.

All the known extraction methods for plutonium may be classified into extractions as molecular compounds of plutonium with the extractant, and extractions as inner complexes or salts of complex plutonium anions with amines.

EXTRACTION OF PLUTONIUM AS A MOLECULAR COMPOUND WITH THE EXTRACTANT

Most extraction methods are based on the solubility of plutonium nitrates in various organic solvents.

In nitric acid solutions tetravalent plutonium forms a number of nitrate complexes from $Pu(NO_3)^{3+}$ to $Pu(NO_3)_6^{2-}$ [3, Chap. 9]. It has been repeatedly demonstrated that plutonium is extracted as molecular compounds between $Pu(NO_3)_4$ or $Pu(NO_3)_6^{2-}$ and the extractant. Plutonium nitrates are readily extracted with oxygenated extractants: alcohols, ethers, ketones and esters of inorganic acids of the tributyl phosphate type.

It is known that in aqueous solutions the stability of the nitrate complexes of tetravalent and hexavalent actinide elements is much higher than that of the complexes of ions of other metals, including trivalent lanthanides and actinides.

The methods of extraction of plutonium and other tetravalent and hexavalent actinides from nitrate media are quite selective, since many bivalent and trivalent elements are not extracted at all under these conditions.

Extraction of plutonium with ethers

Ethers are employed in the extraction of tetravalent and hexavalent plutonium. The extraction of Pu(VI) with diethyl ether was used in the first stages of preparation of actinide elements for their purification and separation.

The extractive capacity of diethyl ether is not very high, so that the distribution coefficient of plutonium nitrate between the extractant and water is low. Sometimes the extent of extraction of plutonium is increased by using a salting-out agent, which acts by increasing the concentration of nitrate ions in the aqueous solution. As a result, the equilibrium in the aqueous phase shifts towards the formation of the neutral complex of plutonium. The efficiency of a salting-out agent depends on its identity and concentration. Vdovenko and Kovaleva [36], who studied the extraction of uranyl nitrate with diethyl ether, showed that the salting-out effect increases with the decrease in the radius and with the increase in the charge on the cation. Thus, the salting-out capacity for monovalent cations increases in the sequence $NH_4^+ < Na^+ < Li^+$, while for bivalent cations it increases in the sequence $Sr^{2+} < Ca^{2+} < Mg^{2+}$.

Nitric acid may sometimes act as a salting-out agent. In such a case the increase in the nitrate ion concentration and the decrease in the activity of water are not the only effects involved. Nitric acid, especially in the presence of other salting-out agents, is itself readily extracted by the organic extractants. The molecules of the acid, as well as those of plutonyl nitrate, are solvated with a certain number of the molecules of the extractant. Thus, if the nitric acid concentration is high, the salt and the acid compete with each other for the addition of the organic extractant, with the result that the distribution coefficient of the salt of the metal being extracted decreases. It will be seen from Figure 91 that the increase of the total nitrate ion concentration increases the degree of extraction of Pu(VI) with diethyl ether [231]. At the same time the large increase in the acidity of the solution also produces an increase in the distribution coefficient of the fission products [35]. It is therefore recommended to extract plutonyl nitrate with diethyl ether using aqueous solutions saturated with the salting-out agent and to work at low acidities (0.5–2.0 M HNO_3).

In the process of extraction by diethyl ether, hexavalent plutonium passes from the nitrate solution into the organic phase as the molecular compound $PuO_2(NO_3)_2 \cdot xH_2O \cdot yS$. As regards the number of molecules of water and the molecules of the extractant S in the compounds, no agreement has as yet been reached. It may be assumed, by analogy with U(VI), that plutonyl nitrate binds four molecules of water and is dissolved with 4 ± 2 molecules of the organic extractant [430].

The separation of plutonium by ether extraction is distinguished by the ready

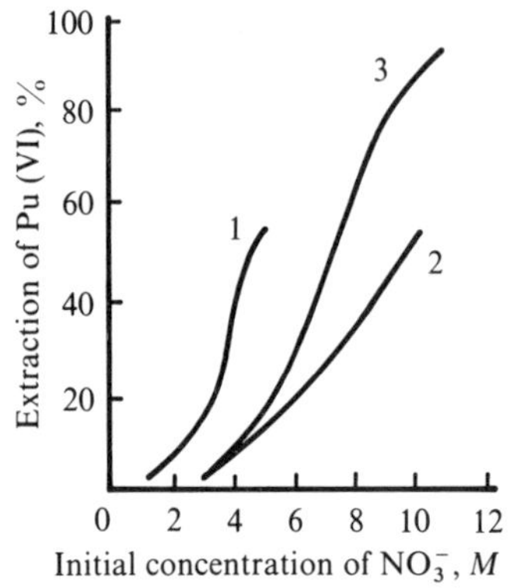

Figure 91

Effect of nitrate ion concentration on the extraction of Pu(VI) with diethyl ether:

1) HNO_3; 2) 1 M HNO_3 and NH_4NO_3; 3) 1 M HNO_3 and $Ca(NO_3)_2$.

extractability of hexavalent plutonium and a poor extractability of most of the impurities and fission products. Table 27 shows the distribution of certain metal nitrates during the extraction with diethyl ether [35].

The only elements other than Pu(VI) which are extracted to a significant extent are U(VI), Ce(IV), Zr(IV), Th(IV) and free halogens. Anions which form stable complexes with plutonium (sulfates, phosphates, etc.) interfere with the extraction of plutonium. The presence of iron nitrate suppresses the interference of phosphates and sulfates. The interference by fluoride ions is eliminated by adding aluminum nitrate.

Potassium persulfate, potassium bichromate and other oxidants are employed to oxidize plutonium to hexavalent state. It must be pointed out, however, that the presence of potassium bichromate results in an increased extraction of individual fission fragments.

As is well known, the bulk of the β- and γ-activity of irradiated uranium originates from the fission elements: zirconium, niobium, ruthenium, cesium, strontium, barium and the rare-earth elements. As a result of the ether extraction plutonium is satisfactorily purified from niobium, cesium and rare-earth elements.

Most difficulties are caused by zirconium and ruthenium. In order to achieve a satisfactory separation of plutonium from zirconium, the extraction should be carried out at relatively low concentrations of both nitric acid and salting-out agent. If ruthenium is present in the starting solution in the tetravalent state, it is separated without difficulty because of the poor extractability of Ru(IV); however, during the dissolution of the irradiated material in nitric acid nitrosylruthenium compounds,

Table 27

DISTRIBUTION OF NITRATES BETWEEN AQUEOUS SOLUTIONS AND DIETHYL ETHER [35]

Element	Distribution coefficient in the presence of		Element	Distribution coefficient in the presence of	
	NH_4NO_3*	$Al(NO_3)_3$**		NH_4NO_3*	$Al(NO_3)_3$**
U	2.3	208	Zn	$< 1 \cdot 10^{-4}$	—
Pu(VI)	1.5	—	Na	$< 1 \cdot 10^{-4}$	—
Th	$3.4 \cdot 10^{-3}$	0.32	K	$< 1 \cdot 10^{-4}$	—
Zr	$1.1 \cdot 10^{-3}$	$1.1 \cdot 10^{-2}$	Mg	$< 1 \cdot 10^{-4}$	—
Bi	$3.1 \cdot 10^{-4}$	$2.1 \cdot 10^{-2}$	Ga	$< 1 \cdot 10^{-4}$	—
Fe(III)	$< 1 \cdot 10^{-4}$	$1.7 \cdot 10^{-3}$	Al	$< 1 \cdot 10^{-4}$	—
Ag	$< 1 \cdot 10^{-4}$	$1.1 \cdot 10^{-3}$	La	$< 1 \cdot 10^{-4}$	—
Cu(II)	$< 1 \cdot 10^{-4}$	$8.2 \cdot 10^{-4}$	Cr(III)	$< 1 \cdot 10^{-4}$	—
Hg(II)	$< 1 \cdot 10^{-4}$	$3.0 \cdot 10^{-4}$	Sr	$< 1 \cdot 10^{-4}$	—
Gd	$< 1 \cdot 10^{-4}$	$2.0 \cdot 10^{-4}$	Mn	$< 1 \cdot 10^{-4}$	—
Li	$3 \cdot 10^{-4}$	—	Be	$2 \cdot 10^{-3}$	—
Rb	$1.4 \cdot 10^{-4}$	—	Sm	$< 9 \cdot 10^{-4}$	—
Co	$< 1 \cdot 10^{-4}$	—	Y	$< 9 \cdot 10^{-4}$	—
Ni	$< 1 \cdot 10^{-4}$	—	Ce(III)	$< 9 \cdot 10^{-4}$	—
Ba	$< 1 \cdot 10^{-4}$	—	Nd	$< 9 \cdot 10^{-4}$	—
			Pu(III)	$< 1 \cdot 10^{-3}$	—

* Composition of aqueous phase: 0.1 *M* solution of the nitrate under study, containing 8 *M* NH_4NO_3 and 1.4 *M* HNO_3.

** Composition of aqueous phase: 0.1 *M* solution of the nitrate under study, containing 1.3 *M* $Al(NO_3)_3$ and 1.3 *M* HNO_3.

which are much more readily extracted, are usually formed. The best extractable form of nitrosylruthenium is the trinitrate complex $RuNO(NO_3)_3 \cdot (H_2O)_2$. The rate of formation of the nitrate complexes of nitrosylruthenium is much less than that of plutonyl and uranyl complexes. This difference in the kinetics of complex formation is utilized to purify the material from ruthenium [34]. The purification of the organic phase improves if it is washed with weakly acid aqueous solutions of the salting-out agent.

The following procedure has been proposed [618] for the purification of plutonium from uranium and fission products by diethyl ether extraction.

The irradiated uranium (about 800 mg), containing plutonium and the fission products, is dissolved in hydrochloric acid. Plutonium chloride is converted into the nitrate by evaporating to dryness. The residue is dissolved in 5 ml of a solution containing 700 g NH_4NO_3 and 100 ml 14 *N* HNO_3 per liter. Plutonium is oxidized to the hexavalent state by potassium persulfate in the presence of silver and the solution is extracted with ether. Uranium and plutonium pass into the ether extract, while the fission products remain in the aqueous phase. Uranium and plutonium are reextracted from the organic phase with acidulated water. Lanthanum nitrate is introduced into the aqueous phase as carrier, and the plutonium is reduced to Pu(III) by sulfurous acid. The solution is evaporated to dryness, dissolved in a solution of ammonium nitrate (700 g/liter) and the extraction with ethyl ether is repeated. Uranium passes into the ethereal layer, while trivalent plutonium remains in the aqueous phase. The aqueous phase, containing plutonium and lanthanum, is then evaporated to dryness, when ammonium nitrate is volatilized. The residue is dissolved in nitric acid and lanthanum fluoride is precipitated together with plutonium.

As a rule, satisfactory purification is achieved in 3–4 extractions.

Vdovenko et al. [35, 37] studied the extraction of Pu(VI) and U(VI) with a mixture of 85 vol.% dibutyl ether and 15 vol.% of carbon tetrachloride. Since the mixture has a low extractive capacity, it is employed only when the nitrate concentration in the initial solution is high. Since the coefficient of purification from the fission products varies considerably with the acidity of the initial solution, it is not recommended to work in solutions more acid than 1 *M* HNO_3. The high nitrate concentration in the initial solution is produced by adding calcium nitrate (up to 5.5 *M*) as salting-out agent. Table 28 gives data on the distribution of hexavalent plutonium and uranium between aqueous solutions and a mixture of 85 vol.% dibutyl ether (DBE) + 15 vol.% CCl_4.

Plutonium is oxidized to the hexavalent state by potassium bichromate.

The experimental procedure includes 3–5 extractions of Pu(VI) together with U(VI) from nitric acid solutions saturated with calcium nitrate into the DBE + CCl_4 mixture (85 vol.% and 15 vol.% respectively). The volume ratio between the organic phase and the aqueous solution is 16:30. These operations separate the bulk of fission products, chromium, iron and other elements. A supplementary purification from the fission products is effected by washing the organic phase with 5 *M* calcium nitrate, or, better, with an aqueous solution of 5.1 *M* $Ca(NO_3)_2$, 0.4 *M* $UO_2(NO_3)_2$, 0.05 *M* $K_2Cr_2O_7$ at equilibrium.

The difference between the extractive properties of plutonyl nitrate and those of trivalent plutonium nitrate is the principle of the separation of plutonium from uranium (Tables 28 and 29).

The separation of uranium from plutonium consists in the reextraction of plutonium in a strongly reducing medium. The reducing agent is $N_2H_4 \cdot HNO_3$ or hydroxylamine nitrate.

After 3–5 successive extraction cycles with DBE, plutonium may be isolated in a yield not inferior to 98%. The overall coefficient of purification for Pu from the γ-active fission products is $8 \cdot 10^4$.

Uranium may also be separated from plutonium in a somewhat different manner. Prior to the DBE extraction, a mild reducing agent, e.g., hydrazine, is introduced into the initial solution in order to reduce plutonium to the trivalent state. Uranium remains hexavalent. As a result of the extraction uranium passes into the organic phase, while plutonium and impurities remain in the aqueous solution. An oxidant

Table 28

DISTRIBUTION COEFFICIENTS OF U(VI) AND Pu(VI) BETWEEN WATER AND DBE:CCl_4 MIXTURE [37]

Composition of initial aqueous solution	Temperature, °C	Distribution coefficient of	
		U(VI)	Pu(VI)
2.5 *M* $UO_2(NO_3)_2$ 2.4 *M* $Ca(NO_3)_2$ 0.53 *M* HNO_3 0.1 g/liter Pu(VI)	60	0.20	0.12
0.76 *M* $UO_2(NO_3)_2$ 4.26 *M* $Ca(NO_3)_2$ 0.53 *M* HNO_3 0.1 g/liter Pu(VI)	35	0.67	0.40
0.01 *M* $UO_2(NO_3)_2$ 5.4 *M* $Ca(NO_3)_2$ 0.05 *M* HNO_3 0.1 g/liter Pu(VI)	20	1.20	0.70

Table 29

DISTRIBUTION COEFFICIENTS OF Pu(III) NITRATE BETWEEN AQUEOUS SOLUTIONS AND A MIXTURE OF 85 vol.% DBE + 15 vol.% CCl_4 AT 20° [37]

Composition of initial aqueous solution	Distribution coefficient of Pu(III)
5.4 *M* $Ca(NO_3)_2$ 0.05 *M* HNO_3 0.2 *M* $N_2H_4 \cdot HNO_3$ 0.1 g/liter Pu(III)	0.003
5.4 *M* $Ca(NO_3)_2$ 0.2 *M* $N_2H_4 \cdot HNO_3$ 0.1 g/liter Pu(III)	0.0005

(usually sodium or potassium bichromate) is added to the aqueous solution to convert plutonium to Pu(VI) and plutonium is extracted by an organic extractant.

Up to 1953 it was considered established that only Pu(VI) may be extracted from nitric acid solutions with oxygenated extractants. However, as early as 1950–

1951 Kuznetsov had studied the extraction of uranium and thorium and had concluded, on the strength of the analogy between thorium and tetravalent plutonium, that Pu(IV) may be extracted from nitrate-saturated nitric acid solutions. This conclusion was confirmed by Chaikhorskii (1953) who observed that Pu(IV) could pass into the ethereal layer from a solution of $\sim 2\ M$ HNO_3 saturated with ammonium nitrate.

In 1953 Palei and Khalkin established the conditions for diethyl ether extraction of tetravalent plutonium from strongly acid solutions with nitric acid as the salting-out agent. The extractability of Pu(IV) increased with the increase in the concentration of nitric acid to 5–6 *N*. At major concentrations of nitric acid the extractability of plutonium decreases. These workers established that plutonium is extracted as the oxonium salt of hexanitroplutonic acid $[Pu(NO_3)_6] \cdot [H\ldots OR'R'']_2$.

The optimum conditions for the extraction of Pu(IV) with diethyl ether are: concentration of nitric acid in aqueous layer 5 *M*, in the organic layer 3.4 *M*; under these conditions the distribution coefficient of Pu(IV) is higher than 10. The degree of extraction of plutonium is independent of the temperature between 18 and 38° and of the initial concentration of plutonium in the aqueous phase between $4 \cdot 10^{-9}$ and $4 \cdot 10^{-3}$ g/ml. The extraction is unaffected by the presence of major amounts of chlorides (up to 2 *M*). The presence of small amounts of sulfates produces a marked decrease in the distribution coefficient of plutonium. Khalkin's investigation paved the way for the extraction of small amounts of plutonium from solutions of complex composition.

Under these conditions the extractability of hexavalent plutonium is much poorer. The distribution coefficient of Pu(VI) is 3. Accordingly, plutonium is converted to Pu(IV) prior to the extraction.

Plutonium is extracted in the extractor shown in Figure 92. The extractor itself is a modified separating funnel, with an outlet tube for draining the organic layer. The extractor is placed in a lead block to protect it from β- and γ-radiations.

To 1 ml of the sample solution are added 0.25 ml of 2 *M* solution of hydrazine hydrate and 2 ml of 13.2 *M* HNO_3 and the mixture is left to stand for 30 minutes. Under these conditions Pu(IV) becomes stabilized and is not reduced to Pu(III). Two ml of water are then added, so that the concentration of HNO_3 in the solution is 5 *M*.

The solution is placed in the extractor and 8–10 ml ether containing 3.4 *M* HNO_3 are added. The extractant is prepared by shaking the ether with an equal volume of 8 *M* HNO_3. The stirrer is switched on and kept in operation for 2–3 minutes. The solution is then left to stand for 4–5 minutes to ensure complete layer separation, after which the ether extract is drained into a washing extractor of similar design and washed. Ether-saturated 5 *M* HNO_3 serves as the wash liquor; this is usually the aqueous layer remaining after the preparation of the extractant. The washed ether extract is evaporated on a water bath. To prevent a possible explosion due to the oxidation of the ether by concentrated nitric acid, 2–3 ml of 2 *M* aqueous hydrazine hydrate are added to the organic layer; hydrazine hydrate is a stronger reducing agent than ether, and will therefore prevent its oxidation. In the solution obtained in this way plutonium is determined either directly by a radiometric method, or only following a supplementary purification with lanthanum fluoride.

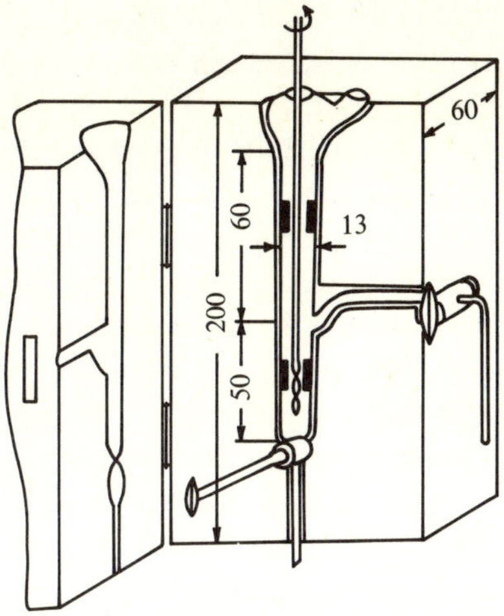

Figure 92
Extractor

Plutonium is extracted practically quantitatively. The experimental error is that of α-radiometric determinations (± 2 rel. %). The method may serve to separate indicator amounts of plutonium from major amounts of chromium, lead, iron and other elements, and also from the fission products.

Palei and Milyukova (1954) used a similar method to separate semimicro- and macroamounts of plutonium from iron, chromium, lanthanum, lead, manganese, calcium, barium, silicon and other elements. Pu(IV) could be quantitatively extracted by treating the aqueous solution (5 *M* in HNO_3) with ten times the volume of ether containing 3 *M* HNO_3. Under these conditions it is unnecessary to wash the ethereal extract. The plutonium-containing organic extract is treated as described for indicator amounts of plutonium.

Other ethers were also tested as potential extractants of plutonium. These included tetraethylene glycol dibutyl ether, diisopropyl ether and ethylene glycol dibutyl ether [31]. They failed to find use, since their distribution coefficients for plutonium were lower than in diethyl ether; the first-named ether, moreover, was decomposed by nitric acid.

Extraction of plutonium with β,β′-dibutoxydiethyl ether

The extractant used for the separation and purification of uranium and plutonium is β,β′-dibutoxydiethyl ether, which is also known under the technical name of Butex [233]. Butex is very stable to nitric acid, owing to which nitric acid alone may be employed as the salting-out agent. Butex has a high flash point and is sparingly soluble in water. Its disadvantages are a high viscosity and a density which is close to that of water. Butex is decomposed when heated in nitric acid, yielding carbon dioxide and water as the final products.

Butex is a good extractant for U(VI), Pu(VI), U(IV), Pu(IV), Zr(IV), Ce(IV) and $RuNO \cdot (NO_3)_3$.

The distribution coefficients of these elements strongly depend on the concentration of nitrate ions in the system. If the nitrate concentration is high, plutonium or uranium may be quantitatively extracted into the organic phase by performing the requisite number of extraction cycles; if the nitrate concentration is low, these elements can be transferred from the organic to the aqueous phase, i.e., they can be reextracted.

The distribution coefficient of trivalent plutonium at all nitrate ion concentrations is much lower than that of tetravalent and hexavalent plutonium and that of hexavalent uranium. This difference may serve to separate uranium from plutonium. Table 30 gives the approximate values of the distribution coefficients of plutonium and uranium in butex extraction.

Table 30

DISTRIBUTION COEFFICIENTS OF URANIUM AND PLUTONIUM BETWEEN THE AQUEOUS PHASE AND BUTEX [233]

Element	Aqueous phase	
	3 *N* HNO_3	0.2 *N* HNO_3 and 8 *M* $Al(NO_3)_3$
U(VI)	1.5	3
Pu(VI)	1.8	2.5
Pu(IV)	7.0	7.0
Pu(III)	< 0.01	< 0.002

Plutonium is extracted, mainly as Pu(VI), together with U(VI) by butex from nitric acid solutions. The optimum acidity of the solution of HNO_3 is 3 *N*. Under these conditions the fission products which are extracted include mainly ruthenium and, to a lesser extent, zirconium, niobium and cerium. To improve the purification from the fission products, the organic phase is washed with 3 *N* HNO_3. The nitric acid in the organic phase is neutralized with aqueous ammonia, after which plutonium is reextracted into the aqueous phase containing iron sulfamate as the reducing agent.

In the extraction of weighable amounts of uranium and trace amounts of plutonium from 3 *N* nitric acid solutions by butex, 99.9% of uranium, 99.8% of plutonium and about 0.5% of the fission products pass into the organic phase. For more thorough purification, the extraction of Pu(VI) with butex may be repeated.

Butex can be used to separate plutonium from zirconium. Ruthenium is not extracted together with plutonium. Tributylphosphate, on the contrary, is an efficient agent for purifying plutonium from ruthenium. By extracting successively with Butex and TBP, plutonium may be quantitatively separated from the fission products.

Extraction of plutonium with methyl isobutyl ketone

Methyl isobutyl ketone $C_4H_9COCH_3$ (hexone) is used to isolate plutonium from the irradiated uranium and for its purification from a large number of other elements [139–141, 217, 527, 529, 632, 650].

Tetravalent and hexavalent plutonium are extracted with methyl isobutyl ketone from solutions of nitric acid. Pu(III), like other triply charged ions, is extracted by methyl isobutyl ketone only to an insignificant extent. This difference may be used for the separation of plutonium from many elements by a combination of the extraction of Pu(IV) and Pu(VI) with the reextraction of Pu(III) with reducing media.

The efficiency of the extraction of plutonium will markedly depend on the concentration of the nitric acid and salting-out agents in the aqueous phase [217, 632]. The salting-out agents employed are ammonium, sodium, calcium, magnesium and aluminum nitrates. Their salting-out effects are similar, except for ammonium nitrate, in the presence of which a better separation from the fission products may be attained [31]. The effect of the concentration of HNO_3 and $Ca(NO_3)_2$ is shown in Figures 93 and 94. According to these data Pu(IV) and Pu(VI) have higher distribution coefficients in the presence of calcium nitrate. Better purification from the fission products is attained at low acidity values. Since Pu(IV) may polymerize under these conditions and thus fail to be extracted, the extraction of Pu(VI) is to be preferred. The oxidants employed include sodium and potassium bichromate, potassium bromate, sodium bismuthate [632] and potassium permanganate [527].

When extracted three times with an equal volume of methyl isobutyl ketone, Pu(IV), Pu(VI) and U(VI) are 99.9% extracted from a solution which is 1.5 *M* in HNO_3 and 3 *M* in $Ca(NO_3)_2$.

In order to achieve complete extraction of plutonium from solutions not containing reducing agents, the volume of the organic phase must be increased, or a greater number of successive extractions performed. A positive effect of the salting-out agents is to improve phase separation; the separation of plutonium from the accompanying elements is, however, poorer in their presence.

The effect of the acid concentration and the salting-out agent on the extractability of the fission products is considerable (Figures 93 and 94). Ruthenium is the element most difficult to separate from uranium and plutonium. Diphenylthiourea and thiourea are added [31] to complex ruthenium in the aqueous phase. In the extraction of plutonium from the oxidizing medium and when washing the organic phase, the concentrations of the salting-out agent and of the acid should be kept at the lowest possible values in order to achieve a better purification from ruthenium. The effect of the concentrations of the salting-out agent and acid on the purification from zirconium and cerium is similar (Figures 93 and 94).

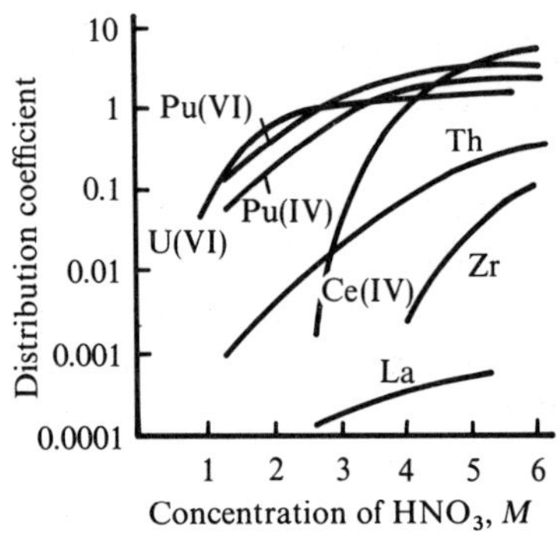

Figure 93

Distribution coefficients of elements in dependence of the equilibrium concentration of HNO_3 in the aqueous phase during extraction with methyl isobutyl ketone

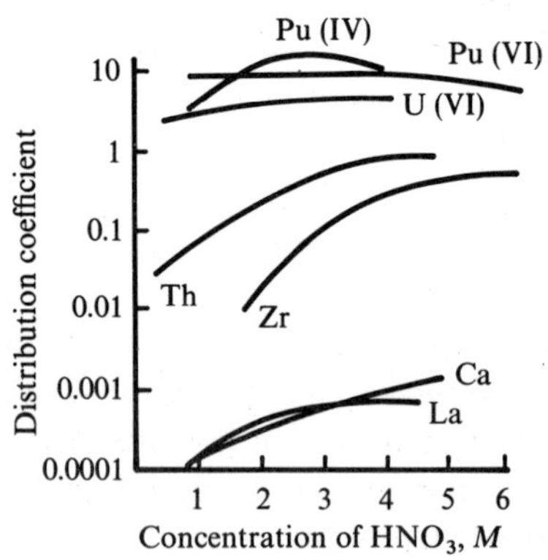

Figure 94

Distribution coefficients of elements in dependence of the equilibrium concentration of HNO_3 in the aqueous phase in the presence of $Ca(NO_3)_2$ as the salting-out agent during extraction with methyl isobutyl ketone. $Ca(NO_3)_2$ concentration, 2 *M*.

The impurities (mesityl oxides, methyl isobutyl carbinol, mesitylene) present in methyl isobutyl ketone have been reported to affect the extraction of the fission products from oxidizing media ($K_2Cr_2O_7$). The purification of the extractant has been described [139].

The separation of plutonium from uranium is effected by a reductive reextraction of plutonium with the aid of hydroxylamine or any other suitable reducing agent. When plutonium is isolated from irradiated uranium, a solution 2 *M* in HNO_3 and 0.5 *M* in hydroxylamine was used for the purpose [527].

Extraction of plutonium with diethyl ketone

The separation of plutonium by extraction with diethyl ketone resembles to a considerable extent the extractions with methyl isobutyl ketone.

The conditions for the extraction of plutonium, uranium and fission fragments were studied by Pal'shin (1958). The concentration of nitric acid and salting-out agent affects the results to a large extent, just as in the case of extraction with methyl isobutyl ketone (Table 31). In the absence of salting-out agents Pu(VI) is extracted to a lesser extent than Pu(IV). The reason for the different behavior of Pu(IV) and Pu(VI) is their unequal ability to form the complex acids $H_2[Pu(NO_3)_6]$ and, probably, $H[PuO_2(NO_3)_3]$, which are readily extractable.

Table 31

EFFECT OF SALTING-OUT AGENTS ON THE DISTRIBUTION COEFFICIENTS OF ELEMENTS BETWEEN 2% SOLUTION OF HNO_3 AND DIETHYL KETONE

(Pal'shin, 1958)

Element	Distribution coefficient					
	concentration of NH_4NO_3 used as salting-out agent, %			concentration of $Ca(NO_3)_2$ used as salting-out agent, %		
	15	30	50	15	30	50
Pu(IV)	$9 \cdot 10^{-2}$	2.4	12	$5 \cdot 10^{-1}$	8.7	125
Pu(VI)	$8.7 \cdot 10^{-1}$	3.7	22	1.2	12	> 250
U(VI)	$2.5 \cdot 10^{-1}$	$9 \cdot 10^{-1}$	3.7	$5 \cdot 10^{-1}$	4.3	> 200
Sr^{89}	—	—	$1 \cdot 10^{-5}$	—	—	$9 \cdot 10^{-4}$
Y^{91}	$8 \cdot 10^{-5}$	$1 \cdot 10^{-4}$	$2 \cdot 10^{-4}$	$2 \cdot 10^{-4}$	$5 \cdot 10^{-4}$	$1 \cdot 10^{-2}$
Zr(IV)	$1 \cdot 10^{-4}$	$2.3 \cdot 10^{-4}$	$6 \cdot 10^{-4}$	$2.5 \cdot 10^{-4}$	$2.5 \cdot 10^{-3}$	$2.6 \cdot 10^{-1}$
Ce(III)	$1 \cdot 10^{-5}$	$4 \cdot 10^{-5}$	$9 \cdot 10^{-5}$	$5 \cdot 10^{-5}$	$3 \cdot 10^{-4}$	$4 \cdot 10^{-3}$
Ru	$1.5 \cdot 10^{-3}$	$6 \cdot 10^{-3}$	$1.6 \cdot 10^{-3}$	$4 \cdot 10^{-3}$	$1.3 \cdot 10^{-2}$	$8 \cdot 10^{-2}$

Note: 1) Concentration of plutonium in aqueous phase 0.2 mg/ml, concentration of uranium 60 mg/ml. 2) Distribution coefficients of Y^{91} and Zr(IV) were determined by Seryakova (1958).

The distribution coefficients of Pu(IV) and Pu(VI) remain practically constant at concentrations of plutonium between 0.002 and 3.0 mg/ml.

It is shown by the data of Pal'shin that one full extraction cycle* will result in a high degree of purification of plutonium and uranium from the fission products. Plutonium is more fully extracted with diethyl ketone than with methyl isobutyl ketone (Table 32).

Table 32

EXTRACTION OF Pu(IV), U(VI) AND FISSION PRODUCTS WITH DIETHYL KETONE AND METHYL ISOBUTYL KETONE FROM 2% SOLUTION OF NITRIC ACID CONTAINING 50% NH_4NO_3 (Pal'shin, 1958)

Solvent	Extraction, %		
	Pu(IV)	U(VI)	fission products (γ-emitters)
Diethyl ketone	99.7	99.8	0.005
Methyl isobutyl ketone	92.0	97.0	0.004

Note: Concentration of plutonium in the aqueous phase 0.2 mg/ml; concentration of uranium 100–150 mg/ml.

Extraction of plutonium with tributyl phosphate

Tributyl phosphate (tri-n-butyl ester of orthophosphoric acid, TBP) is widely employed in the extractional separation of plutonium.

Tributyl phosphate is a nonionized solvent and is therefore a very good extractant for nitrates, which readily associate into molecules. The main advantage of TBP as extractant of plutonium is the fact that the distribution coefficients are sufficiently high for plutonium to be extracted practically quantitatively into the organic phase by one single extraction in the absence of any salting-out agents. The solvent is not volatile (b.p. 289°) and has a low solubility in water and in nitric acid solutions [267].

Temperature, °C	16	17	19	22
Solubility of TBP, g/liter	0.420	0.410	0.397	0.380

TBP is slowly hydrolyzed by water. In acid medium, both in aqueous and in organic phases, the hydrolysis of TBP passes through a few stages, with the successive

* A full extraction cycle is understood to comprise the following operations: 1) a number of successive extractions from the aqueous solution by the organic extraction; 2) washing the organic extract with an aqueous solution of the salting-out agent; 3) supplementary extraction of the aqueous washing liquor; 4) reextraction of element into the aqueous phase.

formation of dibutyl phosphate (DBP), monobutyl phosphate (MBP) and phosphoric acid. Commercial TBP usually contains all these compounds as well as butanol. These impurities interfere with the extraction of plutonium. Butanol reduces Pu(IV) and Pu(VI) to Pu(III) and thus reduces the extent of the extraction. Dibutyl phosphate forms very stable complexes with tetravalent plutonium, which are readily extracted but poorly reextracted; also, the extractability of the impurities is increased in the presence of dibutyl phosphate. Monobutyl phosphate, when dissolved in water, forms with plutonium precipitates which are not extracted.

Shevchenko and Smelov [248, 249] showed that the highest permissible concentrations of monobutyl phosphate and dibutyl phosphate in TBP, which still do not affect the extraction of plutonium, are 0.001 and 0.0001 mole/liter respectively. Accordingly, TBP must be purified before use.

To do this, a portion of TBP is shaken with an equal volume of 5% aqueous solution of Na_2CO_3 or 1–2% solution of alkali. After phase separation TBP is washed a few times with a weak solution of nitric acid and then with distilled water until neutral. TBP is then brought into contact with charcoal and filtered off. To obtain purer material, tributyl phosphate is redistilled.

Tributyl phosphate is very resistant to thermal decomposition and to oxidation by chemical reagents. Nevertheless, above 135° it reacts vigorously with nitric acid [142]. Another advantage of TBP is its resistance to radiation.

The main drawbacks of this solvent are its high viscosity (3.41 cp at 25°) and density (0.973) which is close to that of water and renders phase separation difficult. These drawbacks can be overcome by using inert diluents: kerosene, synthine, chloroform, carbon tetrachloride, benzene and paraffin hydrocarbons. The distribution coefficients of plutonium decrease as a result, but still remain sufficiently high. The selectivity of the extraction increases.

The extraction of tetravalent elements improves in the sequence Th < Np(IV) < Pu(IV), while for hexavalent elements it improves in the opposite sequence Pu(VI) < Np(VI) < U(VI). TBP extraction of trivalent actinides is poor [153, 193, 266].

According to a number of workers [31, 65, 153, 224, 249, 297, 336], metal nitrates are present in TBP as neutral molecules solvated by a certain number of TBP molecules. The following formulas have been proposed for complexes of trivalent, tetravalent and hexavalent plutonium: $Pu(NO_3)_3 \cdot 3TBP$; $Pu(NO_3)_4 \cdot 2TBP$; $PuO_2(NO_3)_2 \cdot 2TBP$ [65, 153, 226, 336]. As distinct from the solvates of plutonium with ethers and ketones, water does not form part of complexes with TBP. This indicates that TBP is a sufficiently strong competitor, which is able to displace all the water from the coordination sphere of plutonium ion.

The TBP extractability of the above complexes varies. Figure 95 gives the distribution coefficients of plutonium in various states of valency.

The complex-forming action of TBP depends on the concentration of nitric acid (cf. Table 33) in the aqueous phase.

Table 33

EFFECT OF ACIDITY ON THE DISTRIBUTION COEFFICIENT OF PLUTONIUM NITRATES BETWEEN TBP AND THE AQUEOUS PHASE

Concentration of HNO_3, M	Distribution coefficient			References
	Pu(III)	Pu(IV)	Pu(VI)	
0.5	—	1.05	—	[65]*
1.0	0.015	1.3	0.7	[225]**
1.4	—	2.3	—	[65]
2.8	—	10.0	—	[65]
6.0	0.01	20.0	3.0	[225]
0.1 M HNO_3 + 6 M $NaNO_3$	—	high	high	[225]

* Extractant: 20% solution of TBP in hydrocarbons.
** Extractant: 33% solution of TBP in hydrocarbons.

It will be seen from Table 33 and Figure 95 that the distribution coefficient of plutonium increases with the increase in acidity, due to the salting-out action of nitrate ions. At acidities above 7 M, the value of K_d decreases. The probable reason for it is that nitric acid forms compounds with TBP, e.g., $HNO_3 \cdot TBP$, which decrease the concentration of free TBP and thus also the extractability of plutonium. The effect of the acid is the same as in the ether extraction, but plutonium is extracted in different forms. Ethers dissolve the nitrate complexes of plutonium, while TBP dissolves molecular compounds.

The distribution coefficients of Pu(IV) and Pu(VI) increase considerably if a part of nitric acid in the aqueous phase is replaced by sodium nitrate or another nitrate which is insoluble in TBP. In this way the reaction between nitric acid and TBP is prevented. A large decrease in acidity may reduce the distribution coefficient of Pu(IV) owing to the hydrolysis of this ionic species.

The value of the distribution coefficient depends only to a small extent on the concentration of plutonium in the initial aqueous solution (Figure 96). At low acidities the distribution coefficient increases with the increase in the concentration of plutonium. This effect is similar to salting-out by common ion effect. At relatively high acidities the effective concentration of TBP decreases owing to complex formation between HNO_3 and TBP. As a result, the distribution coefficient decreases with the increase in the concentration of plutonium.

The efficiency of the extraction of plutonium nitrates with TBP depends to a large extent on the identity of the inert solvent (Figure 97) [192].

Shevchenko et al. [249–251] studied the effect of aliphatic and aromatic hydrocarbons, saturated monohydric alcohols and ethers on the extraction of several

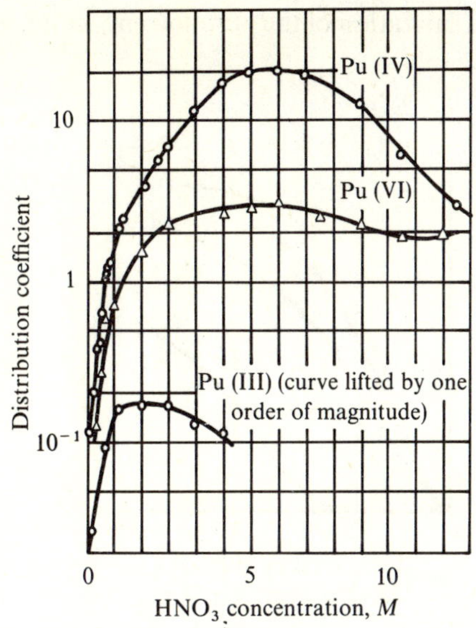

Figure 95

Extraction of plutonium with TBP as a function of the concentration of HNO_3 in aqueous phase

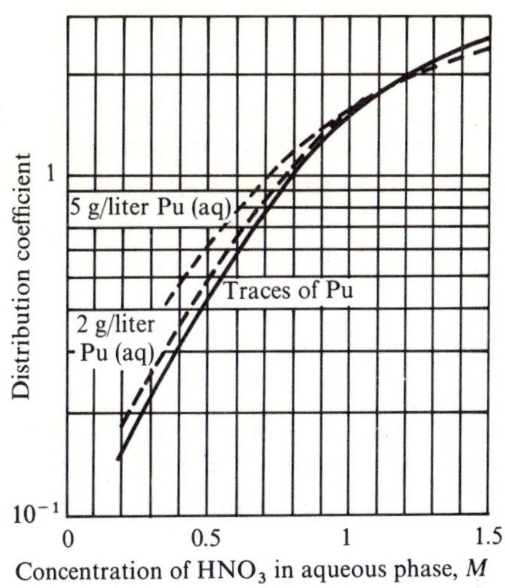

Figure 96

Dependence of the distribution coefficient of plutonium on the concentration of Pu(IV) and HNO_3 during extraction with 19% solution of TBP in kerosene

elements, including the extraction of Pu(IV) with tributyl phosphate. It is considered that the distribution coefficients of Pu(IV) and other elements decrease with the increase in the molecular polarization of the pure solvent, within a given homologous series.

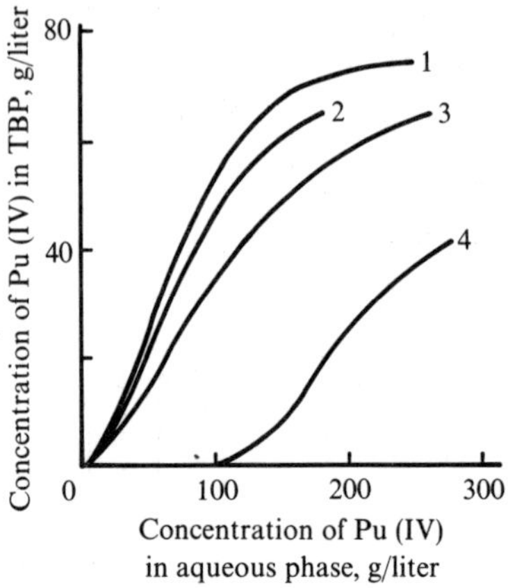

Figure 97

Effect of solvent on the extraction of Pu(IV) with 20% TBP:

1) benzene; 2) synthine; 3) carbon tetrachloride; 4) chloroform. Temperature: 25°.

When a number of metals—Zr, Pu(IV), Ce(IV), Th(IV)—are extracted with tributyl phosphate, a second organic phase occasionally appears. Healy and MacKay [453], who studied this effect in detail, found that the new phase, which was of intermediate density, is a solvate of the metal nitrate, e. g., $Pu(NO_3)_4 \cdot 2TBP$, which is formed as a result of the limited solubility of the compound in the solvent. A second organic phase is seen to appear only when aliphatic hydrocarbons are used as solvents.

The distribution of plutonium during extraction with TBP is a function of the temperature [2, 162, 192] (cf. Figure 98).

The distribution coefficient of plutonium is greatly affected by the concentration of uranium in the initial aqueous solution (Table 34). Hexavalent uranium, which is readily extracted, binds the TBP and thus decreases the distribution coefficient of plutonium.

Tributyl phosphate extracts from nitrate solutions the rare-earth elements, Th, Zr, Ce(IV), Ru(VI), Am(VI), Np(IV) and Np(VI) [225, 233]. Nevertheless, plutonium may be separated from most of these elements by a suitable variation of the extraction conditions. Table 35 gives the distribution coefficients of a number of elements

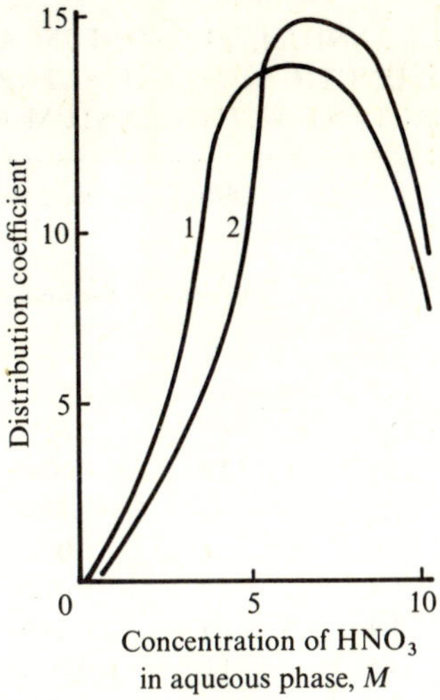

Figure 98
Distribution of Pu(IV) between nitric acid and 20% solution of TBP in saturated hydrocarbons at 20°(1) and 70°(2).

during the extraction with 20% solution of TBP in kerosene. Just as in the extraction with ethers and ketones, the main difficulty in the separation of plutonium from fission fragments is the separation from ruthenium, zirconium and niobium. For an efficient purification of plutonium the experimental parameters of the extraction must be strictly adhered to. The most important are a low acid concentration and the use of a TBP free from hydrolysis products, which would increase the extraction of niobium and zirconium. Thus, for instance, an increase in the concentration of HNO_3 in the aqueous phase from 1 to 5 *M* results in an increase in the distribution coefficient of the total β- and γ-activity by a factor of 5 and 21 respectively. The distribution coefficient of zirconium between the two phases increases almost by a factor of ten [31].

The distribution coefficient of β-activity (ruthenium, zirconium and niobium) increases with increasing temperature, as distinct from the extraction with methyl isobutyl ketone, where the opposite dependence of the distribution coefficient on temperature is noted.

The extraction of the fission products into the organic phase decreases on the introduction of complex-forming reagents: oxalates, phosphates, etc. The presence of phosphates in the concentration of 0.1 mole/liter reduces the distribution coefficient of zirconium almost 100 times; the presence of 0.1 *M* fluorosilicate reduces it 10 times, and the presence of 0.1 *M* sulfate 6 times [31]. This is accompanied by a deterioration in the extraction of plutonium. Thus, when the solution contains 40%

Table 34

DISTRIBUTION OF URANIUM, PLUTONIUM AND FISSION PRODUCTS AS A FUNCTION OF THE SATURATION OF THE EXTRACTANT WITH URANIUM [224]

Percent saturation of extractant with uranium	Distribution coefficient			
	U	Pu(IV)	β-emitters	rare-earth elements
28.0	16.7	4.0	0.013	0.0096
37.0	13.7	3.7	0.011	0.0073
45.6	12.1	2.3	0.0065	0.0048
54.9	10.2	1.3	0.0050	0.0032
61.7	7.9	1.6	0.0035	0.0021
70.2	6.4	1.3	0.0025	0.0014
72.0	5.4	1.1	0.0018	0.0011
77.2	4.5	1.0	0.0015	0.0007
82.4	3.6	0.79	0.0011	0.004
86.8	2.3	0.57	0.007	0.002

Note: Aqueous solution: nitric and solution of uranium, plutonium and fission products. Extractant: TBP in Amsco 123-15; equilibrium established within 5 minutes, temperature 25°C.

of phosphoric acid, the distribution coefficient of Pu(IV) in the absence of salting-out agents decreases by a factor of more than 10, while the distribution coefficient of uranium decreases approximately by a factor of $1\frac{1}{2}$ [247].

The difference in the extractabilities of individual elements with TBP (cf. Table 35) makes it possible to separate them from each other.

A continuous process for the extraction of plutonium with tributyl phosphate was developed by American scientists [224]. The initial 4–5 *N* nitric acid solution, containing uranium, plutonium and fission products, was treated with sodium nitrite to convert plutonium into Pu(IV) and extracted with a solution of TBP in the hydrocarbon Amsco 123-15. The result of this operation was to separate plutonium and uranium from fission products. Plutonium was then reextracted into dilute nitric acid, containing a reducing agent. This method of separating plutonium from the accompanying materials may also be used in analytical practice.

Hyde [231] extracted plutonium with 20% solution of TBP in benzene, CCl_4 and other solvents.

Goldschmidt, Regnaut and Prevot [65] proposed a method, in which plutonium is extracted with TBP diluted to 40% in high-boiling paraffinic hydrocarbons. The first step is to extract a mixture of plutonium and uranium from 1.7 *N* nitric acid solution. The ratio between the volumes of the extractant and the aqueous solution

Table 35

DISTRIBUTION COEFFICIENTS OF PLUTONIUM AND URANIUM NITRATES AND FISSION PRODUCTS DURING THE EXTRACTION WITH 20% SOLUTION OF TBP IN KEROSENE AT 25° [225]

Element	Concentration of element, *M*	Distribution coefficient		
		1 *M* HNO_3*	6 *M* HNO_3*	0.1 *M* HNO_3,* 6 *M* $NaNO_3$
U(VI)	10^{-2}–10^{-3}	5	30	Very high
Pu(VI)	10^{-2}–10^{-3}	0.7	3	” ”
Pu(IV)	10^{-2}–10^{-3}	1.3	20	” ”
Zr(IV)	Traces	0.01	1	1**
Ru(NO) (as trinitrate)	10^{-2}–10^{-3}	10	0.01	Very high
Ce(IV)	Traces	0.01	0.2	—
Pu(III)	10^{-2}–10^{-3}	0.015	0.01	—
Ce(III)†	Traces	0.04	0.02	10
Y(III)	”	0.01	0.02	—
Nb(V)	”	0.01	0.01	—

* Equilibrium concentration in aqueous phase.

** Composition of aqueous phase: 1 *M* HNO_3 + 5 *M* $NaNO_3$.

† The data for Ce(III) refer to a 50% TBP solution.

is kept at 2.8. After phase separation the organic solution is washed with 3–4 *N* HNO_3 to ensure a more complete separation from the fission products. The organic phase contains all the uranium, not less than 99.8% plutonium, about 5% of β-active and about 1% γ-active fission products. The organic extract is then treated with an aqueous solution of a reducing agent, when plutonium is reduced to Pu(III) and passes into the aqueous phase, while uranium remains in the organic phase. When the concentration of plutonium is low (up to 1 μg/ml), the reducing agent is 0.3 *N* solution of hydrazine containing 0.2 *N* HNO_3; if the concentration of plutonium is higher than 2 μg/ml, the reducing agent is an 0.3 *N* solution of hydrazine containing 0.0025 mole of iron sulfamate per liter in 0.24 *N* HNO_3. The authors showed that with the increase in the concentration of plutonium it is no longer fully reduced by hydrazine. The acidity of the reextracting solution must not be below 0.1 *N* in order to prevent hydrolysis of Pu(IV).

In the presence of phosphoric acid plutonium may be separated from uranium and from fission products by extraction with TBP. Shevchenko, Povitskii and Solovkin [247] described the processing of irradiated fuel elements of the first nuclear power station in the Soviet Union. The nitric acid solutions, obtained by dissolving the fuel elements, contained uranium (100–120 grams per liter), plutonium, molybdenum, magnesium, fission elements and phosphoric acid (up to 46 grams per liter). The acidity of the solutions was about 5 *M* HNO_3. The method consisted in a selective extraction first of uranium and then of Pu(IV) with 20% TBP in hydrogenated kerosene.

The first extraction of uranium was performed at an aqueous phase: organic phase volume ratio of 1.5. The aqueous nitric acid phase after the extraction of uranium contained all the plutonium, about 1 g/liter of uranium, molybdenum, magnesium, fission elements and up to 40 g/liter of phosphoric acid. The extraction of plutonium from such solutions was carried out after the addition of a salting-out agent, which was iron or aluminum nitrate. Ferric nitrate proved to be the most efficient salting-out agent in this separation. The addition of the agent up to an approximate concentration of 1 *M* made it possible to extract Pu(IV) in the presence of phosphoric acid, with an approximate distribution coefficient of 10. Plutonium was extracted with a 20% solution of TBP, taken in a volume equal to that of the aqueous phase. More than 98% of plutonium could thus be extracted into the organic phase. The purification coefficient from β-active fission products is 300, from γ-active fission products it is 50. If the organic phase is washed with 3 *M* HNO_3 and plutonium is then reextracted, the purification coefficient from γ-active admixtures can be increased to 150.

Plutonium was then extracted a second time and the organic phase was washed. The coefficient of purification of plutonium from γ-activity in the second extraction stage was about 40.

Plutonium may be reextracted from the organic phase with a mixture of oxalic and nitric acids. The coefficient of purification of plutonium from γ-active fission elements during reextraction is 20, while the overall coefficient of purification from γ-activity up to the final purification stage is 10^4.

Larsen and Siles [518] utilized TBP in the extractional purification of hydrochloric acid solutions of plutonium from uranium and fission products. Carbon tetrachloride served as diluent for TBP, while HCl was employed as salting-out agent.

The advantage of the system HCl–TBP over the nitrate system is its high separation coefficient and the high values of its distribution coefficients (Table 36).

The separation of plutonium from uranium is as follows [518].

A portion of the solution containing 10–20 mg Pu is placed in a conical flask. The hydrochloric acid solution is prepared by evaporating the sample almost to dryness three times with 12 *M* HCl. The volume is made up to 10 ml so as to render the concentration 2 *M* in HCl. About 0.1 g metallic magnesium is added to convert plutonium to Pu (III). The solution is filtered and collected in a flask. The solution is then transferred to a 60 ml separating funnel. The filter and the flask are washed 3 times with 5 ml of 12 *M* HCl each time, and the washings are also placed in the separating funnel. Fifteen ml of 30% TBP in CCl_4 are added, and uranium is extracted during one minute. The phases are separated and the organic phase is transferred to another 60 ml separating funnel. The extraction is repeated with another 10 ml of 30% TBP in CCl_4, and the uranium-containing organic phases are combined. Trivalent plutonium is separated from uranium at this stage.

To the plutonium-containing aqueous phase about 100 mg $NaNO_2$ are added to oxidize plutonium to Pu(IV). Twenty ml of 30% solution of TBP in Amsco-140 (inert diluent) are added and the mixture extracted for one minute. The lower aqueous phase is discarded. To the organic phase 15 ml of 0.2 *M* HCl are added and the mixture shaken for 1 minute. After phase separation, the aqueous phase containing plutonium is transferred to a 50 ml conical flask. To the organic phase 10 ml of 0.2 *M* HCl are added and the reextraction of plutonium is repeated. The aqueous solutions are combined, 2 ml of 12 *M* HCl are added and the solution is evaporated on a sand bath to ~2 ml residual volume. The solution is transferred to a 15 ml centrifuge tube and diluted to 7 ml with acidulated water. Plutonium is then precipitated as the hydroxide, converted to nitrate and determined spectrophotometrically at 475 mμ.

Table 36

DISTRIBUTION COEFFICIENTS OF PLUTONIUM AND URANIUM BETWEEN HCl AND 30% TBP IN CARBON TETRACHLORIDE [518]

Concentration of HCl, *M*	Distribution coefficient			
	Pu(IV)	Pu(VI)	U(IV)	U(VI)
1.0	0.0001	—	—	0.008
2.0	—	—	—	0.9
3.0	0.01	—	—	0.54
4.0	0.54	3.0	—	2.5
5.0	11	12	0.5	3.6
6.0	90	52	15	9.0
7.0	560	104	50	15
8.0	1400	—	102	21

Note: Concentration of plutonium in aqueous phase 10^{-4} *M*; concentration of uranium 10^{-2} *M*. Volume ratio between the phases 1:1.

Solovkin, Ivantsov and Renard [208] studied the extractive capacity of TBP for plutonium (IV) perchlorate. The distribution coefficients are much smaller than in the extraction from nitric acid solutions, but are sufficiently high for the effect of perchlorate ion to be neglected (Table 37).

As distinct from nitrate [208], in the extraction of plutonium (IV) perchlorate, the distribution coefficient increases continuously with the increase in the acid concentration. The extracted complex has the composition $Pu(ClO_4)_4 \cdot 2TBP$.

Extraction of plutonium with dibutylcarbitol

A number of workers [296, 336, 538, 581] extract plutonium from nitrate solutions with dibutylcarbitol, which is close in its properties as extractant to methyl isobutyl ketone. The disadvantages of dibutylcarbitol are its higher price as compared with methyl isobutyl ketone, and the fact that it is not stable to concentrated nitric acid.

Table 37

DISTRIBUTION COEFFICIENTS OF Pu(IV) IN THE SYSTEM TBP–$HClO_4$

Concentration of $HClO_4$, *M*		Increase in volume of organic phase, %	Distribution coefficient of Pu(IV)	Concentration of $HClO_4$, *M*		Increase in volume of organic phase, %	Distribution coefficient of Pu(IV)
in aqueous phase	in organic phase			in aqueous phase	in organic phase		
0.4	0.082	1	0.0045	2.8	0.24	4	0.021
0.9	0.1	2	0.0061	3.7	0.28	7	0.028
1.85	0.14	3	0.0065	6.4	0.57	8	0.9

The distribution coefficients of tetravalent and hexavalent plutonium in the extraction with dibutylcarbitol, as well as those of Th, Np(IV), Np(VI) and U(VI), are close to those in the extraction with methyl isobutyl ketone. The extractability of tetravalent actinides is higher than that of the hexavalent.

Tetravalent plutonium is extracted as a mixture of neutral complexes, strongly solvated with dibutylcarbitol. Hexavalent plutonium is extracted as a compound analogous to that of uranium, $HPuO_2(NO_3)_3 \cdot x\mathrm{DBC} \cdot yH_2O$. The concentration of nitric acid in solution should be maintained at 2–4 *M*, since the extractant is decomposed in more acid solutions.

The separation of plutonium consists in the extraction of Pu(IV) or Pu(VI) together with U(VI) with subsequent reductive reextraction of plutonium.

Extraction of plutonium as benzenesulfinate

Tetravalent plutonium in aqueous solutions forms a precipitate (see p. 92) with sodium benzenesulfinate, which is soluble in certain organic solvents.

It was found (Milyukova, 1958) that plutonium(IV) benzenesulfinate can be extracted into benzene, isobutyric acid and isovaleric acid. The extraction efficiency is markedly dependent on the identity of the solvent and on the concentration of nitric acid in the aqueous solution (Figure 99).

The concentration of sodium benzenesulfinate is a major factor in the extraction of plutonium (Figure 100). The distribution coefficient remains almost constant at 25 when plutonium(IV) benzenesulfinate is extracted from 0.15 *N* HNO_3 into isovaleric acid in the concentration range 0.5–5 mg Pu per ml. At plutonium concentrations below 0.5 mg/ml the distribution coefficient decreases.

The extraction of plutonium is conducted as follows.

One ml of 0.15 *N* nitric acid solution, containing 0.5–1 mg Pu(IV) per ml, is placed in a separating funnel, and 0.2 ml of 15% aqueous solution of sodium benzenesulfinate are added. The solution is stirred, an

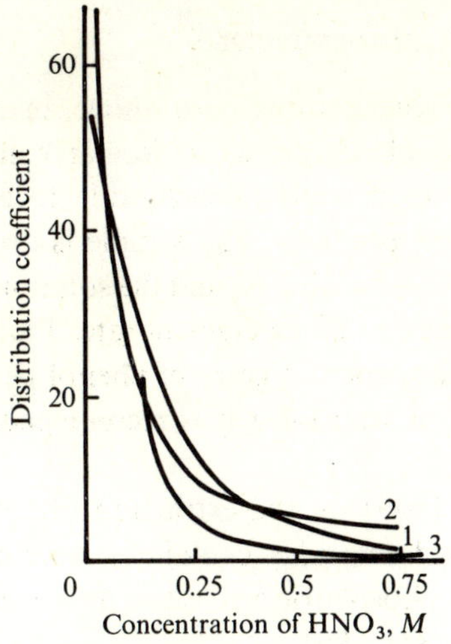

Figure 99

Effect of solvent and HNO_3 concentration on the distribution coefficient of plutonium (IV) benzenesulfinate:

1) isovaleric acid; 2) benzene; 3) isobutyric acid.

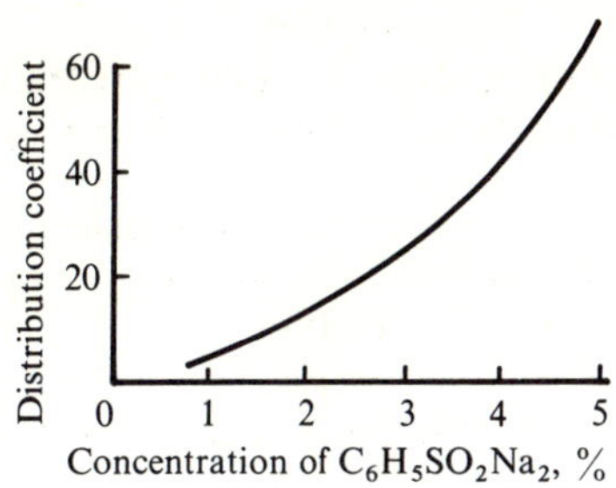

Figure 100

Dependence of the distribution coefficient of Pu (IV) as a function of benzenesulfinate sodium in aqueous phase

equal volume of isovaleric acid is added and the mixture shaken for 2 minutes. Plutonium is reextracted from the organic phase by a warm solution of 0.15 N HNO_3 saturated with sulfur dioxide.

Chromium, manganese, lanthanum and alkali metals do not interfere with the extraction of plutonium. Hexavalent uranium and ferric iron are partially extracted together with plutonium.

Extraction of plutonium with other extractants

In addition to the organic reagents discussed above, many other compounds are utilized in the extraction of plutonium as molecular compounds.

Fitch and Russell [403] used triglycol dichloride to separate plutonium from uranium and from the fission products. The sample is dissolved in nitric acid, the acidity is adjusted to 1–3 *N* (best of all 2 *N*) and the solution is saturated with ammonium nitrate, magnesium nitrate or calcium nitrate. Plutonium is extracted with triglycol dichloride containing about 1 gram of phenol per liter. In the presence of traces of hydrochloric acid or iron salts, it is recommended to add hydroquinone rather than phenol.

Information is available [649] on the extraction of hexavalent plutonium with ethyl sulfide from nitric acid solutions containing ammonium nitrate [649].

In addition to TBP, other phosphoric acid esters have been tried out as extractants for plutonium [21, 249, 279, 331, 699].

Berger [331] determined the distribution coefficients of tetravalent and hexavalent plutonium during extraction with phosphorus-organic compounds and found that the increase in the electronegativity of the phosphoryl group produces an increase in the extractive capacity of the compound in the sequence phosphate, phosphonate, phosphinate and phosphine oxide. Table 38 gives the distribution coefficients of

Table 38

DISTRIBUTION COEFFICIENTS OF PLUTONIUM(IV) NITRATE AND PLUTONIUM(VI) NITRATE AT 25°C [331]

Extractants	Distribution coefficient			
	Pu(IV)		Pu(VI)	
	0.2 M $UO_2(NO_3)_2$, 2 M HNO_3	2 M HNO_3	0.2 M $UO_2(NO_3)_2$, 2 M HNO_3	2 M HNO_3
Dibutyl methyl phosphate	0.27	0.71	0.23	0.63
Tributyl phosphate	0.33	1.57	0.28	1.1
Dibutyl decyl phosphate	0.39	2.3	0.33	1.7
Dibutyl methyl phosphonate	3.0	29	1.0	18
Dibutyl butyl phosphonate	3.7	32	1.3	16
Dibutyl decyl phosphonate	4.4	35	1.5	18
Butyl dibutyl phosphinate	16	170	5.2	99
Ethyl dihexyl phosphinate	16	200	12	140

Note: The composition of the aqueous phase given in the table is the value before extraction. The concentration of plutonium in the aqueous phase is 0.9 g/liter. Composition of organic phase: 0.5 *M* solution of ester in CCl_4.

tetravalent and hexavalent plutonium in the extraction with a number of phosphorus organic solvents.

Boden, Nikitina and Pushlenkov [42] showed that the composition of the complexes of uranyl nitrate with the compounds listed in Table 38 is analogous to that of TBP complexes, viz., $UO_2(NO_3)_2 \cdot 2S$, where S is the phosphorus-containing organic molecule. It may be expected, in analogy with uranium, that plutonium also forms complexes of a similar composition.

The above compounds proved to be the most efficient extractants for nitrates. The extraction of elements from chloride, sulfate and phosphate media is much poorer. The value of the distribution coefficient strongly depends on the concentration of the extractant in the solvent [21].

Umezawa [699] extracted tetravalent plutonium from irradiated uranium with tri-n-butylphosphine oxide (TBPO) in carbon tetrachloride. This worker studied the distribution of Pu(III) and Pu(IV) between the aqueous phase (0.5–10 M HNO_3) and TBPO in the concentration range $1 \cdot 10^{-4}$ to $3 \cdot 10^{-1}$ M.

Pu(IV) is extracted with 0.01 M solution of TBPO in CCl_4 in the presence of sodium nitrate. To separate Pu from uranium, plutonium is reextracted with 5% solution of hydroxylamine in 0.5 M HNO_3.

A number of papers [21, 216, 667, 668] deal with the extractive properties of alkylphosphoric and other phosphorus-organic acids.

Palei, Zemyanukhina and Nikol'skaya (1958) studied the possible utilization of alkyl pyrophosphates in the extractive separation of plutonium. It was found that isoamyl pyrophosphate and isobutyl pyrophosphate in toluene, and also in mixtures with CCl_4 and chloroform have a high extracting capacity for plutonium.

Tri-, tetra- and hexavalent plutonium can be extracted with isoamyl prophosphate in toluene from solutions in HNO_3, HCl, H_2SO_4, $H_2C_2O_4$, CH_3COOH and H_3PO_4. Tetravalent plutonium is extracted with 0.3 M solution of isoamyl pyrophosphate in toluene from all these media to the extent of 99–99.9%. In nitric acid and hydrochloric acid solutions the extraction of plutonium increases with the increase in acid concentration up to 7 N. The extraction of Pu (IV) from sulfuric and acetic acid media remains practically constant in the 0.5–8 N range, and from oxalate solutions in the 0.03–2 N range.

Tetravalent and hexavalent uranium behave similarly. For this reason the use of these reagents cannot bring about a separation of plutonium from uranium, but may be employed to concentrate plutonium in dilute solutions. Ruthenium, cesium and strontium are not extracted into the organic phase.

Plutonium is reextracted from the organic phase by solutions of $(NH_4)_2CO_3$, $NaHCO_3$, CH_3COONH_4 and $(NH)_2C_2O_4$.

Siddall [667, 668] determined the distribution coefficients of plutonium and other elements during their extraction with trialkyl phosphates (Table 39). Nevertheless, most phosphorus–organic reagents have yet to find a practical application.

Table 39

EXTRACTION OF TETRAVALENT AND HEXAVALENT ACTINIDES WITH 1.9 *M* SOLUTIONS OF TRIALKYL PHOSPHATES IN n-DODECANE FROM 3 *N* HNO_3 AT 30° [667, 668]

Extractant	Distribution coefficient*					
	Th	Np(IV)**	Pu(IV)†	U(VI)	Np(VI)††	Pu(VI)††
Tri-n-butyl phosphate	2.9	3.2	16.1	26	15.6	3.5
Triisobutyl phosphate	2.4	2.7	11.8	22	15.9	3.4
Tri-n-amyl phosphate	2.9	4.2	15.6	32	19.3	4.1
Triisoamyl phosphate	4.2	4.7	17.8	34	18.9	4.4
Tri-n-hexyl phosphate	3.0	3.6	15.6	38	20.0	4.5
Tri-n-octyl phosphate	2.4	3.4	15.3	33	15.7	3.9
Tri-(2-ethylhexyl) phosphate	2.5	4.3	25	58	23	5.7
Tri-(2-butyl) phosphate	0.45	4.9	28	42	20	4.6
Tri-(3-amyl) phosphate	0.22	3.5	18.1	49	22	5.0
Tri-(3-methyl-2-butyl) phosphate	0.18	3.0	24	47	25	5.4
Tri-(4-methyl-2-amyl) phosphate	0.047	3.5	22	38	24	4.9

* Investigations were carried out on trace amounts of elements.
** Aqueous phase contained 0.01 *M* iron sulfamate.
† Aqueous phase contained 0.01 *M* $NaNO_2$.
†† Aqueous phase contains 0.01 *M* of ammonium cerium double nitrate.

EXTRACTION OF PLUTONIUM AS CHELATES

Many organic reagents are known to react with plutonium with the formation of chelate compounds. The thenoyltrifluoroacetonates, 8-hydroxyquinolates, cupferronates, etc. are insoluble in water, but are readily extracted with benzene, chloroform, carbon tetrachloride and other solvents.

The extractability of actinide elements as chelate complexes diminishes as a function of the valencies in the sequence IV > VI > III, V. For this reason actinides are usually extracted in tetravalent state.

The efficiency of the extraction of elements as chelate complexes depends on the acidity of the solution. By varying the acidity of the aqueous phase, selective extraction of metals which form chelates of different stabilities with the same chelating agent can be effected.

For the theory of extraction of chelation complexes and the effect of various factors on the extraction of metals the reader is referred to [78, 102, 134, 175, 226, 565, 566].

Extraction of plutonium with thenoyltrifluoroacetone

Of all the chelating agents which have been studied, thenoyltrifluoroacetone (TTA) is the most widely used.

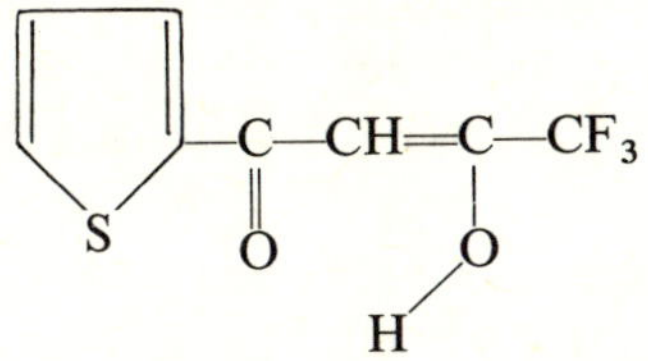

This compound is solid at room temperature, and sparingly soluble in water, but readily soluble in a large number of organic solvents.

The equilibrium constants of plutonium complexes in the system water–TTA–solvent differ widely, depending on the valency of plutonium. In the system plutonium–TTA–benzene the equilibrium constant is about 3 for Pu(III) and $1 \cdot 10^{-5}$ for Pu(IV); the value of the constant for PuO_2^{2+} is very small (by analogy with UO_2^{2+}, it is probably about 10^{-1} or 10^{-2}) [365].

The distribution coefficients of plutonium depend on the nature and concentration of the acid. Plutonium (IV) is most efficiently extracted from nitric acid, hydrochloric acid and perchloric acid solutions. The maximum extraction of plutonium with 0.1 *M* TTA in benzene is noted in the acidity range 0.5–1.0 *M*. At lower acid concentrations the extractability of plutonium decreases, possibly owing to the formation of hydrolyzed polymers [121]. If, on the other hand, the extraction is conducted with 0.5 *M* TTA, the maximum extraction range for plutonium extends up to 3 *M* HNO_3. At higher concentrations of nitric acid plutonium chelate is unstable and the equilibrium shifts towards the formation of nitrate complexes of Pu(IV), which are insoluble in benzene. Figure 101 illustrates the dependence of

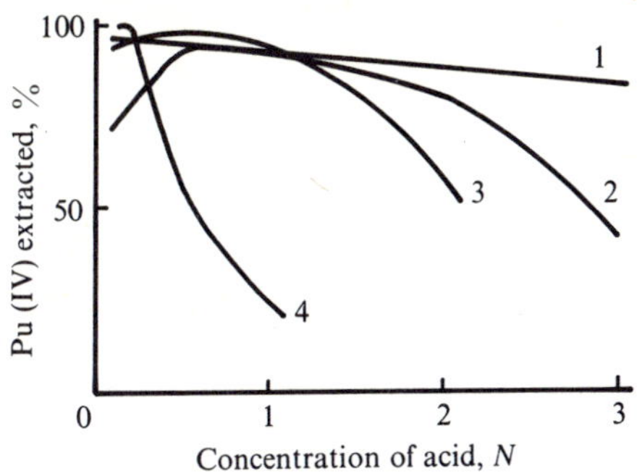

Figure 101

Extraction of Pu (IV) with 0.1 *M* solution of TTA in benzene from solutions of various acids:

1) $HClO_4$; 2) HCl; 3) HNO_3; 4) H_2SO_4.

the extractability of plutonium on the nature and the concentration of the acid (Palei and Milyukova, 1962).

The increase in the concentration of thenoyltrifluoroacetone results in an increase in the Pu(IV) distribution coefficient. Table 40 gives data on the effect of TTA concentration on the extraction of plutonium.

The rate of the extraction of Pu(IV) into the organic phase (cf. Table 41) also depends on the concentration of thenoyltrifluoroacetone [365, 561; Palei and Milyukova, 1962].

The efficiency of the extraction of plutonium with TTA depends to a certain extent on the nature of the organic solvent. This problem was studied by a number of workers [102, 365, 689].

Milyukova (1962) used certain aromatic hydrocarbons as solvents for thenoyltrifluoroacetone. Best extraction of Pu(IV) is achieved when the first few members of

Table 40

EFFECT OF CONCENTRATION OF TTA ON THE DISTRIBUTION COEFFICIENT OF TRACE AMOUNTS OF Pu(IV) IN THE SYSTEM 1 *M* HNO_2–TTA IN CCl_4 [365]

Concentration of TTA, *M*	Distribution coefficient	Concentration of TTA, *M*	Distribution coefficient
0.02	0.063	0.2	155
0.05	2.28	0.5	269
0.1	31.8	1.0	236

Table 41

RATE OF EXTRACTION OF Pu(IV) IN THE SYSTEM 1 *M* HNO_3–TTA IN BENZENE

Time, minutes	Pu(IV) extracted, %			Time, minutes	Pu(IV) extracted, %		
	0.1 *M* TTA*	0.5 *M* TTA*	1.0 *M* TTA**		0.1 *M* TTA*	0.5 *M* TTA**	1.0 *M* TTA**
0.5	—	97.2	97.3	10.0	94.7	—	—
1.0	—	97.3	97.8	12.0	—	99.7	99.7
2.0	80.7	98.0	98.5	15.0	94.7	—	—
5.0	86.7	98.4	99.8	16.0	—	99.7	99.7
7.0	93.0	—	—	20.0	94.7	—	—
8.0	—	99.1	99.6				

* According to Palei and Milyukova (1962). Solutions were manually stirred.

** According to Moore and Hudgens [561]. Solutions were mechanically stirred.

the aromatic series are employed. Of the chlorinated aliphatic hydrocarbons, carbon tetrachloride proved to be the most effective solvent (Table 42).

Table 42

EFFECT OF THE IDENTITY OF SOLVENT ON THE EXTRACTION OF Pu(IV) IN THE SYSTEM 1 *M* HNO_3–0.1 *M* TTA IN BENZENE

(Milyukova, 1962)

Solvent	Pu(IV) extracted, %	Solvent	Pu(IV) extracted, %
Benzene	95.8	Carbon tetrachloride	95.0
Toluene	94.8	Chloroform	88.5
Xylene	89.5	Dichloroethane	85.0
Mesitylene	86.9		

Note: Concentration of plutonium in the aqueous phase $1.3 \cdot 10^{-4}$ *M*. Volume ratio of organic to aqueous phase 1:1.

Thenoyltrifluoroacetone is a strong complexing agent for many ions, especially for multiple-charged ions. TTA extracts Zr(IV), Hf(IV), Ce(IV), Sn(IV), Th(IV), U(IV), Np(IV), Nb(V), Fe(III) and other ions from acid solutions. However, by adjusting the acidity of the solution, it is possible to separate Pu(IV) from a number of elements. Only Zr(IV), U(IV), Np(IV), Ce(IV), Pa(V) and Fe(III) are extracted from solutions in which the acidity is 0.5 *M* or higher [556].

Plutonium is readily separated from thorium, since at acidities above 0.1 *N* the extraction of Th(IV) is negligible. Zirconium is readily extracted [555] together with Pu(IV) from 1 *M* solution of HNO_3, but remains in the organic phase during the reextraction of plutonium with 10 *M* nitric acid. A similar behavior is displayed by Fe(III) and Pa(V). In another method for the separation of plutonium from zirconium, plutonium is reduced to Pu(III) and zirconium is extracted. Trivalent plutonium is not extracted with thenoyltrifluoroacetone from 1 *M* HNO_3 (quoted from [561]). Americium(III), Cm(III) and rare-earth elements behave in a similar manner [567]. Various methods for the separation of plutonium, uranium, neptunium and fission products are based on these considerations.

Sulfate, phosphate, fluoride and oxalate ions interfere with the separation of plutonium by extraction with thenoyltrifluoroacetone [556, 561].

Since the distribution coefficients of trivalent and hexavalent plutonium in thenoyltrifluoroacetone extraction are much lower than the distribution coefficient of Pu(IV), plutonium must be first converted to the tetravalent state.

If plutonium is present in the solutions to be extracted in polymeric or colloidal forms, these must be decomposed.

Plutonium may be reextracted from the organic phase into a weakly acid solution containing a reducing agent. The reducing agents employed include hydroxylamine hydrochloride, stannous chloride, ferrous chloride and potassium iodide [556, 561, 203]. All these agents slowly reduce plutonium to Pu(III). A rapid reextraction of plutonium is attained with the aid of 8–10 *M* nitric [556, 561] or perchloric [628] acid.

The concentration of plutonium and its separation from uranium, neptunium and fission products was described in [139, 231, 556, 558, 573]. Moore and Hudgens [561] recommend the following procedure for the concentration and purification of plutonium from neutron-irradiated uranium.

An aliquot of the sample, containing approximately 150 grams of uranium per liter, is placed in a 10 ml flask, 3 ml of 2 *M* HNO_3 and 1 ml of 1 *M* hydroxylamine hydrochloride are added. The solution is thoroughly mixed, heated at 80° for 5 minutes, the volume adjusted to about 4 ml, and a few drops of 1 *M* HNO_3 added. The solution is quantitatively transferred to a 30 ml separating funnel, 2 ml of 1 *M* $NaNO_2$ solution are added, the solution is mixed and left to stand until the evolution of nitrogen oxides has ceased. During the extraction the aqueous phase should be approximately 1 *M* in HNO_3. The solution is shaken for 10 minutes with an equal volume of 0.5 *M* solution of TTA in xylene. After phase separation, the aqueous layer is discarded. The organic phase is washed with an equal volume of 1 *M* HNO_3 for 3 minutes. Plutonium is then extracted from the organic phase by shaking for 2 minutes with an equal volume of 10 *M* HNO_3. Traces of zirconium and protactinium are removed by treating the plutonium-containing 10 *M* nitric acid solution with an equal volume of 0.05 *M* solution of TTA in xylene. All the plutonium remains in the aqueous phase.

Peppard et al. [590] used extraction with 0.1 *M* solution of TTA in benzene for the final purification of microgram amounts of plutonium, isolated from uranium plant wastes.

Culler [139] described a full cycle of nuclear fuel processing with the use of TTA dissolved in benzene.

The difference between the stabilities of tetravalent plutonium and neptunium have been utilized in their separation from one another with the aid of thenoyltrifluoroacetone [231, 556, 558].

Magnusson, Hindman and La Chapelle (quoted from [230]) developed the following method of separation of Np(IV) from Pu(IV).

The solution, containing major amounts of uranium and trace amounts of neptunium and plutonium, is evaporated with concentrated HCl to remove nitrates and the solution is adjusted to be 5 *M* in HCl. Potassium iodide and hydrazine hydrochloride are then added to a concentration of 0.1 *M*. The solution is heated for 2–3 minutes on a boiling water bath. Under these conditions hexavalent uranium is reduced by the iodide ions only very slowly, while neptunium and plutonium are reduced to Np (IV) and Pu (III) respectively. The solution is now diluted ten times, to obtain the required acidity of the solution. In order to prevent back-oxidation of neptunium by free iodine, the latter is reduced to iodide by hydrazine: this is done by heating the solution for one minute. An equal volume of 0.15 *M* solution of TTA in benzene is added and the solution extracted for 20–30 minutes. Neptunium is extracted almost quantitatively, while uranium and plutonium are extracted to an extent less than 1%.

To separate uranium from plutonium, the latter is oxidized to Pu (IV) by 0.02 *M* solution of potassium iodide. After holding at room temperature for 20 minutes, plutonium is almost quantitatively converted into the tetravalent state. Pu (IV) is then extracted with 0.1 *M* solution of TTA in benzene. Uranium remains in the aqueous phase. Plutonium is reextracted from the organic phase with 5 *M* HCl or with 0.5 *M* HCl containing 0.2 *M* hydroxylamine hydrochloride. This ensures a supplementary purification of plutonium from neptunium and other captured impurities, which now remain in the organic phase. The disadvantage of the method is the long duration of the reduction.

Moore [556, 558] separated plutonium from neptunium by extracting plutonium with 0.5 *M* TTA in xylene. The procedure is similar to that already described [231].

Nebel [573] extracted plutonium (IV) into a benzene or xylene solution of TTA, and reextracted it into 8 *M* HNO_3 containing 0.3 *M* HF.

Brauns and Schneider [29] describe the use of thenoyltrifluoroacetone to isolate microamounts of plutonium from highly active products.

Yakovlev, Vorob'ev and Shvetsov (1957) separated microgram amounts of Pu(III) from weighable amounts of neptunium by extracting Np(IV) from 0.5 *M* solution of HCl with 0.4 *M* TTA in toluene. Less than 1% plutonium passed into the organic phase.

Extraction of plutonium (IV) cupferronate and neocupferronate

In acid solutions plutonium (IV) forms precipitates with cupferron. These precipitates are soluble in various organic solvents.

It was found that plutonium (IV) cupferronate is readily extracted with chloroform. The extractional separation of metal cupferronates is most often carried out from hydrochloric acid or sulfuric acid media. In nitric acid solutions cupferron is decomposed, but high distribution coefficients may be attained. The efficiency of the extraction depends on the concentration of the complexing agent and the acidity of the solution. The extraction of plutonium as cupferronate is best carried out from solutions not more than 1.5 *N* in HNO_3. The degree of extraction of plutonium is a function of the time elapsed between the addition of cupferron to the solution and the beginning of the chloroform extraction of the plutonium complex (Figure 102). For cupferron, this time should clearly be not longer than 2 minutes [234].

In the presence of iron the extraction of plutonium greatly deteriorates (Figure 102). Moreover, the iron passes into the organic layer as cupferronate. Zirconium, vanadium and titanium are extracted together with plutonium. Oxalates, tartrates and citrates do not interfere.

Moiseev (1947) proposed a method for the separation of plutonium based on the extraction of plutonium (IV) cupferronate with chloroform. The method gives a separation of uranium from plutonium when present in the ratio Pu : U $= 1 \cdot 10^{-8}$–$1 \cdot 10^{-10}$. Plutonium in the amount of $1 \cdot 10^{-6}$ mg may be quantitatively isolated from 25–50 ml of solution which is 1.5 *N* in nitric acid or 1–4 *N* in sulfuric acid and which contains chromium, lanthanum, calcium and other elements. The bulk (more

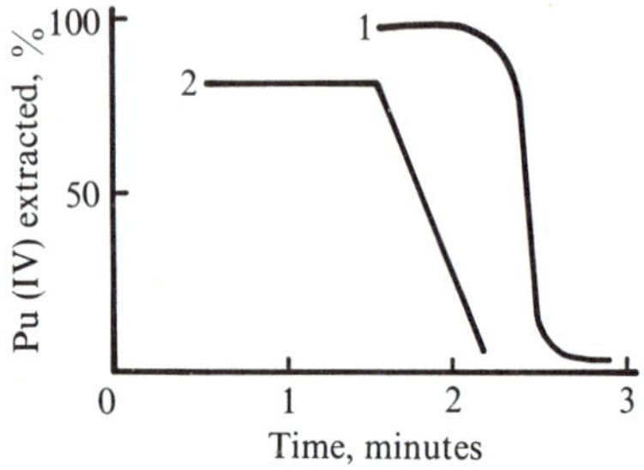

Figure 102

Extraction of plutonium (IV) cupferronate as a function of the time elapsed between the addition of the reagent and the beginning of the extraction:

1) solution of Pu (IV); 2) solution of Pu (IV), containing 0.0025 *M* Fe (III). (Concentration of nitric acid 1.9 *M*; concentration of cupferron 0.014 *M*).

than 98 %) of the fission products is eliminated at the same time. Lanthanum sulfate precipitation (see p. 250) is used to separate plutonium from iron, zirconium, titanium and vanadium which are also extracted.

The extraction of plutonium (IV) cupferronate is conducted as follows.

In a 100 ml separating funnel are placed 10–20 ml of the sample solution, which should be 1.5 *M* in HNO_3 and contain microgram amounts of plutonium and weighable amounts of uranium. The solution is then diluted with an equal volume of 1.5 *M* HNO_3. Water (5–10 ml), saturated with SO_2, is then added, then 3.5 ml of 5% aqueous solution of cupferron, 5–10 ml of chloroform and the mixture is shaken for one minute. The organic layer is separated and the extraction repeated. The organic extracts are combined, transferred to a separating funnel and washed once with 20–25 ml of 0.5 *M* HNO_3 saturated with SO_2 until a faint odor of gaseous SO_2 can be detected. The organic phase is poured into a beaker. To the wash liquid in the separating funnel are added 1–2 ml of 5% solution of cupferron, 3–5 ml of chloroform and the extraction is carried out. The organic extract is combined with the bulk extract in the beaker. Concentrated nitric acid (5–10 ml) is then added to the beaker, the beaker is placed on a water bath and heated until chloroform is fully expelled. The solution is cooled to room temperature, quantitatively transferred into a volumetric flask with the aid of 3 *M* HNO_3 and made up to the mark. Plutonium is then determined radiometrically.

Smales, Airey, Walton and Brooks [672] extracted plutonium (IV) cupferronate with chloroform and developed a method for the determination of plutonium in urine. They achieved a satisfactory separation of Pu from uranium and thorium.

Neagon and Penneman (quoted from [465]) separated plutonium from Am(III) by extraction of plutonium (IV) cupferronate with chloroform from dilute hydrochloric acid solutions. To ensure a quantitative extraction of plutonium, Mohr's salt was added to the solution.

Warren (quoted from [465]) separated plutonium from uranium by extracting plutonium (IV) cupferronate from a dilute hydrochloric acid solution, containing iron as carrier and hydroxylamine.

Kui* (quoted from [465]) utilized the extraction of Pu(IV) as cupferronate from dilute hydrochloric acid solutions to determine traces of plutonium in water.

James and Thompson [478a] proposed a method for the isolation of plutonium from its mixture with lanthanum at a molar ratio La : Pu $\geqq 5$ from phosphate solutions at pH 3–0.2 by extracting plutonium (IV) cupferronate with chloroform.

Neocupferron, which forms chelate complexes with plutonium, behaves in most cases similarly to cupferron. It will be seen from Figure 103 that the extraction efficiency of plutonium with cupferron is somewhat better than with neocupferron. On the other hand, cupferron is more sensitive to the action of nitric acid than neocupferron. The time period within which plutonium is satisfactorily extracted with neocupferron is almost 20 times longer than when cupferron is employed. An increase of the nitric acid concentration above 1.9 *M* and the presence of iron in solution produce a deterioration in the extraction of plutonium with neocupferron. During the extraction of plutonium (IV) neocupferronate from 1–4 *M* HNO_3 into chloroform, less than 1% of uranium (VI) passes into the organic phase [234].

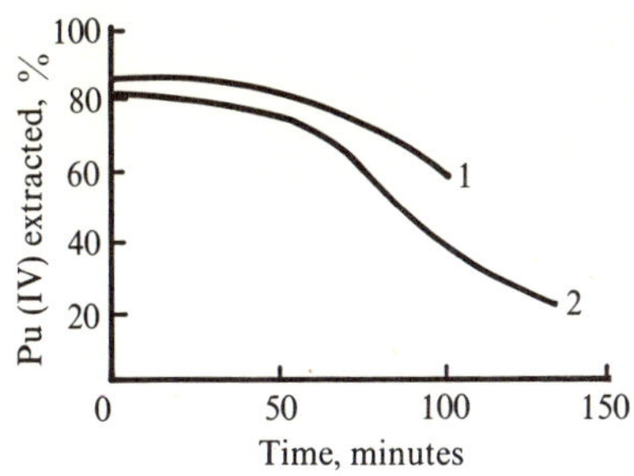

Figure 103

Extraction of plutonium (IV) neocupferronate as a function of the time elapsed between the addition of the reagent and the beginning of the extraction:

1) in the absence of Fe (III); 2 in the presence of 0.0023 *M* Fe (III). (Concentration of nitric acid 1.9 *M*; concentration of neocupferron 0.014 *M*.)

Extraction of plutonium with N-benzoylphenylhydroxylamine

N-benzoylphenylhydroxylamine (BPHA) is an analog of cupferron, but differs from it in that it is stable to mineral acids. A marked decomposition of the reagent is

* [Transliterated.]

noted only in nitric acid solutions above 4 *M*. The reagent is also stable to heat and light.

Chmutova, Petrukhin and Zolotov [242] investigated the extraction of the Pu(IV) compound of N-benzoylphenylhydroxylamine from solutions of nitric, hydrochloric and sulfuric acids. The effect of the identity and the concentration of the acid is shown in Figure 104. Plutonium is readily extracted with a solution of the reagent in chloroform from 1–6 *M* HNO_3. Equilibrium is established within 15 seconds. The extraction of plutonium from sulfuric acid solutions is poor. Plutonium is extracted to a considerable extent from hydrochloric acid solutions on brief shaking (up to 30 seconds), but remains in the aqueous phase almost entirely, if the duration of contact between the phases is increased to one hour. This is accompanied by the reduction of Pu(IV) to Pu(III), BPHA probably acting as reducing agent. The poor extractability of plutonium from sulfuric and hydrochloric acid media may be employed in the reextraction of plutonium into aqueous solutions of these acids.

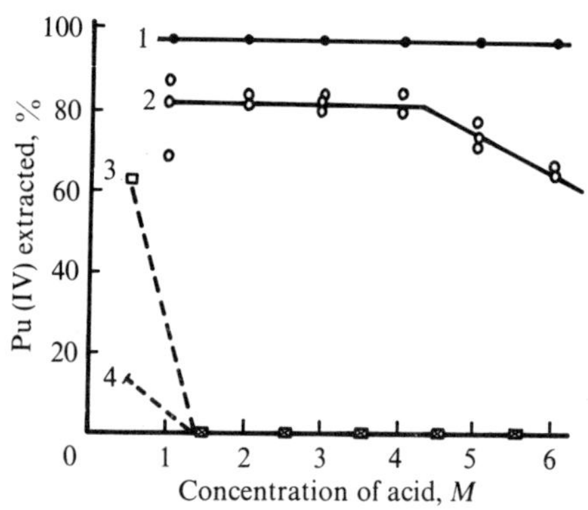

Figure 104

Extraction of Pu (IV) with a solution of BPHA in chloroform from nitric and sulfuric acid solutions:

1) HNO_3, 0.4 *M* BPHA; 2) HNO_3, 0.1 *M* BPHA; 3) H_2SO_4, 0.4 *M* BPHA; 4) H_2SO_4, 0.1 *M* BPHA.

The distribution coefficient of plutonium during extraction from nitric acid solutions strongly depends on the concentration of BPHA (Figure 105).

The distribution coefficient is independent of the concentration of plutonium between 0.002 and 1.5 mg/ml.

Uranium (VI), Np(V) and Am(III) do not interfere with the extraction with BPHA in chloroform. Of the fission products, niobium and zirconium are the only ones extracted. Table 43 shows the results of the extraction of a number of elements from 3 *M* HNO_3 with a solution of BPHA in chloroform.

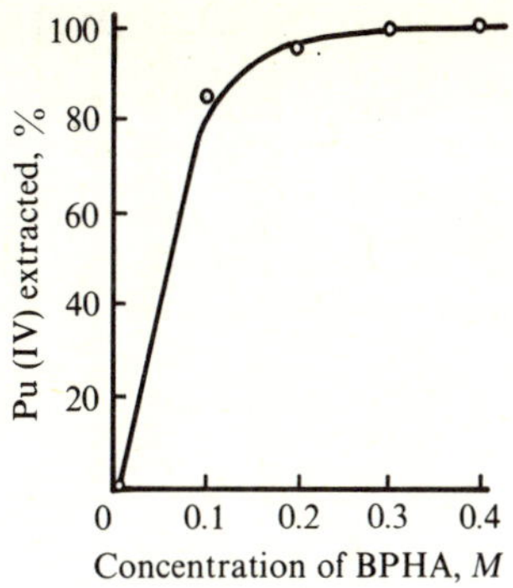

Figure 105
Extraction of Pu (IV) from 3 M HNO_3 as a function of the concentration of BPHA in chloroform

Table 43
EXTRACTION OF ELEMENTS WITH 0.4 M SOLUTION OF BPHA IN CHLOROFORM FROM 3 M HNO_3

Element	Extraction, %	Element	Extraction, %
Pu(IV)	98.0	La^{140}	0.10
U(VI)	0.02	Ce^{144}	0.30
Np(V)	1.47	Zr^{95}	97.5
Ru^{106}	0.06	Nb^{95}	92.3

The following procedure for the separation of plutonium from uranium and the fission products has been proposed [242].

The sample solution (3–5 ml), containing 7–10 μg plutonium and 7–10 mg uranium, is introduced into a 10–20 ml separating funnel, and the acidity is adjusted to 3 M HNO_3. An equal volume of 0.4 M solution of BPHA in chloroform is added and the mixture is shaken for 0.5–1 minute. The extraction separates plutonium from uranium and the fission products except zirconium and niobium. The phases are separated and plutonium is reextracted from the organic phase with 3 M H_2SO_4 at the volume ratio of 2.5 : 1 between the organic and the aqueous phases. Zirconium and niobium remain in the organic phase. The reextract is washed with pure chloroform.

A few results obtained by this method are shown in Table 44. The method may be employed to separate plutonium from americium.

Table 44

ISOLATION OF PLUTONIUM FROM IRRADIATED URANIUM AND ITS RADIOMETRIC DETERMINATION

Pu taken, μg	Pu found in extract, μg	Error, rel. %	γ-Activity in reextract, % of the initial
7.71	7.18	6.8	0
7.71	7.71	0	0
7.71	7.55	2.0	0.9

Extraction of plutonium in the form of other chelate compounds

Pu(IV) can be extracted into benzene as acetylacetonate. The distribution coefficient of the compound increases from 0.5 at pH ~1 to 80 at pH 5–10 [451]. In addition, plutonium acetylacetonate is extracted into benzene, chloroform or hexone from 0.1 *M* solutions of sodium perchlorate [630].

Plutonium (IV) 8-hydroxyquinolate is extracted with amyl acetate from solutions at pH 8, while plutonium (VI) 8-hydroxyquinolate is extracted from solutions at pH 4–8 [451].

EXTRACTION AS SALTS OF COMPLEX PLUTONIUM ANIONS WITH ORGANIC CATIONS

High molecular amines have recently found extensive use in the extractional separation of plutonium. Amines, like TBP, are excellent extractants and may be employed in the absence of salting-out agents. The advantage of amines is their high resistance to radiation [41], which makes it possible to isolate plutonium from highly radioactive wastes, and to employ the reagent repeatedly without purification and regeneration.

The extraction properties of amines were studied by a number of workers [111, 268, 269]. Free amines are readily soluble in organic solvents, but their salts are less so. As a rule, the solubilities of amine salts can be arranged in the following decreasing sequence: sulfates > bisulfates > chlorides > nitrates, and also tertiary > secondary > primary ammonium salts. To increase the solubility of amine salts and occasionally to accelerate phase separation [28] a small amount (3–5 vol. %) of a high molecular aliphatic alcohol is added to the solvent. The solvents used most often for the amines are benzene, xylene, chloroform, etc.

Extraction with amines will take place if the element being extracted is present in the aqueous solution as a complex anion or a neutral complex.

For plutonium (IV) sulfate this process may be represented by the following equations [40]:

$$Pu^{4+} + 3SO_4^{2-} \rightleftharpoons Pu(SO_4)_3^{2-},$$

$$2(RNH_3)_2SO_4 + Pu(SO_4)_3^{2-} \rightleftharpoons (RNH_3)_4 \cdot Pu(SO_4)_4 + SO_4^{2-}; \quad (3)$$

$$Pu^{4+} + 2SO_4^{2-} \rightleftharpoons Pu(SO_4)_2,$$

$$2(RNH_3)_2SO_4 + Pu(SO_4)_2 \rightleftharpoons (RNH_3)_4 \cdot Pu(SO_4)_4. \quad (4)$$

The selectivity of the extraction of plutonium with various amines will mainly depend on their structure and on the medium [111, 497].

Weaver and Horner [719] and Knoch [496] established the following sequence of extraction of plutonium with various amines from nitric acid solutions: quaternary > tertiary > secondary > primary amines; the sequence of extraction from sulfuric acid media is primary > secondary > tertiary. For most reagents the extractability increases in the sequence Pu(IV) > Pu(VI) > Pu(III). Tetravalent plutonium is extracted with amines to a greater extent from nitric acid than from sulfuric acid media, except for primary amines.

The extraction of plutonium with amines is also strongly affected by the nature of the solvent. For this reason the efficiency of an extractant is determined by the combination of the amine with the solvent and not by the amine alone.

Under certain conditions, a second organic phase is formed in the extraction. This is caused by the limited solubility of the compound being extracted in the organic phase.

Wilson [41] pointed out the importance of the solvent in the determination of the maximum admissible concentration of plutonium in the organic phase, at which a second phase still fails to be formed. Thus, for instance, in solutions of xylene and carbon tetrachloride the maximum admissible concentration of plutonium at 25° in 0.22 *M* solution of tri-n-octylamine is approximately 100 mg/liter, whereas it is 8.5 g/liter in chloroform, as much as 12 g/liter in dibromoethylene and more than 26 g/liter in nitromethane.

In some cases the extraction efficiency will depend on the concentration of the acid and of the metal being extracted.

A suitable selection of the reagent, solvent and medium makes it possible to separate various elements from one another.

Extraction of plutonium with primary amines

Vdovenko, Lipovskii and Nikitina [40] studied the extraction of tetravalent plutonium from sulfuric acid solutions with a mixture of normal primary aliphatic amines, containing 7–9 carbon atoms (density 0.7846). Chloroform was employed as solvent in this work.

The amine extraction of tetravalent plutonium from sulfuric acid solutions may be represented—in terms of the concept of the anion-exchange character of such extraction—as an exchange of the sulfate ion for the anionic complex of plutonium, formed in the aqueous phase (cf. equation (3)).

The distribution coefficients of Pu(IV) increase with the increase in the concentration of the amine (Figure 106).

Weaver and Horner [719] also noted the high degree of extraction of plutonium from sulfuric acid solutions with primary amines. Elements such as Zr(IV), Ce(IV) and U(IV) are also readily extracted with primary amines from sulfuric acid solutions.

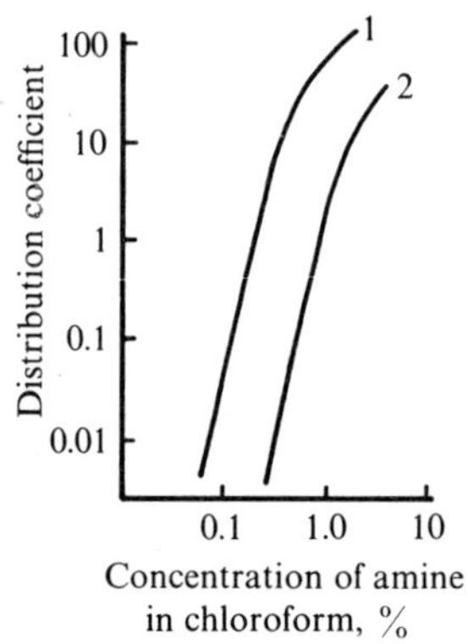

Figure 106

Dependence of the distribution coefficient of trace amounts of Pu(IV) on the concentration of the amine in chloroform.

Aqueous phase: 1) 0.125 *N* H_2SO_4; 2) 0.2 *N* H_2SO_4 and 0.3 *N* HNO_3.

Extraction of plutonium with secondary amines

Baker and Leary [20] noted that secondary amines could be used for extractional separation of plutonium from nitric acid solutions containing zirconium, molybdenum, ruthenium, lanthanum and cerium as fission products.

Winchester and Maraman [729] used secondary amine extraction in processing and purification of an alloy containing 300 grams of plutonium, fission products and iron and cobalt. The best extractant proved to be a solution containing 35% by volume of the secondary amine, 10% by volume decyl alcohol and 55% by volume of the solvent Gulf BT.

The alloy is dissolved in boiling 16 M HNO_3 containing 0.05 M HF. The acidity of the solution is reduced to about 1 M by adding a solution of NaOH and hydroxylamine nitrate is added to a concentration of 0.05 M. Sodium nitrite is then added to decompose the residual hydroxylamine and to stabilize plutonium in the tetravalent state. The acidity of the solution is then adjusted to 8 M with the aid of concentrated HNO_3. The solution is then extracted six times with an equal volume of freshly prepared extractant containing 35 vol. % of the secondary amine, 10 vol. % decyl alcohol and 55 vol. % of Gulf BT solvent. The aqueous phase is discarded. The organic phase is shaken first with water (one-fifth the volume of the organic phase) to reduce the acidity of the organic phase, and then three times with 0.1 M hydroxylamine nitrate at the same volume ratio of aqueous to organic phase. The aqueous reextracts, containing Pu (III), are combined and plutonium (III) is precipitated as oxalate.

The degree of purification of plutonium which can be attained by this method is shown in Table 45.

Table 45

CHEMICAL ANALYSIS OF PLUTONIUM [729]

Element	Content, grams per 10^6 grams of Pu	
	initial solution prior to extraction	final product
Co	37,300	< 500
Zr	15,400	< 30
Mo	8,200	< 100
Ru	11,600	< 100
Ce	8,770	< 100

Extraction of plutonium with tertiary amines

A number of workers separate plutonium by extraction with tertiary amines from solutions of mineral acids and from acetate solutions.

Tri-n-octylamine (TOA) efficiently extracts tetravalent plutonium from hydrochloric acid [254], nitric acid [41, 488] and sulfuric acid [246] solutions. Trivalent plutonium is not extracted with tri-n-octylamine. The distribution coefficient of Pu(IV) depends on the concentration of the acid and of tri-n-octylamine.

The maximum extraction of Pu(IV) with a solution of TOA in xylene from nitric acid solutions corresponds to an acid concentration of 4–6 M, while the value of the distribution coefficient is about one order of magnitude higher than the value obtained on extracting with 19% TBP under analogous conditions.

In the extraction of Pu(IV) from sulfuric acid solutions the distribution coefficient increases with the increase in the acid concentration to 0.1 M, and then decreases (Table 46). The increase in the distribution coefficient is explained by a partial suppression of the hydrolysis of plutonium. The decrease in the degree of extraction

Table 46

EFFECT OF THE SULFURIC ACID CONCENTRATION ON THE EXTRACTION OF Pu(IV) WITH TOA [246]

Equilibrium concentration of H_2SO_4 in aqueous phase, *M*	Distribution coefficient of Pu(IV)	Equilibrium concentration of H_2SO_4 in aqueous phase, *M*	Distribution coefficient of Pu(IV)
0.01	11.3	0.88	0.415
0.1	43	2.44	0.008
0.323	18.2		

Note: Initial concentration of TOA, 0.1 *M*. Concentration of Pu(IV), $2.34 \cdot 10^{-4}$ *M*.

of plutonium is probably due to the fact that with the increase in the concentration of sulfuric acid in the aqueous phase above 0.1 *M*, there is a considerable increase in the concentration of amine bisulfate, with consequent deterioration in the extraction of plutonium (Figure 107).

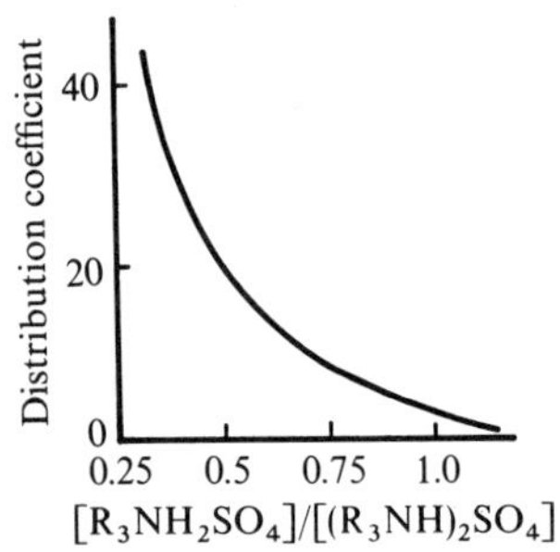

Figure 107

Dependence of the distribution coefficient of Pu(IV) on the molar ratio of the bisulfate to trioctylamine sulfate in solution

The effect of the concentration of the amine on the extraction of plutonium is shown in Table 47.

Shevchenko, Shmidt and Mezhov [254] studied the extraction of plutonium with tri-n-octylamine in xylene from hydrochloric acid media and found that Pu(III) is not extracted, while Pu(IV) is quantitatively extracted as the complex $PuCl_4 \cdot 2TOA \cdot 2HCl$.

Table 47

DISTRIBUTION COEFFICIENT OF PLUTONIUM(IV) SULFATE AS A FUCTION OF THE CONCENTRATION OF TOA [246]

Initial concentration of TOA, *M*	Distribution coefficient of Pu(IV)	Initial concentration of TOA, *M*	Distribution coefficient of Pu(IV)
0.01	0.9	0.07	36
0.02	5.9	0.10	43
0.05	15	0.25	250

Note: Concentration of Pu(IV) in the aqueous phase: $2.34 \cdot 10^{-4}$ *M*.

Bertocci (quoted from [465]) studied the extraction of Pu(IV) and uranium from nitrate solutions with triisononylamine in xylene and found that the distribution coefficient of plutonium increases with the increase in the concentration of nitric acid in the aqueous phase. Tetravalent plutonium is readily reextracted from the organic phase with a solution of hydroxylamine.

Triisooctylamine has also found wide application in the extractional separation of plutonium [496, 497, 557, 559]. Moore [557] studied the extraction of Pu(VI) from 4.8 *M* HCl with triisooctylamine (TIOA) in xylene and methyl isobutyl ketone. Pu(VI) was obtained by oxidation with potassium bichromate with heating. The effect of the concentration of potassium bichromate on the extent of the extraction is shown in Table 48.

Table 48

EFFECT OF $K_2Cr_2O_7$ CONCENTRATION ON THE EXTRACTION OF TRACE AMOUNTS OF Pu(VI) WITH 5% SOLUTION OF TRIISOOCTYLAMINE IN XYLENE FROM 4.8 *M* HCl [557]

Concentration of $K_2Cr_2O_7$, *M*	Pu(VI) extracted, %	Concentration of $K_2Cr_2O_7$, *M*	Pu(VI) extracted, %
0	60.8	0.02	98.9
0.01	99.8	0.05	83.2
0.01	99.8	0.09	1.2

Plutonium and uranium are extracted with an equal volume of 5% triisooctylamine in xylene. After phase separation, the elements are reextracted with 0.1 *M* HCl. Plutonium and uranium are separated from thorium, ruthenium, niobium, rare-earth elements, americium and curium. After reducing plutonium to Pu(III), hexavalent uranium is extracted with triisooctylamine. The degree of purification of plutonium with the use of TIOA is higher than by the extraction with TBP [20].

Triisooctylamine fully extracts hexavalent plutonium from acetate solutions [559]. Under these conditions Pu(III) is practically unextracted, while Pu(IV) is extracted to not more than 1.9%.

The extraction of Pu(VI) is conducted from a solution 0.5 *M* in HNO_3 and 0.5 *M* in CH_3COOH with a 5% solution of triisooctylamine in xylene. In this way Pu(VI) is separated from Fe(III), Th(IV), Pb(II), Ni(II), Co(II), Mn(II), Ce(IV), Cr(III), Al(III) and other elements. The extraction of ruthenium, zirconium and niobium varies from 10 to 50%. These elements are removed from the organic phase by washing with 5 *M* HCl. A reductive reextraction of plutonium is then carried out; uranium remains in the xylene layer.

The separation of uranium from plutonium may also be effected by extracting hexavalent uranium from Pu(III)-containing solution with TOA. Plutonium is additionally purified by extraction with triisooctylamine after oxidation to Pu(VI).

Knoch and Sinder [497] reported a selective extraction of tetravalent plutonium with triisooctylamine in xylene and trilaurylamine in kerosene from nitric acid solutions. The extraction of Pu(IV) is carried out from 6.5 *M* HNO_3, in which plutonium is partly present as the anion $[Pu(NO_3)_6]^{2-}$. As the result of the extraction the complex-formation equilibrium is shifted and plutonium is quantitatively extracted (similarly to ion exchange). Plutonium can thus be separated from uranium and fission products with satisfactory selectivity (Table 49). The purification of plutonium conducted in this matter is more efficient than when TBP is employed.

Table 49

DISTRIBUTION COEFFICIENTS OF ELEMENTS DURING AMINE EXTRACTION AS A FUNCTION OF NITRIC ACID CONCENTRATION [497]

Extractant*	HNO_3, *M*	Distribution coefficient			
		U(VI)**	Pu(IV)†	Ru	Zr (IV)
0.2 *M* solution of TIOA in xylene, containing 5% octanol	1	0.10	2	0.016	0.011
	3	0.095	3.6	0.035	0.021
	6.5	0.096	7	0.046	0.025
0.5% solution of TLA in kerosene, containing 5% octanol	1	0.065	5	0.008	0.006
	3	0.060	16	0.010	0.013
	6.5	0.058	35	0.021	0.018

* The organic phase is in equilibrium with HNO_3 of the given concentration.

** Concentration of uranium in aqueous phase 180 g/liter.

† Trace amounts of Pu(IV).

Wilson [41] utilized the high selectivity of the extraction of Pu(IV) with tertiary amines to isolate plutonium from uranium slugs. The extractant was a solution of trilaurylamine (TLA) in Amsco (kerosene). Plutonium may be separated from uranium and from the fission products in one cycle of operations (cf. Table 49).

The solution, containing 4 *M* HNO_3, 0.95 *M* U(VI), 0.03 *M* $NaNO_2$, $4 \cdot 10^{-4}$ *M* Pu(IV) and the fission products is shaken with a threefold volume of 0.15 *M* solution of TLA in Amsco, containing 2 vol.% n-octanol. The octanol is added to prevent the formation of a third liquid phase at high concentrations of uranium and plutonium in the aqueous phase. The organic extract is washed twice with $\frac{1}{3}$ of the volume of 4 *M* HNO_3. Plutonium is reextracted from the organic phase with 1 *M* solution of sulfamic acid containing 0.05 *M* iron sulfamate.

The results obtained are shown in Table 50.

Table 50

DISTRIBUTION COEFFICIENTS OF COMPONENTS DURING THE EXTRACTION OF PLUTONIUM FROM NITRIC ACID SOLUTIONS [41]

Operation	Phase	Concentration, *M*		Distribution coefficient of components			
		U(VI)	HNO_3	Pu	Ce, Pr	Ru, Rh	Zr, Nb
Extraction	Aqueous Organic	1.0 0.04	4.80 0.30	6.2	0.0002	0.0033	0.042
1st washing	Aqueous Organic	0.04 0.02	3.90 0.30	91	Not detd*	0.28	1.5
2nd washing	Aqueous Organic	0.02 0.005	3.90 0.29	190	"	0.1	63
Reextraction	Aqueous Organic	0.005 0.0005	— 0.20	0.004 Not detd*	"	Not detd*	85

* Distribution coefficients were not determined owing to the very low specific activity of solutions.

The purification coefficients of plutonium from the fission products are $2.3 \cdot 10^7$ (Ce), $2.7 \cdot 10^5$ (Ru), $2 \cdot 10^4$ (Zr, Nb). The purification from zirconium and niobium is improved if the extraction is carried out at a temperature higher than 25°. This is accompanied by a decrease in the distribution coefficients of zirconium and niobium. The extraction of plutonium also becomes poorer, but the deterioration is not as great. The change in the temperature is without significant effect on the extraction of ruthenium.

Trentinion and Chesne [697] utilized a solution of trilaurylamine in dodecane for the purification of plutonium.

Plutonium (IV) may be extracted with solutions of other tertiary amines in xylene. Some of the extractants are listed in Table 51.

Table 51

EXTRACTION OF PLUTONIUM WITH SOLUTIONS OF TERTIARY AMINES IN XYLENE

Amine	Distribution coefficient of Pu(IV)	Amine	Distribution coefficient of Pu(IV)
Tri-n-hexylamine	85	Triisooctylamine	300
Tri-n-heptylamine	92	Tridecylamine	115
Tri-n-octylamine	250	Trilaurylamine	140

Note: Composition of aqueous phase: 2 M HNO_3, 0.03 M $NaNO_2$, trace amounts of Pu(IV). Extractant—10% solution of amine in xylene. Conditions of extraction: equal volumes were contracted for 5 minutes at 25°; the phases were separated by centrifugation.

Extraction with quaternary ammonium salts

Quaternary ammonium salts are of major interest for the extractional separation and purification of plutonium.

Maeck et al. [528] studied the behavior of 57 metal ions including plutonium, uranium and neptunium during extraction with tetrapropylammonium, tetrabutylammonium and tetrahexylammonium salts. The authors assume that the compound being extracted is an associated ionic complex, containing a quaternary ammonium cation and one charged anion $[(R_4N)^+(M^{x+}A_{x+1})^-]$. It was found that the extraction of the elements depends on the nature and the concentration of the acid. Pu(IV) and Pu(VI) are readily extracted with these compounds from nitric acid media, while the extraction from hydrochloric acid solutions is less effective. There is no extraction of plutonium from sulfuric acid and hydrofluoric acid media. The degree of extraction increases with the increase in the acid concentration. The efficiency of extraction of plutonium depends on the identity of the quaternary ammonium salt employed (Figure 108). Tetrahexylammonium salts may be employed to separate plutonium from uranium, since hexavalent plutonium is, to all intents and purposes, not extracted by this reagent from solutions with nitric acid concentration below 1 M, while hexavalent uranium is readily extracted under these conditions.

Plutonium (IV) is fairly readily extracted from nitric acid solutions with hydrocarbons in the form of complexes with quaternary ammonium salts of the type $(CH_3)_2R^1R^2NX$, where R^1 and R^2 are aliphatic groups with 8, 10 and 12 carbon

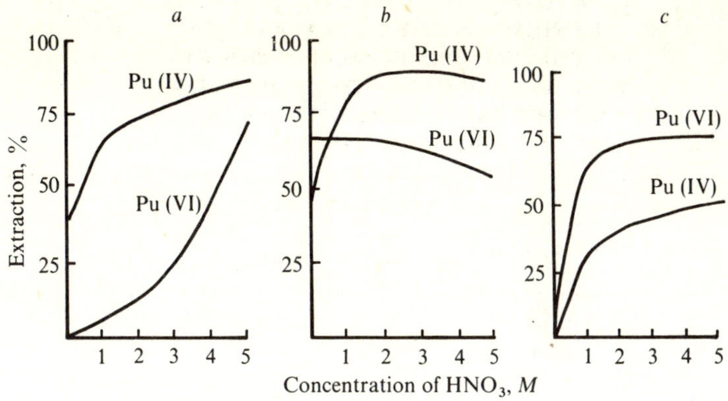

Figure 108

Extraction of Pu (IV) and Pu (VI) from nitric acid solutions with tetrahexylammonium (*a*), tetrabutylammonium (*b*) and tetrapropylammonium (*c*) salts

atoms [65]. As for other amines, the distribution coefficient is a function of the concentration of the reagent (Table 52).

Plutonium may be separated from uranium, but if the latter is present in large amounts, the degree of extraction of plutonium becomes much lower (Table 53).

With the increase in the acidity the distribution coefficients of Pu(IV), U(VI) and of the fission products increase. In a neutral medium all plutonium and uranium remain in the aqueous phase.

Table 52

EFFECT OF THE CONCENTRATION OF DIDECYLDIMETHYLAMMONIUM NITRATE IN BENZENE ON THE EXTRACTION OF Pu(IV) AND U(VI) FROM 8 *M* HNO_3

Concentration of reagent, %	Distribution coefficient	
	Pu(IV)	U(VI)
1.0	0.4	0.012
2.0	0.5	0.022
3.0	2.2	0.034
4.0	5.0	0.044
5.0	8.0	0.053

Note: Aqueous phase: plutonium(IV) nitrate, concentration of uranium 300 g/liter, acidity 0.8 *M* HNO_3.

Table 53

EFFECT OF THE CONCENTRATION OF URANIUM ON THE EXTRACTION OF PLUTONIUM AND FISSION PRODUCTS WITH 10% SOLUTION OF DIDECYLDIMETHYLAMMONIUM NITRATE IN BENZENE [65]

Content of uranium, g/liter*	Distribution coefficient of	
	Pu(IV)	β-activity
125.0	50.0	$5.5 \cdot 10^{-3}$**
250.0	45.0	$3 \cdot 10^{-3}$
375.0	6.1	$1 \cdot 10^{-3}$
500.0	1.4	

* Aqueous phase: solution of uranium irradiated for one year and then held for 6 months, in 0.08 *M* HNO_3.

** Residual β-activity originated mainly from ruthenium.

Goldschmidt, Regnaut and Prevot [65] extracted plutonium with quaternary ammonium salts under technological conditions, as one of the stages of concentration of plutonium extracted from uranium slugs. Plutonium was extracted from 0.8 *M* nitric acid solution with 5% solution of dimethyldidecylammonium nitrate in benzene. The extraction of plutonium was quantitative.

Boase, Foreman and Drummond [302] described an accurate and reproducible method for the determination of plutonium in nitric acid solutions of irradiated uranium. Tetravalent plutonium was extracted with a solution of Hyamine 1622 in benzene. Hyamine 1622 is p-diisobutylphenoxyethoxyethylbenzyldimethylammonium chloride [599]. The distribution coefficient of Pu(IV) is more than 200, that of Pu(VI) is 0.2 and that of Pu(III) less than 0.2. The acidity which the solution must have for the maximum possible extraction of Pu(IV) depends on the concentration of uranium in the sample (Table 54).

If the uranium concentration is high, and the acidity is higher than 4 *M*, the extraction of plutonium deteriorates considerably. A certain amount of uranium is extracted together with plutonium, but most of uranium and most β-activity may be removed by washing the organic phase with 4 *M* HNO_3.

Two ml of the sample solution, containing up to 300 mg/ml of uranium, 50–150 μg/ml of plutonium and the fission products, is placed in a separating funnel and the acidity is adjusted in accordance with the concentration of uranium. In order to convert plutonium to Pu (IV), 0.15 ml of 1 *M* sodium nitrite solution and 0.3 ml of freshly prepared 10% solution of iron sulfamate are added. The solution is carefully mixed and shaken twice with an equal volume of 1 % solution of the amine in benzene. The extracts are combined and washed with a solution of 4 *M* HNO_3.

Losses of plutonium are 1.5–2 % in the aqueous phase after the first extraction, 0.1 % in the aqueous phase after the second extraction and 0.15–0.2 % in the wash liquor. The purification factor from uranium is 100, while from the fission products it is 10^5.

Table 54

OPTIMUM ACIDITY FOR THE EXTRACTION OF Pu(IV) WITH HYAMINE 1622 IN DEPENDENCE OF THE CONCENTRATION OF URANIUM IN THE INITIAL SOLUTION [302]

Concentration of uranium, mg/ml	Concentration of HNO_3, *M*	Pu(IV) extracted, %
< 10.0	4.0	99.6
100	2.0	98.0
300	1.0	97.0

Heffner, Nilsson and Hultgren [234] studied the effect of several factors on the extraction of plutonium (VI) with tetrabutylammonium nitrate in chloroform. The extraction of plutonium is directly proportional to the concentration of TBAN. Thus, for instance, when the concentration of TBAN is increased from 0.2 to 25 mg/ml, the extraction of plutonium increases from 0.3 to 97.7 %. The dependence of the extraction of plutonium and uranium on the acidity of the water phase is shown in Figure 109. Under certain conditions the TBAN extraction may be used to separate Pu(VI) from the bulk of uranium.

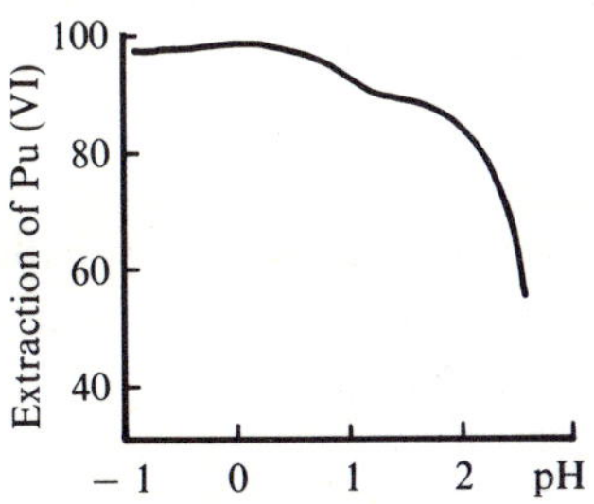

Figure 109

Extraction of Pu (VI) with a solution of tetrabutylammonium nitrate from HNO_3 and NH_4NO_3 (7 *M*). Concentration of TBAN 25 mg/ml. Concentration of Pu (VI) 4700 cpm/ml.

Taube [689a] studied the effect of the concentration of nitric acid on the extraction of the complex of plutonium (IV) with tetrabutylammonium nitrate with mixed solvents. The author found that, depending on the identity of the solvent, the distribution coefficient of plutonium reaches a maximum value at a certain concentration of nitric acid. Thus, for instance, the maximum values of the distribution coefficient of plutonium correspond to 5–6 *N*, 4–5 *N*, 4 *N* and 5 *N* nitric acid when the solvents used are, respectively, benzene or benzene–toluene mixture, mixture of benzene with chloroform or dibromomethane, chloroform or mixture of chloroform with dibromomethane, and mixture of toluene with chloroform or carbon tetrachloride with chloroform.

CHROMATOGRAPHIC SEPARATION OF PLUTONIUM

Methods based on ion-exchange chromatography and partition chromatography have become very important in plutonium technology and in analytical control. A recent development is the separation of plutonium on inorganic sorbents, which are attractive because of their stability to radiation and the simplicity of preparation.

ION-EXCHANGE CHROMATOGRAPHY

The synthesis and the investigation of ion exchangers and the theory and practice of ion-exchange chromatography have been repeatedly described [63, 176, 195, 196, 199, 219].

In ion-exchange separation of elements, ions are selectively adsorbed from solution by replacement of the active ionic groups of the ion exchanger; they are then eluted (desorbed, stripped) from the ion exchanger.

The sorptive capacity of an ion exchanger is characterized by its total exchange capacity, defined as the number of milligram-equivalents which can be sorbed on one gram of the dry exchanger. For practical purposes another parameter is more important, viz., the dynamic break-through capacity, which will depend on the total capacity and the experimental conditions (concentration of element in solution, temperature, height of ion-exchanger column, rate of percolation of sample solution, etc.). For methods of determination of exchange capacities, see [196].

One of the important parameters of the absorption of the element on the ion exchanger is the distribution coefficient K_d:

$$K_d = \frac{\text{amount of element adsorbed by 1 g of exchanger}}{\text{equilibrium amount of element in 1 ml of solution}}.$$

The value of K_d is a measure of the capacity for the separation of different ions from one another under the given conditions. The conditions of sorption are usually so chosen that the distribution coefficient of plutonium is high, while the coefficients of extraneous elements are low. Less frequently, plutonium is isolated by adsorption of the impurities.

The separation of elements markedly improves if different techniques are employed in their elution from the exchanger. Thus, for instance, in cation-exchange methods, desorption is often carried out with solutions of complex-forming reagents. In anion-exchange methods reductive desorption is also frequently employed.

The published papers on the ion-exchange behavior of plutonium largely refer to investigations carried out under industrial conditions. Nevertheless, these results are undoubtedly of use also in the analytical work performed in the laboratory.

Cation exchange

Plutonium in all its valency states is readily sorbed on cation exchangers from dilute acid solutions. Ion-exchange resins with strongly acid sulfonic groups are mainly employed. The selectivity of the sorption on such ion exchangers is poor. Separation from most elements takes place at the elution stage, carried out with specially selected eluents.

Of the cation-exchange resins manufactured in the Soviet Union, KU-1 and KU-2 brands were tried out; the foreign resins tested include Dowex-50 and C50. Their total exchange capacity is between 4–6 mg-equiv/g.

It will be seen from Figure 110 that the sorption of plutonium strongly depends on the brand of the resin and also on the ionic strength of the solution [448]. Clearly, cation exchange will not take place if the concentration of the electrolyte is such that plutonium is present as anionic complex which is not sorbed on a cation-exchange resin. This is the reason for the decrease in K_d which is noted when the concentration of nitric or hydrochloric acid is increased (Figure 110).

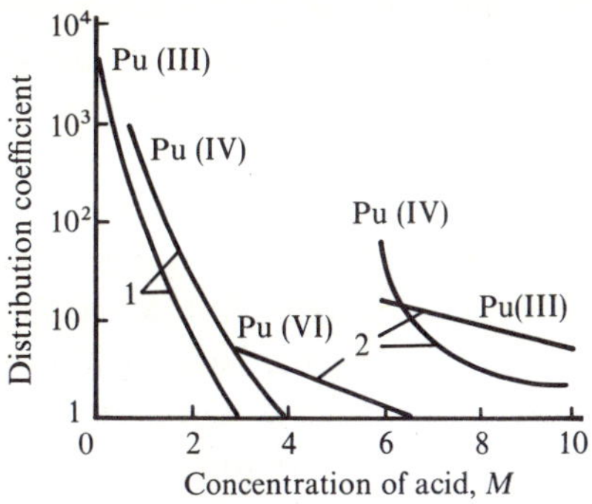

Figure 110

Distribution coefficients of plutonium ions as a function of acid concentration:

1) C50 cation exchanger, nitric acid; 2) Dowex-50 cation exchanger, hydrochloric acid.

Khalturin (1958–1959) studied the sorption of plutonium on the KU-2 cation-exchange resin from nitric acid solution and showed that in the HNO_3 concentration range from 0.2 to 1 *M* tetravalent plutonium is more readily adsorbed than Pu(III). The sorption of Pu(IV) deteriorates at acidities of less than 0.2 *M* or more than 1 *M* HNO_3. This is caused, respectively, by hydrolysis of plutonium and by complex formation with nitrate ions. According to Khalturin, the ionic strength of solutions in sorption from nitrate media should not exceed 0.5–0.7. The conditions for the sorption from hydrochloric acid solutions are similar.

In practice, plutonium is most frequently sorbed in its trivalent state. Even though tetravalent plutonium is more fully adsorbed, it is difficult strip from Dowex-50 or KU-2 even with 5.7 *M* solution of HNO_3 [696]. It is more convenient to sorb Pu(III) during its separation from uranium. When U(VI) is selectively eluted with dilute sulfuric acid, Pu(III) is almost totally retained by the resin, while Pu(IV) is rapidly eluted together with U(VI), due to the formation of stable complexes.

Khalturin also studied the effect of temperature on the time required for equilibrium to be attained during the sorption of Pu(III) from nitric acid solutions. When the temperature was raised from 20 to 60°, the time of establishment of equilibrium decreased from 100 to 15 minutes.

Cation-exchange resins of KU-2 and Dowex-50 type usually sorb from moderately acid (about 0.5 *M*) solutions of perchloric, nitric and hydrochloric acids, in addition to Pu(III), Pu(IV) and Pu(VI), the ions of other elements usually present together with plutonium, viz., Th^{4+}, U^{4+}, UO_2^{2+}, Np^{4+}, NpO_2^{+}, NpO_2^{2+}, Am^{3+}, fission products, Fe^{3+} and Cr^{3+}. Solutions of complex-forming reagents are usually employed to effect a selective elution of plutonium in the presence of various extraneous elements.

The elution of plutonium and other actinides with hydrochloric acid of various concentrations was studied by Diamond, Street and Seaborg [380]. Trace amounts of elements were sorbed on Dowex-50 in a 10 cm tall column. The elution was carried out with a solution of HCl at a linear rate of 0.1 cm per minute. The location of the elution peaks (amount of milliliters of eluate from the beginning of the elution to the maximum ion concentration in the eluate) is shown in Figure 111 as a function of the concentration of hydrochloric acid. All ions were desorbed by the first portions of HCl when its concentration was above 9 *M* (Th^{4+} was the only one not to be eluted on passing 1000 ml 12.2 *M* HCl). At lower acidities the peaks separate out. It will be seen from Figure 111 that the best separation of plutonium from thorium, uranium and neptunium is attained when the elements are present as Pu(III), Th(IV), U(VI) and Np(V), and 3–4 *M* HCl is used as diluent.

The desorption of trivalent transplutonides and of lanthanides by hydrochloric acid was studied in [340, 683]. Actinide elements are the first to be eluted.

The conditions for the stripping of trivalent plutonium with nitric acid are similar to those used with hydrochloric acid. Plutonium must be stabilized in nitric acid

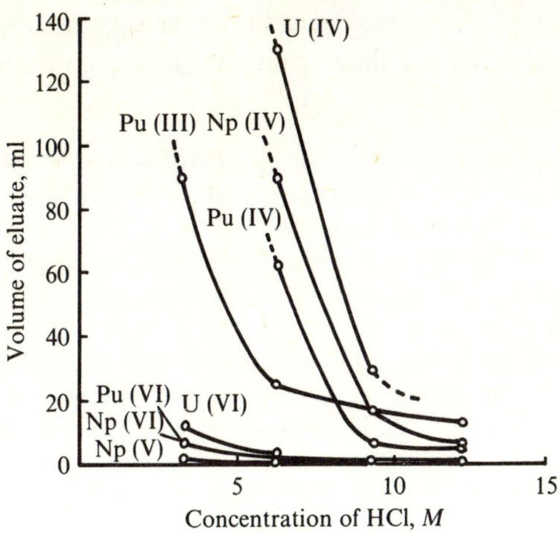

Figure 111

Location of elution peaks of actinide elements in dependence of the HCl concentration in the eluent. Cation exchanger: Dowex-50.

solutions as Pu(III) by the addition of excess hydroxylamine. The oxidation of Pu(III) during the elution with concentrated solutions of nitric acid is prevented by adding sulfamic acid to the eluent.

The technological method of Tober [696] for the purification and concentration of plutonium from nitric acid solutions is given below; its principles are also applicable to analytical work.

The sample solution, containing 0.5 g/liter of Pu (III), 0.5 g/liter of U (VI), fission products and 0.025 *M* hydroxylamine in 0.25 *M* HNO_3, is passed at a rate of 30 ml/min · cm^2 through a 36 cm tall column with a diameter of 17.5 cm, containing Dowex-50 × 12 in the H-form (50–100 mesh). Plutonium, uranium and radioactive zirconium and niobium are totally adsorbed, as well as about 40% ruthenium. The resin is washed with 0.25 *M* H_2SO_4 (12 volumes of the column) passed at a rate of 5 ml/min · cm^2, when the bulk of uranium and about 90% each of Zr^{95} and Nb^{95} are eluted. After 24 column volumes of 0.25 *M* H_2SO_4 have been passed, the purification coefficient from these fission products is about 25, i.e., increases by a factor of 2.5. Trivalent plutonium is then eluted with 5.7 *M* HNO_3 (containing 0.3 *M* sulfamic acid) at a rate of 0.5 ml/min · cm^2. The eluate contains, in addition to plutonium, considerable amounts of desorbed ruthenium. The overall coefficient of purification from Ru^{106} is ~ 2–3. The elution of Zr^{95} and Nb^{95}, which have been adsorbed by the resin together with plutonium, is insignificant, and these isotopes accumulate in the resin phase. After a few cycles have been effected, these elements are removed by passing five column volumes of 0.5 *M* oxalic acid at a rate not faster than 0.2 ml/min · cm^2.

The solution obtained contains 50–60 g/liter of plutonium, 4.7 *M* HNO_3, 0.3 *M* sulfamic acid and 0.35 *M* hydroxylamine nitrate. The amount of uranium is less than 0.001 g per gram of plutonium.

A similar method of purification of plutonium was employed by Groh [436] and Bruce [322].

Dizdar and DenBoer [382] employed cation exchange on Dowex-50 as one of the stages in the processing of irradiated uranium for the concentration and purification of plutonium.

The first stage consisted of the extraction of uranium and plutonium (5 mg plutonium per 1 kg uranium) with a mixture of 30% TBP and 70% kerosene. As a result of the reductive reextraction with hydrazine, 50 ml aqueous solution were obtained, containing $2.8 \cdot 10^{-6}$ *M* Pu (III); $3.2 \cdot 10^{-2}$ *M* U (VI); 0.23 *M* hydrazine and 1 *M* HNO_3. The solution also contained some β-active fission products. The solution was diluted to pH 0.27 and passed through a column containing one gram of air dry cation-exchange resin in the H-form (160-200 mesh) at a rate of 1 ml/min · cm^2. The solution at the issue from the column contained 82% of the original amount of hydrazine and 69% uranium. After washing the column with 0.1 *M* HNO_3, a solution of 1.5 *M* HNO_3 was passed through it. Hydrazine and uranium were stripped with the first portions of the eluate (about 30 ml), while the following 100 ml contained plutonium and some β-activity (Ru^{106}–Rh^{106}). When the resin was then washed with 5 *M* HNO_3, most fission products were desorbed. Similar results were obtained in the elution of Pu (III) with 1.5 *M* HCl.

Zolotov and Nishanov [103] described the chromatographic isolation of microgram amounts of neptunium and plutonium from irradiated uranium.

The irradiated U_3O_8 is dissolved in the minimum volume of hot 4*M* HNO_3 and the solution is diluted with water until the nitric acid concentration is 1 *M*. Np (V), Pu (IV) and U (VI) are sorbed on KU-2 cation-exchange resin (average grain size 0.1 mm) in a 35 × 0.5 cm column. Thirteen ml of 1 *M* HNO_3 are then passed through the column at a rate of 0.2 ml/min, when Np (V) and a small part of the fission products are eluted. Hexavalent uranium is eluted with 1 *M* HNO_3 only much later (after passing 19 ml 1 *M* HNO_3). Tetravalent plutonium is eluted with 3 *M* HNO_3.

The authors also showed that trace amounts of neptunium can be separated in this way from weighable amounts of plutonium and uranium. The original solution (0.5–0.8 *M* HNO_3) is first treated with sodium nitrite with prolonged heating, in order to convert the plutonium to Pu(IV) and neptunium to Np(V).

Zagrai and Sel'chenkov [96] developed a method for the separation of plutonium and neptunium in microamounts on KU-1 and KU-2 cation exchangers by eluting with dilute solutions of hydrofluoric acid. They found that when the concentration of HF is increased from 0.02 to 0.05 *M*, the distribution coefficient under stationary conditions decreases from 10^5 to 10^3 for Pu(III) and from 10^3 to 8 for Np(IV).

In a conical test tube are placed 6–8 ml sample solution, 0.25 *M* HCl containing microgram amounts of neptunium and plutonium, about one-half of KU-1 or KU-2 resin in the H-form (0.3–0.4 mm fraction) from a 9.0 × 0.25 cm Plexiglas column and 1–2 ml water. The test tube is placed on a boiling water bath and SO_2 is passed through the solution for 15–20 minutes in order to convert plutonium to Pu(III) and neptunium to Np(IV). The mixture is cooled, the resin is transferred to the column with a pipet and washed a few times with 10 ml portions of 0.25 *M* HCl and with water. Neptunium is eluted first by passing 40 ml (for KU-1) or 60 ml (for KU-2) of a 0.02 *M* solution of HF. Plutonium is removed from the resin with 4–5 ml of 0.5 *M* HF. Each fraction is collected in a platinum dish or a Teflon beaker and is analyzed radiometrically.

Fritz et al. [422] studied the elution of the ions of various elements from Dowex-50 × 8 with hydrofluoric acid. By varying the acid concentration and the ionic form of the resin (H-form or $EDTA^{2+}$ form) the elements may be separated into a number of groups.

The methods given above are based on the selective elution of the sorbed ions with solutions of mineral acids. The scope of the method may be considerably extended if organic complexing agents are used instead.

Tober [696] considers that Pu(III) should be eluted with ammonium lactate rather than with nitric acid. Before passing the solution containing 3.25 *M* ammonium lactate at pH 5.3, the column should be washed free from hydrogen and hydroxylamine ions. This is achieved by selective desorption of U(VI) with a mixture of $(NH_4)_2SO_4$ (0.085 *M*) and ammonium sulfamate (0.015 *M*) which acts similarly to 0.25 *M* H_2SO_4. Trivalent plutonium is 99.95% eluted at 60° if the ammonium lactate eluent is passed at a rate of 0.4 ml/min · cm^2.

Bubernak et al. [325] separated a mixture of trace amounts of Pu(III), americium, zirconium, yttrium and rare-earth elements by eluting with a saturated solution of EDTA at pH 2.30, which contained 5% ammonium nitrate, in a 19 × 0.5 cm column filled with Dowex-50 × 4 (200–400 mesh) in the H-form. Plutonium and zirconium are eluted together with the first portions and then yttrium, americium, samarium, promethium and neodymium, in that order. Praseodymium, cerium and lanthanum remain on the column and may be eluted with a solution of the same composition at pH 2.60.

Selective sorption of Pu(III) from solutions containing oxalates, tannates, citrates and tartrates at pH 2.4–4 has been briefly reported [291]. The fission products remain in solution.

Methods involving a combined utilization of cation and anion exchangers will be discussed in the section after the next.

Anion exchange

The ability of plutonium ions to form negatively charged complexes is employed in ion-exchange separation of plutonium from the accompanying elements. Studies of the anion-exchange behavior of plutonium and other metals were reported by Hyde [230], Kraus and Nelson [123], Ryan and Wheelright [623–627], and Nikol'skii et al. (1956–1960). The exchangers mostly used in these studies were the strongly basic anion exchangers of Soviet manufacture (AM-1 and AV-17) and of foreign manufacture (Dowex-1, Dowex-2, Amberlite IRA-400, De-Acidite FF). They are copolymers of polystyrene and divinylbenzene, with active tetramethylammonium groups introduced into the network.

Especially popular is the sorption of tetravalent plutonium from solutions of nitric and hydrochloric acids and their salts, in which Pu(IV) may add up to six nitrate or chloride ions. The behavior of hexavalent plutonium is similar, since in concentrated solutions of nitric and hydrochloric acids it is mainly present as the anions $[PuO_2(NO_3)_3]^-$ and $[PuO_2Cl_3]^-$ [624]. However, anion-exchange methods are not extensively applied to hexavalent plutonium, because of the difficulties involved in converting plutonium to the hexavalent state, and also because of its incomplete separation from impurities. In these media trivalent plutonium is present mainly as cation, and its sorption on anion exchangers is insignificant. This property of trivalent plutonium makes it possible to separate plutonium from other tetravalent actinides—Th(IV), U(IV) and Np(IV), which behave similarly to Pu(IV).

Plutonium may be separated from numerous elements by sorption on anion exchangers from solutions of nitric or hydrochloric acid.

Sorption from nitric acid solutions

The mechanism and kinetics of sorption of tetravalent plutonium on Dowex-1 × 4 from nitrate media have been described by Ryan and Wheelright [623, 626].

The process of complex formation and sorption in solutions of nitric acid may be written down in a simplified manner as follows:

$$Pu^{4+} + 6NO_3^- \rightleftharpoons [Pu(NO_3)_6]^{2-}, \tag{1}$$

$$[Pu(NO_3)_6]^{2-} + 2H^+ \rightleftharpoons H_2[Pu(NO_3)_6], \tag{2}$$

$$[Pu(NO_3)_6]^{2-} + 2NO_{3\,(c)}^- \rightleftharpoons [Pu(NO_3)_6]_{(c)}^{2-} + 2NO_3^-, \tag{3}$$

where *c* denotes the phase of the ion exchanger.

In solutions of 4 *M* HNO_3 the amount of the hexanitrate complex is still so small that it cannot be detected by spectroscopic or electromigratory methods. As the concentration increases, a successive addition of nitrate groups, up to a maximum of six, takes place. With the formation of the doubly charged anionic complex at higher acidities, reaction (2) begins to be more important, while in concentrated nitric acid solutions (more than 9 *M*) the hexanitrate complex is almost fully present as $H[Pu(NO_3)_6]^-$ and $H_2[Pu(NO_3)_6]$. This notwithstanding, the sorption of Pu(IV) takes place both in moderately acid and in concentrated nitric acid solutions.

It was found by special experiments that the absorption spectrum of Pu(IV) in the anion exchanger phase is identical with the spectrum of $[(C_2H_5)_4N]_2Pu(NO_3)_6$ in acetonitrile. Moreover, the capacity of the anion exchanger for plutonium is close to its capacity for a bivalent ion. It is therefore most probable that the ion $[Pu(NO_3)_6]^{2-}$, and not $[Pu(NO_3)_5]^-$ or $H[Pu(NO_3)_6]^-$ displays the highest tendency to be sorbed. Clearly, the affinity of ion $[Pu(NO_3)_6]^{2-}$ for the anion exchanger results in its considerable sorption in relatively weakly acid, as well as in strongly acid, solutions, even when its equilibrium concentration is low.

The effect of the concentration of nitric acid and calcium nitrate on the distribution coefficient is shown in Figure 112. As the nitrate concentration increases, so does the distribution coefficient in accordance with reaction (1). The maximum K_d value in nitric acid solutions is attained at an acidity of about 7.7 *M*, even though, according to Ryan [623], the proportion of the hexanitrate forms under these conditions is about 40%. At acidities higher than 7.7 *M* the distribution coefficient decreases owing to a decrease in the activity of the ion being sorbed. In calcium nitrate solutions containing a small amount of nitric acid, the effect of reaction (2) is much smaller, and the K_d-value is much higher as compared to solutions of pure nitric acid. However, the conversion of Pu(IV) into a complex form with the aid of salts such as calcium nitrate is not used in anion exchange, because of the low rates of sorption of plutonium under these conditions.

These authors [623, 626] also studied the effect of the temperature on the sorption process. It will be seen from Figure 112 that the distribution coefficient at room temperature is higher than at 60°, but that the rate of sorption increases at higher temperatures.

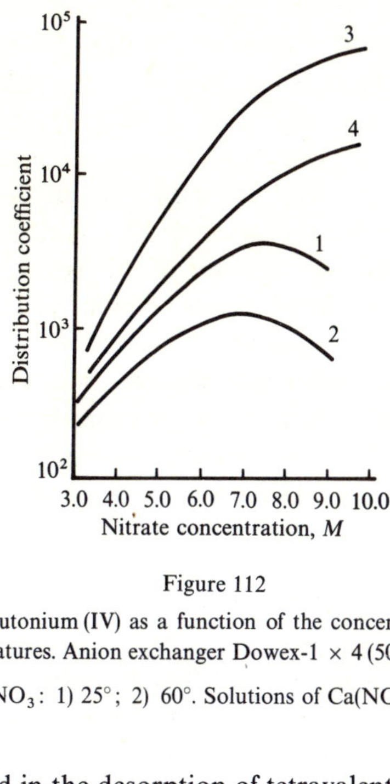

Figure 112

Distribution coefficients of plutonium (IV) as a function of the concentration of nitrate ions at various temperatures. Anion exchanger Dowex-1 × 4 (50–100 mesh).

Solutions of HNO_3: 1) 25°; 2) 60°. Solutions of $Ca(NO_3)_2$; 3) 25°; 4) 60°.

Two methods are used in the desorption of tetravalent plutonium from the anion exchanger. The more widespread technique is to shift the complex formation equilibrium by decreasing the concentration of nitrate ions in the eluent. It is preferable to elute at 60–70° rather than at room temperature in order to achieve a higher rate of desorption and a higher degree of concentration of plutonium. The second method consists in a reduction of Pu(IV) to Pu(III) by hydroxylamine. This method is more rarely employed because of the gas evolution which accompanies the reduction, because of the swelling of the resin in the column, and also because of the danger of decomposition of the resin by the reducing agent [626].

The first successful attempts at utilization of strongly basic anion exchangers of Dowex-1 type for the sorption of Pu(IV) from technological solutions were made by Campbell [334] and Aikin [265]. The reaction rates and the degrees of concentration attained were low.

The work of Ryan and Wheelright [626, 627] established the optimum conditions of isolation and purification of plutonium from impurities. The best ion exchanger for the purpose proved to be Dowex-1 × 4 in the nitrate form, with a grain size of 50–100 mesh. The acidity of the solution to be sorbed is 7.2–7.5 *M*; this corresponds to the maximum value of K_d. The temperature at this stage is kept at 50–60°, while the rate of flow of the feed solution should not exceed 20 ml/min · cm^2 at a concentration of plutonium ≧0.5 g/liter. The resin is washed with 20 column volumes of 7.2–7.5 *M* HNO_3, which is passed at approximately the same rate. The elution is carried out with 0.35 *M* HNO_3, since at a lower acidity there is the danger of hydrolysis of tetravalent plutonium. The eluent is passed at the rate ≦1.5 ml/min · cm^2 at 50–60°. The solution thus obtained contains up to 63 g Pu(IV) per liter. If the temperature is considerably raised, a better purification from U(VI) is achieved. Thus, when the concentration of uranium in the feed solution was about 20 g/liter, its concentration in solution after purification did not exceed $4 \cdot 10^{-5}$ g/liter.

These authors also studied the behavior of the fission products. It was shown [626, 627] that the degree of purification of plutonium from the fission elements depends on the acidity at which the dissolution was conducted and on the treatment of the sample prior to the sorption. It would appear that when the nitric acid content is low, the radioactive products are not fully converted into ions, as a result of which they are mechanically retained by the resin and subsequently contaminate the eluate during the desorption. This is evidenced by the fact that a better purification is attained in the case of solutions of samples in strongly acid media, and also when the feed solution is filtered. The coefficient of purification from γ-activity (mainly from the sum $Zr^{95} + Nb^{95}$) for the filtered solution is $6.4 \cdot 10^4$. The purification coefficients from other elements had the following values: $>7.7 \cdot 10^3$ (Al); $>5 \cdot 10^4$ (Ag); $>2 \cdot 10^4$ (Ca); $>2 \cdot 10^4$ (Cr); $1 \cdot 10^4$ (Cu); $4.4 \cdot 10^4$ (Fe); $>1 \cdot 10^4$ (Ni); $\sim 5 \cdot 10^3$ (Mg, Mn); $2 \cdot 10^4$ (K); $5 \cdot 10^2$ (Na). The behavior of plutonium and other elements during sorption from nitric acid solutions on anion exchangers may be approximately evaluated from the distribution coefficients given in Table 55.

Ryan and Wheelright [626, 627] also tested other ion exchangers—Permutite SK, Amberlite IRA-400 and IRA-401, as well as Dowex-1 with 1 and 8% cross-linking. Permutite SK has properties similar to Dowex-1 × 4. The effect of the exchanger brand on the distribution coefficient of Pu(IV) is shown in Figure 113.

Table 55

DISTRIBUTION COEFFICIENTS OF ELEMENTS IN HNO_3 SOLUTIONS AT 20–25°

Element	Concentration of HNO_3, M	Anion exchanger	K_d	References
Pu(IV)	8	Dowex-1 × 10	$9 \cdot 10^3$	[326]
	7.5	Dowex-1 × 4	$6 \cdot 10^3$	[626]
U(VI)	7.2	Dowex-1 × 4	7.8	[626]
	8	Dowex-1 × 10	11	[326]
	7	Dowex-2 × 8	8	[330]
Np(IV)	8	Dowex-1 × 10	$5 \cdot 10^3$	[326]
	7.5	AM-1 × 10	$1.5 \cdot 10^3$	Gel'man et al. (1960)
Th	7	Dowex-2 × 8	150	[330]
	8	Dowex-1 × 10	600	[326]
Pa(V)	8	Dowex-1 × 10	70	[326]
	7	Dowex-2 × 8	10.0	[330]
Zr	7	Dowex-2 × 8	0.9	[330]
	8	Dowex-1 × 10	3	[326]
Zr^{95}	11	De-Acidite FF	17	[594]
	7	"	2	[594]
Zr^{95}–Nb^{95}	7	"	24	[594]
Nb^{95}	11	"	30	[594]
	7	"	2	[594]
Ru	7	Dowex-2 × 8	10	[330]
	8	Dowex-1 × 10	3	[326]
Ru^{106}	11	De-Acidite FF	8	[594]
	7	"	12	[594]
Ce^{144}	11	"	3	[594]
	7	"	4	[594]
Au(III)	8	Dowex-1 × 10	700	[326]
Pd(II)	8	"	15	[326]
Tl(III)	8	"	12	[326]
Bi(III)	8	"	10	[326]
Re(VII)	8	"	3	[326]
Hg(II)	8	"	7	[326]
$\sum$REE	8	"	1–5	[326]

Note: Mo(VI), Rh(III), Hf, Pt(IV), Pb(II), and Nd(III) are weakly sorbed. Remaining elements are not sorbed at all [326].

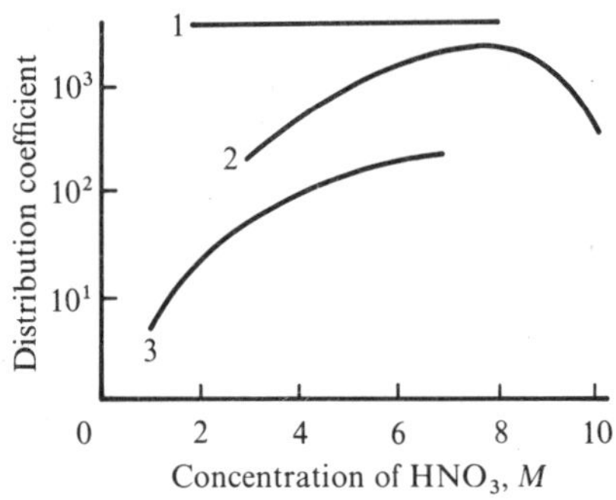

Figure 113

Distribution coefficients of plutonium (IV) as a function of the concentration of HNO_3:

1) De-Acidite FF; 2) Dowex-1; 3) Amberlite IRA-400.

Phillips and Jenkins [594] made a detailed study of the separation of Pu(IV) from fission elements under conditions of anion exchange on the resin De-Acidite FF (60–100 mesh). As a result of the sorption of Pu(IV) from 7 *M* HNO_3 onto 2 ml of the ion exchanger and washing with 200 ml of 7 *M* HNO_3, americium and the actinides following americium were fully removed, as well as all fission elements (Sr^{90}, Mo^{99}, Ru^{103} and Ru^{106}, Te^{132}, I^{131}, Cs^{132}, Ce^{141} and Ce^{144}), except for Zr^{95}–Nb^{95} which were retained by the anion exchanger to the extent of 10–15%. These isotopes can be eluted to the extent of 97–100% by a supplementary washing of the ion exchanger with solutions containing 0.001–0.1 *M* HF, 0.001–0.002 *M* $H_2C_2O_4$ or 0.002–2.5 *M* H_2SO_4. This operation may be accompanied by a 1–5% desorption of plutonium.

Tober [696] separated plutonium on Dowex 21K (50–100 mesh) anion exchanger from a solution containing 1 g of Pu(IV) per liter and 0.4 *M* Al(III), 1.3 *M* Mg(II), 0.2 *M* Ca(II), 5.5 *M* H^+, ~9.0 *M* NO_3^-, ~0.2 *M* F^-. Plutonium alone was sorbed on the column. Its elution with 0.3 *M* HNO_3 yielded a solution containing 45 gram of plutonium per liter; the content of other elements in solution did not exceed 10–20 mg/liter.

Davis et al. [80] sorbed Pu(IV) on Dowex-1, thus separating it from aluminum, iron and uranium, which form part of plutonium alloys. Less than 1% of each of these elements is retained on the resin.

Nikol'skii, Popik, Mordberg, Khalturin et al. (1956–1959) studied the ion exchange of Pu(IV) and other elements in nitrate media on USSR-made ion exchangers of AM-1 × 10 and AV-17 × 6 brands. The work was carried out with 8–10 cm long columns, 0.68–0.8 cm in diameter. A charge of an ion exchanger with a grain size of 0.5–1 mm weighed about 2 grams. The procedure for sorption and desorption (temperature, acidity, rates of passage of solutions) was similar to that described by

Ryan and Wheelright [626]. The capacity of the resin for plutonium at 60° and at an acidity of 7 *M* HNO_3 was 60–100 mg/g. Desorption carried out with a solution of 0.5 *M* HNO_3 at 50–70° is preferable to a reductive desorption with a weakly acid solution of hydrazine or Rongalite at room temperature. The coefficients of purification from β- and α-activity which could be attained were about $1 \cdot 10^3$ for ruthenium, zirconium and rare-earth elements. The extent of purification from niobium is smaller. The contamination of plutonium by fission elements decreases considerably if oxalic acid is added to the feed solution (~2 g/liter). It was also shown that the presence of oxalic acid (up to 20 g/liter) has no effect on the sorption of plutonium. A supplementary purification by a factor of 10–20 may be attained by filtering the solution through a silica gel column prior to the sorption on anion exchanger.

Gel'man, Kazantsev, Kondratov, and Kalinichenko (1957) isolated plutonium and neptunium from a solution of the following composition: 200 mg/liter Pu(IV), 17.5 mg/liter Np(V), 25 g/liter U(VI), 40 mg/liter Fe(III), 130 g/liter $NaNO_3$, fission products (Ce, Zr–Nb, Sr) and 7.5 *M* HNO_3.

One hundred ml of sample solution is passed at the rate of ~1 ml/min through a column containing one gram of AM-1 × 10 anion exchanger (grain size 0.25–0.5 mm) in the nitrate form. The column is then washed with 25 ml of 7.5 *M* HNO_3. Np(V), U(VI), Fe(III) and most of the fission products pass into the filtrate. The sorbed plutonium is stripped with 22 ml of 0.2 *M* HNO_3 to the extent of 98 %. The solution contains about 0.1 % of the initial amount of uranium and about 5 % of the original β- and γ-activity. The neptunium-containing solution, after reduction of Np(V) to Np(IV) by heating with hydrazine, is passed through another anion exchange column and the sorption and desorption of Np(IV) is conducted under the same conditions; the element is thus separated from uranium, iron and fission elements.

In all the above methods dilute solutions of HNO_3 were used as stripping solutions. In a number of studies [449, 594, 711, 729] reductive stripping was employed.

Winchester and Maraman [729] used the anion-exchange method to separate plutonium from long-lived fission products, present in plutonium-enriched nuclear fuel. The initial solution (2.7 *M* HNO_3) contained: 60.58 g/liter Pu, 1.26 g/liter Fe, 1.86 g/liter Co, 0.34 g/liter Zr, 0.17 g/liter Mo, 0.98 g/liter Ru, 0.36 g/liter Ce and a small amount of F^- and Na^+ ions.

Twenty ml of sample solution is evaporated to a volume of 10 ml. The residue is treated, successively, with hydroxylamine nitrate and sodium nitrite to stabilize Pu(IV) and the solution is acidified with 16 *M* HNO_3 to a final acidity of 8.3 *M*. Twenty-five ml of the solution thus obtained are placed in a flask with 10 grams of Dowex A-1 anion exchanger (250–500 mesh) in the nitrate form, and the contents of the vessel are stirred for one hour. The anion exchanger is separated by filtration, washed with three volumes of 8 *M* HNO_3 by 25 ml portions, and with one portion of water. The sorbent is then treated three times with a solution of hydroxylamine nitrate to desorb plutonium, filtering each time.

The analysis of the combined filtrate showed that the coefficients of purification of plutonium from impurities are: 111 (Fe); 300 (Co); 52 (Zr); 47 (Mo); >16 (Ru) and 44 (Ce). The purification from these elements by the same method using Amberlite IR-4B is much poorer. It was shown by the determination of the light absorption of

Pu(III) at 602 mμ [449] that the desorption of plutonium from Dowex-1 anion exchanger with 10% solution of hydroxylamine nitrate proceeds to the extent of 99.7–99.8%.

A better extraction yield of plutonium could be achieved [595] by using anion exchange to separate plutonium from cerium and cobalt in solutions of ternary alloys. When Pu(IV) was stripped with 5% solution of hydroxylamine hydrochloride in 4 *M* HCl, the anion exchanger De-Acidite FF was found to retain less than 0.025% of plutonium.

Sorption from hydrochloric acid solutions

In all probability, the sorption of Pu(IV) from hydrochloric acid solutions takes place by the mechanism described above for nitric acid media. The absorption spectrum of Pu(IV) in 12 *M* HCl is similar to that of $[(C_2H_5)_4N]_2PuCl_6$ in nitromethane and to that of Pu(IV) sorbed from 12 *M* HCl onto Dowex-1 × 1 [625]. This would appear to show that the anion-exchange resin sorbs plutonium as the anion $[PuCl_6]^{2-}$. Small amounts of Pu(IV) are captured by the anion-exchange resin Dowex 2 × 8 even from 2 *M* HCl (Figure 114), and in 4 *M* HCl solutions the distribution coefficient sharply increases [731]. The maximum on the curve giving the dependence of the distribution coefficient on the concentration of the acid, which is noted in 7.5 *M* solutions of nitric acid (Figure 112), is not noted in the case of hydrochloric acid up to 12 *M*. This is probably due to the fact that Wish [731] failed to attain the HCl concentration above which the formation of acid complexes, analogous to hexanitrates, is important (pp. 347–348). This suggestion is confirmed by the following results. In 8 *M* HCl solutions the amount of hexachloride is so small that

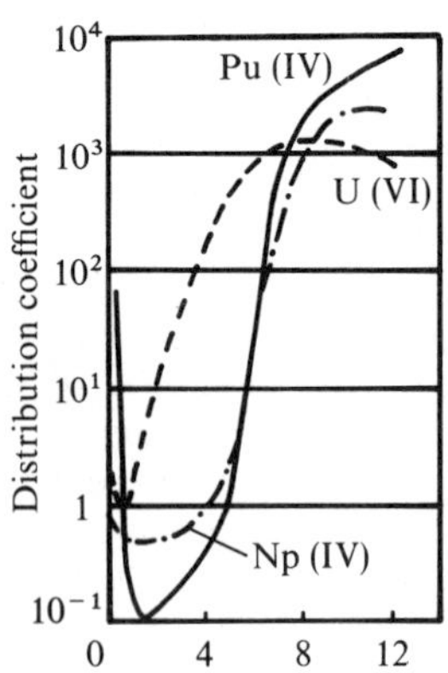

Figure 114

Distribution coefficients of Pu(IV), Np(IV) and U(VI) as a function of the concentration of HCl. Anion exchanger: Dowex-2 × 8

it cannot be detected spectrophotometrically, whereas in 8 M HNO_3 the content of the hexanitrate is as high as 40%. In 12 M HCl and 12 M HNO_3 these concentrations increase to 75 and 100% respectively. Thus, the maximum on the K_d–C_{HA} curve, which is noted for HNO_3 at 7.5 M, is shifted in the case of HCl towards higher acid concentrations.

The sorption on Pu(IV) on anion-exchange resins is usually conducted from 12 M HCl. There are no available literature data on the effect of the temperature, but it may perhaps be assumed that it will be similar to that in the sorption of Pu(IV) from 7–8 M HNO_3.

The elution is conducted with dilute solutions of HCl (about 0.5 M) and hydrochloric acid solutions of the reducing agents hydroxylamine and hydrogen iodide.

The distribution coefficients of elements in the sorption from hydrochloric acid solutions are given in Table 56.

Table 56

DISTRIBUTION COEFFICIENTS OF ELEMENTS IN CONCENTRATED SOLUTIONS OF HCl AT 20–25°

Element	Concentration of HCl, M	Anion exchanger	K_d	References
Pu(IV)	12	Dowex-2 × 8	$7 \cdot 10^3$	[731]
	12	Dowex-2 × 4	$8 \cdot 10^3$	[551]
	8	"	$1.4 \cdot 10^3$	[551]
U(VI)	10	Dowex-2 × 8	500	[330]
	12	"	800	[731]
	12	Dowex-1 (2)	$\sim 1 \cdot 10^3$	[123]
U(IV)	12	"	$\sim 10^3$	[123]
Np(IV)	12	Dowex-2 × 8	$2.5 \cdot 10^3$	[731]
Zr(IV)	12	"	37	[731]
	12	"	10	[330]
	12	Dowex-1 (2)	~ 100	[123]
Pa(V)	10	Dowex-2 × 8	200	[330]
Nb(V)	10	"	200	[330]
	12	"	$8 \cdot 10^3$	[731]
Mo(VI)	10–12	Dowex-1 (2)	85–100	[123, 731]
Ru(IV)	10–12	"	20–25	[123, 330]
Fe(III)	9	Dowex-1 × 4	$6 \cdot 10^3$	[551]
	12	Dowex-1 (2)	$4 \cdot 10^4$	[123]
Fe(II)	12	"	< 10	[123]
Cr(III)	12	Dowex-1 (2)	10	[123]
Co(II)	12	"	20	[123]
Cu(II)	12	"	10	[123]
Zn(II)	12	"	20	[123]
Ga(III)	12	"	$4 \cdot 10^4$	[123]

Table 56 (continued)

Element	Concentration of HCl, M	Anion exchanger	K_d	References
Ge(IV)	12	Dowex-1(2)	100	[123]
As(III)	12	”	10	[123]
Tc(VII)	12	”	~ 100	[123]
Rh(IV)	12	”	>1	[123]
Pd(II)	12	”	~ 20	[123]
Cd(II)	12	”	~ 10	[123]
In(III)	12	”	~ 1	[123]
Sn(IV)	12	”	$>10^3$	[123]
Sn(II)	12	”	~ 10	[123]
Sb(V)	12	”	$>10^5$	[123]
Sb(III)	12	”	~ 10	[123]
Te(IV)	12	”	>1	[123]
Hf(IV)	12	”	$\sim 10^3$	[123]
Ta(V)	12	”	$>10^2$	[123]
Re(VII)	12	”	~ 1	[123]
W(VI)	12	”	>10	[123]
Os	12	”	>100	[123]
Pt(IV)	12	”	~ 10	[123]
Au(III)	12	”	$>10^3$	[123]
Hg(II)	12	”	>10	[123]
Tl(III)	12	”	>100	[123]
Bi(III)	12	”	>1	[123]
V(IV)	12	”	$\sim 10^3$	[123]

Note: The following elements are sorbed weakly or not at all: Li, Na, K, Rb, Sr, Cs, Mg, Ba, Ra, Al, Be, Fr, Ac, Sc, $\sum$REE(III), Y, Mn(II), Ni and Th.

The behavior of the nearest neighbors of plutonium in the actinide series [230] is of interest. Tetravalent thorium does not form anionic chloride complexes and passes unchanged through a column filled with an anion-exchange resin [80]. This means that the separation of Pu(IV) from Th(IV) on an anion-exchange resin presents no difficulties.

Hexavalent uranium and neptunium are readily sorbed from HCl solutions above 6 M. To sorb Np(V) the HCl concentration in solution must be above 4 M. The sorption of Np(IV) is considerable even from 1 M HCl (Figure 114), whereas U(IV) is sorbed at acidities above 7 M. In concentrated hydrochloric acid solutions the sorptional properties of Pu(IV) are very similar to those of U(IV) or U(VI). The separation of these elements from one another may only be attained if plutonium is present as Pu(III), which is not readily sorbed on ion exchangers [600]. Similar considerations apply to the separation of plutonium from neptunium.

Hyde [230] points out two techniques for the isolation of plutonium and neptunium from hydrochloric acid solutions.

In the first variant, Pu(IV) and Np(V) or Np(VI) are sorbed from 12 *M* HCl onto Dowex-1 in the Cl-form. The solution in the column is then replaced by a solution of 0.1 *M* HI in 12 *M* HCl, which reduces Pu(IV) to Pu(III) and Np(V) to Np(IV) during 20 minutes at room temperature. At 80–90° the reduction is complete within one minute.* Pu(III) is removed by washing the column with concentrated acid, when Np(IV) is strongly retained by the resin and may be eluted with 0.5 *M* HCl.

In the other separation variant, Np(IV) is sorbed from a solution preliminarily treated with HI. Pu(III) is not sorbed at any concentration of HCl.

Jackson and Short [477] used this variant to separate 2.3 g of Np^{237} from 50 mg Pu^{239}.

Neptunium and plutonium hydroxides are precipitated from an acid solution, the precipitate is centrifuged and dissolved in 210 ml conc. HCl saturated with NH_4I. After 30 minutes reduction of plutonium to Pu (III), the solution is passed at 1 ml/min through 20 × 0.25 cm column of Deacidite FF. Traces of plutonium and other impurities are removed by passing 100 ml conc. HCl through the column. Neptunium is desorbed with 2 *M* HCl.

Anion-exchange separation of Pu(IV) from americium and transamericium elements (with a main valency of three) presents no special difficulties, owing to their low tendency to be sorbed on anion exchangers from concentrated solutions of HCl. Pu(IV), Am(III) and Cm(III) are sorbed on the strongly basic Dowex-1. Americium and curium are eluted by washing the resin with 12 *M* HCl, since their K_d-values are small. Tetravalent plutonium is eluted by any suitable method. Brody et al. [316] separated Pu(IV) from the impurities present in metallic plutonium by sorbing it onto Dowex-1 from 12 *M* hydrochloric acid (p. 365). The method cannot be used to separate and purify plutonium and americium (or curium) in solution together with U(VI), Fe(III) and certain other elements whose behavior is similar to that of Pu(IV). In such cases anion exchange is combined with precipitation and extraction methods [180, 318, 321, 466].

Gavrilov, Myasoedov and Khlebnikov [45] attained a high degree of purification of plutonium from small amounts of thallium, mercury, platinum, bismuth lead and iron by conducting an extraction with diethyl ether and an ion exchange on Dowex-1 from a solution of HCl. They noted that small amounts of nitrates (of the order of 0.01 *M* NO_3^-) decrease the coefficient of purification of plutonium from lead and bismuth by a factor of more than 100 in the ion-exchange stage.

Hoffman [465] gave a short survey of the separation methods for plutonium, uranium, neptunium, americium and fission products by sorption from hydrochloric acid media.

* The sorbed Pu(IV) may also be reduced to Pu(III) by sulfurous acid [585].

The separation of Pu(III) and transplutonium trivalent actinides by elution with 10 *M* LiCl was described in [473].

Wish [731] studied the behavior of U(VI), Np(IV), Nb(V), Zr(IV) and Mo(VI) which, because of their high distribution coefficients in 12 *M* HCl (Figure 114, Table 56), cannot be fully separated from Pu (IV). These elements could be separated by elution with a mixture of hydrochloric and hydrofluoric acids. A similar technique was used by Kraus and Nelson [122, 515] to separate U(VI) from W(VI) and Mo(VI). Hydrofluoric acid displays different competing effects on the formation of anionic chloride complexes and their sorption, depending on the relative concentrations of HCl and HF. Thus, for instance, the distribution coefficient of Pu(IV) in 8 *M* HCl (Dowex-2 × 8) is about 10^3, while in a mixture of 8 *M* HCl with 0.3 *M* HF this value is much less than unity.

In a method proposed by Wish [731], Pu(IV), Zr(IV), Np(IV), Nb(V), U(VI), Mo(VI) and Tc(VI) are sorbed on a column (15 × 0.2 cm) of the anion-exchange resin Dowex-2 × 8 (200–400 mesh) and eluted with solutions of HCl with HF of various compositions at a rate of ~ 0.3 ml/min in the following sequence: Zr (12 *M* HCl + 0.06 *M* HF), Np (6.5 *M* HCl + 0.004 *M* HF), Nb + Pu (6 *M* HCl + 0.06 *M* HF), U (0.1 *M* HCl + 0.06 *M* HF), Mo + Tc (12 *M* HNO_3).

A comparison of these methods shows that anion-exchange techniques in nitric acid solutions are the more selective. However, the inconvenience of working with highly radioactive substances (resulting from the decomposition of HNO_3 by radiation) and the need for a special stabilization of the valency forms of plutonium in nitric acid often make it necessary to have recourse to anion exchange in solutions of hydrochloric acid.

Sorption from other media

Negatively charged complexes of plutonium with sulfate, oxalate and other ions are known. However, anion exchange from sulfate, oxalate and other media is not frequently employed, since many other elements which usually accompany plutonium also form complexes with these ions.

Plutonium, neptunium and uranium are sorbed from sulfuric acid solutions onto strongly basic anion-exchange resins. According to Parker et al. [585], the values of K_d in 0.5 *M* H_2SO_4 (Dowex-1) are for Pu(III) less than unity, for U(VI) about 50. Clearly, the difference in the distribution coefficients is not large enough to attain a separation of U(VI) from Pu(III) by the sorption of uranium from this medium. Plutonium cannot be separated from neptunium by sorption from dilute sulfuric acid [465].

The sorption of Pu(III) and Am(III) onto Dowex-1 from 5 *M* NH_4SCN makes it possible to separate these metals from the rare-earth and from some fission elements [159, 344]. The method was employed for the analysis of samples of irradiated Pu [544]. The elution of plutonium and americium was effected with 0.1 *M* HCl.

Interesting results were obtained on studying the sorption of Pu(IV) on ion exchangers from organic solutions containing nitric acid [698a; Mikhailov, 1961].

Tuck and Welch [698a] showed that Pu(IV) is sorbed on the anion-exchange resin Amberlite IRA-400, with $K_d = 6 \cdot 10^3$ from Butex containing 0.4 M HNO_3. For the cation exchanger IR-120 K_d is $5 \cdot 10^4$ under these conditions. That the sorption on the anion-exchange resin is indeed of the ion exchange type is confirmed by the decrease in the value of K_d for sorption from Butex, containing 0.4 M $HClO_4$, to 10. It is assumed that the sorption reaction is

$$Pu(NO_3)_{4(b)} + 2NO^-_{3(i)} \rightleftharpoons [Pu(NO_3)_6]^{2-}_{(i)},$$

where b is the organic phase (Butex), while i is the ion-exchanger phase.

According to Mikhailov (1961), Pu(IV) may be sorbed on strongly basic anion exchangers from TBP solutions in CCl_4, containing 10–20% TBP, 0.25–0.6 M HNO_3 and $1.2 \cdot 10^{-5}$–$2 \cdot 10^{-2}$ M Pu(IV). Plutonium is sorbed to the extent of 99.3–99.8% by 0.5 gram of Dowex-1 × 4 (Cl-form, grain diameter 0.1 mm) from 10–12 ml solution after a contact time of 120–250 hours. The sorption on Dowex-2 × 8 is 99.0–99.5% complete; on AV-17 anion exchanger it is about 95%. This technique probably may be employed for concentrating plutonium.

Combined ion exchange

Purification of plutonium by successive ion exchange on cation and anion exchange resins was reported in a number of papers [600, 669, 711].

Aikin [265] reported an effective industrial purification scheme of plutonium, based on the sorption of Pu(IV) from nitrate solution first on an anion exchanger, and then, after desorption of plutonium with hydroxylamine, on a cation exchanger. A second anion-exchange cycle may be carried out. The method was employed by Watanabe [711] to isolate plutonium from irradiated uranium.

Sikkeland [669] proposed a somewhat different procedure for the separation of microamounts of plutonium from uranium and fission products.

The solution of the irradiated uranium, containing 0.19 M of U(VI), 0.1 M $NaNO_2$ to stabilize Pu(IV) and 0.5 M HNO_3, is passed through a small column charged with Dowex-50 (H-form, 200–250 mesh) at a rate of 0.4 ml/min · cm^2. Uranium and sodium are desorbed with 2 M HCl. Pu(IV) and some of the fission products are eluted with 5 M HCl. The plutonium-containing eluate is evaporated to a final acidity > 8 M. From this solution, Pu(IV) is sorbed on a column of Dowex-1 (Cl-form, 100–140 mesh), and the resin is washed with concentrated HCl. Plutonium is desorbed with a dilute HCl solution, or with a mixture of concentrated HCl and ammonium iodide; radiochemically pure Pu is obtained with a 99% yield.

Prevot and Regnaut [600] described a laboratory scale isolation of plutonium from a solution containing 2–10 grams Pu(III) per liter (stabilized with sulfamic acid), 0.5 g/liter U(VI), 30–40% Fe(III) calculated on the amount of plutonium, 0.5–0.7 M HNO_3 and Zr^{95}–Nb^{95}. The method comprises three ion-exchange cycles. In the first cycle, plutonium, uranium, iron and some Zr–Nb sorbed on C50 cation exchanger. The desorption of uranium and other elements is conducted with 5 *M* HCl, and the eluate obtained is run through a column charged with A300D anion exchanger to absorb uranium and iron (second cycle). The solution which has passed through the column and which contains Pu (III) is treated with sodium nitrite to oxidize Pu(III) to Pu(IV), the acidity is adjusted to 7 *M* HCl, and Pu(IV) is sorbed on a second anion-exchanger column. The resin is washed with 7 *M* HCl, and the Pu(IV) is desorbed with 0.7 *M* HCl (third cycle). The overall purification coefficients thus attained are: >2000(U), >700(Fe), ~ 100(Zr–Nb).

SEPARATION OF PLUTONIUM ON INORGANIC SORBENTS

Attention has been recently directed to the sorptional and exchange properties of sparingly soluble inorganic compounds (hydroxides, oxides and phosphates) of elements in Groups IV–VI of the periodic system. General information on the utilization of inorganic sorbents will be found in [30, 124, 391, 706]. As far as plutonium is concerned, most of these publications contain merely brief summaries.

The compounds tested as plutonium absorbents were SiO_2, TiO_2, Nb_2O_5, MnO_2, ZrO_2, CaF_2, zirconium phosphates and pyrophosphates, and mixed compounds of the zirconium phosphosilicate type.

Ahrland et al. [263, 264] consider silica gel as a weakly acid cation exchanger:

$$\geq Si{-}OH + Me^+ \rightleftharpoons \geq Si{-}OMe + H^+.$$

Silica gel was saturated with hydrogen ions by subjecting SiO_2 samples of Kebo brand (100–200 and 50–100 mesh) to a double treatment with solutions of 5 *M* HCl or 5 *M* HNO_3 at 80–90°. The gel was washed with water and dried for 20 hours at 100°.

It was found that the distribution coefficients of the various ions between the gel and the solution strongly increase with an increase in pH. For the same acidity value, the sorption decreases in the sequence Zr(IV) > U(IV) ≈ Pu(IV) > U(VI) > Ca ≈ Ba ≈ Na. The rate of sorption is a function of the ionic charge: it is slower for Pu^{4+} and U^{4+} than for UO_2^{2+}, Ca^{2+} and Na^+. It was shown [263, 264] that the K_d value of Pu(IV) in 0.1 *M* HNO_3 is much higher than in solutions of 0.5 *M* HNO_3. A solution of 4.5 *M* HNO_3 elutes Pu(IV) quantitatively, while zirconium is sorbed on silica gel from this medium. Tetravalent zirconium is held by the sorbent when the latter is washed with 5 *M* HNO_3. The behavior of Pu(IV) is sufficiently different from that of Zr(IV) for the separation of the two elements to be possible. U(VI) is not sorbed from 0.1 *M* HNO_3, but it is absorbed from weakly acid (pH ~ 3) solutions.

The isolation of uranium and plutonium from 100 ml of a solution of 4 *M* HNO_3, containing 9.17 millimoles Zr per liter, $2.39 \cdot 10^{-3}$ millimoles Pu(IV) per liter, and 199 millimoles U(VI) per liter, was carried out as follows.

The solution was passed through a 100 × 1 cm column, filled with silica gel (100–200 mesh), at a rate of 3.5 ml/hour · cm^2 to sorb zirconium. The sorbent was washed with 4 M HNO_3 to effect complete removal of U(VI) and Pu(IV) from the column. Zirconium was then eluted with 0.1 M solution of oxalic acid. The combined filtrate from the sorption of zirconium and washing of silica gel, containing uranium and plutonium, was evaporated to dryness, the volume is made up to 100 ml to obtain an acidity of ~ 0.1 M. This solution was run through the column, when Pu(IV) was sorbed. Hexavalent uranium passed into the filtrate and was completely removed from the column by washing it with 0.1 M HNO_3 (rare-earth elements are washed out of the column and pass into the filtrate together with uranium). Tetravalent plutonium was eluted with 4.5 M HNO_3.

A similar technique was employed by the authors to isolate plutonium from irradiated uranium, with the difference that the stage of sorbing Pu(IV) from 0.1 M HNO_3, desorbing it with 4.5 M HNO_3 and evaporating the eluate was repeated three times. In the evaporation process Pu(IV) seems to be partly oxidized, and the treatment of the residue with sodium nitrite proved insufficient to convert all the plutonium to Pu(IV). The result was that the yield of plutonium in the three-stage sorption did not exceed 50%. The coefficients of purification of the final product from γ- and β-activity were $5.4 \cdot 10^3$ and $2.65 \cdot 10^4$ respectively. If plutonium had been preserved in the tetravalent state, its yield would have been higher than 99%.

Beaton [292] sorbed Pu(IV) on pelleted titanium dioxide from a sulfuric acid solution at pH ~ 3.5. The completeness of separation from the fission activity depends on the technique of pelleting. Good performance was given by a mixture of 95% hydrated TiO_2 and 5% mica, ignited at 900–1000° for one hour. The sorbent when ignited at 600–800° sorbs less selectively. It is recommended to sorb plutonium on TiO_2 after cation-exchange purification on Amberlite-IR.

According to other data [373], TiO_2 may sorb fission products from acid solution, while Pu(VI) remains in solution.

Maddock and Boothe [526] reported the use as sorbent of calcium fluoride, obtained by sintering at 800°. Plutonium in a nitric acid solution of irradiated uranium is oxidized to Pu(VI) with potassium bichromate and the fission products are sorbed on CaF_2. The solution issuing from the column is treated with SO_2 to reduce Pu(VI) and Pu(IV) and Pu(III) is sorbed on a second column with CaF_2. Plutonium is desorbed with a saturated solution of ammonium oxalate or with 10–20% solution of Na_2CO_3. Prior to the sorption, soluble calcium salts are added to the solution to prevent dissolution of sorbent.

The report [370] on the utilization of manganese dioxide as sorbent is also of interest. Manganese dioxide was prepared by adding manganese sulfate (101.4 g $MnSO_4 \cdot H_2O$ in 1.0 liter of water) to a solution of permanganate (63.2 g $KMnO_4$ in 5 liters of water). After filtration the precipitate was washed first with hot water to remove hydrogen ions, then with 3 M HNO_3 and then again with water. The filtered precipitate was dried for 20 hours at 110°. Manganese dioxide was boiled for 5 minutes in 3 M HNO_3. Boiling was discontinued by adding cold water. The precipitate was filtered, washed with water and dried for 80 hours at 110°.

The 100–140 mesh fraction of MnO_2 was used in the work. The fraction contained a certain amount of finer particles, which were easily removed by washing the sorbent with water and decanting. Manganese dioxide was placed in a 2 ml column and washed first with 50 ml 1 *M* HNO_3, and then with water until free from hydrogen ions. The capacity of MnO_2 for zirconium is 24 ± 1 mg/g, calculated on the dry dioxide.

The authors carried out experiments on the sorption of trace amounts of UO_2^{2+}, PuO_2^{2+}, Pu^{4+}, Zr^{95}–Nb^{95}, Ru^{106}, Sr^{90}–Y^{90} and Cs^{137}–Ba^{137}. Portions of solution (100–200 μl, 0.001 *M* in HNO_3) were applied to the top of the column and MnO_2 was washed with nitric acid of various concentrations passed at a rate of 3 ml/min · cm^2. Microgram amounts of U(VI) and Pu(VI) are eluted almost quantitatively with $\sim$ 10 ml of 0.1 – 1 *M* HNO_3. Zr^{95} and Nb^{95} are eluted to within a few hundredths of one percent with 1 *M* HNO_3. The desorption of Ru^{106} increases from 0.2 to 7–17% as the acidity is increased from 0.1 to 5 *M* HNO_3. Cs^{137} and Ba^{137} are not eluted with 1 *M* HNO_3. Sr^{90} and Y^{90} are less firmly held by the sorbent. The overall coefficient of purification of Pu(VI) and U (VI) from Zr^{95}, Nb^{95}, Ru^{106} and Cs^{137} is about 130, if the elution of uranium and plutonium from the column is conducted with 10–20 ml of 0.1 *M* HNO_3.

A few papers deal with the sorptional properties of the compounds of zirconium: phosphate [311, 622], pyrophosphate [682] and phosphosilicates [313, 572].

According to Naumann [572], a mixed gel of zirconium hydroxide and silicic acid displays the properties of a cation exchanger with a large exchange capacity, which is unstable to acids. This disadvantage could be overcome by treating the mixture with phosphoric acid as stabilizer.

The gel obtained by adding ammonia to the weakly acid solution of Na_2SiO_3 and $ZrOCl_2$ is treated with a mixture of dilute HNO_3 and H_3PO_4, owing to which a phosphate-containing sorbent is formed. The properties of the sorbent do not significantly depend on the conditions of drying. The product of the composition $SiO_2:ZrO_2:P_2O_5 = 4:1:0.8$ behaves as a strongly acid cation exchanger with a capacity of 2.5 mg-equiv/g. Its properties do not alter as a result of being treated with $\sim$ 6 *M* HNO_3, and the reproducibility of its ion-exchange properties is satisfactory. Experiments on the separation of plutonium, uranium and fission elements were conducted in a 12.0 × 0.3 cm column, filled with the sorbent (grain size 0.06–0.1 mm).

A weighed sample (10 mg) of irradiated uranium is dissolved in 2 *M* HNO_3 or in a mixture of 2 *M* HNO_3 and 0.01 *M* H_3PO_4, and the solution is passed through a column. Hexavalent uranium is contained in the first portions of the filtrate, as well as most of the fission product. Pu(IV) remains on the sorbent and is desorbed with 7 *M* HNO_3 solution. The purification of plutonium markedly improves in the presence

The coefficient of purification from the sum total of fission products may reach $0.5 \cdot 10^3$; the purification coefficient from uranium is more than 10^4. The main

β-active impurity in the plutonium fraction is Cs^{137}. A satisfactory separation was also attained for the mixture of 10 mg uranium, 0.1 mg cerium and 0.005 mg plutonium in 4 ml of 2 M HNO_3. Qualitative elution of plutonium required about 10 ml of 7 M HNO_3. A similar technique was employed in [313].

The sorptional and exchange properties of inorganic compounds depend to a large extent on the manner of their preparation. They have a number of advantages over the organic ion-exchange resins: a higher stability to high-energy γ-radiation and to high temperatures.

The disadvantages of most inorganic sorbents include their relatively low capacities, low rates of sorption, and instability to prolonged action of concentrated mineral acids.

PARTITION CHROMATOGRAPHY

The methods of partition chromatography have much in common with the extractional separation of elements.

Extraction by partition chromatography is effected on special columns filled with a porous carrier (silica gel, paper pulp and similar chemically stable substances). A part of the carrier is impregnated with pure solvent (called the stationary phase); this should not be washed out by the moving solvent passed through the column, which is immiscible with the stationary solvent. This layer of the carrier is called the purification layer. The other part of the carrier (partition layer) is impregnated with the stationary solvent, containing the mixture to be analyzed, and is introduced into the top part of the column if the mobile solvent is passed downwards, and into the bottom part of the column if the solvent is passed upwards. The extraction of the components of the mixture on passing the mobile solvent will be more rapid, the smaller their distribution coefficients between the mobile and the stationary solvents.*

Partition chromatography on silica gel

Methods in which silica gel serves as carrier are the most frequently used. Other carriers are not employed, probably because the process in too slow. The theoretical principles of the method and the calculation of the efficiency of a column were developed by Markov et al. [156, p. 106]. These authors also studied the application

* Usually, aqueous solutions are used as the stationary solvent, while an organic solvent—diethyl ether, TBP, etc.—is employed as the mobile solvent. The corresponding expression for the distribution coefficient is:

$$K_d = \frac{\text{concentration of component in stationary (aqueous) phase}}{\text{concentration of component in mobile (organic) phase}}$$

of partition chromatography to the purification of plutonium from uranium and from other impurities. The high extractability of Pu(IV) and Pu(VI) from aqueous nitric acid solutions with oxygen- and phosphorus-containing organic solvents makes it possible to separate plutonium from Fe, Cr, Al, Mn, alkali and alkaline-earth metals and lanthanides. The separation of plutonium from uranium is based on the low extractability of Pu(III) as compared with U(VI).

The preparation and the quality of silica gel and the technique employed in charging it into the column are very important. The preparation of pelleted silica gel under laboratory conditions has been described in detail [9, p. 334; 156, pp.111–118].

According to Markov [156, p. 119], the large-pore silica gel, manufactured by the Voskresenskii plant, may also be employed. It should be crushed prior to use, and purified by washing first with a hot mixture of HNO_3 and HCl, and then with a 5–10% solution of redistilled HNO_3, containing some H_2O_2, until the reaction for Ti(IV), Fe(III) and Cl^- is negative. The silica gel is then washed with water, held for 1.5–2 hours in 2–2.5% solution of ammonia, and dried at 100–150°. The dry silica gel is ignited for 30–60 minutes at 400–450° to remove organic matter and packed while hot into ground-glass stoppered jars. It is particularly important to ensure that the surface of silica gel, which is treated with the most suitable stationary solvent, be dry. The use of moist silica gel impairs the efficiency of the process. The impregnation of dry silica gel with a stationary solvent is carried out in the proportion of 1 ml solvent to 3 ml sorbent (large-pore, pelleted sorbent, grain size about 0.5 mm). One or two minutes after mixing, the silica gel again becomes loose. After the column has been charged with silica gel (purification and partition layer) the mobile phase is passed through. The same technique of gel impregnation and column charging was also adopted by Cvjeticanin [369].

Hultgren and Heffner [238] used a simplified technique for the preparation of the column and carrier. Chromatographic grade silica gel (50–100 mesh) is held prior to use in hot 5 *M* HNO_3 for about 2 hours, washed 8–10 times with water, and dried for 24 hours at 110° until water content becomes as low as 4–5%. Before use the column is filled with silica gel and water and the stationary aqueous phase is passed through; the excess of this phase is displaced by the organic solvent.

Markov's technique of charging the column with silica gel would appear to be more suitable.

The dimensions of the column and the volume of the mobile solvent required for the separation of two or more substances may be calculated from the values of distribution coefficients [156, p. 113]. It was found that with the increase in the surface area of the porous carrier and with the decrease in the rate of passage of the mobile phase the concentration maximum of the element on the elution curve becomes higher, while the volume of solvent required for quantitative elution becomes smaller.

Chromatography of plutonium may be of the oxidative or of the reductive type. In the former method, Pu(IV) and Pu(VI) are extracted into the mobile solvent phase, while elements with a higher K_d-value are sorbed. In the latter method, on the contrary, plutonium as Pu(III) is held in the stationary aqueous phase, while elements such as U(VI) are removed with the mobile solvent. Occasionally the two techniques must be applied in succession.

The oxidative variant (Markov, 1956) for the isolation of microgram amounts of plutonium from a solution containing 0.1–0.5 mg Pu/liter and up to 0.5 g/liter of U, Cr, Fe, Ca, Al, Mg, and Mn is given below. The β- and γ-activity of the solution was high.

A 0.2 ml portion of the sample is treated with 2–3 drops of concentrated HNO_3 and evaporated to dryness. To the dry residue a few $KMnO_4$ or $K_2Cr_2O_7$ crystals are added, together with 0.3 ml of 0.15 *M* HNO_3 saturated with NH_4NO_3. The solution is heated on the water bath for 3–5 minutes to oxidize plutonium to Pu(VI) and is mixed with 1–1.5 cm^3 of dry, pelleted, large-spore silica gel (grain size 0.5 mm). The impregnated silica gel is placed in a 20 × 0.8 cm column, and the beaker is washed with a small volume of the stationary solvent (saturated solution of ammonium nitrate in 0.15 *M* HNO_3). This solution is used to impregnate 0.5 cm^3 silica gel, which is also transferred to the column. This is the partition layer. The purification layer is prepared by wetting 3 cm^3 of the silica gel with 1 ml of the stationary solvent. Since the mobile phase (diethyl ether) is passed through the column in an upward direction (to effect a better elimination of air from the column), the partition layer is placed in the bottom of the column, while the purification layer is placed in the top of the column. The silica gel layer in the column is plugged from both sides with glass wool. Diethyl ether, which has been treated with a mixture of 0.15 *M* HNO_3 and 10 *M* NH_4NO_3, is fed through a tube joined with the bottom of the column through a ground glass joint. The mobile phase (30 ml) is passed at about 1 ml/min. Uranium passes into the ether phase together with plutonium.

Plutonium is determined radiometrically with an error not exceeding $+ 5\%$, since the contribution of uranium to the α-activity measured is small. In this way more than 99.5% of fission elements may be eliminated.

Markov et al. recommend the use of the same technique to isolate plutonium from solutions of Pu–Be and Pu–Be–Al alloys.

Radiometric determination of plutonium isolated from samples containing ~ 0.1 g/liter U and ~ 1 μg/liter Pu will be much less accurate and must be corrected for the α-count of uranium the concentration of which is found by an independent method such as photometry.

Solutions with a larger U:Pu ratio are analyzed by reductive chromatography.

To 0.5 ml of a solution of 0.5 *M* HNO_3 saturated with NH_4NO_3 are added 0.1 ml of 3 *M* $N_2H_4 \cdot HNO_3$ and 0.1 ml of 10% solution of $FeSO_4$. The mixture is stirred and then used to impregnate 2–2.3 cm^3 of dry silica gel (purification layer), which is placed in the column. Nitric acid solution of sample (0.5–5 ml) is evaporated to dryness and the residue mixed with 2–3 ml of acidulated diethyl ether. The organic extract is transferred to the column and filtered. The sorbent is washed 2–3 times with 3 ml portions of diethyl ether. During the percolation plutonium is reduced to Pu(III) and passes into the stationary solvent phase, while all the uranium remains in the ether phase. The sorbed Pu(III) is eluted with 10 ml of a warm solution of 2–2.5 *M* HNO_3.

The separation of uranium, plutonium and impurities is possible only if the oxidative and the reductive cycles are combined. The procedure includes an oxidative stage, during which the impurities which are not extracted with ether are retained in the stationary solvent, and a reductive stage during which plutonium is separated from uranium. The wide scope of the procedure may be illustrated by its application to exhaustive purification of large amounts of Np^{237} and Pu^{239} (Np:Pu $\cong$ 6) from americium, uranium, iron and aluminum. As a result of the first (reductive) cycle, an almost complete separation of U(VI) and Np(IV) takes place, since the the first portions of ethyl ether used as the extractant contain only uranium, while the following portions contain only neptunium. The content of plutonium, sorbed as Pu(III) on the column, in the neptunium fraction is 2000 times smaller than the original ratio. A second purification of neptunium, conducted in the same manner, yields a product from which even traces of plutonium are absent (Pu^{239} was absent from the α-spectrogram). Trivalent plutonium, sorbed on the silica gel, as well as other elements, were eluted by warm ~ 2.5 *M* HNO_3 and, after Pu(III) has been oxidized to Pu(VI) by potassium permanganate, Pu(VI) was extracted in an oxidative stage, during which Am(III) and Fe(III) remained on the silica gel.

Hultgren and Heffner [238] used partition chromatography in their separation of plutonium and uranium from 30% solution of TBP in kerosene.

Silica gel (50–100 mesh) is placed in a 25.7 × 2.2 cm column and a solution containing 1 g/liter Fe(II), 0.3 *M* $N_2H_4 \cdot HNO_3$ and 0.1 *M* HNO_3 which serves as the stationary phase, is passed through. When the organic solution containing 79.3 g/liter U, 22.3 μg/liter Pu and 0.049 *M* HNO_3 is now run through, plutonium is sorbed on the silica gel. Uranium is retained almost entirely in the organic phase, and is quantitatively removed from the column by washing with 400 ml of a solution of TBP containing 0.021 *M* HNO_3. Trivalent plutonium is eluted with 500 ml of 0.98 *M* HNO_3; its content in the eluate is 99.7 ± 0.5% of the original amount. The coefficients of purification of plutonium from radioactive isotopes of zirconium and ruthenium are 100 and 14 respectively.

For weakly acid TBP solutions, containing 0.05 *M* HNO_3 and 80–100 times less uranium than in the preceding example, a variant of the method was tried in which the stationary aqueous phase (0.1 *M* HNO_3) contained no reducing agents. Uranium was eluted with 880 ml of a solution of TBP containing 0.026 *M* HNO_3, while plutonium was eluted with 780 ml of a TBP solution containing 0.18 *M* HNO_3. The bulk of the ruthenium (~ 70% was held on the silica gel and could be eluted only by 0.95 *M* HNO_3. The plutonium fraction contained some 7% of the original amount of Ru^{106}–Rh^{106}.

The same authors [238] separated gram amounts of uranium from microgram amounts of plutonium by taking advantage of the poor extractability of Pu(IV) with very weakly acid (0.005 *M* in HNO_3) solutions of TBP. Uranium is eluted first, and then, after acidification of TBP, plutonium is also eluted. The purification coefficient of plutonium from Zr^{95} is almost 100.

In another work, Heffner and Hultgren [439] reported a method for the separation of plutonium from uranium with the aid of the mobile phase–ethylene glycol dibutyl ether or its mixture with kerosene (3:1). Similarly to the preceding method, it is based on the relatively low extractability of Pu(IV) with the organic solvent, containing about 0.35 *M* HNO_3. Plutonium is eluted with a new portion of the extractant, which is 1 *M* in HNO_3.

A method of the purification of Pu (IV), Pu (VI) and U (VI) from fission elements ($Zr^{95}-Nb^{95}$) with the utilization of a nitric acid solution of methyl isobutyl ketone as eluent has been described [369]. Zirconium and niobium are held by silica gel, but the filtrate issuing from the column is contaminated with ruthenium.

Gwozdz and Siekierski [437] separated the various valency forms of plutonium, using TBP as the stationary phase. Nitric acid (0.5–1.0 *M*) served as the eluent. About 0.3 g of carrier—Kieselguhr—preliminarily treated with dichloromethylsilane, and then 0.18 ml TBP are placed in a 11 × 0.3 cm column. A 0.02–0.04 ml volume of the acid solution containing ~25 mg Pu per liter is introduced into the top of the column, and the column is washed with 0.5–1 *M* HNO_3. If the acidity is less, the solution peaks of Pu(IV), Pu(III) and Pu(VI) are located quite close to one another, while if the acidity is higher, the elution is unduly prolonged and the peaks are diffuse (see Figure 115). The acid is passed at about 4 drops per minute. Trivalent plutonium is eluted first; the location of its peak is not affected by changes in the acidity. The next to be eluted is Pu(VI), the elution curve of which is not significantly dependent on the acidity of the eluent. Tetravalent plutonium is the species last issuing from the column, and its elution is delayed if the nitric acid concentration of the eluent is increased. It was noted that a part of polymerized Pu(IV) is eluted

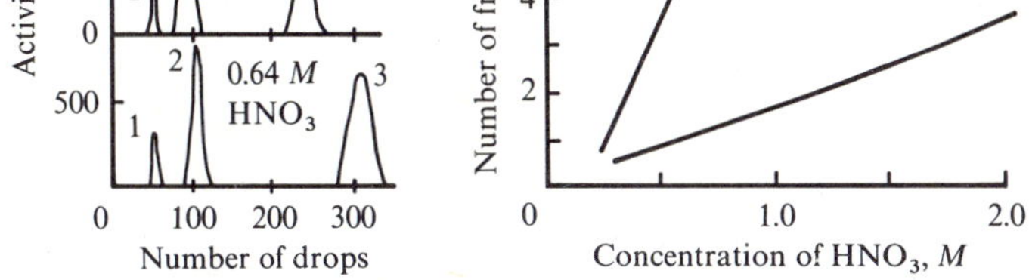

Figure 115

Location of the elution peaks of plutonium ions as a function of the concentration (*a*, *b*) and of the amount of nitric acid (*a*). Stationary phase: TBP.

1) Pu(III) and polymerized Pu(IV); 2) Pu(VI); 3) Pu(IV).

together with trivalent plutonium. It was also shown in the study that plutonium can be separated from major amounts of uranium (U:Pu = 10^6) if a mixture of 3 M HNO_3 with 0.05 M NH_2OH is passed through the column. Trivalent plutonium is eluted with one free column volume of the eluent, while six volumes are required for the uranium to appear in the eluate.

Paper chromatography

Partition chromatography on cellulose columns is not employed in the case of plutonium, probably because the process is too long and large volumes of eluent are needed; even so, the method has been studied [9, p. 331]. There are also references in the literature to paper chromatography. This method is simple and makes it possible to separate very small amounts of substances (fractions of a microgram). A small portion of the aqueous solution to be analyzed is applied to a paper strip near one end and is dried. This end of the strip is dipped into a vessel with the mobile solvent, which moves along the strip, entraining with it the substances to be separated.

These substances move at different speeds, depending on their distribution coefficients. Their rates of motion are described by the parameter R_f (rate of flow), which is the ratio between the shift of the concentration maximum of the substance and the shift of the solvent front during the same time. It is considered that if the R_f-values of two components differ by 0.1 or more, they may be separated and their concentration ranges are localized.

Keller [489] developed in more detail the theory of separation of actinide elements by paper chromatography.

The spots are revealed radiometrically or by color reactions with specific reagents. The quantitative determination of the element may be carried out by any suitable highly sensitive method, after the element has been extracted from the paper.

The separation of plutonium by paper chromatography can only infrequently compete with the numerous rapid methods which are available, since the time required for such separation may be 20 hours or more. The method is employed when only small volumes of a dilute solution are available, and when the constituents of the solution cannot be determined radiometrically or photometrically without separation. Accordingly, the published studies on paper chromatography of plutonium deal either with the separation of microamounts of plutonium in various valency states, or with the separation of plutonium from other actinide elements.

Cowan and Foreman [362] studied the behavior of Pu(III), Pu(IV) and Pu(VI), when these species are eluted with mixtures of n-butyl or n-amyl alcohol with HCl and mixtures of ethyl acetate with HCl. The difference in the R_f-values of the ion couples is sufficient to achieve a sharp separation. The separation of Pu(IV) from Pu(VI) is carried out as follows.

A 0.005–0.05 ml volume of an acid solution of plutonium (up to 100 μg in the sample) is evenly applied with a micropipet onto the end of a Whatman No. 1 paper strip, about 2.5 cm wide. The paper is dried for one hour, and the end with the applied sample is dipped into a vessel with n-butanol or n-pentanol, saturated with 3 *M* HCl. The development takes 16 hours, during which the solvent front moves about 30 cm. The R_f-values for the butanol mixture are: Pu(IV), 0.09; Pu(VI), 0.31; for n-pentanol mixture they are 0.01 and 0.17 respectively. The composition of the mixture is determined by radiometry after the components have been reextracted with dilute nitric acid or else subjected to wet digestion or ignition.

To separate Pu(III) and Pu(IV) these authors use mixtures of ethyl acetate with 11 *M* HCl (volume ratio 7:3). The respective R_f-values are 0.10 and 0.79. The time of the determination is as short as five hours; the solvent front moves through a distance of 15 cm.

The method of Cowan and Foreman does not permit separation of all three valency states in the presence of each other.

Clanet [341] succeeded in achieving a sharp separation of zones of Pu(III), Pu(IV), Pu(VI) and Am(III) on Whatman No. 1 filter paper, using a mixture of n-butanol with 3–10 *M* HCl (1:1). The R_f-values, determined by this worker in the above acidity range, make it possible to select the experimental conditions for the separation of mixtures which contain, in addition, U(IV) and U(VI).

Keller [489] chromatogrammed mixtures of actinide ions on Schleicher and Schull filter paper No. 20436. He separated U(VI), U(IV) and Pu(III), and also U(VI), Np(VI) and Pu(III) by elution with mixtures of methanol (ethanol, n-propanol and n-butanol) with hydrochloric acid, in the volume ratio of 1:1. A sharp zone separation was obtained for two preparations containing Ac(III), Th(IV), Am(III), Pu(IV), and also U(VI), Am(III) and Pu(IV) by developing with a mixture of methanol with nitric acid (1:1). The determination takes between 10 and 24 hours, depending on the nature of the organic component of the eluent.

CHAPTER VI

Analysis of Plutonium Preparations and Alloys

Depending on the nature of the sample, the concentration of the element and the accuracy requirements, radiometric, spectroscopic and physicochemical methods are employed.

Radiometric methods are the most sensitive. They are used to determine fission products in plutonium fuel and neptunium and americium in pure plutonium. In each case a high degree of purification from plutonium is required; this is achieved by chemical methods.

Spectroscopic methods are used mainly in the analysis of impurities present in plutonium or in plutonium dioxide. The development of such methods involves serious difficulties due to the fact that the spectrum of plutonium itself is very complex. The result is that it is impossible to determine the impurity lines directly, since the lines of plutonium overlap them. The ways of eliminating these interferences will be briefly discussed below.

Physicochemical methods (spectrophotometry, polarography, etc.) are employed for more accurate determination of impurities present in plutonium, and for analysis of alloys.

DETERMINATION OF IMPURITIES IN PREPARATIONS OF PURE PLUTONIUM

SPECTROSCOPIC METHODS

These methods are usually employed when it is desired to analyze plutonium or plutonium compounds simultaneously for a number of impurity elements. In one such method [412, 739] the separation of plutonium is not required. The analysis is carried out on a small portion of sample solution (total amount of plutonium 0.05–01 mg). The determination is carried out with a spark source between copper electrodes. The sensitivity of the determination of Be, Cr, Cu, Fe, Mg, Si, Ti, Al, Pb, Ba, Ni, Sr, Zn and La is low (0.01–0.03%). The error is ± 10–15%. The method is mainly used to determine nonvolatile impurities.

The low sensitivity of the method is caused by the spectral background produced by plutonium. An increase in sensitivity may be attained if the sample is physically enriched.

Methods of fractional distillation are based on the fact that the vapor pressure of the impurities is higher than that of the base compound—plutonium oxide. In these methods the impurities are distilled and their spectra excited at the same time. The methods involve the use of carriers which are introduced to stabilize the arc.

In the method of Scribner [547, 643], the carrier is Ga_2O_3 (4 wt. % on the weight of plutonium dioxide in the 100 mg sample which is analyzed). It is considered [98] that when the carrier vapors enter the discharge zone, the concentration of the impurities in the plasma is temporarily increased, because in a steadily burning arc the diffusion and convectional efflux of the impurities to be determined is slower. In this way the impurities become concentrated and their lines are intensified.

In another method of fractional distillation [480a], sodium fluoride is used as carrier. The sample taken for analysis consists of 0.5 mg of plutonium as a nitrate solution, which is mixed with 0.8 mg NaF in the crater of the graphite electrode. The solution is evaporated to dryness and the residue excited in an AC arc. Since the sample is small, the sensitivity of the method is small as compared with the gallium oxide method. Its advantages are the simplicity of sample preparation and the short duration of the determination.

Recourse to the spectroscopic techniques described above involves grave danger to personnel because of the high toxicity of the aerosols of plutonium.

The evaporation method, proposed by Zaidel' et al. [18, 97, 98, 98a, 99] is undoubtedly preferable. The authors completely separated the processes of evaporation of impurities and of the excitation of their spectra. The impurities were separated from the base by vaporization in vacuo and not in the air. The vacuum evaporation results in the formation of a dense impurity layer, which condenses on the end of a cooled copper capsule, with a consequent increase in the accuracy and sensitivity of the determination, as well as a diminished danger of accidental contamination of samples and penetration of radioactivity into the laboratory. The copper capsule serves as the electrode in spectroscopic analysis.

A 10–50 mg sample of metallic plutonium is converted into the dioxide by heating in a muffle furnace. The temperature is raised slowly, in order to prevent the sample from catching fire. The dioxide obtained is placed in an ignited graphite crucible which is placed in a small graphite furnace. The copper capsule is placed exactly above the crucible. The space is then evacuated. When the residual pressure is ~ 0.01 mm Hg, the heating oven is switched on and the temperature is slowly raised to 1800°, while following the temperature rise on a pyrometer. At this temperature ~ 0.001% of the plutonium sample is vaporized per minute.

Heating to 1900° results in a considerable increase in the evaporation of plutonium. At 1800° the sample lasts for 2 minutes. The condensation of impurities on the copper capsule is 85–98%. The capsule is transferred to the spectrum excitation chamber. The spectra of the impurities are taken simultaneously with the aid of three instruments (ISP-51, ISP-22 and KS-55), so that 15 or 16 elements can be determined. Standard spectra are photographed on the same plates. The exposure time is 10 seconds at the following parameters of the spark generator IG-2: capacity 0.012 μF, inductance 0.15 mh, current intensity 2A, interelectrode gap 2 mm.

It was found [97] that thorium dioxide may sometimes be substituted for PuO_2 as the base of the standards.

The method serves to determine Li, Be, B, Mn, Cr, Na, K, Zn, Fe, Co, Ni, Cu and certain other elements, with sensitivities ranging from $2 \cdot 10^{-3}$% for Fe to $1 \cdot 10^{-5}$% for Li, Be, B, Mn and Cr. A comparison between the fractional distillation and the vaporization methods shows that for most impurities the sensitivity of the latter method is 5–10 times higher. The mean square error of a single determination of impurities in plutonium by the vaporization method is ± 15–20%.

The methods of Scribner and Zaidel' cannot be employed to determine nonvolatile substances: Th, Zr, rare earths and other elements.

In order to raise the sensitivity of the spectroscopic determination of admixtures in plutonium, chemical methods have been developed for the preliminary separation of plutonium and concentration of impurities (extraction, ion-exchange and precipitation methods). To determine Al, Ba, Be, Ca, Cd, Co, Cr, K, La, Li, Mg, Mn, Na, Ni, Pb, Sr and Zn, extraction of plutonium cupferronate with an ether–chloroform mixture is employed [547]. A three-stage method of extraction of impurities is also known.

The solution (0.1 ml) containing 50 mg Pu per ml in 6 *N* HCl is placed in a 1 ml measuring tube. One drop of 0.5 *M* solution of hydroxylamine is added and the solution is heated until bright blue. Hexone (0.5 ml) is added to the tube and the tube is shaken for two minutes. The organic phase is analyzed by the copper spark method. The residual aqueous solution containing Pu(III) is shaken for 2 minutes with 0.5 ml of 0.5 *M* solution of TTA in hexone. Zirconium and hafnium pass into the organic layer. The aqueous phase is acidified with HNO_3 to 6 *N* tetravalent plutonium is extracted twice with 0.5 *M* TTA in hexone for 2 minutes each time. The remaining aqueous phase, which is sent for spectroscopic analysis, contains alkali metals, alkaline-earth metals, lanthanides and B, Al, Co, Ni, Pb, Sc, Th and U. Other elements pass into organic phase in the first stage of the extraction.

Brody et al. [316] employed the sorption of Pu(IV) on an anion exchanger to separate it from a large number of other elements.

A 0.2 gram sample of metallic plutonium is dissolved in a polythene beaker of 250 ml capacity in 15 ml of 12 *M* HCl in the presence of 1 ml concentrated HNO_3 (to convert plutonium into Pu(IV)). The solution is passed through a 200 × 10 mm polythene column, charged to half capacity with Dowex-1 × 10 (200–400 mesh). The column is first washed with concentrated HCl. The impurities are washed out of the column first with two free column volumes of 12 *M* HCl, then with one volume each of 11, 10, 9, 8, 7, 6, and 5 *M* HCl. The combined eluate contains Al, B, Be, Mg, Ca, Cr, Cu, K, Zn, Mn, Na, Ni, Pb and Ti. Five volumes of 8 *M* HNO_3 are then passed through to elute Fe, Co, Mo, Ga, Zn and Cd, which form stable chloride complexes and are retained by the anion exchanger. Prior to analysis the solution is evaporated to dryness to remove HNO_3 and the residue is dissolved in 1 ml of 6 *M* HCl. The impurities are determined by the copper spark method.

Impurities may also be separated from plutonium by sorbing plutonium on strongly basic anion exchangers from nitric acid solutions.

Chemical enrichment ensures an increase in the sensitivity of determination of most impurities as compared with the vaporization and fractional distillation methods.

CHEMICAL METHODS

Hydrogen

Hydrogen is determined by vacuum melting in an iron or a platinum bath [690]. The determination of nitrogen and oxygen is carried out at the same time.

Boron

The spectrophotometric determination of boron is essentially the same as that employed for the determination of boron in uranium [9]. Methyl borate is distilled from weakly acid solution. Boron is determined photometrically by its reaction with curcumin. The experimental error is ± 3 rel. %, if the boron content is 1 μg.

Carbon

Carbon in metallic plutonium is determined by combustion of 30–50 mg of the metal in oxygen, followed by electroconductivity determination of a barium hydroxide solution before and after it has absorbed the carbon dioxide evolved (Palei et al.,

1948). The method is similar to the determination of carbon in uranium [9]. The experimental procedure and description of the apparatus for the determination of carbon in plutonium are given below.

1. The oxygen from a gasometer, used for the combustion of the metal, passes into a quartz tube 9 filled with copper oxide placed in an electric furnace (600–700°). In this way carbon monoxide, hydrocarbons and vapors of organic impurities from lubricants, rubber tubing, etc. are removed by combustion. The heated copper layer is 10 cm long and 15–20 mm in diameter. Tube 9 is positioned vertically (Figures 117 and 118). The temperature of the tube with the copper oxide, which is controlled by means of a thermocouple with a galvanometer, should not exceed 700°, since above this temperature copper oxide begins to sinter and cake with the quartz. The top of the tube, which is fitted with a ground glass joint, should be 10–15 cm above the oven to prevent heat-up of the joint.

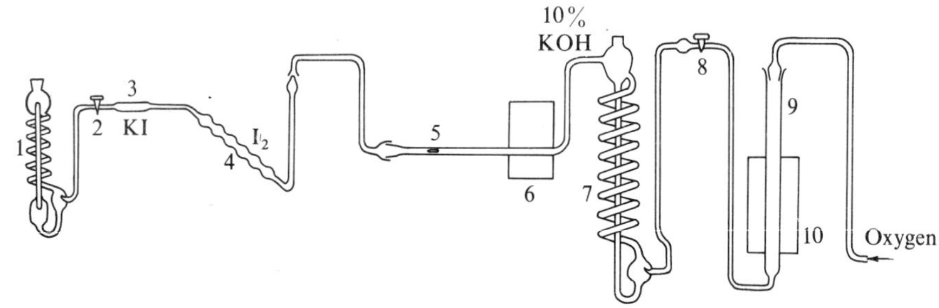

Figure 116

Apparatus for the microdetermination of carbon:

1) coil-type cell for determination of electrical conductivity; 2) head control stopcock; 3) KI tube; 4) bulb tube; 5) boat; 6) sample combustion furnace; 7) oxygen purification coil; 8) stopcock; 9) tube with copper oxide; 10) furnace.

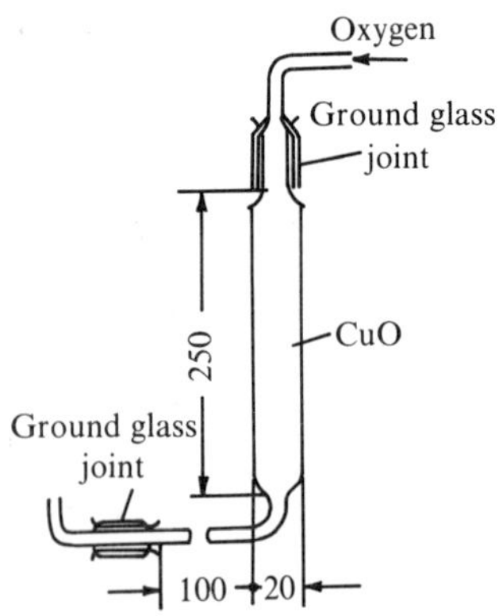

Figure 117

Tube with copper oxide for the purification of oxygen

2. Oxygen is purified from traces of CO_2 in the coil 7 in Figure 116 (see also Figure 118), which contains 40 ml of 40% solution of KOH. The joints between the outlet end of the tube with copper oxide and stopcock 8 and between stopcock 8 and the end of the coil inlet should be aligned.

3. The quartz tube with combustion boat for the sample (Figure 116, 5 and Figure 119) should be 2.2–2.5 times longer than the furnace, i.e., 25–30 cm long if the length of the furnace is 10–11 cm. The internal furnace diameter is 10 mm larger than the external diameter of the tube, so as to allow free insertion of the end of the thermocouple, permanently attached to furnace 6. The furnace moves freely from one end of the tube to another. The internal diameter of the tube is 9–10 mm to permit ready insertion and removal of the quartz boat, which is 5–6 mm wide and 15–20 mm long. The inlet end of the tube is bent at a right angle to prevent heat-up of its joint with the coil by the radiation emanating from the furnace. The outlet end of the tube widens and is covered with a ground glass cap. The joint is lubricated with the minimum possible amount of good quality high-vacuum grease.

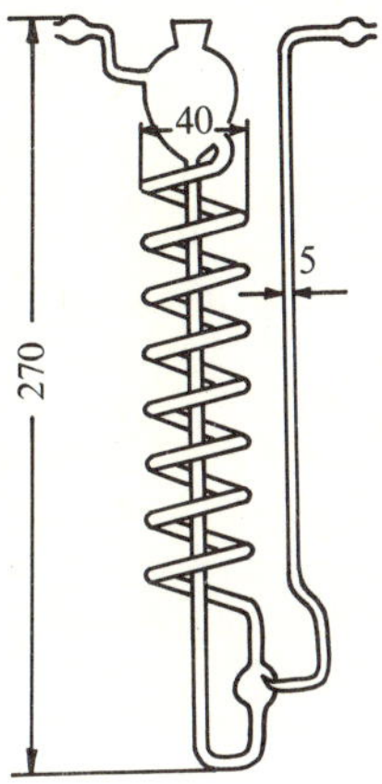

Figure 118

Coil for the purification of oxygen

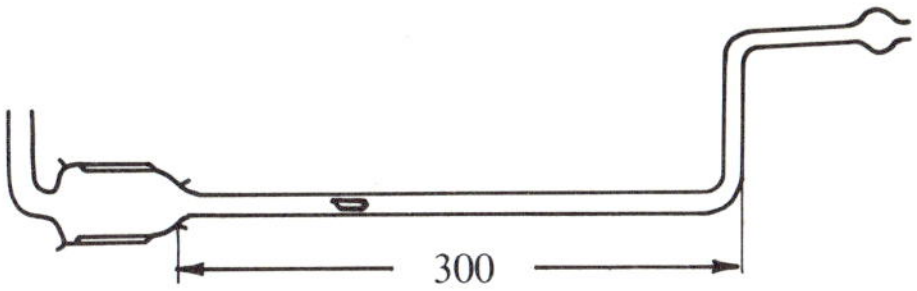

Figure 119

Quartz tube with boat for combustion of sample

4. Stopcock 2 serves for fine adjustment of the stream of oxygen during combustion. It has a long turning head and two notches in the body of the stopcock. These notches, whose length should be a quarter of the circumference, are made with a thin file or a needle file and gradually taper off. This stopcock is also greased with the minimum amount of high-vacuum or Apiezone grease. The stopcock is permanently fixed on the lid of a thermostat. The outlet tube of the stopcock and the inlet tube of the absorption coil (Figure 116, 1) in front of the narrow nozzle have a very small volume (0.3–0.5 ml) in order to attain a regular bubbling rate of the gas in the coil 1.

5. The coil, containing 4 ml of 0.01–0.002 *N* $Ba(OH)_2$, has a wider bottom part, into which two platinized platinum electrodes are sealed. The motion of the gas bubbles along the coil ensures a continuous circulation of the solution. The top, outlet part of the coil is covered with a soda lime tube to prevent penetration of atmospheric carbon dioxide. During the initial adjustment of the instrument it is advisable to place a small rheometer, designed for a bubbling rate of 10–15 ml of gas per minute, at the outlet of the tube. The absorption coil is immersed up to the middle of the top bulb into a water thermostat at $25 \pm 0.1°$.

6. If the metal contains sulfur, a bulb absorber with 0.01 *N* solution of iodine and a tube with solid KI are inserted between stopcock 2 and tube 5 to trap sulfur compounds and iodine vapors.

7. The electrical conductivity of the solution of $Ba(OH)_2$ is determined by the conventional Wheatstone bridge technique. The conductometer designed by Gusev and Sentyurin [76] which measures the resistance of the solution to within $\pm 0.1\%$ is very convenient. More bulky is the unit which consists of a 3G-2 audio-frequency oscillator and electronic oscillograph of the EO-4 or EO-5 type as the zero indicator. The bridge is fed with AC of a frequency of 1000 sec^{-1}. The determination takes place as follows. The boat with the weighed sample for combustion is placed in the left-hand side of the tube, but not too near the joint; furnace 10 with tube 9, filled with copper oxide, is switched on. Furnace 6 is placed on the extreme right and switched on. The individual parts of the apparatus are assembled and oxygen is passed through at a rate of 10–15 ml/min. The apparatus is checked for leaks. All adjustments of the gas flow rate are effected with the aid of stopcock 2. The gas is passed through the apparatus for 20 minutes; when the temperature of the copper oxide furnace has reached 600–700°, the absorption coil is disconnected, without interrupting the stream of oxygen, and filled with 4 ml of 0.01–0.02 *N* $Ba(OH)_2$. The $Ba(OH)_2$ solution is introduced into the coil with the aid of a microburet with a long drawn-out tip. When introducing the solution, the tip should be inserted into the lower, wider part of the coil, or an air seal will be formed which is difficult to remove.

After filling, the coil is reconnected to the apparatus and the initial conductivity of the $Ba(OH)_2$ solution is determined while the oxygen stream continues to be bubbled through. The determination is repeated a few times at 3–5 minute intervals, until a constant conductivity value is obtained.

The furnace 6 is now slipped over the boat. The temperature of the combustion furnace is maintained at about 900° throughout. After 0.5–1 minutes there is a flash, and after 1–2 minutes the oscillograph (or galvanometer) begins to record the variation in the concentration of $Ba(OH)_2$ solution. The evolution and absorption of carbon dioxide are considered complete when the conductivity value remains constant for for not less than ten minutes. The carbon content is calculated by the formula

$$\%\,C = \frac{\Delta n \cdot 4.31 \cdot v \cdot 6 \cdot 100}{m \cdot 1000},$$

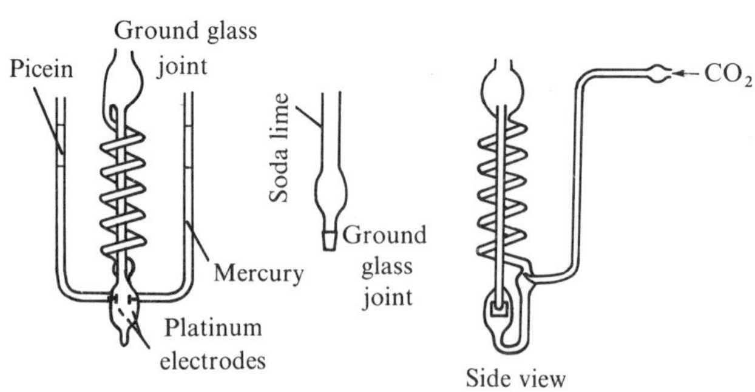

Figure 120

Circulation coil for CO_2 absorption and electrical conductivity measurements

where Δn is the difference between the initial and the final conductivities, 4.31 is the conversion factor between the conductivity of the solution and the normality of the $Ba(OH)_2$ solution (this factor is constant between concentrations 0.01 N and 0.002 N), v is the volume of $Ba(OH)_2$ solution in ml, 6 is half the atomic weight of carbon and m is the sample weight in grams.

At $v = 4$ ml, the formula may be written as follows:

$$\%C = \frac{\Delta n \cdot 10.3}{m}.$$

The sensitivity of the method is to within a few micrograms of carbon, the accuracy is about ± 3 rel. % at a carbon content of ~0.1%.

Nitrogen

Nitrogen in plutonium is determined by the classic method of Kjeldahl [328]. The apparatus employed has been designed with the specific features of working with plutonium in view (Figure 121).

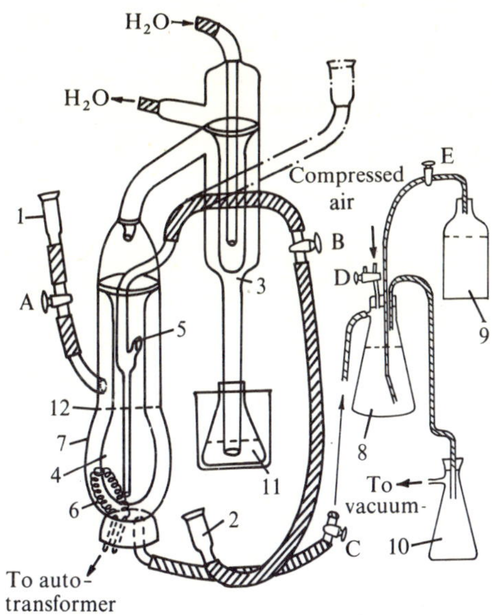

Figure 121

Diagram of apparatus for the distillation of nitrogen:

1) water funnel; 2) funnel for sample solution; 3) condenser; 4) inner flask; 5) steam inlet; 6) heater; 7) steam generator; 8) two-liter vessel for spent liquid; 9) spare vessel for spent liquid; 10) trap; 11) receiver (25 ml conical flask); 12) level; A, B, C, D, E: Teflon stopcocks.

The weighed sample is dissolved while working in a glove box, from which all nitrogen-containing reagents have been removed the day before. About 0.5 g of the metal is weighed out and dissolved in 20% redistilled HCl in a 40 ml tube. The residue is dissolved on heating by adding a few portions of 30% H_2O_2. If a small residue still remains, it is treated with a few drops of 1 *M* HF. After the sample has dissolved, the test tube, which contains about 5 ml of solution, is transferred to the distillation apparatus. The apparatus should have been steamed through not less recently than two days before, until the titration in a blank run consumes not more than 0.06 ml 0.01 *M* HCl.

Stopcocks A and B are opened; stopcock C is closed. Through funnel 1 distilled water is poured into steam generator 7 to the level 12. Stopcock A is left open. The sample is transferred from the test tube into flask 4 through funnel 2 held in the upright position, with the aid of a syringe. The test tube and the syringe are washed twice with water (total volume 7 ml).

Nine ml of 50% solution of NaOH are added to the flask and the solution diluted with water to the level 12 (25 ml in all). Stopcocks A and B are closed and a 25 ml conical flask 11, containing 5 ml of 2% H_3BO_3 and three drops of mixed indicator (40 mg methyl red and 60 mg bromocresol green in 100 ml ethanol), is placed under condenser 3. The lower end of the condenser dips below the surface of the liquid in the receiver. The autotransformer is then switched on and the distillation begun. When 10 ml distillate have been collected (after about 15 minutes), the receiver flask is lowered and the condenser is steamed through for about two minutes. The flask is placed in a fume cupboard and its contents titrated against 0.01 *N* HCl until the color changes from blue to pink-yellow. The solution is then removed from the distillation flask. Stopcock C is opened, stopcock D is opened to the atmosphere, stopcocks A, B, and E are left closed and the system connected to the vacuum line. By adjusting the opening of stopcock D, the solution is aspired from the steam generator 7 into vessel 8. The apparatus is carefully rinsed with water, the water drained by aspiration and the stopcock C is closed. The apparatus is then ready for the next determination. The spent solutions are transferred into the spare vessel 9 with the aid of compressed air. Each series of determinations must include blank experiments and determinations of nitrogen in standard specimens. A convenient standard is a solution containing 25–100 μg nitrogen as NH_4Cl in 5 ml of redistilled 6 *M* HCl. After dissolution of sample, one determination takes one man-hour.

The calculations are carried out using the formula

$$\mu\text{g N} = \frac{(v_1 - v_2) \cdot M \cdot A \cdot 10^3}{G},$$

where v_1 is the volume of standard solution of HCl consumed in sample titration, v_2 is the volume of standard solution of HCl consumed in blank titration, ml; M is the molarity of the standard solution of HCl, A is the atomic weight of nitrogen (14,007) and G is the sample weight in grams.

The experimental error of the determination is ± 5 μg of nitrogen. The lower limit of the determination by this method is 5–10 μg of nitrogen in a 0.5 gram sample.

Fluorine

The determination of fluorine [397] involves a steam distillation of HF from a sulfuric acid solution of the sample at 150° and its titration with thorium nitrate at pH 3.2 in the presence of an indicator. Prior to the distillation the bulk of plutonium is separated by precipitation as sulfate. It would appear that the method differs but little from that employed for the determination of fluorine in nitric acid solutions of uranyl [700], in which the titration is conducted in the presence of hydroxylamine hydrochloride which is added to decompose nitrogen oxides. The procedure employed to determine fluorine in solutions of uranium is given below.

Reagents. Solutions: H_2SO_4 (concentration 1:1), NaOH (40 g/liter) and hydroxylamine hydrochloride (250 g/liter).

Indicator solution: Chrome Azurol S (0.2 g/liter), filtered before use.

Monochloroacetate buffer solution (0.24 *M*): fifty ml of monochloroacetic acid (22.4%) are neutralized with a solution of NaOH (220 g/liter) to phenolphthalein. Fifty ml of monochloroacetic acid are then added to the solution and the solution is then diluted with water to one liter. The pH is adjusted to 3.0 ± 0.05 with the aid of solutions of NaOH and monochloroacetic acid.

Nitrite-free solution of nitric acid: two hundred ml of CP HNO_3 (sp. gr. 1.42) are evaporated to 100 ml. The solution is cooled and diluted with water to 200 ml.

Standard fluoride solution: two or three grams of NaF are placed in a platinum dish, dried at 110° for 30 minutes and cooled in a desiccator. The sodium fluoride (1.105 g) is dissolved in water and made up to 500 ml in a measuring flask (concentration of fluorine in solution is 1000 μg/ml). A dilute standard solution, containing 10 μg fluorine per ml, is obtained by diluting 10 ml of the stock solution to one liter.

Standard solution of 0.001 *N* $Th(NO_3)_4$: the titer of the solution is checked as follows. Portions of standard fluoride solution, containing 0, 5, 10, 20, 30 and 50 μg F, are placed in six 150 ml flasks containing 4 ml NaOH (40 g/liter) and 0.2 ml of $NH_2OH \cdot HCl$ (250 g/liter). The volume of the contents of each flask is adjusted with water to 40 ml. The pH is adjusted to 3.0 ± 0.05 by adding nitrite-free nitric acid, with the aid of a pH-meter. To each solution is added 1.0 ml of the buffer solution and 2 ml indicator solution. The solution not containing fluoride is titrated first against 0.001 *N* $Th(NO_3)_4$ until the color becomes red-purple. The remaining solutions are then titrated until their color matches that of the blank experiment (volumes of titrant consumed are, respectively, v_2, v_3, v_4, v_5, and v_6 ml). The fluorine titer of the thorium nitrate solution is now calculated (μg/ml) by the expression

$$F = \frac{115}{(v_2 + v_3 + v_4 + v_5 + v_6 - 5b)},$$

where b is the volume of the thorium solution consumed in the blank titration.

Procedure. The distillation apparatus (Figure 122) is assembled. Fifty ml H_2SO_4 (1:1) are placed in a 250 ml flask, heated to 150°, and steamed out for 15 minutes at 150°. In the steamed flask are placed 50 ml sulfuric acid (1:1) and a portion of the sample solution (for uranium, nitric acid solution; for plutonium, the filtrate from the precipitation of sulfate). Twenty ml of NaOH (40 g/liter) are measured out into a tall 250 ml beaker and placed under the outlet tube of the condenser, so that the tube dips into the solution. The contents of the flask are heated to 150° and steam is passed through the apparatus. The temperature is kept at 148–152° and the steam is passed through until 200 ml distillate are collected. To the distillate 1 ml of $NH_2OH \cdot HCl$ solution (250 g/liter) is added, the solution transferred to a 250 ml measuring flask, diluted to the mark with water and mixed. Fifty ml of the solution thus obtained are placed in a tall 150 ml beaker and the pH is adjusted to 3.0 ± 0.05, using solutions of nitrite-free nitric acid and NaOH. Buffer solution (1.0 ml) and 2 ml of indicator solution are added, the latter solution immediately before titration. All these operations are also performed in a blank determination; the blank solution is titrated first, and then the sample solution.

About 87% fluorine are determined in samples of plutonium containing $(1\text{–}15) \cdot 10^{-4}\%$ fluorine. The 95% expectancy of error is $\pm 6.5 \cdot 10^{-5}\%$.

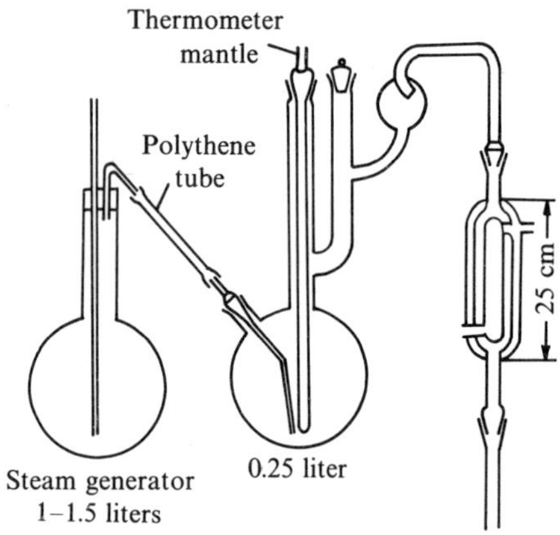

Figure 122
Fluorine distillation apparatus

Silicon

Holt [468] developed a colorimetric distillation method for the determination of silica in plutonium, based on the distillation of silicon tetrafluoride followed by the measurement of the color intensity of silicomolybdic acid.

The diagram of the distillation apparatus is shown in Figure 123. Distillation vessel 2 is made entirely of platinum (diameter 2.25 cm, length 10 cm, volume 25 ml, diameter of outlet tubes ~ 0.6 cm). Heat is supplied uniformly to the entire vessel, except for the outlet tubes, with the stream of hot air ejected by a heat gun with adjustable electric heater. The air temperature is held at 170–280°. The cylindrical part 4 of the gun is extended, and has an 0.8 × 3.3 cm slit in its top part so as to be able to lower the outlet tube of the distillation vessel into the position shown in Figure 123. The distillation vessel is connected to the vessel 6, containing a solution of the reagent. Air is aspired through the system by means of a vacuum pump. Vessels 4 and 6 are clamped to a stand (not shown in the figure) so as to be able to lift or to lower the distillation vessel in the heat gun. The stand also carries a tubular furnace (3.1 cm diameter, 10 cm length) designed so as to produce a temperature of 300°. It is used to heat the platinum vessel prior to distillation, to dissolve the sample and to treat the sample with perchloric acid.

The preparation of solutions and the experimental procedure are described below.

Reagents. Absorbing solution: aqueous saturated solution of boric acid (200 ml) is mixed with 40 ml of 10% solution of molybdic acid and the solution is diluted to one liter. To prepare a 10% solution of molybdic acid, 25 grams $(NH_4)_2MoO_4 \cdot 4H_2O$ are dissolved in 200 ml water, the resulting solution is mixed with 20 ml concentrated H_2SO_4 and is diluted to 250 ml.

Mixture of nitric and hydrofluoric acids. 5 ml 50% HF are mixed with 25 ml conc. HNO_3.

Mixture of sulfuric and tartaric acids. 125 ml conc. H_2SO_4 are mixed with a solution of 8 g tartaric acid and the solution is diluted with water to 500 ml.

Reducing solution: 0.1 g of 1-amino-2-naphthol-4-sulfonic acid is dissolved together with 0.4 g NaOH in 25 ml water; 5.4 g $NaHSO_3$ are added and the mixture diluted to 50 ml.

Procedure. About 0.3 g metallic plutonium is placed in the distillation vessel and 1 ml of 10% HCl, 2 ml $HClO_4$ (1 : 1) and 4 ml of 72% $HClO_4$ are added, in that order. The vessel is partly inserted into the tubular

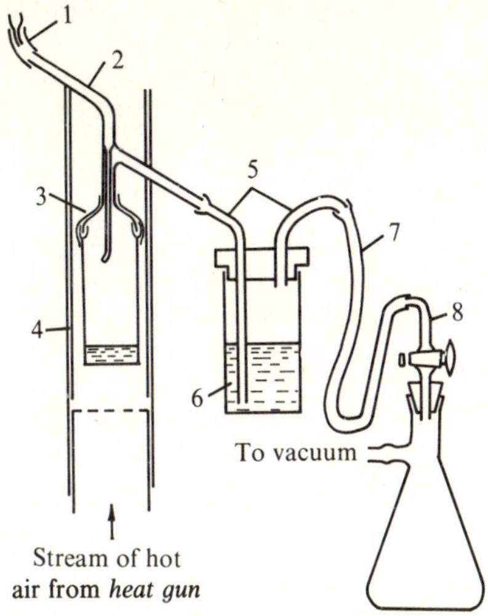

Figure 123

Apparatus for the distillation of silicon:

1) funnel; 2) platinum distillation vessel with outlet tubes; 3) aluminum foil cap; 4) barrel extension of heat gun; 5) Teflon tubes; 6) receiver, made of semitransparent plastic material; 7) rubber tube; 8) Teflon stopcock.

furnace and cautiously heated to attain full decomposition of sample. It is then lowered into the furnace almost completely, heated until abundant evolution of perchloric acid fumes, the vessel taken out and cooled to room temperature. The platinum lid with the outlet tubes is then placed on the vessel and the joint is carefully sealed with 7.5 × 12.5 cm aluminum foil to prevent air motion through the gap between the vessel and the lid. The distillation vessel is now suspended from a stainless steel wire inside the cylinder of the heat gun as shown in Figure 123. The gap in the slit of the cylinder between the outlet tube and the cylinder is plugged with a small piece of aluminum foil in order to prevent penetration of cold air inside the cylinder, which could produce a premature vapor condensation in the distillation vessel and losses of SiF_4. The platinum outlet tube is connected with vessel 6 (Figure 123), containing 25 ml of the absorbing solution, with the aid of a Teflon tube 0.6 cm in diameter. They are connected during heating. Vessel 6 is connected to the vacuum line via Teflon tube 5 and rubber tube 7. With the aid of stopcock 8, the rate of flow of the air drawn through the system is adjusted at 150 ml/min. Through funnel 1 the mixture of HNO_3 and HF (0.3 ml) is drawn into the distillation vessel. The distillation is begun by switching on the heat gun for 10 minutes (timer). The temperature of the hot air in the top part of the gun cylinder is about 200°. During the distillation, the thermometer readings in the heat gun are noted, and the distillate flow is followed by observing through the transparent Teflon tube. As soon as the heating is discontinued, the Teflon and the platinum tubes are disconnected, and then the Teflon and the rubber tube. The tubes are removed from vessel 6 and the Teflon tube dipping into the solution is rinsed with water. The pH of the solution is adjusted to 1.2–1.3 with concentrated solution of ammonia (free from silicon). The solution is left to stand for 10 minutes to allow the β-form of silicomolybdic acid to form. Five ml of sulfuric acid–tartaric acid mixture are added to the solution, the solution is stirred and 0.8 ml of the reducing solution are added immediately. The solution is then diluted to 50 ml and, after 20 minutes, placed in a cell (layer thickness 1 cm for samples with 0–50 μg Si, 5 cm for samples with 0–10.0 μg Si), and the optical density D is measured at 815 mμ

against water. From the value D thus obtained is subtracted the value D_{bl} obtained in the blank experiment, and the silicon concentration is calculated from the formula

$$C = \frac{(D - D_{bl})}{D_{st} \cdot W},$$

where C is the concentration of silicon, in μg/g Pu, D_{st} is the optical density corresponding to 1 μg Si in 50 ml of the solution used for the photometric determination (average value obtained by making photometric determinations on standard solutions of silicon of various concentrations) and W is the weight of the plutonium sample in grams.

The average error of the determination of 5–30 μg Si in samples of plutonium weighing about 0.3 g is $\pm$ 0.8 μg.

The completeness of separation and the accuracy of the determination of silicon were found to be markedly affected by the concentration of $HClO_4$, HF and HNO_3.

It was reported [680] that the method may be used to determine silicon in amounts between 25 and 100 μg per one gram of plutonium. Plutonium is preliminarily separated by precipitation of plutonium (III) fluoride. Prior to the photometric determination fluoride ion is bound by adding $AlCl_3$.

Iron

The determination of iron by o-phenanthroline is described on p. 384. The method [409, 474] may be applied to the analysis of metallic plutonium without preliminary separation, provided the iron content is not lower than 0.01 %. At lower concentrations of iron, plutonium must be separated; this is achieved by extraction or ion-exchange methods.

Nickel

The determination of nickel is carried out after its preliminary separation from plutonium by ion exchange [406]. The hydrochloric acid solution, containing milligram amounts of plutonium and microgram amounts of nickel, is passed through a column of Amberlite IR-120 (30–60 mesh) in the H-form and the sorbed nickel is selectively eluted with a solution of potassium cyanide. Nickel is determined photometrically by the reaction with dimethylglyoxime.

Taylor [690a] employed α-furildioxime to extract nickel from plutonium solutions into chloroform, and determined the complex in the organic phase photometrically at 435 mμ.

An aliquot of the weakly acid solution (pH not below 1) containing 1–20 μg Ni is placed in a 50 ml separating funnel. To the solution are added 0.05 ml of 1 N potassium bichromate, 5 ml of 10 % sodium citrate, 0.06 ml of 1 % solution of α-furyldioxime in 50 % ethanol and 0.5 ml aqueous ammonia (sp.gr. 0.88).

Nickel is extracted with three 7 ml portions of chloroform (in each extraction the funnel is shaken 200 times). The combined organic phases are transferred to a 25 ml volumetric flask and diluted to the mark with chloroform. The optical density of the solution is determined immediately, using 4 cm cells, at 435 mμ. A blank determination is run in the same manner. A calibration curve is constructed using standard solutions of nickel in sulfuric acid, containing 5, 10, 15 and 20 μg Ni in the sample.

Zirconium

Spectrographic determination of zirconium (up to $2 \cdot 10^{-3}\%$) is carried out after separating plutonium by an anion-exchange method [500], similar to that used by Brody et al. [316] (see p. 365).

Fifty mg of metallic plutonium are dissolved in 1 ml conc. HCl. To this solution is added 0.1 ml solution of cobalt in 4 *M* HNO_3 (concentration 0.01%, internal standard), 1 drop of 0.1 *M* HF and 1 ml conc. HNO_3. The solution is evaporated, 1 ml of 7.2 *M* HNO_3 is added and the solution is passed through a 10 × 0.6 cm column of Dowex-1 × 4 (50–100 mesh), pretreated with 6 ml of 7.2 *M* HNO_3. The column is washed with 6 ml of 7.2 *M* HNO_3. The combined solution is evaporated, dissolved in 0.5 *M* HCl, and 0.1 ml portions of the solution thus obtained are placed in graphite electrodes for subsequent determination in the spark.

The accuracy of a single determination is ± 17% to a 95% confidence limit.

Niobium and tantalum

Methods for the determination of both elements in plutonium alloys by hydroquinone will be described below (pp. 387 and 395). The method may be employed in the analysis of metallic plutonium since $\geqq 0.008\%$ niobium and tantalum may be determined by it [714, 715].

Molybdenum

A sensitive reaction of molybdenum with chloroanilic acid [713] is employed for the determination of the element when alloyed with plutonium (p. 388). The method may also be employed in analyzing plutonium for molybdenum as impurity.

Tungsten

Tungsten in plutonium is determined colorimetrically (Palei and Deberdeeva, 1951). Plutonium is converted into Pu(III) by treating a strongly acid solution of plutonium in HCl with 5% solution of $SnCl_2$ in the heat. The solution is cooled and an equal volume of 5% solution of KCNS is added, the mixture is left to stand for two hours, after which tungsten is extracted as the thiosulfate complex with two portions of ether (3 and 2 ml) preliminarily treated with a mixture of $SnCl_2$ and KCNS solutions. Colorimetric determination is carried out on the organic phases in a Duboscq colorimeter with a blue filter, using a standard scale. Tungsten in concentrations between 0.005 and 0.1% may be determined to within ± 5 rel. %. Equal or lower amounts of iron and uranium do not interfere. In the presence of molybdenum high results are obtained.

Lanthanum

The determination of lanthanum in plutonium samples [294] involves the precipitation of lanthanum as fluoride from solutions containing hexavalent plutonium. In the last stage of purification lanthanum 8-hydroxyquinolate is precipitated, the precipitate dissolved in dilute HCl and the photometric determination carried out at 365 mμ.

The sample of metallic plutonium, containing about 50 mg plutonium and 25–200 μg lanthanum, is placed in a quartz tube (12 × 100 mm). To prevent spattering, the tube with the solution is dipped in water and 0.4 ml of 70% $HClO_4$ are added. If the sample consists of a nitric or hydrochloric acid solution, to an aliquot volume (not more than 5 ml) 0.4 ml 70% $HClO_4$ is added and the solution evaporated to about 0.5 ml residual volume. After dissolution (or evaporation), the test tube with the solution is placed in a special aluminum heating block with holes for test tubes, designed to furnish a temperature of up to 225°. The oxidation of plutonium at 225° is prolonged for 10 minutes.* The tube is then taken out of the block, the solution is cooled to room temperature and 3 ml water are added while rinsing the inner walls of the tube. The residue is dissolved by stirring with a platinum rod, and the rod is left in the tube. Concentrated HF (0.4–0.8 ml) is gradually introduced into the tube by pouring along the rod (the HF solution should not come in contact with the walls of the tube). The solution is stirred, the rod is washed with a few drops of 1 *N* HF and removed from the tube. After 10 minutes the mixture is centrifuged for 5 minutes at the maximum velocity. The solution is cautiously removed, 1–2 ml 1 *N* HF are added to the tube, the contents are stirred, the stirrer washed and the mixture is again centrifuged. To the solution is added 0.3 ml 70% $HClO_4$ and the tube is heated in the heating block at 225° for 7–8 minutes. Acid fumes are evolved during the last 3–4 minutes. The tube is cooled to room temperature, its inner walls are rinsed down with 0.5 ml water and the solution is centrifuged. The solution is transferred into another identical-shaped quartz tube. The first tube is successively rinsed three times with 0.7–0.8 ml water each time, and the washings are poured into the second tube. One hundred mg ammonium persulfate, and 2 drops of 0.05% solution of thymol blue are added and dissolved by stirring with a platinum rod. The solution is neutralized by adding first 1.5 *N* and then 0.15 *N* solution of ammonia (final volume of solution should not exceed 5 ml). The stirrer is washed with a few drops of water, the solution heated at 85° for 15–20 minutes and 0.4 ml conc. HF is cautiously added, pouring along the rod. The mixture is stirred, and the stirrer washed with a few drops of HF and taken out of the tube. After 10 minutes the mixture is centrifuged, the solution withdrawn, the precipitate washed with 1–2 ml 1 *N* HF and the washings discarded. The precipitate is dissolved in 0.3–4 ml 70% $HClO_4$ at 225° as described above. The inner surface of the test tube is rinsed with 1.5–2.0 ml water and to the solution thus obtained 0.5 ml of 1% 8-hydroxyquinoline in 2 *N* acetic acid are added. The solution is then neutralized with 1.5 *N* NH_4OH while stirring.

When the neutralization point is attained, the solution turns intense yellow and the lanthanum pre-

* The perchlorate acid oxidation of liquid samples, containing major concentrations of sulfates, is slow. In such cases ammonium persulfate is used as oxidant.

cipitate separates out. A few more drops of ammonia are added and the mixture is stirred. The stirrer is rinsed with water and removed from the tube. The solution with the precipitate is heated for 2 minutes on a boiling water bath and is left to stand during one hour. It is then filtered through a glass filter of medium porosity, the tube rinsed with the minimum amount of water, and the washings also passed through the filter. The tip of the filter funnel is wiped, after which it is connected to a vacuum vessel containing a 10 ml volumetric flask. The tube is washed with a few ml 0.1 N HCl and the solution poured into the funnel. The vacuum is cautiously connected and the solution collected in the volumetric flask. The funnel is again washed with 0.1 N HCl. The total volume of added HCl should be 8–9 ml. The solution in the flask is diluted to the mark with 0.1 N HCl, mixed, and photometric determination is carried out at 365 mμ against water, in a 1 cm cell.

The content of lanthanum in the alloy is found from a calibration curve which is constructed as follows. Samples of high purity (99.9%) metallic plutonium (50 mg) are dissolved in 0.4 ml 70% $HClO_4$ in quartz tubes. To each solution an accurately known amount of standard lanthanum nitrate solution is added and the solution treated as above. The calibration curve is plotted from the optical densities found. The content of lanthanum in the standard solution is determined by precipitation of lanthanum 8-hydroxyquinolate from acetate solution. Lanthanum 8-hydroxyquinolate is dried at 130° (gravimetric factor 0.2433).

When analyzing samples containing up to 100 mg Pu, three precipitations of lanthanum fluoride from the oxidized solution are carried out.

The average deviation of 15 determinations in samples containing 50 mg Pu and 40–220 μg La is $\pm$ 3 μg.

The method was tested on relatively pure plutonium preparations. Ce(IV), Am(III), Th(IV), etc. are precipitated together with lanthanum and interfere with its determination.

Thorium

Spectroscopic determination of thorium is conducted after its separation from plutonium. This operation is based on the precipitation of thorium benzenesulfinate $Th(C_6H_5SO_2)_4$. Trivalent plutonium is not precipitated by this reagent (Palei and Milyukova, 1951). Zirconium is used as thorium carrier.

To 5 ml solution of sample in 0.2 M nitric acid, the carrier solution (2 mg Zr) is added. Sulfur dioxide is passed through the solution during 10 minutes and the solution is left to stand for 15–20 minutes. Aqueous 15% sodium benzenesulfinate (0.4–0.5 ml) is then added to the solution. The mixture is stirred and, after 10–15 minutes, centrifuged. The precipitate is washed twice with 1 ml of double-distilled water and dissolved in concentrated HNO_3. The solution is evaporated a few times with concentrated HNO_3 to decompose organic matter. The solution is then evaporated to dryness with concentrated HCl and the residue is dissolved in 1 ml of 1% solution of HCl. Four drops each (~0.25 ml) of the resulting solution are applied to copper electrodes and thorium is determined spectrographically, using the 2832.32 Å line. Zirconium (lines at 2833.908 Å and 2834.395 Å) serves as the comparison element.

The sensitivity of the method is 0.003%, the error is $\pm$ 10%. 10–300 μg Th can be determined in 25–100 mg Pu.

Uranium

Uranium in plutonium is determined mainly by luminescence methods. The known methods differ from each other in the manner of separation of plutonium and other quenchers. Buchanan et al. [328] analyzed samples containing 10^{-5}–10^{-4}% of uranium. Hexavalent uranium is preliminarily separated from the reduced aqueous solution by extraction with 30% solution of TBP in benzene.

Reagents. 2 *M* solution of iron (II) sulfamate: 21 g sulfamic acid are dissolved in 30 ml water, preheated to about 100°, and 5.5 g metallic iron powder are added. After the dissolution of the iron, the solution is filtered and diluted to 50 ml with water. The reagent is stable for 5–6 days if kept cool.

30% solution of TBP in benzene: TBP is preliminarily washed with saturated solution of $NaCO_3$ and then with water. The purified TBP is dissolved in benzene. Prior to use, the solvent is shaken with an equal volume of a mixture of the solution of iron sulfamate and 1 *N* HNO_3 (1:15).

Procedure. A weighed sample of metal is dissolved in 20% redistilled HCl, and 30% H_2O_2 and 1 *N* HF are added if an insoluble residue remains. A portion of the solution, containing ~ 5 μg Pu, is transferred with a micropipet to a 15 ml centrifuge tube and evaporated to dryness under a lamp. The residue is dissolved in 5 drops of 1 *N* HNO_3, 5 drops of iron (II) sulfamate solution are added and the solution is left to stand during 5 minutes. Thirty drops of TBP solution are introduced into the test tube and the extraction carried out during 10 minutes. Stirring is effected by aspiration and expression of the two phases through a pipet placed in the tube, with the aid of a pump. The aqueous phase is removed and the organic phase is washed five times, using 8 drops of 1 *N* HNO_3, and 2 drops of iron (II) sulfamate for each washing. Iron sulfamate is not used in the last washing, since iron is a quencher of the fluorescence of uranium. The organic phase is evaporated to 1 ml. Portions of organic solution (100 μl) are applied to tablets of flux (98% NaF and 2% LiF) on platinum plates (1.26 cm diameter) and dried under the lamp for 20 minutes. The tablets are fused, cooled and fluorimetric determination is made (LYuF-57 instrument may be used). Standards for comparison are prepared by adding to the tablet 100 μl solution of uranium (concentration 2 μg/ml).

The method determines 85–95% of uranium with an error of ~ 10%.

Boase and Foreman [301] recommend a preliminary anion-exchange separation of the extraneous elements in fluorimetric determinations of uranium in solutions containing 0.001–0.005 *M* of Pu, 0.02–0.03 *M* of iron and 0.01 *M* of aluminum.

Reagents. Solutions of 11 *M* HCl, containing 7% HI, and of 4 *M* HCl, containing 7% HI: iodine is removed from the HI serving for the preparation of the mixture by passing the acid through a 20 × 0.6 cm column of De-Acidite FF in the Cl-form. Most of the iodine may be removed in advance by shaking the HI with a few grams of the anion exchanger in a flask. Solution of 11 *M* HCl, containing 7% HI, is prepared immediately before use.

Chromatographic column (10 × 0.4 cm) is fitted at the top with a 10 ml reservoir. Air-dry De-Acidite FF (60–80 mesh) (0.5 g) is placed in the column, converted to the Cl-form and treated with 11 *M* HCl immediately before the sample solution is passed.

If the column is used for a long time, there will be an accumulation of iodine in the resin. Iodine may be removed by washing the resin with a solution of NaOH, but it is preferable to fill the column with a new batch of resin.

Procedure. In the reservoir of the column are placed 2 ml of 11 *M* HCl containing 7% HI and 1 ml the hydrochloric acid sample solution; the mixture is stirred and passed through the column at a rate of 0.4 ml/min. The column is washed with 1 ml of 11 *M* HCl containing 7% HI, and then with 2 ml of

4 M HCl containing 7% HI, at the same rate. Uranium is eluted with 4 ml of 0.1 M HCl, passed at a rate of 0.4 ml/min. The eluate is divided into two equal parts. Each portion is placed in a platinum capsule, to one of which is added a standard solution of uranium (0.2 μg). The content of the capsules is evaporated to dryness under an IR lamp, and is then fused at 850 ± 20° with 0.6 g flux (90% $NaHCO_3$ + 10% NaF) during five minutes. The melt is cooled for 5 minutes and the fluorescence measured on Harwell 1080 A fluorimeter. Blank determinations in the absence of uranium are periodically run in the same manner, and the results obtained on samples are corrected accordingly. The correction is usually ~ 0.1 μg/ml of uranium.

Solutions in which the content of plutonium, iron and aluminum is much smaller may be analyzed without separating uranium from other elements, but a correction for the quenching of the fluorescence of uranium must be introduced [10]. For the various aspects of fluorimetry, see [9, p. 143].

Neptunium

Np^{237} is determined by α-spectrometry after extractional separation from plutonium [328].

A weighed sample of metal is dissolved in 6 M HCl. The undissolved residue is treated with 30% H_2O_2 and a small amount of 1 M HF. Concentrated HCl is added to a final acidity of 12 M, and an aliquot part, containing ~ 250 mg Pu, is withdrawn into a separating funnel. Hydroquinone is added to a concentration of 0.1 M and the solution is cautiously stirred for 30 minutes. Potassium iodide is then added to a concentration of 0.1 M and the mixture is left to stand for 30 minutes. Tetravalent neptunium is extracted with an equal volume (~ 10 ml) of 0.5 M mono-(2-ethylhexyl) orthophosphoric acid in toluene for 30 seconds. Pu(III) remains in the aqueous phase. The organic layer is washed 10–20 times with 25 ml portions of freshly prepared (not more than 2 hours before use) solution of 12 M HCl containing 0.1 M hydroquinone. An aliquot part of the organic phase is applied to a platinum disc, dried, ignited and the α-activity is measured in an α-analyzer at 4.8 MeV.

The method will determine 10^{-4}% neptunium in plutonium. The yield of neptunium is approximately 90%.

Khramtsova and Mikhailov (1962) developed a photometric determination of neptunium, when present in plutonium in concentrations above 0.003%.

Tetravalent Pu is preliminarily extracted three times from 0.5–1 M HNO_3, containing 4.5 M $NaNO_3$ with some nitrite, with a solution of 20% diisoamyl ester of methylphosphoric acid (DAMPA) in carbon tetrachloride. Pentavalent Np, stabilized with sodium nitrite, remains in the aqueous phase. The aqueous phase is washed with carbon tetrachloride, acidified with HNO_3 to ~5 M and treated with excess hydrazine on a boiling water bath for 30 minutes to convert Np(V) to Np(IV). After cooling, a small amount of urea solution (200 g/liter), a solution of Arsenazo III (1.5 g/liter) and water are added to the solution so that its final content is about 2 grams of urea, 30 mg of Arsenzo III and 3 moles of HNO_3 per liter. If the content of plutonium in the aqueous phase exceeds that of neptunium by a factor of 30–200, 50 mg ascorbic acid are introduced. A similar analysis is performed on a parallel sample to which a known amount of neptunium has been added (since the losses of the element on extraction are ~ 50%) as well as a blank determination on a dummy sample not containing neptunium. The optical densities of all three solutions are determined with the aid of SF-4 or SF-10 spectrophotometers at 665 and 750 mμ, using 2 cm layer cells. The former wavelength corresponds to the maximum absorption of the complex of Np(IV) with Arsenazo III, while the measurement at the latter determines the so-called scattering background.

The amount of neptunium in the sample is determined by the formula:

$$n = \frac{\Delta D_s - \Delta D_0}{\Delta D_a - \Delta D_s} \cdot m,$$

where n is the amount of neptunium in micrograms, m is the amount of added neptunium in micrograms, ΔD_s is the difference between the optical densities of the sample solution at 665 and 750 mμ, ΔD_a is the corresponding difference for the solution with added neptunium, and ΔD_0 is that difference for the blank solution.

The method will determine 0.2–1.0 micrograms of neptunium in the sample with an accuracy of $\pm$ 20–30 rel. %.

The solution used in the spectrophotometric determinations may contain uranium (2 : 1), plutonium (up to 200 : 1), lanthanum (300 : 1), iron (300 : 1), nickel and chromium (1000 : 1) and major amounts of aluminum and manganese.

Microgram amounts of neptunium in plutonium are determined by square-wave polarography, after neptunium has been extracted with 2-thenoyltrifluoroacetone [671]. The polarographic determination is carried out on a dropping mercury electrode in the presence of EDTA at pH 5.5–6.5. The wave peak is at -0.8 V. The addition method or the calibration curve may both be used.

The sensitivity of the method is 0.4 μg/ml of neptunium. In the polarography of a solution containing 50 μg Np(IV) per ml, the following may be present: 50 μg/ml of U(VI), 10 μg/ml of Cu(II), Pb(II) and Cd(II), 50 μg/ml of Pu(IV) and 200 μg/ml of Fe(II). If the metallic plutonium sample contains 0.003 % Np, the error is usually $\pm$ 10 %. If the neptunium content is $\cong$ 0.05 %, the error does not exceed $\pm$ 3 %.

Americium

Am^{241} in plutonium is determined radiometrically. The element is first separated from plutonium by various precipitation and ion-exchange methods, depending on the radioactive elements which are present in plutonium.

Buchanan et al. [328] found the content of Am^{241} from the difference between the 5.5 MeV α-activity of (Am^{241} + Pu^{238}) before and after the separation of americium from plutonium.

To a 25 microliter portion of the hydrochloric acid solution are added HCl and HNO_3 to a concentration of 12 *M* and 0.1 *M* respectively. The resulting solution is passed through a 6×0.6 cm column of AG 1×10 anion exchanger (200–400 mesh, Cl-form) which is first washed with a mixture of 12 *M* HCl and 0.1 *M* HNO_3, at a rate of 1–2 ml/min. The column is washed with 12–20 ml of the mixture. Americium passes through the column. Plutonium is desorbed with 1 *M* HCl and the elution is continued until 25 microliters of the eluate give a count of (2–3) $\cdot 10^4$ cpm. Twenty-five microliters of the solution are placed on a platinum disc to determine the 5.5 MeV activity in a 256-channel α-analyzer. The content of Pu^{238} is found in this way. The sum of Am^{241} and Pu^{238} is determined in the same way prior to the separation of americium from plutonium. The content of americium is found by difference.

The experimental error is $\pm$ 10% for samples containing ~ 0.01 wt. % Am^{241}. Methods in which Am^{241} in Pu^{239} is determined from α-radiation energy, when present in contents of 0.032–3.6%, have been described. Small amounts of Am^{241} (> 0.001%) can be determined from the energy of γ-radiation [534].

The separation of americium from plutonium is also possible by the precipitation of the former on a carrier (cerium fluoride) from an oxidizing medium [180, p. 48].

A portion of the solution with a total α-activity of about $1 \cdot 10^6$ cpm is placed in a conical 3 ml centrifuge tube and diluted to 2 ml with 2 *M* HNO_3. The tube is heated on a water bath at 85–95° for 15 minutes. Ceric sulfate (0.1 *M*, 0.25 ml) in 1 *M* H_2SO_4 is added and the solution is held for another 15 minutes at the same temperature. The solution is cooled to room temperature and ~ 0.5 ml of 27 *M* HF are added.* The precipitate is left to stand for 5 minutes, the solution is then centrifuged and the supernatant separated. To the precipitate are added 1 drop of 0.05 *M* potassium bichromate, 3 drops of 0.1 *M* solution of $ZrO(NO_3)_2$ and 2 ml of 2 *M* HNO_3. The mixture is stirred until the precipitate is dissolved. In order to reprecipitate cerium fluoride 10 drops of 27 *M* HF are added.* After five minutes the solution is centrifuged and the supernatant is separated. The precipitate is washed three times with a solution of 1 *M* HNO_3 and 1 *M* HF.* The precipitate is suspended in a few drops of the washing liquid and placed on a platinum disc for radiometric determinations.

ANALYSIS OF ALLOYS

The methods of determination of metals which are constituents of plutonium alloys will now be discussed. Some of them are so sensitive that they may be employed in the determination of these elements when present as mere impurities. Plutonium, when present as an alloy component, is usually determined by spectrophotometry of the absorption by Pu(III) in hydrochloric acid solutions. Plutonium may also be determined by potentiometric titration of Pu(III) with ceric sulfate, radiometric and coulometric methods.

Aluminum

The content of aluminum in plutonium alloys may be determined by photometry (reaction with 8-hydroxyquinoline or with aluminon), gravimetry (as 8-hydroxyquinolate) or volumetry (titration of a solution of aluminum 8-hydroxyquinolate with bromine or complexonometric titration) [551]. In all these methods plutonium must first be separated; this is best done by anion-exchange methods. Photometric

* Solutions of HF are preliminarily treated with potassium bichromate to oxidize impurities, and are stored in platinium vessels.

[4] and gravimetric [114, p. 342] methods are less accurate and more awkward to handle. The complexonometric method is best.

A weighed sample of the alloy, containing about 5 mg Al, is dissolved in 2 ml of 8 *M* HCl. Two ml conc. HNO_3 are then cautiously added in order to oxidize plutonium to Pu(IV); the end of the oxidation is indicated by the cessation of gas evolution and by the clarification of the intensely brown-colored solution. The solution is transferred to a 35 × 2.2 cm column with Dowex-1 × 4 anion exchanger (50–100 mesh, Cl-form), which has been preliminarily treated with 8 *M* HCl, and is passed through the column at the rate of 2 ml/min. The column is then washed with 60 ml of 8 *M* HCl. To the eluate are added 5.00 ml of 0.1 *M* solution of Complexone III, 2 drops of 0.1% solution of bromocresol green indicator and the solution is carefully neutralized with 12 *N* NaOH, until the color turns from yellow to blue (pH 4–5). Five ml of an acid buffer solution (108.9 g sodium acetate and 23 ml conc. acetic acid are dissolved in water and the volume made up to one liter) are then added as well as 5 drops of 0.2% xylenol orange indicator solution and the mixture is titrated against 0.02 *M* solution of $Zn(CH_3COO)_2$ until purple. (The aqueous solution of zinc is prepared by dissolving the salt in water, to which 2–3 drops of 8 *M* HCl have been added to prevent hydrolysis. The titer of the solution is determined with the aid of 0.1 *M* Complexone III, with xylenol orange as indicator.)

The aluminum content is found by the formula

$$\text{mg Al} = 26.98\,(a - b),$$

where a is the amount of zinc equivalent to the added amount of Complexone III solution, mg-equivalents; and b is the amount of zinc consumed in the titration, in mg-equivalents.

Trivalent chromium and nickel cannot be separated from aluminum by this method. Cr(III) may be oxidized to chromate. Nickel, when present in amounts larger than 0.3 mg in the sample, interferes with the determination of aluminum and may be removed by dimethylglyoxime precipitation.

The error of the determination of aluminum in dummies, containing 1% aluminum, is about ± 2.0 rel. %. The same results were obtained in the titration of aluminum 8-hydroxyquinolate.

Titanium

The optical absorption of the peroxide complex of titanium is determined at 410 mμ after the separation of the solution from the precipitate of plutonium peroxide by the method proposed in [294a].

A weighed sample of the alloy, containing not more than 50 mg Pu and 50–500 μg Ti, is placed in a 13 × 100 mm test tube and 0.25 ml of 70% $HClO_4$ is added, care being taken not to wet the glass near the top end of the tube. The tube is closed with a cloth stopper and cooled. When the sample has dissolved, the stopper is removed (if many undissolved particles remain, the tube is heated at 140° during 10 minutes).

To the solution obtained 1 ml of 1.5 *N* H_2SO_4 and a few drops of 0.1% *Aerosol* solution are added. A mixture of 1 ml of 1.5 *N* H_2SO_4 and 1 ml of 30% H_2O_2 is added and the solution vigorously stirred. The tube is again stoppered, clamped at a 30° angle, and left overnight. The stopper is removed and the contents of the tube are centrifuged for a few minutes at maximum speed. The clear supernatant is immediately aspired into a 10 ml volumetric flask. To the tube are added 1.5–2.0 ml of the washing solution (mixture of 95 ml of 1.5 *N* sulfuric acid and 5 ml of 30% H_2O_2) and the contents of the tube are thoroughly stirred with the aid of a rod made of a plastic material. The rod is rinsed with a few drops of the washing solution, and the contents are centrifuged for 30 seconds at the maximum speed. The washings are also aspired into the 10 ml volumetric flask. The peroxide precipitate is washed twice more, without, however, using more liquid than would make the total volume of the solution larger than 10 ml. The solution is diluted to volume with the washing solution. Most of the solution is poured into a dry 13 × 100 mm tube and is then centrifuged to remove the last particles of plutonium peroxide. The solution is cautiously decanted and its optical density measured at 410 mμ against 1.5 *N* H_2SO_4, using a 1 cm cell.

To construct the calibration curve, synthetic mixtures of pure (99.9%) plutonium with titanium(IV) sulfate are prepared.

Milligram amounts of thorium do not interfere with the determination. In the presence of uranium the results are high. Niobium also strongly interferes. The method is applicable to the analysis of binary alloys containing 0.1–1.0% Ti. The standard deviation is ± 2 μg when the content of titanium in the sample is 50–500 μg.

Iron

Waterbury et al. [159, 718] developed spectrophotometric methods for the determination of a number of elements, including iron, in synthetic alloys, the composition of which corresponds to a 10% burn-up of plutonium in fast neutron reactors. The synthetic mixtures were prepared by dissolving the stable isotopes in a melt, containing 9.5 at. % Fe or 12 at. % Co.

The alloys were dissolved in acids according to one of the following two techniques.

In the first technique, the alloy is subjected to a high temperature and high pressure treatment. A 1–2 g sample of the alloy is placed in a quartz tube with 10 mm internal diameter and ~ 30 cm length. Seven ml of 12 *M* HCl and 4 drops of 70% perchloric acid are added, after which the tube is sealed and placed in a steel cartridge (Figure 124). Before the lid of the tube is screwed down, a calculated amount of dry ice is placed in it in order to equilibrate partially the pressure inside and outside the tube. The tube is then heated at 300–325° for 2 hours. The method is convenient for the dissolution of samples containing major amounts of plutonium dioxide.

The second, more frequently used method is to treat the sample with excess concentrated HCl in a quartz distilling apparatus. The excess acid and water are removed by evaporation, ~ 10 ml of 60% $HClO_4$ are added and the mixture is boiled, while passing through a slow air stream. The distillate containing ruthenium tetroxide passes through an air condenser and collects in a trap with 3 *M* NaOH.

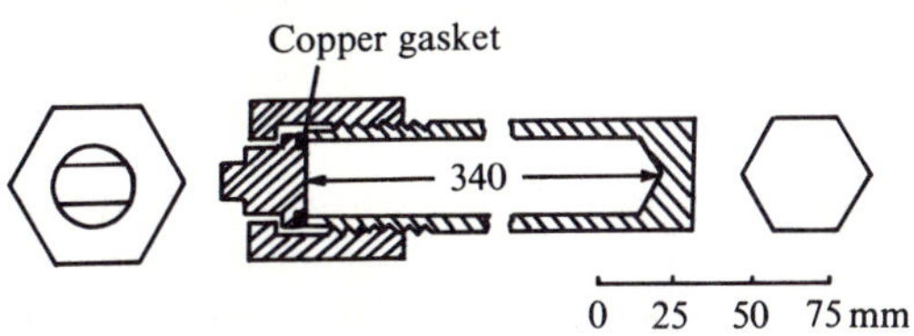

Figure 124

Steel cartridge for the protection of the sealed tube

About 40 minutes are required to dissolve a 1–2 gram sample. If the sample contains major amounts of plutonium dioxide, the dissolution will not be complete. In such a case the mixture is centrifuged, and the residue is dissolved by heating with HF, HNO_3 and H_2SO_4.

The iron is determined as the o-phenanthroline complex of ferrous iron [409, 474]. Plutonium must first be converted into Pu(III).

To a portion of the solution, containing 2–40 μg Fe and not more than 50 mg Pu, 1 ml of 25% solution of hydroxylamine hydrochloride is added. After one hour, 1 ml of 20% sodium acetate and 1 ml of 0.5% o-phenanthroline solution are added to the reduced solution. The solution is diluted to 10 ml and the photometric determination carried out after 1 hour.

The experimental error is 1–2%.

If the Pu : Fe ratio is higher than 1000, a correction for the absorption by Pu(III) must be introduced. To do this, the optical density of the solution, obtained by dissolving the sample as above, is determined; the solution should contain all the reagents except for o-phenanthroline. In addition, iron may be separated by extracting Fe(III) with hexone from 6 *M* HCl solution, or by precipitating plutonium as oxalate.

Cobalt

Cobalt in ternary alloys with plutonium and cerium is determined by the reaction with nitroso-R-salt, after plutonium has been separated by sorption on an anion exchanger [595].

The weighed sample is placed in a beaker, covered with water, and conc. HCl is cautiously added. The solution is heated under a watch-glass until reaction ceases, boiled for a few minutes, cooled and an equal volume of conc. HNO_3 is added. The solution is passed through a 7 × 0.6 cm De-Acidite FF (100–200 mesh) column, which has been preliminarily washed with 8 *M* HNO_3, at a rate of 5–6 drops per minute. Cobalt is, practically speaking, not sorbed at all, while cerium is eluted by passing 80 ml of 8 *M* HNO_3 at 10–12 drops a minute. The solutions issuing from the column are combined for the determination of cobalt and cerium.

A portion of the solution containing 100–300 μg Co is diluted to 10 ml with water and 30 ml of 0.2% solution of nitroso-R-salt and 10 ml of 50% sodium acetate solution are added. The solution is boiled for 3 minutes, 3 ml conc. HNO_3 are added, the solution is again boiled for 15 minutes and cooled. The volume is made up to 100 ml with water and the optical density at 510 mμ is measured against a comparison solution.

The presence of 14-fold amounts of cerium does not interfere with the determination of cobalt.

In the method proposed by Kressin (quoted from [718]) no separation of plutonium is required. The optical density of Co(II) is determined in acetone–hydrochloric acid solution. It was found that the absorption spectrum of Co(II) has a maximum at 665 mμ, which coincides with the absorption maximum of Pu(III). However, the molar extinction coefficients of Pu(III) at 665 and 785 mμ are equal, while Co(II) does not absorb at 785 mμ at all. The difference between the optical densities of solutions of Pu(III) at 665 and 785 mμ is due to the absorption of Co(II) at 665 mμ. The method was applied to the analysis of synthetic alloys of plutonium with cobalt, containing 0.5–3.0% Co and stable isotopes of fission products.

The sample is dissolved (p. 384) and a portion of the solution containing 3–18 mg cobalt is withdrawn. Perchloric acid (0.2 ml, 70%) is added and the solution evaporated almost to dryness. Four ml of 25% solution of hydroxylamine hydrochloride are added and the solution is heated for 15 minutes. Five ml of 12 *M* HCl are added, the solution diluted with acetone to 25 ml, and photometric determinations made at 665 and 785 mμ. The cobalt content is determined from the difference between the optical densities at 665 and 785 mμ, with the aid of a calibration curve.

The error does not exceed ± 1.5%. Three samples can be analyzed in one hour. Fission products, present in weighable amounts (Mo, Ce, Ru, Zr, La), do not interfere.

The photometric determination of Co(II) may be carried out in the same solution at 695 mμ, i. e., at a wavelength at which the absorption by Pu(III) is small, and no correction for its presence need be introduced. In this case the accuracy is somewhat lower.

Zirconium

Buchanan and co-workers [327] developed a spectrophotometric determination of zirconium by sodium alizarinsulfonate in alloys of plutonium and uranium, used as nuclear fuel. These alloys have the following composition: 20–40% Pu, 50–75% U, 0.05–4.5% Zr, 1.25–6.0% Mo, 1.25–6.0% Ru, 0.25–0.9% Rh and 0.7–2.5% Pd. Zirconium and other long-lived fission elements accumulate in the alloys as a result of fission processes and are not fully removed during pyrometallurgical regeneration of the fuel.

The weighed sample (a few milligrams) is dissolved in 6 *M* HCl containing 30% H_2O_2. One or two drops of 1 *M* HF may be added to ensure full dissolution. The solution is diluted to a suitable volume, and an aliquot containing 40–100 μg Zr is withdrawn and placed in a test tube. Concentrated HNO_3 is added to a final acidity of 10 *M*. Prior to the anion-exchange separation of zirconium from plutonium the solution is left to stand overnight. A 6 × 0.6 cm column is filled with AG-1 × 10 (Dowex-1 × 10 type) anion-exchange resin, of 200–400 mesh grain size. The anion exchanger layer is treated with 5 ml portions of 10 *M* HNO_3 and water, the last washing being carried out with 10 *M* HNO_3. The total volume of the wash liquor should be 25–30 ml. The sample solution is passed through the column at a rate of 1–2 ml/min and the column is immediately washed with 25–30 ml of 10 *M* HNO_3 at a rate of 6 ml/min (displacing the solution by vacuum). The eluate contains all the zirconium and some plutonium. Plutonium does not interfere with the determination of zirconium if its content in the zirconium fraction does not exceed 60 μg. The amount of plutonium is found by the radiometric method (the sum of Pu^{239}, Pu^{240} and Am^{241} is determined; the last mentioned isotope is the daughter product of Pu^{241}). The solution is then evaporated to 5–10 ml, cooled and 5 drops of 70% $HClO_4$ are added. The solution is heated for 5 minutes to eliminate ruthenium and is then cooled. If HF has been added during the dissolution of the sample, the solution must be evaporated to dryness, since traces of HF interfere with color formation. To the residue 0.2 ml concentrated HNO_3 is added and the mixture is boiled under reflux for about 30 minutes to attain a full dissolution of zirconium. The cooled solution is transferred to a 25 ml volumetric flask and is diluted to 15 ml with water. Five ml of 0.0125% aqueous sodium alizarinsulfonate solution are added and the contents of the flask are mixed. The solution is then diluted to the mark with water, mixed and left to stand for $1\frac{1}{2}$ hours. The optical density of the solution is determined at 520 mμ using a Beckman spectrophotometer (Model B) and a 1 cm cell, the comparison solution being 0.12 M HNO_3 containing the reagent.

Plutonium does not affect the optical density value at 520 mμ up to a concentration of 60 μg in 25 ml of solution. At higher concentrations of plutonium, the optical density decreases. In such a case a correction for plutonium must be introduced; the value of the correction is obtained from the preliminarily plotted calibration curve giving the ratio μg Zr/D as a function of Pu content (in μg). Th, Mo and Pd do not interfere if present in sample solutions in amounts not greater than that of zirconium. Uranium does not interfere even at U : Zr = 2000 : 1. The concentration of HNO_3 should not exceed 0.15 M. The experimental error is not more than 3 rel. % under these conditions.

Another spectrophotometric determination of zirconium involves the separation of zirconium from Pu(III) by precipitation with p-bromomandelic acid with subsequent photometric determination of the complex of zirconium with chloroanilic acid [310a, 718]. The method is recommended for the analysis of alloys with 0.01–1.0% of zirconium.

The alloy sample is dissolved as described above (p. 385). An aliquot part of the solution, containing not more than 10 μg Zr, is placed in a 15 ml centrifuge tube and diluted with water to 3 ml. Using concentrated HCl, the acidity is adjusted to 1–2 N. Two ml of 10% solution of hydroxylamine hydrochloride are added and the solution heated on a boiling water bath for 30 minutes. To the hot solution are added 5 ml of p-bromomandelic acid solution (2.0 grams of reagent dissolved in 100 ml warm water). The resulting solution is heated for 30 minutes on a boiling water bath and is then cooled to room temperature. The walls of the tube are washed down with 8–12 drops of *Aerosol* solution (5% solution in water) and the mixture is centrifuged for 5 minutes. The supernatant is separated by aspiration. The precipitate is washed with three drops of *Aerosol* solution mixed with 2–3 ml water, the mixture is centrifuged and the supernatant is again removed by aspiration. To the washed precipitate are added 7 drops of 70% $HClO_4$ and 2 drops conc. HNO_3. The tube is heated to 200° in a heating block, and left at this temperature for 3–5 minutes, to decompose organic matter. The cooled solution is diluted with water to 3 ml, two ml of hydroxylamine solution are added and the solution is heated for 5 minutes on a boiling water bath. Five ml of p-bromomandelic acid solution are added and the solution is heated on a water bath for 30 minutes. The solution is cooled, centrifuged and the precipitate washed as described above. To the precipitate 3 drops of 70% $HClO_4$ and 2 drops of conc. HNO_3 are added and the precipitate dissolved as above.

The solution is transferred to a 25 ml volumetric flask with the aid of 2 M $HClO_4$ and the solution made up to the mark with the acid. Three ml of chloranilic acid are placed in a 10 ml volumetric flask (for preparation of reagent see p. 388) and the solution is diluted to the mark with the sample solution from the 25 ml flask. The optical absorption is measured at 350 mμ using a 1 cm cell; the comparison solution is 3 ml of standard chloranilic acid diluted with 2 M $HClO_4$ to 10 ml. The calibration curve is constructed with the aid of standard solutions of zirconium as described above.

If the amount of zirconium in the sample is less than 20 μg, the sensitivity of the determination is increased by transferring zirconium p-bromomandelic solution immediately after the second precipitation to a 10 ml volumetric flask containing 3.00 ml chloranilic acid (eliminating the stage of dilution in 25 ml flask). If the content of zirconium in the sample is higher than 80 μg, an aliquot is withdrawn from the 25 ml flask and the photometric determination performed.

If the sample contains 25 μg Zr or more, the average deviation is 2.7% for 20 determinations. Twentyfold amounts of Al, As, Bi, Cd, Ce, Cr, Co, Cu, Pb, Mg, Mn, Mo, Hg, Nb, Ni, Ag, Sr, Sn, Ti, V and Zn do not interfere with the determination. Thousandfold amounts of uranium give results which are low by 5–7%. Hafnium, which is precipitated by p-bromomandelic acid, interferes.

Niobium

Niobium in plutonium and in plutonium alloys is determined colorimetrically, using the reaction with hydroquinone in sulfuric acid medium. Niobium is preliminarily extracted with hexone from a mixture of 6.3 *M* H_2SO_4 and 1.6 *M* HF [715].

Reagents. Solution of hydroquinone (55 mg/ml): 5.5 g of hydroquinone are dissolved in concentrated H_2SO_4 in a 100 ml volumetric flask. The solution is placed in a 5 ml Koch microburet, which carries a silica gel tube at the top to absorb moisture.

Standard solution of niobium: ~ 0.3 g of pure metallic niobium are accurately weighed out (tantalum content must be less than 0.1%) and placed in a 50 ml platinum dish, containing a few ml HNO_3 (1 : 1). Ten ml H_2SO_4 are added and then HF drop by drop so that the reaction proceeds at a fairly slow rate. After the sample has dissolved, the solution is evaporated to dryness until abundant fumes of sulfuric acid are evolved, and the cooled solution is transferred to a 100 ml volumetric flask (insoluble residue is treated with 10 ml of 4 *M* HF). The dish is washed with water, which is also introduced into the flask and the volume is made up to the mark. The solution is mixed and placed in a polythene beaker. Ten ml of this stock solution are diluted to 100 ml with water; the resulting solution contains 0.3 mg Nb per ml.

Procedure. A weighed sample of the alloy is placed in a centrifuge tube and 2–3 ml water are added. Concentrated HCl is added drop by drop; the reaction should not be allowed to become too vigorous. When the hydrogen evolution has ceased, the tube is heated. The mixture is centrifuged, the solution transferred to a volumetric flask and the residue is placed in a platinum dish, in which it is treated with concentrated solutions of HF, HNO_3 and H_2SO_4. The mixture is evaporated until copious H_2SO_4 fumes appear. If the residue is fully soluble in water, it is also transferred to the volumetric flask; otherwise it is treated with 2 drops of conc. HF in 1–2 ml water with heating. The solution is added to the volumetric flask. If the dissolution is still incomplete, the residue is fused with a mixture of $NaNO_3$ and NaOH (p. 389).

An aliquot of the solution, containing 25–1000 μg Nb and not more than 300 μg Pu, is placed in a platinum dish, conc. H_2SO_4 is added to a volume of 3.5 ml and the solution evaporated to the appearance of abundant sulfuric acid fumes. The solution is transferred to an extractor (p. 389), the dish is washed with 2 ml water, and the washings are also placed in the extractor. The solution is cooled in an ice bath, the platinum dish is washed with 4 ml of 4 *M* HF, and then with 0.5 ml water. These solutions are placed in the extractor. Ten ml of hexone are also added and the mixture is stirred for 5 minutes. After layer separation, two 3 ml portions of the organic phase are withdrawn and transferred into 30 ml platinum dishes containing 2–3 drops of 5 *N* NaOH.* The mixture is slowly evaporated to dryness under an IR lamp with stirring, the dish is cooled and 10 drops of water and 5 drops of 48% HF are added, then 2–3 ml conc.

* The extraction is repeated in the presence of milligram amounts of Au, Ir, Pt, Mn, Ti, V and Mo.

H_2SO_4, and the solution is evaporated under the lamp until copious H_2SO_4 fumes appear. The solution is cooled, 5–10 mg crystalline ammonium persulfate are added and the mixture heated on an electric heater until gas evolution ceases. The persulfate treatment is repeated until the sulfuric acid solution clears up. The solution is then heated until abundant sulfuric acid fumes are evolved, during 3–5 minutes. The dish is cooled and the solution transferred to a dry 10 ml volumetric flask, washing the dish with small portions of sulfuric acid. The total volume should not exceed 7.0 ml. The solution is cooled and 3.0 ml hydroquinone solution are added from the buret, and the volume is made up to the mark with conc. H_2SO_4. The optical density D_s is measured at 410 mμ in a 1 cm cell relative to hydroquinone (3.0 ml reagent and 7.0 ml H_2SO_4). In order to introduce the blank correction, the H_2SO_4 solution is subjected to the same treatment as the sample and D_b is obtained. Using the standard solution of niobium, D_{st} is also determined. All three D values are determined on equal volumes of organic phase. The niobium content in the sample is calculated by the formula:

$$\text{Nb (mg)} = \frac{a(D_s - D_b)}{D_{st} - D_b},$$

where a is the niobium content in the standard, mg.

At a niobium content in the sample of 300–600 μg, the following do not interfere: U, Pu, Na (up to 500 : 1), Al, Bi, Cd, Ce, Cr(III), Cu, Fe, Pb, Mg, Hg, Nd, Ni, Ru, Tl, Ag, Sr, Sn, Zn and Zr (up to 30 : 1) and Ti (7 : 1). In the presence of 20- to 30-fold amounts of Au, Ir, Mn, Pt, Ti and V and 4- to 6-fold amounts of molybdenum a second extraction is carried out. The interfering effect of Cr(VI) may be reduced by adding HCl prior to sulfuric acid evaporation. Tantalum and molybdenum strongly interfere. Water strongly affects the optical density value; in the presence of 1 mg water in 10 ml solution, the value of D decreases by 0.1%. If the sample contains 1 mg of phosphate ions, the optical density decreases by 0.07%. The analyst must ensure that the acidity of the aqueous solution prior to extraction is 6.3 M in H_2SO_4 and 1.6 M in HF. Unless this is the case, the extraction efficiency of niobium decreases considerably. Under the conditions here indicated, 99.9% Nb can be found with a standard deviation of ± 2.0%. The analysis of 15–20 dissolved samples can be performed in one day. The method has been employed to determine 0.03 wt. % Nb in alloys with plutonium and other elements (Ce, Ru, Zr, etc.), which simulate the composition of plutonium after prolonged irradiation [718].

Molybdenum

Microgram amounts of molybdenum in plutonium and its compounds [713] are determined by extracting molybdenum with methyl isobutyl ketone (hexone) from a solution of 0.4 M HF in 6 M HCl and determining the optical absorption of molybdenum complex with chloranilic acid in the aqueous reextract.

Reagents. Chloranilic acid: 8 g of chloranilic acid are dissolved in one liter of boiling water and the hot solution is filtered. The filtrate at 50° is treated with two successive 200 ml portions of benzene. The organic phase is discarded, while the aqueous phase is cooled in ice water. The precipitated red crystals are filtered and washed on the filter with three 10 ml portions of water. The crystals are dried at 115°. Ninety mg chloranilic acid are dissolved in 100 ml water.

Hexone: to a solution of 20 ml water, 20 ml conc. HCl and 5 ml of 4 *M* HF, about 100 ml hexone are added. The mixture is stirred for 10–15 minutes, and the phases separated after standing. The organic phase thus prepared may be utilized during one day.

Standard molybdenum solution: a weighed amount of pure metallic molybdenum is dissolved in aqua regia and the solution evaporated to dryness. A few milliliters of water are added to the residue, then 1 ml concentrated NH_4OH and the solution is again evaporated to dryness. The residue is transferred to a volumetric flask with the aid of water and the solution is made up to the mark with water. By suitable dilution of the resulting solution, more dilute solutions with any desired molybdenum concentration may be prepared.

Procedure. To a weighed sample of plutonium or plutonium alloy in a centrifuge tube are added first, a little water, and then, gradually, concentrated HCl. In this way, foaming and spraying are avoided. When hydrogen evolution has ceased, the tube is heated until the sample is fully dissolved, the solution is transferred to a volumetric flask and made up to the mark with HCl solution. A portion of this solution (not more than 4 ml), containing 10–100 μg Mo, is placed in the extractor (Figure 125a). The solution is diluted to 4 ml with water and 5 ml conc. HCl and 1 ml 4 *M* HF are added.

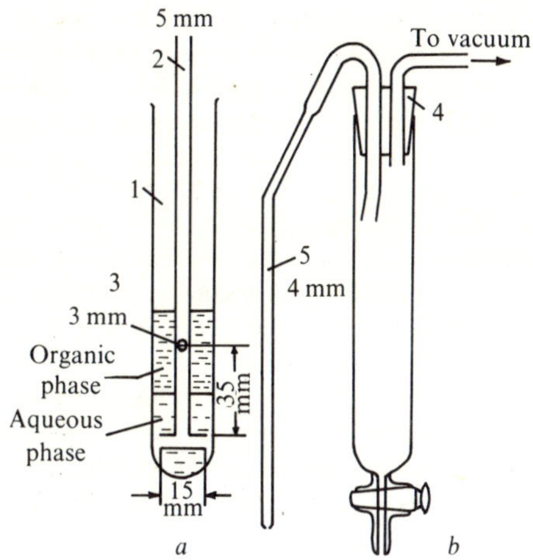

Figure 125

Diagram of extractor (*a*) and reextractor (*b*):

1) test tube (diameter 25 mm, length 150 mm); 2) hollow glass stirrer; 3) hole; 4) rubber bung; 5) tube.

If the initial sample is not completely soluble in hydrochloric acid, the residue is treated with HF, HNO_3 and H_2SO_4 with heating until abundant H_2SO_4 fumes are evolved. The residue is dissolved in water and transferred to the same volumetric flask. If the residue is not fully dissolved, it is placed in a gold crucible and fused with a mixture of 0.1–0.5 g NaOH and 0.2–1.0 g $NaNO_3$ (depending on the amount of the residue). The melt is dissolved in water, the solution acidified with HCl, transferred to a volumetric flask and made up to the mark with water. The portion required for the analysis is then treated with a few drops of H_2SO_4 in a platinum dish and evaporated to dryness. The residue is dissolved in 4 ml water with the addition of 5 ml HCl and 1 ml 4 *M* HF. The solution obtained is placed in the extractor. Since the extraction of molybdenum should be carried out from solutions containing 6 *M* HCl and 0.4 *M* HF, evaporation with sulfuric acid is performed if the aliquot for the analysis is larger than 4 ml or if the HCl concentration in the dissolved sample is not known.

Ten ml of hexone are added to the extractor and the mixture is stirred for 5 minutes. After layer separation, the organic phase is cautiously aspired into the reextraction vessel (Figure 125b). The walls of the extractor are washed with 5 ml of hexone, without stirring, and the washings are transferred to the reextractor. The aqueous phase is then treated with two 5 ml portions of hexone while stirring, and the organic phase is cautiously removed into the reextractor each time. To the combined organic solution 5–8 ml of water are added, the mixture is stirred for 3–5 minutes and the aqueous phase is drained through a stopcock into a platinum dish. The hexone layer is treated with two more 5 ml portions of water, the extract being poured into the platinum dish each time. The reextract is evaporated to dryness under an infrared lamp. To the residue are added 5–6 drops of conc. HCl and about 1 ml water and the mixture is again evaporated. The residue is wetted with 2 drops of conc. HCl and the solution is quantitatively transferred to a 25 ml volumetric flask with the aid of water and 5 ml of 0.2 *M* NaOH. The solution in the flask is heated under an IR lamp during 30 minutes to coagulate the hydroxides (e. g., iron hydroxide). The mixture is cooled, made up to the mark with water and filtered through Whatman No. 42 filter paper into a dry flask. Twenty ml of the filtrate are placed in a platinum dish and evaporated to dryness under the lamp. A smaller volume of the filtrate is taken for samples containing more than 100 μg Mo. The residue is dissolved in 3 ml water, the solution is cautiously acidified with HCl and again evaporated to dryness. The residue is transferred to a 10 ml volumetric flask with the aid of 2 *M* $HClO_4$ and made up to the mark with the perchloric acid. At least 3 ml of this solution are placed in a cell and the optical density D_1 is determined at 350 mμ relative to 2 *M* $HClO_4$. Exactly 3.00 ml of the chloranilic acid solution is added to the solution which has remained in the 10 ml flask, and the solution is made up to the mark with solution in the cell. The optical density D_2 of the solution is now measured at 350 mμ against the solution of the reagent (3.00 ml of chloranilic acid solution and 7.00 ml of 2 *M* $HClO_4$). The contribution of molybdenum to the optical density D_{Mo} is calculated by multiplying D_2 by $\frac{10}{7}$ and subtracting D_1 from the result.

The extraction and all subsequent operations are now carried out on 10 ml of blank solution, whose concentration is 6 *M* in HCl and 0.4 *M* in HF. The correction D_b is introduced into the value of D_{Mo} found.

The calibration curve is preliminarily determined as follows. Portions of solution with a known content of molybdenum are evaporated to dryness in platinum dishes, with the addition of 2–3 drops of conc. HCl. The residues are quantitatively transferred with 2 *M* $HClO_4$ into 10 ml volumetric flasks containing 3.00 ml of the solution of chloranilic acid, the volumes are made up to mark with 2 *M* $HClO_4$ and the photometric determination is made against the solution of the reagent (3.00 ml of chloranilic acid solution and 7.00 ml of 2 *M* $HClO_4$). The molybdenum concentration sought is found on the calibration curve thus prepared.

The color of the molybdenum complex is stable for several hours (a small increase in the optical density was noted after 15 hours). Many elements (Fe, Zr, U(VI), Bi, Sn, Cr(VI), V and W) strongly interfere with the color determination of the molybdenum complex. Mn(II), Ti, Ni, Pb, Al and certain other elements interfere to a lesser extent. If the complete experimental procedure is carried out, Sn, Bi, and W are the only elements which accompany molybdenum and interfere with the determination. The method was tested out on synthetic mixtures. The average standard deviation for 32 determinations is 1.6%.

The method was employed in determining molybdenum in alloys of plutonium with a number of elements, which simulated the composition of plutonium after irradiation in a reactor [718].

Phillips et al. [596] described the analysis of ternary alloys of plutonium, uranium and molybdenum (14.7% Mo, 0–85.3% Pu, 0–85.3% U). The determination of molybdenum consists in a spectrophotometric determination of the color intensity of the thiocyanate complex, without preliminary separation of U and Pu.

A weighed sample of an alloy containing not less than 10 mg Mo is treated with 20 ml of 6 *M* HCl and the solution is left to stand until the reaction has ceased. Twenty ml conc. HNO_3 are then added and the mixture is cautiously heated for one hour. Seven ml conc. $HClO_4$ are then added and the mixture is heated for one hour. If the sample is not fully dissolved, 0.1 ml conc. HF is added, the mixture heated for another hour and is finally boiled to expel HCl, HF and most of HNO_3. During this operation abundant $HClO_4$ fumes must not be allowed to evolve for too long a period of time because of the danger of precipitation of molybdenum oxide. The solution is diluted to 30 ml, filtered into a 50 ml volumetric flask and made up to the mark with water. Five ml of the resulting solution is evaporated with 10 ml conc. H_2SO_4 to remove nitric acid, cooled and diluted to 25.0 ml. A 5 ml portion of this solution is placed in each of two 50 ml volumetric flasks, and the following are added to each flask: 20 ml 6.4 *N* H_2SO_4, 5 ml 0.2 *M* $Fe(NH_4)_2(SO_4)_2 \cdot 6H_2O$, 1 ml $HClO_4$ (sp. gr. 1.54), 5 ml 1.5 *M* NH_4CNS, 5 ml 0.5 *M* $SnCl_2$ and the mixture is made up to the mark with water. The comparison solution, containing a known, but somewhat lower amount of molybdenum, is prepared in the same manner from a standard solution of molybdenum. The solutions are left to stand for 10 minutes and the optical density at 500 mμ is determined with a spectrophotometer (slit width 0.06 mm).

Uranium (up to 250-fold amounts) and plutonium (up to 15-fold amounts) do not affect the determination of molybdenum. The experimental error does not exceed ±2 rel. % for synthetic mixtures.

Ruthenium

The determination of ruthenium (~1.3%) in synthetic plutonium–cobalt alloys (or plutonium–iron alloys), containing stable isotopes of fission products, is based on the determination of the light absorption of the chloride complex of Ru(IV) at 485 mμ ($\varepsilon_{485} \cong 5300$) [718]. The preparation of the sample for analysis varies in accordance with the composition of the dissolved sample.

After the dissolution of the alloy (p. 382) the solutions are treated with SO_2 to reduce ruthenium to Ru(III), which is then quantitatively oxidized to Ru(IV) by chlorine in 6 *M* HCl solution. Solutions containing up to 15 μg Ru per ml obey Beer's law. The method is not applicable to the analysis of solutions containing major amounts of nitrates and iron. In such cases ruthenium is preliminarily distilled from a nitrate-free solution and is absorbed in 3 *M* NaOH (p. 391). The solution is diluted to a known volume, a portion containing 25–400 μg Ru is withdrawn, 10 ml of 12 *M* HCl are added, the solution made up to 25 ml and the solution heated on a boiling water bath for 20 minutes. The solution is cooled and its optical density measured at 485 mμ.

The absorption spectra of Ru(IV), obtained by the two methods just mentioned, are identical. In 16 determinations of ruthenium in amounts between 70–280 μg, the deviation from the average is 0.9% after distillation. Variations in the volumes of 3 *M* NaOH and 12 *M* HCl not exceeding ± 1 ml do not affect the results. None of the fission products interferes with the determination.

Rhodium and platinum

In order to determine rhodium and platinum in alloys of these elements with plutonium, Smith [678] used the method of Ayres [281], based on the formation of colored compounds with stannous tin by these elements.

The absorption spectra of rhodium and platinum have maxima at 470 and 399 mμ respectively. The optical density values obtained at these wavelengths must be corrected for the absorption by Pu(III). At these wavelengths the absorption by Pu(III) is insignificant. The concentration of plutonium is found from its absorption at 665 mμ.

Calibration curves are constructed with the aid of standard solutions of rhodium and platinum.

Standard solution of rhodium: about 100 mg of pure rhodium foil accurately weighed, and a mixture of 4 ml 12 M HCl and 5 drops of 16 M HNO_3 are placed in a thick-walled test tube made of borosilicate glass (the volume of the solution should not exceed half the volume of the test tube). The test tube is sealed, placed in a protective steel jacket (see p. 383) and heated at 200° for 4 hours. The tube is then cooled, opened, and the solution is transferred to a 100 ml volumetric flask. The tube is washed with water, and the washings are also added to the flask. The volume of the flask is made up to mark with water.

Standard solution of platinum: about 100 mg of platinum foil, accurately weighed, is dissolved in aqua regia. The solution is evaporated to dryness on a steam bath and the residue is dissolved in 10 ml of 6 M HCl. The evaporation and dissolution is repeated three times. The final residue is dissolved in water and the solution is transferred to a 100 ml volumetric flask.

Solution of stannous tin (1 M): 22.6 g $SnCl_2 \cdot 2H_2O$ are dissolved in 21 ml of 12 M HCl, transferred to a 100 ml volumetric flask and made up to the mark with water. The solution must be freshly prepared each week.

Procedure. Rhodium and platinum alloys, which contain up to 5% of each element, are treated as follows. A weighed (not more than 100 mg) sample is placed in a thick-walled, carefully annealed borosilicate glass tube. Four ml of 12 M HCl are slowly added and, when the reaction has ended, 2 drops of 72% $HClO_4$ (the total volume should not exceed half the tube volume). The test tube is sealed with an oxygen burner, and the seal is annealed in a gas flame. The tube is placed in a steel cartridge and 40 g of dry ice (see p. 383) are added. The lid of the cartridge is immediately screwed on and tightened to prevent CO_2 from escaping. The tightness of the screw is tested by immersing the cartridge in water. The cartridge is then heated in a furnace at ~ 300° for 3 hours. The cartridge is taken out, cooled to room temperature and the test tube is extracted. If the sample is fully dissolved, the solution is cooled with dry ice and the tube opened. The solution is transferred to a 50 ml flask, and the tube is rinsed out with water, the washings also being added to the flask. The solution is now evaporated down to ~ 5 ml on a boiling water bath. It is then transferred to a 10 ml volumetric flask (if the total amount of rhodium and platinum does not exceed 200 μg). If this is not the case, a smaller volume is taken, or the solution is suitably diluted prior to the determination. One ml of 12 M HCl and 1 ml of 1 M $SnCl_2$ are added. The flask is placed on a boiling water bath, heated for one hour, cooled and made up to mark with water. After one hour, the optical density of the solution is determined at 399, 470 and 665 mμ in a 1 cm cell against a solution of the reagents. The contribution of plutonium to the absorption at 470 and 399 mμ is found from the relationships:

$$D_{Pu(470)} = 0.216 \cdot D_{Pu(665)} - 0.004,$$

$$D_{Pu(399)} = 0.157 \cdot D_{Pu(665)} - 0.004.$$

The calculated values are subtracted from the experimental values of the optical densities at 470 and 399 mμ:

$$D_{470} - D_{\mathrm{Pu}(470)} = D_{\mathrm{Rh\,+\,Pt}(470)};$$

$$D_{399} - D_{\mathrm{Pu}(399)} = D_{\mathrm{Rh\,+\,Pt}(399)}.$$

To find the concentration of rhodium and platinum, the specific optical densities a of both elements are determined at both wavelengths. To do this, a few portions of various volumes of standard solutions of rhodium and platinum are each treated with hydrochloric acid and stannous chloride in the same manner as the sample solutions, the optical densities of each solution are measured, and calibration curves at 470 and 399 mμ are plotted for each element. From these curves the values a are found, which are given by the slope of the calibration curves or by the ratio between the optical density and concentration. The concentrations of rhodium and platinum are calculated from the equations:

$$C_{\mathrm{Rh}} = \frac{a_4 \cdot D_{\mathrm{Rh\,+\,Pt}(470)} - a_2 \cdot D_{\mathrm{Rh\,+\,Pt}(399)}}{a_1 \cdot a_4 - a_2 \cdot a_3},$$

$$C_{\mathrm{Pt}} = \frac{a_1 \cdot D_{\mathrm{Rh\,+\,Pt}(399)} - a_3 \cdot D_{\mathrm{Rh\,+\,Pt}(470)}}{a_1 \cdot a_4 - a_2 \cdot a_3},$$

where a_1 and a_2 are specific optical densities at 470 mμ of rhodium and platinum respectively, and a_3 and a_4 are their specific optical densities at 399 mμ.

The method was checked on simulating solutions with rhodium and platinum concentrations of 5–14 μg/ml and with plutonium concentrations of 10 mg/ml, as well as on ternary Pu—Rh—Pt alloys containing 4–5 wt. % of rhodium and platinum. The standard deviation is about ± 2 rel. %. Ru, Pd, Os, Au and Cr interfere. Rhodium and platinum can be determined in the presence of 15-fold amounts of Zr and U.

Cerium

As a result of prolonged irradiation of plutonium by fast neutrons, the plutonium collects major amounts of fission products including cerium. In synthetic alloys, the composition of which corresponds to a 10% burn-up of plutonium, cerium (up to 0.74 wt. %) is determined by its own coloration when in the form of Ce(IV) [718]. The method for the dissolution of such samples has been described above (see p. 382).

A portion of solution containing 50 mg Pu is diluted to 10 ml with 1 *N* sulfuric acid. Plutonium is oxidized to Pu(VI) by persulfate in the presence of Ag^+ ions, a solution of carrier (lanthanum, 1 mg) and 0.5 ml of 48% HF are added. The precipitate is centrifuged and washed with 2 ml of 1 *M* HF. Sulfuric acid (0.1 ml, 18 *M*) is added to the precipitate and the mixture is heated at 200° for 20 minutes. One ml of 1 *N* H_2SO_4 is added, 0.2 ml of 0.25% solution of $AgNO_3$ and 1 ml of 2.4% solution of ammonium persulfate. The solution is heated on a boiling water bath, cooled, diluted to 10 ml and the optical density is determined at

320 mμ. In order to determine the contribution of extraneous ions to the value determined, 2 drops of 3% H_2O_2 are added, when Ce(IV) is reduced to Ce(III), and the optical density at 320 mμ is again measured.

The experimental error is 5% and about 1% for cerium contents in plutonium of 0.01% and 0.3% respectively. Small amounts of cerium (less than 0.1%) are much more accurately determined in 10 cm than in 1 cm cells. The determination may be carried out in the presence of 12-fold amounts of Mo, Zr, Ru; 24-fold amounts of Nd; 36-fold amounts of Co, Ta, Fe and 44-fold amounts of K. Major amounts of ruthenium are eliminated by evaporation with perchloric acid.

Cerium in ternary alloys with plutonium and cobalt is determined by oxidative titration of Ce(III) against permanganate after plutonium has been removed by ion exchange [595].*

A portion of the solution, containing not less than 25 mg Ce is evaporated to 5 ml residual volume, after plutonium has been separated on an anion exchanger. Ten ml of 0.5% H_2SO_3 are added, diluted with water to 50 ml and the solution boiled for 15 minutes. The cooled solution is placed in a photometric cell, 100 ml of 11% solution of $K_4P_2O_7$ are added, the pH adjusted to 5.5–7 with HNO_3 or NH_4OH and the solution titrated against 0.02 *N* $KMnO_4$ (Ilford 604 filter).

Cerium may be determined in the amounts of 6–90 mg cerium in the presence of twofold amounts of cobalt.

Lanthanum and cerium

These elements, when present in alloys with plutonium, iron and other fission products (2.5% Fe; 0.7% Ce; 1.3% La; 0.7% Mo; 1.3% Ru and 0.8% Zr) are determined spectrographically after separation of plutonium [307].

To a portion of the solution of the alloys (for the dissolution, see p. 382) containing 40–100 μg each of lanthanum and cerium, 1 mg Eu is added as coprecipitant, and as spectroscopic internal standard. The solution is strongly heated with 1–3 *M* $HClO_4$ to oxidize plutonium to Pu(VI) and cerium, and lanthanum and europium are precipitated by adding HF. The precipitate is dissolved in conc. $HClO_4$ and the solution evaporated to dryness to remove excess acid. The residue is taken up in HCl, portions of the resulting solution are applied to copper electrodes and spectroscopic determination is carried out in the spark.

The accuracy of the determination is ± 6%. The sensitivity of the determination of lanthanum is 5 μg when the content of plutonium is up to 100 mg; the sensitivity of the determination of cerium is half as high.

Tantalum

The determination of tantalum (0.01–2%) in alloys with plutonium [714] is carried out according to a method which is similar to the one described for niobium.

* For a description of the dissolution of the alloy and anion-exchange elimination of plutonium see p. 384.

Standard solution of tantalum (0.6 mg/g): about 0.1 g of pure metallic tantalum is accurately weighed out (the sample should not contain more than 0.1 wt. % Nb) and placed in a 50 ml platinum dish. Four ml water and 2 ml conc. HNO_3 are added. Hydrofluoric acid is then added drop by drop, taking care for the reaction not to become too violent, until the metal is fully dissolved. Thirty-five ml of conc. H_2SO_4 are cautiously added and the solution evaporated to copious evolution of sulfuric acid vapors. The solution is transferred to a tared flask with a ground glass stopper, the dish is washed and the total volume is made up to 100 ml with conc. H_2SO_4. The flask with its contents is weighed and the concentration of tantalum, in mg/g, is calculated.

Procedure. A portion of the alloy solution (for solubilization of alloy see p. 386), containing 0.025–3.0 mg Ta, is placed in a platinum dish. Conc. H_2SO_4 is added to an approximate volume of 3.3 ml and the solution evaporated under a heating lamp or on an electric heater until abundant sulfuric acid vapors are evolved. The solution is transferred to an extractor (for design of extractor see p. 389). The dish is washed with 5 ml water, which is also poured into the extractor and the solution is cooled. The dish is washed with 1 ml 4 *M* HF and 1 ml water, and the washings are added to the extractor. The solution is again cooled, 10.0 ml hexone are added and the mixture stirred for 5 minutes. After phase separation, 8 ml of the organic phase are placed in a 30 ml platinum dish, which contains 3 drops of 5 *N* NaOH. In the case of samples containing more than 0.5 mg Ta, two 2 ml portions are withdrawn for parallel determinations. The mixtures are slowly evaporated to dryness under the lamp, while stirring. After cooling 10 drops of water are added, and then, cautiously, 5 drops of conc. HF. (For samples containing milligram amounts of Ti and Mo, 3.5 ml conc. H_2SO_4 are added, the solution is evaporated to sulfuric acid fumes and the extraction as above is repeated.) Sulfuric acid (2–3 ml) is added, and the solution evaporated under the lamp until sulfuric acid fumes are evolved. The solution is cooled, 5–10 mg of crystalline ammonium persulfate are added, and the solution heated on a plate until gases are no longer evolved. The persulfate treatment is repeated until the sulfuric acid solution becomes colorless. The solution is heated 3–5 minutes until abundant H_2SO_4 fumes are evolved. The dish is cooled, the contents transferred to a 10 ml volumetric flask, using small portions of conc. H_2SO_4 to rinse the dish. The total volume of the solution plus washings should not exceed 7.0 ml. The solution is cooled to room temperature, 3.0 ml solution of hydroquinone (in conc. H_2SO_4, 55 mg/ml) are added from a buret (for preparation and storage of the reagent solution see p. 387). The solution is made up to the mark with conc. H_2SO_4 and the optical density D_s is measured in a 1 cm cell at 375 mμ relative to a solution of the reagent (3.00 ml hydroquinone solution and 7.00 ml sulfuric acid). A blank solution of H_2SO_4 and a portion of the standard solution are analyzed in the same way, equal volumes of solution being taken in all three determinations. The content of tantalum is given by the formula:

$$\text{Ta (mg)} = \frac{a \cdot (D_s - D_b)}{D_{st} - D_b}$$

where D_b, D_{st} and a are, respectively, the optical density of the blank solution, the optical density of the standard solution, and the content of tantalum in the standard solution in mg.

The method makes it possible to determine 1 mg Ta in the presence of Na, K, U(VI), Pu(III) or Pu(IV) (up to 100 : 1); Al, Cr(III), Sr, V(V), Tl(I), Co or Fe(III) (up to 40 : 1); W, Ce, Zr, Ru(IV) or Au (up to 3 : 1). Molybdenum is extracted together with tantalum, and titanium also partly ($\sim$ 0.5%) passes into the organic phase. Both these elements react with hydroquinone. If two extraction cycles are carried out, 1 mg Ta may be determined in the presence of 30 mg Mo or 100 mg Ti. Niobium is extracted to an extent of 60–70%. If niobium is present in a tenfold amount, the result are 10–15% high after two extraction cycles. In the presence of

1 mg water in 10 ml of solution the results are 0.0019% low, in the presence of one mg of phosphate ions they are 0.024% low, while in the presence of 1 mg F^- they are 6% low.

When all the precautions given above are taken, the error is ± 2 rel. %.

Uranium

The determination of uranium alloyed with plutonium is carried out spectrophotometrically, using the absorption of uranyl sulfate in 4 *N* H_2SO_4 at 430 mμ [596]. Interference by Pu(III), Pu(IV) and Pu(VI) makes a preliminary separation necessary. To separate uranium from plutonium it is recommended to sorb Pu(IV) as nitrate complex on strongly basic anion exchanger. Uranium passes through the column and may be determined spectrophotometrically in the sulfate solution.

Uranium in ternary alloys with plutonium and molybdenum is determined in a similar manner [596, 410].

Extraction methods may also be used to separate uranium from plutonium. One efficient method is to extract Pu(IV) with 0.5 *M* TTA in xylene (see p. 316). Uranium remaining in the aqueous phase is determined by titration, after the organic matter has been decomposed by evaporation with sulfuric or perchloric acid and after U(VI) has been reduced to U(IV) in a lead reductor [359a]. Direct titration or back-titration methods may be employed. Good results are obtained in direct titration of U(IV) by cerium(IV) sulfate with ferroin as indicator (at elevated temperatures) or copper phthalocyanintetrasulfonic acid at room temperature [9, p. 92], and also by titration of the excess Ce(IV) with a solution of ferrous sulfate with ferroin as indicator.

Uranium(VI) is determined by polarographic method in 2 *M* hydroxylamine hydrochloride (plutonium is present as Pu(III)) to within ±2 rel. % at the weight ratio Pu : U $\leqq 3 \cdot 10^3$ [679].

Zhuravlev (1961) determined uranium in alloys with plutonium (9–13% U, calculated on the sum of Pu and U) by condensed spark spectrography, using a high dispersion spectrograph. Plutonium served as the internal standard. Calibration curves, constructed for three pairs of lines (Å) were employed: 1) U 3932.026 Å and Pu 3931.3 Å; 2) U 4090.135 Å and Pu 4103.7 Å; 3) U 4241.669 Å and Pu 4229.7 Å. All three curves had approximately the same slope. To improve the accuracy and the reliability of the determination, it is recommended to carry out the determination using the three curves simultaneously. The error is ± 2 rel. %.

DETERMINATION OF FISSION PRODUCTS IN IRRADIATED PLUTONIUM

Systematic radiochemical analysis of irradiated materials has been repeatedly described [187, 361, 364]. The present section contains brief descriptions of the determinations of individual long-lived isotopes obtained in fairly high yields, each method including a different technique for the separation of plutonium.

Phillips and Jenkins [594] recommend that nongaseous fission products (except for Zr and Nb which are not fully separated) be separated together by sorbing Pu(IV) from solutions of 7 *M* HNO_3 on strongly basic anion exchange resin De-Acidite FF (see p. 344). Plutonium is sorbed to an extent of more than 99.9%. Sr, Mo, Ru, Te, I, Cs, Ba and rare-earth elements are 98–100% transmitted by the column. The eluate is subjected to a radiochemical analysis.

In order to determine the yield of any isotope, constituent of the fission products, the time which has elapsed from the moment the irradiation has ceased must be allowed for. In addition, times at which the chemical operations are performed must be so chosen as to reduce to a minimum the effect of other isotopes of the element being determined on the radiochemical measurements. For examples of such calculations, see [187]. After the dissolution of the sample, a known amount (usually 10–50 mg) of a carrier (a stable isotope of the element being determined) is introduced into the solution. The separation and purification of the radioactive product with the carrier will only be efficient if there is full isotopic exchange between the radioactive and the stable isotopes. This applies particularly to elements of variable valency (Mo, Ru, I and Ce).

The separation process must ensure a coefficient of purification from α-activity of not less than $1 \cdot 10^6$, and from extraneous β- and γ-activity of not less than $1 \cdot 10^5$.

The final stage of the separation consists in obtaining the preparation in a form suitable for the determination of the yield. The yield must not be less than 50%.

Strontium

Sr^{89} (half-life period 53 days) and Sr^{91} (half-life period 9.7 hours) are formed from short-lived β-active isotopes of bromine, krypton and rubidium isotopes. When determining Sr^{89}, the irradiated sample is opened only 30 minutes after termination of the irradiation, to ensure that the mother isotopes of bromine and krypton have decayed.

In the method of purification developed by Soviet workers [187, p. 63], strontium and barium sulfates are first precipitated from a weak hydrochloric acid solution containing 0.5% sulfuric acid. To ensure quantitative precipitation, 50% ethanol must be added. The next purification stage is to convert the sulfates into carbonates, dissolve the latter in 2 *M* HCl, neutralize the solution and precipitate barium chromate from an acetic buffer solution. From the strontium-containing filtrate the carbonate is then precipitated, dissolved in hydrochloric acid, barium is added as carrier, and the cycle of operations is repeated, beginning with the precipitation of barium chromate. To separate Sr^{91} from traces of plutonium, iron hydroxide is precipitated five times. $Y(OH)_3$ is also precipitated to achieve a better elimination of Y^{91}. The mother liquors from the hydroxide precipitation and the washings are acidified with HCl, evaporated to a small volume and strontium sulfate is precipitated by adding 5% H_2SO_4 and an equal volume of alcohol. The precipitate is washed, dried, ignited at 600°, weighed, and its β-activity is determined.

The chemical yield of $SrSO_4$ is 60–75%. The purification coefficient is $3 \cdot 10^6$. The determination takes 10 hours.

In the determination of Sr^{89}, its separation from the sample solution is carried out at the end of $2\frac{1}{2}$ hours to ensure a full conversion of Rb^{89} and Sr^{89}. Measurements of β-activity are carried out after 7 days, during which time Sr^{91} will have decayed.

Sr^{91} is determined immediately after purification, correcting for the time elapsed from the last precipitation of the daughter isotope Y^{91}.

According to another method [159], strontium nitrate is precipitated from concentrated solution of HNO_3. Further purification consists in a repeated precipitation of iron hydroxide. Strontium oxalate is precipitated in the final stage. The chemical yield is 60–70%. The standard error of the determination is ± 4 rel. %.

Zirconium

The isotopes Zr^{95} ($T_{\frac{1}{2}}$ = 65 days) and Zr^{97} ($T_{\frac{1}{2}}$ = 17 hours) are formed from the short-lived isotopes of krypton, rubidium, strontium and yttrium [213]. Zr^{91} should be isolated not earlier than 1.75 hours after the termination of the irradiation (this period is equal to 10 half-life periods of the mother Y^{95}). The isolation and purification of Zr^{97} should be carried out immediately after irradiation.

In the method of Tolmachev et al. [187, p. 58], the sparingly soluble barium fluozirconate and zirconium phenylarsonate are precipitated. Advantage is also taken of the ability of zirconium to form soluble fluoride complexes.

The first step is to carry out a double precipitation of the fluorides of zirconium and other elements by HF. $BaZrF_6$ is then precipitated by a saturated solution of $Ba(NO_3)_2$. The precipitate is dissolved in boric acid, to which nitric acid has been added, after which barium is separated by precipitation as sulfate. From the filtrate a double precipitation of $Zr(OH)_4$ is carried out in the cold with the aid of 30% KOH. The hydroxides are dissolved in HCl and a double precipitation of zirconium phenylarsonate is carried out, the precipitate being each time converted into hydroxide by treatment with NaOH. Zirconium hydroxide is dissolved in 6 *N* HNO_3, lanthanum is added and LaF_3 is precipitated once; this is followed by several precipitations of $BaZrF_6$ and $Zr(OH)_4$. In the determination of Zr^{95}, the last precipitation of $BaZrF_6$ should be carried out not earlier than 12 days after the irradiation has been terminated, to ensure a full decay of Zr^{97}. Barium fluorozirconate is converted to the hydroxide, the hydroxide dissolved in HCl and zirconium phenylarsonate is precipitated. In the final stage it is converted into the hydroxide, which is dried and ignited to ZrO_2.

The measurement of the β-activity of Zr^{95} is carried out not later than 28 hours after the separation of the daughter Nb^{95} (in this case the contribution of Nb^{95} to the radioactivity will be about 2%).

The chemical yield is 65–70%. The coefficient of purification from niobium is $\sim 3 \cdot 10^5$, and from the sum total of the fission products it is $1 \cdot 10^7$.

Another method [543] involves repeated precipitation of zirconium by p-bromomandelic acid. Plutonium, when converted to Pu(III), is not precipitated.

To a hydrochloric acid solution of plutonium, carriers—zirconium, cerium and tellurium—are added. The first stage is the precipitation of elementary tellurium by SO_2 from 3 *M* HCl. From the filtrate acidified with HCl zirconium bromomandelate is precipitated. The bromomandelate is converted to the hydroxide, the hydroxide dissolved in HCl, and the precipitation of tellurium, bromomandelate and zirconium hydroxide is repeated. The precipitate is filtered, washed with water, alcohol and ether, dried at 110°, weighed and the measurements carried out.

Four determinations may be run in parallel during 4–5 hours. The approximate chemical yield is 75%. The experimental error is ± 2.3 rel. %.

Niobium

The niobium isotope Nb^{95} ($T_{\frac{1}{2}} = 38.7$ days) is the daughter product of Zr^{95}.

One method of isolation of Nb^{95} [544, 563] is given below.

Niobic acid is precipitated from a solution of 8 *M* HNO_3, containing $NaBrO_3$ and suitable carriers—niobium, zirconium and tellurium. The precipitate is dissolved in a saturated solution of oxalic acid, HCl is added to a concentration of 1 *M*, and certain radioisotopes are precipitated on copper sulfide as carrier. Radioactive tellurium is then separated on carrier after preliminary reduction to the elementary state by SO_2 from 3 *M* HCl. From the filtrate niobic acid is again precipitated, the precipitate is dissolved in concentrated sulfuric acid in the presence of conc. HF. Niobium is extracted with TBP from this solution. The organic layer is treated with concentrated ammonium hydroxide and with petroleum ether to precipitate niobic acid. The precipitate is dried, ignited to at 800° to niobium oxide, suspended in water, filtered, washed with ethanol and ether, dried in vacuo and weighed. The activity measurements are carried out 7.5 days after separation from the mother isotope Zr^{95}.

The yield of niobium is 60–80%. The standard deviation of a single determination is ± 3 rel. %. Four parallel determinations take about 4 hours.

Molybdenum

The longest lived isotope of molybdenum—Mo^{99}—has a half-life period of 67.2 hours. Its isolation from the solution should be carried out not before 4.5 hours, to ensure completion of the decay of Mo^{101} and Mo^{102}, with half-life periods of 14.6 and 11.0 minutes respectively.

In the method of Venediktova, Rodionova and Tolmachev [187, p. 28], the sparingly soluble lead molybdate and molybdenum α-benzoinoximate are precipitated. Since Pu(IV) is also partly precipitated, apparently as molybdate, plutonium (IV) hydroxide is preliminarily precipitated by NaOH (not by NH_4OH or KOH), or else fluorides of trivalent and tetravalent plutonium are precipitated.

The filtrate, which must not contain fluoride ions, is acidified to ~ 0.5 *M* HNO_3 and molybdenum α-benzoinoximate is precipitated in the presence of oxalic acid in order to keep niobium in solution. The precipitate is treated with concentrated HNO_3 and $HClO_4$ to remove daughter technetium and organic matter. The residue is dissolved in concentrated solution of alkali, iron is added and iron hydroxide precipitated. This operation is then repeated. The filtrate is acidified and α-benzoinoximate precipitated; this is followed by two precipitations of iron hydroxide (the time of the last precipitation is noted, since it is the moment of separation of the daughter technetium). The filtrate is neutralized, acetic acid and ammonium acetate are added and $PbMoO_4$ is precipitated by the addition of lead acetate. The residue is dissolved in HNO_3 and the precipitation of lead molybdate is repeated. The precipitate is washed, ignited at 500° and weighed.

The chemical yield is 75% (molybdenum is added in an amount not less than 80 mg calculated as lead molybdate). All these operations should take not more than 8 hours. The coefficient of purification from the sum total of fission products is $1 \cdot 10^6$.

The activity of Mo^{99} is measured at equilibrium with Tc^{99*}, which has a half-life period of 6 hours. The first activity measurements are carried out 17.5 hours after the separation of the daughter technetium; the measurements should be terminated at the end of two half-life periods of Mo^{99}.

An ion-exchange method for the isolation of molybdenum has been briefly described [159]. The precipitation of lead molybdate again serves as the final stage.

Ruthenium

The isolation of the short-lived isotope Ru^{105}, with a half-life period of 4.5 hours, must be begun immediately after the decomposition of the mother Mo^{105} and Tc^{105}. The separation of the long-lived Ru^{103} ($T_{\frac{1}{2}}$ = 42.5 days) and Ru^{106} ($T_{\frac{1}{2}}$ = 290 days) is carried out 4 days after the termination of the irradiation, when the activity of Ru^{105} becomes insignificant.

The method of Venediktova et al. [187, p. 81], as well as other methods [159, 361, p. 1549], involve distillation of ruthenium as the tetraoxide.

Prior to distillation a carrier is introduced into the solution—ruthenium, KI, KBr, sodium bismuthate, conc. H_3PO_4—to prevent volatilization of molybdenum. The solution is heated at 80° for one hour, when iodide and bromide are oxidized, respectively, to iodate and bromate and remain in solution. The temperature of the distilling apparatus is then raised to 200° and RuO_4 is distilled off. The absorption is carried out in a receiver with 6 *M* NaOH. The solution is acidified and ruthenium is precipitated by hydrogen sulfide. If the precipitation does not proceed readily, the solution is made alkaline, and hydrogen sulfide is passed through, then acidified, and hydrogen sulfide is again passed through. Ruthenium sulfide is treated with conc. HCl and H_2O_2. The solution thus obtained is again distilled from a solution containing sodium bismuthate, phosphoric and perchloric acids. The alkaline solution of ruthenium in the receiver is treated with ethanol at the boil.

The precipitated ruthenium tetroxide is dissolved in conc. HCl and ruthenium is reduced to the metal with metallic magnesium. The precipitate of ruthenium is centrifuged, washed with HCl, alcohol and ether and dried at 120°.

The chemical yield of ruthenium is 70%; the lowest value of the purification coefficient is $1 \cdot 10^4$.

The β-activity of Ru^{105} is measured immediately after completion of purification. Aluminum filter (300 mg/cm^2) is employed to absorb the radiation of daughter rhodium and Ru^{103}. The activity of Ru^{106} is determined from the activity of daughter Rh^{106} at equilibrium, using an aluminum filter (300 mg/cm^2) to absorb the activity of Ru^{103}.

Another determination [544] involves the extraction of RuO_4 with carbon tetrachloride.

Ruthenium and zirconium are added to the solution of irradiated plutonium as carriers, ruthenium is oxidized by alkaline sodium hypochlorite and RuO_4 is extracted twice with CCl_4 at pH 4. Ruthenium is then extracted from the combined organic phase with an alkaline solution of $NaHSO_3$. The chemical yield of ruthenium is determined spectrophotometrically (see p. 391). For radiometric measurements, a portion of the solution is treated with alcohol in alkaline solution, the resulting hydroxide is filtered, washed and dried at 110°.

Iodine is not extracted under these conditions. The yield of ruthenium is about 95%. The standard deviation of one determination is $\pm 1\%$. Four determinations may be carried out in 3 hours, including one hour which is necessary for the oxidation and isotopic exchange.

Rhodium

A description has been given [544] of the determination of radioactive rhodium Rh^{105}, the daughter product of Ru^{105}, with a half-life period of 36.5 hours.

The first step is to precipitate palladium iodide from 1 *M* HCl. Rhodium iodide is then precipitated from the filtrate (precipitation from $\sim$ 6.5 *M* HCl). The precipitate is treated twice with 72% $HClO_4$ while heating, to decompose the iodide and distill off ruthenium. Rhodium is then sorbed from dilute $HClO_4$ on the cation-exchange resin Dowex-50 × 2 and eluted with hot 3 *M* HCl. The eluate is acidulated and the double potassium rhodium nitrite is precipitated. The precipitate is dissolved in aqua regia and the nitrite, which, when dried at 110°, is the gravimetric form, is reprecipitated.

The chemical yield is $\sim$ 70%. The standard deviation of a single determination is 4% for β- and 3% for γ-radiation measurements; four samples may be analyzed in 4 hours.

Tellurium

The tellurium isotope Te^{132} ($T_{\frac{1}{2}} = 77.7$ hours) is formed from Sb^{132} ($T_{\frac{1}{2}} = 5$ minutes) and decays to I^{132} ($T_{\frac{1}{2}} = 2.4$ hours).

In one method [159] tellurium is precipitated by passing SO_2 through its HBr solution. Tellurium is dissolved in nitric acid and iron hydroxide is precipitated. The precipitation of tellurium and of iron hydroxide is repeated, and in the final stage tellurium is again precipitated. The precipitate is washed, dried at 110°, weighed, and the γ-activity of the precipitate is measured after equilibrium with I^{132} has been attained.

In another method [187, p. 75], which is used in analysis of samples of irradiated uranium, Te^{132} is determined from the daughter I^{132}.

The first stage is to evaporate the sample solution to dryness with conc. HBr to eliminate $SeBr_4$ and nitrates, which interfere with the reduction of tellurium to the metal. This operation is performed three times. After the dissolution of the residue in 3 M HCl tellurium is precipitated by SO_2. The precipitate is dissolved in conc. HNO_3, the solution evaporated to dryness, dissolved in HCl and the precipitation of tellurium is repeated (I^{132} begins to accumulate from this moment on). The precipitate is centrifuged, washed with water and alcohol and dried at 110°.

A portion of the isolated tellurium is weighed out and dissolved in nitric acid with heating. The solution is diluted with water and left to stand for 12 hours, during which time equilibrium between Te^{132} and I^{132} becomes established. The solution is neutralized with saturated sodium carbonate until the precipitate of H_2TeO_3 dissolves, and an excess of sodium carbonate, and KI and NaClO solutions are added to oxidize iodide to periodate. The solution is acidified with nitric acid, IO_4^- is reduced to I_2 with hydroxylamine hydrochloride and the iodine is extracted with carbon tetrachloride. Another two extractions are carried out. These operations should be terminated as rapidly as possible, since they mark the moment of separation of iodine from the mother tellurium. The organic phase is treated with a solution of sodium bisulfite until both phases become decolorized. The aqueous solutions are treated with conc. HNO_3 and sodium nitrate and another three extractions of iodine are carried out, the iodine being reextracted with sodium sulfite each time. In the final stage AgI is precipitated from the aqueous solution, the precipitate is treated with 1:1 nitric acid in the heat, centrifuged, washed with water and ethanol and subsequently dried at 110°.

The time between the separation of I^{132} from the mother Te^{132} and the beginning of the measurement should be as short as possible.

Cesium

The isotopes of cesium formed during the irradiation of plutonium are Cs^{136} ($T_{\frac{1}{2}} = 13.7$ days) and the long-lived isotope Cs^{137} ($T_{\frac{1}{2}} = 33$ years).

The first stage is the precipitation of cesium silicotungstate $Cs_8SiW_{12}O_{42}$ [159; 361, p. 1642] from $\sim 6\,M$ HCl by the addition of silicotungstic acid. After standing overnight, the precipitate is centrifuged and dissolved in 6 M NaOH. The solution is treated with 70% perchloric acid, with heating. After dilution, the precipitate of SiO_2 and tungstic acid is separated out, and the filtrate is evaporated to the appearance of perchloric acid fumes. Cesium is precipitated as $CsClO_4$ by adding three volumes of absolute alcohol. The precipitate is filtered off, dissolved in water and ferric hydroxide is precipitated with 6 M NH_4OH. To the filtrate iron is again added and its hydroxide is precipitated. Sodium hydroxide is added to the filtrate and the filtrate evaporated, acidified with 70% perchloric acid and $CsClO_4$ is again precipitated. The precipitate is dried at 110°, weighed and its γ-activity is measured. About 90% of the activity is due to the radiation of Cs^{136}, the remainder by that of Cs^{137} (provided the sample has not been left to stand for more than one month).

The yield is $\sim 70\%$; the standard error of a single determination is 2%.

Barium

Barium isotopes other than Ba^{140} ($T_{\frac{1}{2}}$ = 12.8 days) have short half-life periods. The isolation of barium from solution should be carried out 6 hours after the irradiation.

Venediktova et al. [187, p. 37] carried out the first separation of plutonium by precipitating barium sulfate from ~ 0.5% H_2SO_4. The precipitate is treated with sodium carbonate, dissolved in HCl with heating, ferric hydroxide is precipitated twice, and then lanthanum hydroxide is precipitated six times, the time of the sixth precipitation being noted. Barium chromate is then precipitated from an acetate buffer solution, after having added inactive strontium to the solution. Barium is precipitated twice more, the last time without the addition of strontium. The precipitate is washed with water, ignited at 600° and the β-radiation of Ba^{140} in equilibrium with La^{140} is measured (the first measurements should not be carried out before 12.8 days have elapsed from the last separation of lanthanum).

Ba^{140} may be determined from the activity of daughter La^{140} ($T^{\frac{1}{2}}$ = 40 hours). To do this, a weighed sample of purified barium chromate after the accumulation of lanthanum is dissolved in HCl and lanthanum oxalate is precipitated from 0.3 *M* HCl solution by the addition of a saturated solution of oxalic acid. The oxalate is converted into the hydroxide by treating with KOH in the heat. The precipitate is then dissolved in HNO_3 and $La(OH)_3$ is precipitated in the presence of inactive barium. Lanthanum hydroxide is precipitated twice more, the last precipitation being carried out without the addition of barium. Lanthanum oxalate is precipitated at the last stage. It is washed, dried, calcined and ignited to La_2O_3 at 1000°.

The chemical yield of lanthanum is 80%.

Barium may be isolated by three successive precipitations of barium chloride from a 4 : 1 mixture of hydrochloric acid with ether [159]. Iron hydroxide is then precipitated. Barium is precipitated as sulfate in the last stage. The yield is 60–70%. The coefficient of purification from α-activity is $1 \cdot 10^6$.

Cerium

The isolation of Ce^{144} ($T_{\frac{1}{2}}$ = 282 days) is best done 290–300 days after the irradiation, when Ce^{141} ($T_{\frac{1}{2}}$ = 33.1 days) and Ce^{143} ($T_{\frac{1}{2}}$ = 33 hours) have decayed [187, p. 24], and so have other isotopes of rare-earth elements. Ce^{143} is determined from the activity of the daughter Pr^{143} (see below).

The preliminary separation of plutonium is conveniently carried out by sorption on strongly basic anion exchanger from 7–8 *M* HNO_3 [159]. Almost all the fission products pass through the column. The separation and purification of cerium may be based on its extraction with hexone after oxidation with sodium bromate, with subsequent reextraction with dilute hydrogen peroxide. The oxalate precipitation is carried out twice in the last stage. The yield of separated cerium is 70–80%.

The isolation of cerium from a mixture of fission products (after elimination of plutonium) may be carried out according to the method of Boldbridge and Hume [361, p. 1693].

From a solution containing cerium, lanthanum and zirconium as carriers, fluorides are precipitated. The precipitate is dissolved in boric and nitric acids. After the oxidation of cerium with sodium bromate, $Ce(IO_3)_4$ is precipitated from $\sim 4\,M$ HNO_3. The precipitate is dissolved in a mixture of HCl, HNO_3 and H_2O_2, and the iodate is again precipitated by the addition of $KBrO_3$. The iodate precipitation is repeated once more. From the reduced solution zirconium iodate is precipitated, which entrains all active zirconium. Cerium (III) hydroxide is precipitated from the filtrate, the precipitate is dissolved in HCl and SO_2 is passed through the solution to reduce iodate to iodide. The hydroxide is again precipitated from the solution, and then cerium oxalate, which serves as the gravimetric form.

Praseodymium and neodymium

These elements are best separated by ion-exchange methods [159, 544].

Plutonium is first converted to Pu(III) by treatment with hydroxylamine. The solution is evaporated to dryness and the residue is dissolved in $5\,M$ ammonium thiocyanate. When this solution is passed through the strongly basic anion-exchange resin Dowex-1, plutonium and americium are sorbed, while the rare-earth elements pass through the column. The lanthanides are then purified by hydroxide precipitation, and by double precipitation of fluorides and hydroxides. The hydroxides are dissolved in dilute $HClO_4$ and the solution is passed through the cation-exchange resin Dowex-50. When the ion exchanger is washed with $1\,M$ solution of lactic acid (pH 3.15), rare-earth elements are separated. The separation of praseodymium and neodymium is indicated by the precipitation of the oxalates from individual eluate fractions (γ-spectrometric analysis is also carried out). Fractions containing Nd^{147} and Pr^{143} are combined, the oxalates are reprecipitated, dried and the γ-activity of Nd^{147} and the β-activity of Pr^{143} are measured.

The chemical yield is not more than 30–50%. The standard deviation of a single determination for each element is $\pm$ 3%.

Americium

To effect a radiometric determination of americium in neutron-irradiated plutonium, americium is separated from radioactive elements by precipitation of fluorides and hydroxides, and also by ion-exchange methods [325].

The sample is dissolved in HCl, a portion of the solution containing 1 mg Pu is withdrawn, 1 mg La is added as carrier and the hydroxides are precipitated with ammonia (the precipitate occludes americium, plutonium and rare-earth fission products). The precipitate is dissolved in HNO_3, 4 drops of $0.1\,M$ solution of cerium (IV) are added and the solution is heated for 10 minutes to convert plutonium to Pu(VI). The solution is cooled and 3 drops of $20\,M$ HF are added, when americium and rare-earth elements are precipitated. The precipitate is dissolved in nitric acid to which one drop of boric acid has been added and the hydroxides are again precipitated by ammonia. The precipitate is dissolved in the minimum amount of conc. HNO_3. The solution, which is 0.3–0.5 N with respect to the acid, is passed through a 19×0.5 cm column of the cation-exchange resin Dowex-50 × 4 (200–400 mesh) in the H-form. The column is eluted with 5% solution of ammonium nitrate (pH 2.30) until the pH of the eluate becomes 2.30. The sorbed elements are then selectively eluted with 5% NH_4NO_3 (pH 2.30), saturated with EDTA. Zirconium, plutonium, yttrium and americium (with gadolinium and europium) are eluted first. Samarium, promethium and neodymium appear in the following fractions, while praseodymium, cerium and lanthanum remain on the cation exchanger. Gadolinium and europium do not interfere with the radiometric determination of americium.

In samples containing ~ 0.25 μg Am per 1 mg Pu, 97.0–98.6% Am on the average are found; the standard deviation is 2.4–3.8%.

Because the americium fraction is contaminated with short-lived γ-activity, the sample is left to cool for 2 days prior to the analysis.

It was reported [701] that americium and curium can be efficiently separated from plutonium by extracting the latter as Pu(IV) with a solution of trioctylphosphine oxide in cyclohexane.

Appendix

COMPLEX COMPOUNDS OF PLUTONIUM IN AQUEOUS SOLUTIONS

Cation	Anion	Reaction of complex formation	Ionic strength, μ	Constant	References
Pu^{3+}	Cl^-	$Pu^{3+} + Cl^- = PuCl^{2+}$	0.5	1.72	[710]
			1.0	0.67	[186]
	NO_3^-	$Pu^{3+} + NO_3^- = PuNO_3^{2+}$	—	5.9	[252]
		$Pu^{3+} + 2NO_3^- = Pu(NO_3)_2^+$	—	14.3	[252]
		$Pu^{3+} + 3NO_3^- = Pu(NO_3)_3^0$	—	14.4	[252]
	SO_4^{2-}	$Pu^{3+} + SO_4^{2-} = PuSO_4^+$	—	10	[186]
		$Pu^{3+} + 2SO_4^{2-} = Pu(SO_4)_2^-$	—	50	[186]
	$C_2O_4^{2-}$	$Pu^{3+} + 2C_2O_4^{2-} = Pu(C_2O_4)_2^-$	—	$2.0 \cdot 10^9$	[54]
			—	$1.4 \cdot 10^9$	[55]
Pu		$Pu^{3+} + 3C_2O_4^{2-} = Pu(C_2O_4)_3^{3-}$	—	$2.4 \cdot 10^9$	[54, 56]
			—	$4.5 \cdot 10^{10}$	[227]
		$Pu^{3+} + 4C_2O_4^{2-} = Pu(C_2O_4)_4^{5-}$	—	$8.3 \cdot 10^9$	[54, 56]
			—	$4.2 \cdot 10^{11}$	[227]
		$Pu^{3+} + 4HC_2O_4^- = Pu(HC_2O_4)_4^-$	1.0	$9.1 \cdot 10^{10}$	[55]
	γ^{4-}*	$Pu^{3+} + H\gamma^{3-} = PuH\gamma^0$	1.0	$1.6 \cdot 10^9$	[165, 59]
		$Pu^{3+} + \gamma^{4-} = Pu\gamma^-$	1.0	$2.3 \cdot 10^{17}$	[4]
			—	$3.5 \cdot 10^{15}$	Palei and Kucha (1957)
Pu^{3+}	$C_6H_5H_7^{3-}$ (citrate ion)	$Pu^{3+} + 3C_6H_5O_7^{3-} = Pu(C_6H_5O_7)_3^{6-}$	1.0	$1 \cdot 10^{11}$	Vinogradov and
	$C_4H_4O_6^{2-}$ (tartrate ion)	$Pu^{3+} + 6C_4H_4O_6^{2-} = Pu(C_4H_4O_6)_6^{9-}$	1.0	$5 \cdot 10^{15}$	Gryzin (1952)

Appendix (continued)

Cation	Anion	Reaction of complex formation	Ionic strength, μ	Constant	References
Pu^{4+}	Cl^-	$Pu^{4+} + Cl^- = PuCl^{3+}$	1.0	1.38	[186]
			1.0	2.10	[603]
			2.0	1.26	[605]
			4.0	1.4	[433]
			—	0.8	[105]
		$Pu^{4+} + 2Cl^- = PuCl_2^{2+}$	4.0	1.2	[433]
			—	0.3	[105]
		$Pu^{4+} + 3Cl^- = PuCl_3^+$	4.0	0.1	[433]
	NO_3^-	$Pu^{4+} + NO_3^- = PuNO_3^{3+}$	1.0	3.48	[611]
			2.0	2.9	[461]
			4.0	5.5	[433]
			6.0	4.7	3, Chap. 9
		$Pu(NO_3)^{3+} + NO_3^- = Pu(NO_3)_2^{2+}$	6.0	0.96	3, Chap. 9
		$Pu^{4+} + 2NO_3^- = Pu(NO_3)_2^{2+}$	4.0	23.5	[433]
Pu^{4+}	NO_3^-	$Pu(NO_3)_2^{2+} + NO_3^- = Pu(NO_3)_3^+$	6.0	0.33	3, Chap. 9
		$Pu^{4+} + 3NO_3^- = Pu(NO_3)_3^+$	4.0	15 ± 10	[433]
		$[Pu(H_2O)_{8-n}(NO_3)_n]^{4-n} + NO_3^- = [Pu(H_2O)_{8-(n+1)}(NO_3)_{n+1}]^{4-(n+1)} + H_2O; n = 0-5$	—	—	[149, 623, 626]
	F^-	$Pu^{4+} + F^- = PuF^{3+}$	1.0	$1.7 \cdot 10^4$	[203, p. 337]
			2.0	$1.0 \cdot 10^4$	[203, p. 337]
	CH_3COO^- (Ac^-)	$Pu^{4+} + Ac^- = PuAc^{3+}$	0.1	$2 \cdot 10^5$	[663a]
		$Pu^{4+} + 2Ac^- = PuAc_2^{2+}$	0.1	$1 \cdot 10^{-9}$	[663a]
		$Pu^{4+} + 3Ac^- = PuAc_3^+$	0.1	$8 \cdot 10^{12}$	[663a]
		$Pu^{4+} + 4Ac^- = PuAc_4^0$	0.1	$2 \cdot 10^{18}$	[663a]
		$Pu^{4+} + 5Ac^- = PuAc_5^-$	0.1	$4 \cdot 10^{25}$	[663a]
	SO_4^{2-}	$Pu^{4+} + SO_4^{2-} = PuSO_4^{2+}$	1.0	$4.6 \cdot 10^3$	[611]
			2.33	$(3.02 \pm 0.4) \cdot 10^2$	[157]
		$Pu^{4+} + HSO_4^- = PuSO_4^{2+} + H^+$	0.5	9.45	[157]
		$Pu^{4+} + 2HSO_4^- = Pu(SO_4)_2^0 + 2H^+$	0.5	20.0	[157]
		$Pu^{4+} + 3HSO_4^- = Pu(SO_4)_3^{2-} + 3H^+$	0.5	$1.25 \cdot 10$	[157]
		$[Pu(H_2O)_{8-n}(SO_4)_n]^{4-2n} + SO_4^{2-} = [Pu(H_2O)_{8-(n+1)}(SO_4)_{n+1}]^{2-2n} + H_2O$, where $n = 0-7$	—	—	[150]
	$C_2O_4^{2-}$	$Pu^{4+} + C_2O_4^{2-} = Pu(C_2O_4)^{2+}$	1.0	$3.6 \cdot 10^8$	[57, 167]
		$Pu^{4+} + 2C_2O_4^{2-} = Pu(C_2O_4)_2^0$	1.0	$8.3 \cdot 10^{16}$	[57, 167]
		$Pu^{4+} + 3C_2O_4^{2-} = Pu(C_2O_4)_3^{2-}$	1.0	$2.5 \cdot 10^{23}$	[57, 167]
		$Pu^{4+} + 4C_2O_4^{2-} = Pu(C_2O_4)_4^{4-}$	1.0	$3.1 \cdot 10^{27}$	[57, 167]
			—	$3.0 \cdot 10^{27}$	[227]

Appendix (continued)

Cation	Anion	Reaction of complex formation	Ionic strength, μ	Constant	References
	HPO_4^{2-}	$Pu^{4+} + HPO_4^{2-} = Pu(HPO_4)^{2+}$	—	$8.3 \cdot 10^{12}$	[84]
		$Pu^{4+} + 2HPO_4^{2-} = Pu(HPO_4)_2^0$	—	$5.6 \cdot 10^{23}$	[84]
		$Pu^{4+} + 3HPO_4^{2-} = Pu(HPO_4)_3^{2-}$	—	$2.7 \cdot 10^{33}$	[84]
		$Pu^{4+} + 4HPO_4^{2-} = Pu(HPO_4)_4^{4-}$	—	$1.7 \cdot 10^{43}$	[84]
	HPO_4^{2-}	$Pu^{4+} + 5HPO_4^{2-} = Pu(HPO_4)_5^{6-}$	—	$1.1 \cdot 10^{52}$	[84]
	CO_3^{2-}	$Pu^{4+} + CO_3^{2-} = Pu(CO_3)^{2+}$	10.0	$9.1 \cdot 10^{46}$	[167]
	H_2O_2	$2Pu^{4+} + H_2O_2 + H_2O =$ $= Pu(OO)(OH)Pu^{5+} + 3H^+$ (brown-colored complex)	0.5	$8.8 \cdot 10^6$	[353]
Pu^{4+}	H_2O_2	$2Pu^{4+} + 2H_2O_2 = Pu(OO)_2Pu^{4+} +$ $+ 4H^+$ (red complex)	0.5	$6.3 \cdot 10^8$	[353]
	γ^{4-}*	$Pu^4 + \gamma^{4-} = Pu\gamma^0$	0.1	$1.58 \cdot 10^{24}$	[407]
			—	$1.4 \cdot 10^{26}$	[110]
			—	$1.38 \cdot 10^{25}$	[241, p. 47]
		$Pu^{4+} + OH^- + \gamma^{4-} = Pu(OH)\gamma^-$	—	$2.9 \cdot 10^{17}$	[407]
	$C_6H_5O_7^{3-}$ (citrate ion)	$Pu^{4+} + 4C_6H_5O_7^{3-} = Pu(C_6H_5O_7)_4^{8-}$	1.0	$1.7 \cdot 10^{27}$	Vinogradov and Gryzin (1951)
	$C_4H_4O_6^{2-}$ (tartrate ion)	$Pu^{4+} + 6C_4H_4O_6^{2-} = Pu(C_4H_4O_6)_6^{8-}$	1.0	$2.0 \cdot 10^{31}$	Ibid.
	$AcAc^-$**	$Pu^{4+} + AcAc^- = Pu(AcAc)^{3+}$	0.1	$3.16 \cdot 10^{10}$	[60]
		$Pu(AcAc)^{3+} + AcAc^- = Pu(AcAc)_2^{2+}$	0.1	$1.58 \cdot 10^9$	[60]
		$Pu(AcAc)_2^{2+} + AcAc^- = Pu(AcAc)_3^+$	0.1	$2.51 \cdot 10^8$	[60]
		$Pu(AcAc)_3^+ + AcAc^- = Pu(AcAc)_4^0$	0.1	$1.0 \cdot 10^6$	[60]
PuO_2^+	Cl^-	$PuO_2^+ + Cl^- = PuO_2Cl^0$	0.05	0.67	[603]
	$C_2O_4^{2-}$	$PuO_2^+ + C_2O_4^{2-} = PuO_2C_2O_4^-$	0.05	$3.3 \cdot 10^4$	[60, p. 79]
		$PuO_2^+ + 2C_2O_4^{2-} = PuO_2(C_2O_4)_2^{3-}$	0.05	$2.4 \cdot 10^7$	[60, p. 79]
	γ^{4-}*	$PuO_2^+ + \gamma^{4-} = PuO_2\gamma^{3-}$	0.05	$1.5 \cdot 10^{10}$	[60, p. 98]
PuO_2^{2+}	NO_3^-	$PuO_2^{2+} + NO_3^- = PuO_2NO_3^+$	—	72	[126]
		$PuO_2NO_3^+ + NO_3^- = PuO_2(NO_3)_2^0$	—	36	[126]
	Cl^-	$PuO_2^{2+} + Cl^- = PuO_2Cl^+$	—	1.25	[578]
			2.0	0.56	[612]
			1.0	0.73	[603]
		$PuO_2^{2+} + 2Cl^- = PuO_2Cl_2^0$	—	0.35	[578]
	CH_3COO^- (Ac^-)	$PuO_2^{2+} + Ac^- = PuO_2Ac^+$	—	$1.9 \cdot 10^3$	[60, p. 98]
		$PuO_2^{2+} + 2Ac^- = PuO_2Ac_2^0$	—	$2.0 \cdot 10^6$	[60, p. 98]
		$PuO_2^{2+} + 3Ac_3^- = PuO_2Ac_3^-$	—	$2.3 \cdot 10^7$	[60, p. 98]
	$C_2O_4^{2-}$	$PuO_2^{2+} + C_2O_4^{2-} = PuO_2C_2O_4^0$	1.0	$4.3 \cdot 10^6$	[51]
		$PuO_2^{2+} + 2C_2O_4^{2-} = PuO_2(C_2O_4)_2^{2-}$	1.0	$3.0 \cdot 10^{11}$	[51]
	γ^{4-}*	$PuO_2^{2+} + \gamma^{4-} = PuO_2\gamma^{2-}$	—	$1.75 \cdot 10^{16}$	[407]

* γ^{4-}—anion of ethylenediaminetetraacetic acid.

** $AcAc^-$—$[CH_3—CO—CH—CO—CH_3]^-$.

BIBLIOGRAPHY

[Note: Items 1–259 in this bibliography mainly refer to publications in Russian (both original and translated) and are given in the Cyrillic alphabetic order. Items 260–739 refer to publications in languages other than Russian. An explanatory list of Russian abbreviations appearing in this bibliography is given on p. 433].

1. ADAMOVICH, L.B. and R.S. DIDENKO.—*Uchenye zapiski KhGU*, **54**. 189. 1954.
2. ADAMSKII, N.M., S.M. KARPACHEVA, I.M. MELNIKOV. and A.M. ROZEN.—*Radio-khimiya*, **2**, 16. 1960.
3. SEABORG, G.T. and J.J. KATZ. *Actinide Elements*.—N.Y., McGraw-Hill. 1954.
4. ALEKSEEVSKII, E.V., R.K. GOL'TS, and A.P. MUSAKIN. *Kolichestvennyi analiz (Quantitative Analysis)*, p. 390.—Leningrad-Moscow, Goskhimizdat. 1948.
5. ALENCHIKOVA, I.F., L.L. ZAITSEV, L.V. LIPIS. V.V. FOMIN, and N.T. CHEBOTAREV.—*Second International Conference on Peaceful Uses of Atomic Energy, Geneva*, **28**, 309. 1958.
6. ALENCHIKOVA, I.F., L.V. LIPIS, and N.S. NIKOLAEV.—*Atomnaya energiya*, **10**, 592. 1961.
7. ALIMARIN, I.P. and S.S. ALIKBEROV.—*Zavodskaya laboratoriya*, **23**, 658. 1957.
8. ALIMARIN, I.P. and L.Z. KOZEL'.—In: *Khimiya redkikh elementov*, Moscow, Izdatel'stvo AN SSSR, **3**, 114. 1957.
9. *Analiticheskaya khimiya urana (Analytical Chemistry of Uranium)*.—Moscow, Izdatel'-stvo AN SSSR. 1962.
10. *Analytical Chemistry of Uranium and Thorium*. [Russian translation. 1956.]
11. ARON, P.M.—*Atomnaya energiya* **5**, 183. 1958.
12. ARTYUKHIN, P.I., A.D. GEL'MAN, and V.I. MEDVEDOVSKII.—*Doklady AN SSSR*, **120**, 98. 1958.
13. ARTYUKHIN, P. I., V.I. MEDVEDOVSKII, and A.D. GEL'MAN.—*Zhurnal neorgan, khimii*, **4**, 1324. 1959.
14. ARTYUKHIN, P.I., V.I. MEDVEDOVSKII, and A.D. GEL'MAN.—*Radio-khimiya*, **1**, 131. 1959.

15. BABKO, A.K. *Fiziko-khimicheskii analiz kompleksnykh soedinenii v rastvorakh (Instrumental Analysis of Complex Compounds in Solutions).*—Kiev, Izdatel'stvo AN UkrSSR. 1956.
16. BARANOV, V.I. *Radiometriya (Radiometry).*—Moscow, Izdatel'stvo AN SSSR. 1956.
17. BAKH, N.A., V.I. MEDVEDOVSKII, and V.V. SARAEVA.—*Second International Conference on Peaceful Uses of Atomic Energy, Geneva,* 1958.
18. BELYAEV, YU. I. and A.N. ZAIDEL'.—*Zhurnal analiticheskoi khimii,* **12**, 30. 1957.
19. BENTLEY, W.C., H.DIAMOND, P.R. FIELDS, A.M. FRIEDMAN, J.E. GINDLER, D.C. HESS, J.R. HUIZENGA, M.G. INGHRAM, A.H. JAFFEY, L.B. MAGNUSSON, W.M. MANNING, J.F. MACH, G.L. PYLE, R. SJOBLOM, K.M. STEVENS, and M.H. STUDIER.—*International Conference on Peaceful Uses of Atomic Energy, Geneva,* **7**, 261. 1955.
20. BAKER, R.D. and J.A. LEARY.—*Second International Conference on Peaceful Uses of Atomic Energy, Geneva,* **17**, 356. 1958.
21. BLAKE, C.A., C.F. BAES, K.B. BROWN, C.F. COLEMAN, and J.C. WHITE.—*Second International Conference on Peaceful Uses of Atomic Energy, Geneva,* **28**, 284. 1958.
22. BOBROVA, V.N.—*Radiokhimiya,* **2**, 364. 1960.
23. BOBROVA, V.N.—*Radiokhimiya,* **2**, 500. 1960.
24. BEAUFAIT, L.J., F.R. STEVENSON, and G.C. ROLLEFSON.—U.S. Patent 2860949.
25. BOCHVAR, A.A., S.T. KONOBEEVSKII, A.S. ZAIMOVSKII, G.Ya. SERGEEV, V.I. KUTAITSEV, N.F.PRAVDYUK, and B.M. LEVITSKII.—*Atomnaya energiya,* **5**, 5. 1958.
26. BOCHVAR, A.A., S.T. KONOBEEVSKII, V.I. KUTAITSEV, T.A. MEN'SHIKOVA, and N.T. CHEBOTAREV.—*Atomnaya energiya,* **5**, 503. 1958.
27. BOCHAR, A.A., S.T. KONOBEEVSKII, V.I. KUTAITSEV, T.A. MEN'SHIKOVA, and N.T. CHEBOTAREV.—*Second International Conference on Peaceful Uses of Atomic Energy Geneva,* **6**, 184. 1958.
28. BROWN, K.B., C.F. COLEMAN, D.J. CROUSE, C.A. BLAKE, and A.D. RYAN.—*Second International Conference on Peaceful Uses of Atomic Energy, Geneva* **3**, 472. 1958.
29. BROUNS, R.J. and R.A. SCHNEIDER.—*Second International Conference on Peaceful Uses of Atomic Energy, Geneva,* **28**, 436. 1958.
30. BRESLER, S.E.—*Uspekhi khimii.* **29**, 993. 1960.
31. BRUCE, F.R.—*International Conference on Peaceful Uses of Atomic Energy, Geneva,* **17**, 49. 1955.
32. BUDAEV, I.V. and A.N. VOL'SKII.—*Second International Conference on Peaceful Uses of Atomic Energy, Geneva,* **28**, 316. 1958.
33. BAGLEY, K. *Plutonium and its alloys.* [Russian translation, 1958.]
34. VDOVENKO, V.M. *Khimiya urana i transuranovykh elementov (Chemistry of Uranium and Transuranides).*—Moscow-Leningrad, Izdatel'stvo AN SSSR. 1960.
35. VDOVENKO, V.M.—*Second International Conference on Peaceful Uses of Atomic Energy, Geneva,* **17**, 175. 1958.
36. VDOVENKO, V.M. and T.V. KOVALEVA.—In: *Sbornik rabot po radiokhimii,* Izdatel'stvo LGU, p. 44. 1959.
37. VDOVENKO, V.M. and M.P. KOVAL'SKAVA.—*Second International Conference on Peaceful Uses of Atomic Energy, Geneva,* **17**, 329. 1958.
38. VDOVENKO, V.M., A.A. LIPOVSKII, and M.G. KUZINA.—*Radiokhimiya,* **2**, 301. 1960.

39. VDOVENKO, V.M., A.A. LIPOVSKII, and M.G. KUZINA.—*Radiokhimiya*, **2**, 307. 1960.

40. VDOVENKO, V.M., A.A. LIPOVSKII, and S.A. NIKITINA.—*Radiokhimiya*, **3**, 396. 1961.

41. WILSON, A.S.—*Second International Conference on Peaceful Uses of Atomic Energy, Geneva*, **17**, 348. 1958.

42. VODEN, V.G., G.P. NIKITINA, and M.F. PUSHLENKOV.—*Radiokhimiya*, **1**, 121. 1959.

43. VOLSKY, A.N., N.V. VYATKINA, N.P. ZVEREVA, I.L. KALTSMAN, and T.S. POSOBILO.—*Second International Conference on Peaceful Uses of Atomic Energy, Geneva*, **28**, 170. 1958.

44. VYAZEMSKII, V.O., I.I. LOMONOSOV, A.N. PISAREVSKII, Kh. V. PROTOPOPOV, V.A. RUZIN, and E.D. TETERIN. *Stsintillyasionnyi metod v radiometrii (Scintillation Methods in Radiometry)*.—Moscow, Atomizdat. 1961

45. GAVRILOV, K.A., B.F. MYASOEDOV, and G.I. KHLEBNIKOV.—*Radiokhimiya*, **3**, 62. 1961.

46. HAISSINSKY, M.N. *La chimie nucléaire et ses applications*.—Paris. Masson et Cie. 1957.

47. GALKIN, N.P., L.A. PONOMAREV, and YU.D. SHISHKOV. *Geksaftorid plutoniya, ego poluchenie i svoistva (Plutonium Hexafluoride, its Preparation and Properties)*.—Moscow, Atomizdat. 1961.

48. HARVEY. B.G., G.C. HEAL, A.G. MADDOCK, and E.L. ROWLEY. *Isotope Chemistry*, Coll. 2 [Russian translations, 1948.]

49. GEL'MAN, A.D., I.P. ARTYUKHIN, and A.I. MOSKVIN.—*Zhurnal neorganicheskoi khimii*, **4**, 1332. 1959.

50. GEL'MAN, A.D. and L.E. DRABKINA.—*Zhurnal neorganicheskoi khimii*, **3**, 1105. 1958.

51. GEL'MAN, A.D., L.E. DRABKINA, and A.I. MOSKVIN.—*Zhurnal neorganicheskoi khimii*, **3**, 1546. 1958.

52. GEL'MAN, A.D. and L.M. ZAITSEV.—*Zhurnal neorganicheskoi khimii*, **3**, 1304. 1958.

53. GEL'MAN, A.D. and L.M. ZAITSEV.—*Zhurnal neorganicheskoi khimii*, **3**, 1551. 1958.

54. GEL'MAN, A.D., N.N. MATORINA, and A.I. MOSKVIN.—*Atomnaya energiya*, **3**, 308. 1957.

55. GEL'MAN, A.D., N.N. MATORINA, and A.I. MOSKVIN.—*Atomnaya energiya*, **4**, 52. 1958.

56. GEL'MAN, A.D., N.N. MATORINA, and A.I. MOSKVIN.—*Doklady AN SSSR*, **117**, 88. 1957.

57. GEL'MAN, A.D. and A.I. MOSKVIN.—*Doklady AN SSSR*, **118**, 493. 1958.

58. GEL'MAN, A.D. and A.I. MOSKVIN.—*Atomnaya energiya*, **3**, 314. 1957.

59. GEL'MAN, A.D., A.I. MOSKVIN, and P.I. ARTYUKHIN.—*Atomnaya energiya*, **7**, 162. 1959.

60. GEL'MAN, A.D., A.I. MOSKVIN, L.M. ZAITSEV, and M.P. MEFOD'EVA. *Kompleksnye soedineniya transuranovykh elementov (Complex Compounds of Transuranides)*.—Moscow, Izdatel'stvo AN SSSR. 1961.

61. GEL'MAN, A.D., A.I. MOSKVIN, and V.P. ZAITSEVA.—*Radiokhimiya*, **4**, 154. 1962.

62. GEL'MAN, A.D. and L.P. SOKHINA.—*Zhurnal neorganicheskoi khimii*, **3**, 1100. 1958.

63. HEFFERICH, H. *Ion exchange*.—N.Y., McGraw-Hill. 1962.

64. GOL'DIN, L.L. and E.F. TRET'YAKOV.—*Izvestiya AN SSSR*, phys. ser., **20**, 859. 1956.

65. GOLDSCHMIDT, B., P. REGNAUT, and I. PREVOT. *Chemistry of Nuclear Fuel*. [Russian translation, 1956.]

66. GOLDSCHMIDT, B., P. REGNAUT, and I. PREVOT.—*International Conference on Peaceful Uses of Atomic Energy, Geneva*, **9**, 492. 1955.

67. HOFMAN, CONNICK and WALL.—U.S. Patent 2871251.

68. Grebenshchikova, V.I. and V.N. Bobrova.—*Radiokhimiya,* **3**, 544. 1961.
69. Grebenshchikova, V.I. and R.V. Bryzgalova.—*Radiokhimiya,* **2**, 264. 1960.
70. Grebenshchikova, V.I., R.V. Bryzgalova, N.B. Chernyavskaya, and V.N. Borrova. —*Radiokhimiya,* **1**, 11. 1959.
71. Grebenshchikova, V.I. and N.B. Chernyavskaya.—*Zhurnal neorganicheskoi khimii,* **3**, 38. 1958.
72. Grebenshchikova, V.I. and N.B. Chernyavskaya.—*Zhurnal neorganicheskoi khimii,* **4**, 941. 1959.
73. Grebenshchikova, V.I. and N.B. Chernyavskaya.—*Radiokhimiya,* **4**, 232. 1962.
74. Grebenshchikova, V.I. and N.B. Chernyavskaya.—*Radiokhimiya,* **3**, 650. 1961.
75. Gusev, N.I., P.N. Palei, I.G. Sentyurin, and I.S. Sklyarenko.—*Zhurnal analiticheskoi khimii,* **14**, 606. 1959.
76. Gusev, N.I. and I.G. Sentyurin.—Soviet patent 133257; *Peredovoi Nauchno-tekhnicheskii Opyt. Pribory dlya khimicheskikh, elektrokhimicheskikh i fiziko-khimicheskikh issledovanii,* No. 8, Moscow, Izdatel'stvo TsITEIN. 1960.
77. Gusev, N.I., I.G. Sentyurin, and I.S. Sklyarenko.—*Peredovoi Nauchno-tekhnicheskii Opyt. Primenenie radioaktivnykh izotopov i yadernykh izluchenii v promyshlennosti,* No. 8. Moscow, Izdatel'stvo TsITEIN. 1961.
78. Diamond, R.M. and D.G. Tuck. *Extraction of Inorganic Compounds into Organic Solvents.*—Interscience. 1960.
79. Dedov, V.B. and N.N. Kosyakov.—*International Conference on Peaceful Uses of Atomic Energy, Geneva,* **7**, 369. 1955.
80. Davis, H.M., F.J.W. Garton, H.I. Shalgosky, E.N. Jenkins, and G.W.C. Milner.—*Second Int. Conf. on Peaceful Uses of Atomic Energy, Geneva,* **28**, 403. 1958.
81. Deichman, E.N. and I.V. Tananaev.—*Radiokhimiya,* **4**, 66. 1962.
82. Delahay, P. *New Instruments and Methods in Electrochemistry.*—Interscience. 1954.
83. Denotkina, R.G., A.I. Moskvin, and V.B. Shevchenko.—*Zhurnal neorganicheskoi khimii,* **5**, 806. 1960.
84. Denotkina, R.G., A.I. Moskvin, and V.B. Shevchenko.—*Zhurnal neorganicheskoi khimii,* **5**, 1509. 1960.
85. Denotkina, R.G. and V.B. Shevchenko.—*Zhurnal neorganicheskoi khimii,* **6**, 1476. 1961.
86. Dzhelepov, B.S., R.B. Ivanov, A.S. Krivokhatskii, and V.G. Nedovesov.—*Radiokhimiya,* **4**, 619. 1962.
87. Dokuchaev, A.P.—*Atomnaya energiya,* **6**, 74. 1959.
88. Drabkina, L.E.—*Zhurnal neorganicheskoi khimii,* **3**, 1109. 1958.
89. Drabkina, L.E.—*Radiokhimiya,* **2**, 377. 1960.
90. Drabkina, L.E. and A.D. Gel'man.—*Radiokhimiya,* **1**, 137. 1959.
91. Drabkina, L.E., A.I. Moskvin, and A.D. Gel'man.—*Zhurnal neorganicheskoi khimii,* **3**, 1934. 1958.
92. Dreher, D.L., D.E. Coshland, S.G. Thompson, and D.E. Willard.—U.S. Patent 2907631.
93. Druin, V.A., V.P. Perelygin, and G.I. Khlebnikov.—*Zhurn. eksperim. teoret. fiziki,* **40**, 1296. 1961.

94. ERMAKOV, A.N., V.K. BELYAEVA, I.N. MAROV, and M.K. CHMUTOVA.—*Zhurnal neorganicheskoi khimii,* **4**, 493. 1959.

95. ZHERNOV, V.S. and S.V. MAMIKONYAN. *Novye radiometricheskie i spektrometricheskie pribory (Modern Radiometric and Spectrometric Instruments).*—Moscow. 1960.

96. ZAGRAI, V.D. and L.I. SEL'CHENKOV.—*Radiokhimiya,* **4**, 181. 1962.

97. ZAIDEL', A.N., N.I. KALITEEVSKII, L.V. LIPIS, and V.M. TARAKANOV.—*Optika i spektroskopiya,* **3**, 16. 1957.

98. ZAIDEL', A.N., N.I. KALITEEVSKII, L.V. LIPIS, and M.P. CHAIKA. *Emissionnoi spektral'nyi analiz atomnykh materialov (Emission Spectroscopic Analysis of Nuclear Materials).*—Moscow–Leningrad, Fizmatgiz. 1960.

98a. ZAIDEL', A.N., N.I. KALITEEVSKII, L.V. LIPIS, and M.P. CHAIKA.—*Zhurnal analiticheskoi khimii,* **12**, 17. 1957.

99. ZAIDEL', A.N., N.I. KALITEEVSKII, L.V. LIPIS, and M.P. CHAIKA.—*Zhurnal analiticheskoi khimii,* **11**, 21. 1956.

100. ZVYAGINTSEV, O.E. and B.N. SUDARIKOV.—*Zhurnal neorganicheskoi khimii,* **3**, 975. 1958.

101. ZIL'BERMAN, Ya. I. *Osnovy khimicheskoi tekhnologii iskusstvennykh radioaktivnykh elementov (Fundamentals of the Technology of Artificial Radioactive Elements),* p. 28.—Moscow, Atomizdat. 1961.

102. ZOLOTOV, Yu. A. and I.P. ALIMARIN.—*Radiokhimiya,* **4**, 272. 1962.

103. ZOLOTOV, Yu. A. and D. NISHANOV.—*Radiokhimiya,* **4**, 241. 1962.

104. KABANOVA, O.L.—*Zhurnal neorganicheskoi khimii,* **6**, 786. 1961.

105. KABANOVA, O.L. and P.N. PALEI.—*Zhurnal neorganicheskoi khimii,* **5**, 31. 1960.

106. CULLER, F.L. and R.E. BLANCO.—*Second International Conference on Peaceful Uses of Atomic Energy, Generva,* **17**, 259. 1958.

107. KAMACK, J. and J.H. BALTIS.—U.S. Patent 2863719.

108. KARTUSHOVA, R.E., T.I. RUDENKO, and V.V. FOMIN.—*Atomnaya energiya,* **5**, 24.1958.

109. KLYGIN, A.E. and V.K. PAVLOVA.—*Zhurnal neorganicheskoi khimii,* **6**, 1050. 1961.

110. KLYGIN, A.E., N.D. SMIRNOVA, and N.A. NIKOL'SKAYA.—*Zhurnal neorganicheskoi khimii,* **4**, 2766, 1959.

111. COLEMAN, C.F., K.B. BROWN, J.G. MOORE, and C.A. ALLEN.—*Second International Conference on Peaceful Uses of Atomic Energy, Geneva,* **28**, 278. 1958.

112. KOLODNY, M.—U.S. Patent 2893928.

113. KOLTHOFF, I.M. and D.D. LINGANE. *Polarography.*—N.Y., Interscience. 1946.

114. KOLTHOFF, I.M. and E.B. SANDELL. Textbook of Quantitative Inorganic Analysis.—N.Y., Macmillan. 1943.

115. KONOBEEVSKII, S.T.—*Sessiya AN SSSR po mirnomu ispol'zovaniyu atomnoi energii. Zasedanie otdeleniya khimicheskikh nauk (Conference of the Academy of Sciences of the USSR on Peaceful Uses of Atomic Energy. Meeting of the Section of Chemical Sciences),* p. 362.—Moscow, Izdatel'stvo AN SSSR. 1955.

116. KONOBEEVSKII, S.T., A.S. ZAIMOVSKII, B.M. LEVITSKII, Yu. N. SOKURSKII, N.I. CHEBOTAREV, Yu. V. BOBKOV, P.P. EGOROV, G.N. NIKOLAEV, and A.A. IVANOV.—*Second International Conference on Peaceful Uses of Atomic Energy, Geneva,* **6**, 194. 1958.

117. COFFINBERRY, A.S. and F.H. ELLINGER.—*International Conference on Peaceful Uses of Atomic Energy, Geneva,* **9**, 138. 1955.
118. COFFINBERRY, A.S., F.W. SCHONFELD, E.M. CRAMER, W.N. MINER, and F.H. ELLINGER.—*Second International Conference on Peaceful Uses of Atomic Energy, Geneva,* **6**, 681. 1958.
119. KRAVTSOV, V.A.—*Uspekhi fiz. nauk,* **78**, 65. 1962.
120. KRAUS, K.A.—*International Conference on Peaceful Uses of Atomic Energy, Geneva,* **7**, 245. 1955.
121. KRAUS, K.A.—*Ibid.*
122. KRAUS, K.A. and F. NELSON.—*Ibid.*
123. KRAUS, K.A. and F. NELSON. *Khimiya yadernogo gorynchego (Chemistry of Nuclear Fuel).*—Moscow, Goshkimizdat, 1956.
124. KRAUS, K.A., H.O. PHILLIPS, T.A. CARLSON, and J.S. JOHNSON.—*Second International Conference on Peaceful Uses of Atomic Energy, Geneva,* **28**, 3. 1958.
125. KREVINSKAYA, M.E., V.D. NIKOL'SKII, and B.G. POZHARSKII.—*Radiokhimiya,* **1**, 554. 1959.
126. KREVINSKAYA, M.E., V.D. NIKOL'SKII, B.G. POZHARSKII, and E.E. ZASTENKER.—*Radiokhimiya,* **1**, 548. 1959.
127. KREVINSKAYA, M.E., V.D. NIKOL'SKII, B.G. POZHARSKII, and E.E. ZASTENKER.—*Radiokhimiya,* **1**, 562. 1959.
128. KRYUKOVA, T.A., S.I. SINYAKOVA, and T.V. AREF'EVA. *Polyarograficheskii analiz (Polarographic Analysis).*—Moscow, Goskhimizdat. 1959.
129. KUZNETSOV, V.I.—*Doklady AN SSSR,* **31**, 895. 1941.
130. KUZNETSOV, V.I.—*Zhurnal analit. khimii,* **14**, 7. 1959.
131. KUZNETSOV, V.I.—*Trudy Komissii po analit. khimii,* **11**, 13. 1960.
132. KUZNETSOV, V.I.—*Trudy po khimii i khimicheskoi tekhnologii (Gor'kovskii unit),* No. 3, 559. 1960.
133. KUZNETSOV, V.I.—*Uspekhi khimii,* **21**, 175. 1952.
134. KUZNETSOV, V.I.—*Uspekhi khimii,* **23**, 654. 1954.
135. KUZNETSOV, V.I. and T.G. AKIMOVA.—*Atomnaya energiya,* **8**, 148. 1960.
136. KUZNETSOV, V.I. and T.G. AKIMOVA.—*Radiokhimiya,* **2**, 357. 1960.
137. KUZNETSOV, V.I. and L.I. BOL'SHAKOVA.—*Zhurnal analit. khimii,* **15**, 523. 1960.
138. KUZNETSOV, V.I. and S.B. SAVVIN.—*Radiokhimiya,* **3**, 79. 1961.
139. CULLER, F.L.—*International Conference on Peaceful Uses of Atomic Energy, Geneva,* **9**, 464. 1955.
140. CULLER, F.L. *Chemistry of Nuclear Fuel.* [Russian translation, 1956.]
141. CULLER, F.L. and F.R. BRUCE.—*International Conference on Peaceful Uses of Atomic Energy, Geneva,* **9**, 484. 1955.
142. COOPER, V.R. and M.T. WALLING.—*Second International Conference on Peaceful Uses of Atomic Energy, Geneva,* **17**, 291. 1958.
143. KURCHATOV, B.V., V.I. GREBENSHCHIKOVA, N.B. CHERNYAVSKAYA, and G.N. YAKOVLEV.—*International Conference on Peaceful Uses of Atomic Energy, Geneva,* **7**, 318. 1955.
144. LAVROVSKII, S.—*Ibid.*

145. LAVROVSKII, S. *Khimiya yadernogo goryuchego (Chemistry of Nuclear Fuel).*—Moscow. Goskhimizdat, p. 152. 1956.

146. LATIMER, W.M. *The Oxidation States of the Elements and Their Potentials in Aqueous Solutions.*—N.Y., Prentice-Hall. 1952.

147. LEVICH, V.G. *Fiziko-khimicheskaya gidrodinamika (Physicochemical Hydrodynamics).* —Moscow, Fizmatgiz. 1959.

148. LIPIS, L.V., B.G. POZHARSKII, M.E. POZHARSKAYA, and V.V. FOMIN.—*Zhurnal neorganicheskoi khimii,* **5**, 2190. 1960.

149. LIPIS, L.V., B.G. POZHARSKII, and V.V. FOMIN.—*Zhurnal strukt. khimii,* **1**, 135. 1960.

150. LIPIS, L.V., B.G. POZHARSKII, and V.V. FOMIN.—*Zhurnal strukt. khimii,* **1**, 417. 1960.

151. LIPOVSKII, A.A. and N.B. CHERNYAVSKAYA.—*Zhurnal neorgan. khimii,* **4**, 2244. 1959.

152. MADDOCK, A.G. and F. SMITH.—U.S. Patent 2863892.

153. MCKAY, H.A. *Chemistry of Nuclear Fuel.* [Russian translation, 1956.]

154. MALY, Ya., I. PEKA, M. TALASH, and M. TYMPL.—*Radiokhimiya,* **3**, 195. 1961.

155. MARGULIS, U. Ya. *Zashchita ot deistva pronikayushchei radiatsii (Protection from the Effects of Penetrating Radiation).*—Moscow, Atomizdat. 1961.

156. MARKOV, V.K., A.V. VINOGRADOV, S.V. ELINSON, A.E. KLYGIN, and I.V. MOISEEV. *Uran, metody ego opredeleniya (Uranium and its Determination).*—Moscow, Atomizdat. 1960.

157. MAROV, I.N. and M.K. CHMUTOVA.—*Zhurnal neorganicheskoi khimii,* **6**, 2664. 1961.

158. *Metody opredeleniya radioaktivnykh elementov v mineral'nom syr'e (Determination of Radioactive Elements in Mineral Materials).*—Moscow, Gosgeoltekhizdat. 1958.

159. METZ, C.F., G.M. MATLACK, and G.R. WATERBURY.—*Second International Conference on Peaceful Uses of Atomic Energy, Geneva,* **28**, 441. 1958.

160. MECHELNIK, F.—*Ibid.*

161. MOISEEV, I.V., N.N. BORODINA, and V.G. TSVETKOVA.—*Zhurnal neorganicheskoi khimii,* **6**, 543. 1961.

162. MOISEENKO, E.I. and A.M. ROZEN.—*Radiokhimiya,* **2**, 274. 1960.

163. MANDLEBERG, C.J., H.K. RAE, R. HURST, G. LONG, D. DAVIS, and K.E. FRANCIS.—*International Conference on Peaceful Uses of Atomic Energy, Geneva,* **7**, 374. 1955.

164. MOSKVIN, A.I.—*Radiokhimiya,* **1**, 430. 1959.

165. MOSKVIN, A.I. and P.A. ARTYUKHIN.—*Zhurnal neorganicheskoi khimii,* **4**, 591. 1959.

166. MOSKVIN, A.I. and A.D. GEL'MAN.—*Zhurnal neorganicheskoi khimii,* **3**, 456. 1958.

167. MOSKVIN, A.I. and A.D. GEL'MAN.—*Zhurnal neorganicheskoi khimii,* **3**, 962. 1958.

168. MOSKVIN, A.I. and A.D. GEL'MAN.—*Zhurnal neorganicheskoi khimii,* **5**, 1013. 1960.

169. MOSKVIN, A.I. and V.P. ZAITSEVA.—*Radiokhimiya,* **4**, 73. 1962.

170. NAST, R. and T. KRAKKAI.—*Uspekhi khimii,* **5**, 583. 1953.

171. NEBEL, D.—Paper read at the Congress of Analytical Chemistry, Budapest, 24–29 April 1961.

172. NEMODRUK, A.A., Yu. P. NOVIKOV, A.M. LUKIN, and I.D. KALININA.—*Zhurnal analit. khimii,* **16**, 292. 1961.

173. NEMODRUK, A.A., Yu. P. NOVIKOV, A.M. LUKIN, and I.D. KALININA.—*Zhurnal analit. khimii,* **16**, 180. 1961.

174. NIKITIN, B.A. *Izbrannye trudy* (*Selected Works*).—Izdatel'stvo AN SSSR. 1956.

175. Nikolaev, A.V.—*Khim. nauka i promyshlennost',* **1**, 548. 1954.
176. Nikol'skii, B.P., Editor. In: *Khromatografiya (Chromatography).*—Izdatel'stvo LGU. 1956.
177. Nikol'skii, V.D., M.E. Pozharskaya, and B.G. Pozharskii.—*Radiokhimiya,* **2**, 320. 1960.
178. Palei, P.N.—*International Conference on Peaceful Uses of Atomic Energy, Geneva,* **8**, 225. 1955.
179. Palei, P.N., Chzhan Ven'-Tsin.—*Zhurnal analiticheskoi khimii,* **15**, 598. 1960.
180. Penneman, R. and T. Keenan. *The Radiochemistry of Americium and Curium.*—Washington, Nat. Acad. Sci. 1960.
181. Popov, N.I.—*Atomnaya energiya,* **6**, 71. 1959.
182. Popov, N.I., V.I. Medvedovskii, and N.A. Bakh.—*Atomnaya energiya,* **4**, 154. 1958.
183. Popov, N.I., V.I. Medvedovskii, and N.A. Bakh.—*Atomnaya energiya,* **4**, 159. 1958.
184. Popov, M.M. and M.I. Ivanov.—*Atomnaya energiya,* **2**, 360. 1957.
185. Přibil, R. *Komplexony v chemicke analyse.*—Prague. Nakl. Československé akad. věd. 1953.
186. Rabideau, S.W., L.B. Asprey, T.K. Keenan, and T.W. Newton.—*Second International Conference on Peaceful Uses of Atomic Energy, Geneva,* **28**, 361. 1958.
187. Tolmachev, Yu. M., Editor. *Radiokhimicheskii analiz produktov deleniya (Radiochemical Analysis of Fission Products).*—Moscow-Lenningrad, Izdatel'stvo AN SSSR. 1960.
188. Ritter, D.M.—U.S. Patent 2868619.
189. Roberts, L.E., L.E. Russell, A.G. Adwick, A. Walter, and M.H. Rand.—*Second International Conference on Peaceful Uses of Atomic Energy, Geneva,* **28**, 215. 1958.
190. Rozen, A.M.—*Atomnaya energiya,* **2**, 445. 1957.
191. Rozen, A.M. In: *Ekstraktsiya,* No. 1, p. 6. A.P. Zefirov and M.M. Senyavin, Editors.—Atomizdat. 1962.
192. Rozen, A.M. and E.I. Moiseenko.—*Ibid.,* No. 2, p. 235.
193. Rozen, A.M. and E.I. Moiseenko.—*Zhurnal neorganicheskoi khimii,* **4**, 1209. 1959.
194. Ryabchikov, D.I., V.K. Belyaeva, and A.N. Ermakov.—*Zhurnal analiticheskoi khimii,* **11**, 658. 1956.
195. Ryabchikov, D.I., M.M. Senyavin, and K.V. Filippova.—*Zhurnal analiticheskoi khimii,* **8**, 220. 1953.
196. Ryabchikov, D.I. and I.K. Tsitovich. *Ionoobmennye smoly i ikh primenenie (Ion-Exchange Resins and Their Application).*—Moscow, Izdatel'stvo AN SSSR. 1962.
197. Savvin, S.B.—*Doklady AN SSSR,* **127**, 1231. 1959.
198. Samartseva, G.I.—*Atomnaya energiya,* **8**, 324. 1960.
199. Samuelson, O. *Ion Exchangers in Analytical Chemistry.*—N.Y., Wiley. 1953.
200. *Sanitarnye pravila raboty s radioaktivnymi veshchestvami i istochnikami ioniziruyuschikh izluchenii (Sanitation Rules for Work with Radioactive Substances and Sources of Ionizing Radiations).*—Moscow, Atomizdat. 1960.
201. Saukov, A.A. *Radioaktivnye elementy Zemli (Radioactive Elements in the Earth's Crust).*—Moscow, Atomizdat. 1961.
202. Seaborg, G.T.—*Atomnaya energiya,* **6**, 21. 1959.

203. SEABORG, G.T. and J.J. KATZ. *Chemistry of the Actinide Elements.*—N.Y., WILEY. 1958.

204. SINITSYNA, G.S., S.A. FADEEV, and G.M. SUKHODOLOV.—*Radiokhimiya,* **1**, 295. 1959.

205. SKLYARENKO, I.S.—*Zhurnal analiticheskoi khimii* (in press).

206. SKLYARENKO, I.S. and T.M. CHUBUKOVA.—*Zhurnal analit. khimii,* **18**, 1963.

207. SKLYARENKO, I.S. and T.M. CHUBUKOVA.—*Zhurnal analiticheskoi khimii,* **15**, 706. 1960.

208. SOLOVKIN, A.S., A.I. IVANTSOV, and E.V. RENARD.—*Zhurnal neorganicheskoi khimii,* **4**, 2826. 1959.

209. SOKHINA, L.P., and A.D. GEL'MAN.—*Zhurnal neorganicheskoi khimii,* **5**, 1013. 1960.

210. STARIK, I.E., A.P. RATNER, M.A. PASVIK, and F.L. GINZBURG.—*Geokhimiya,* **2**, 142. 1957.

211. STARIK, I.E., A.P. RATNER, M.A. PASVIK, and F.L. GINZBURG.—*Radiokhimiya,* **1**, 545. 1959.

212. STAUTEK, R.V. and R.B. DAFFIELD.—U.S. Patent 2856261.

213. STEINBERG, E.R. and L.E. GLENDENNIN.—*International Conference on Peaceful Uses of Atomic Energy, Geneva,* **7**, 3. 1955.

214. STYUNKEL', G.V. and E.M. YAKIMETS.—*Zavodskaya laboratoriya,* **24**, 23. 1958.

215. TARAYAN, V.M. *Merkuroreduktometriya (Mercuroreductometry).*—Izdatel'stvo Erevanskogo universiteta. 1958.

216. TIMASHEV, V.G., K.A. PETROV, A.V. RODIONOV, V.V. BALANDINA, A.A. VOLKOVA, A.V. EL'KINA, and Z.I. NAGNIBEDA. In: *Ekstraktsiya,* No. 1, p. 88. A.P. ZEFIROV and M.M. SENYAVIN, Editors.—Moscow, Atomizdat. 1962.

217. TOMLINSON, R.E.—*International Conference on Peaceful Uses of Atomic Energy, Geneva,* **9**, 511. 1955.

218. THOMPSON, S.G. and M.L. MOOCHA.—*Second International Conference on Peaceful Uses of Atomic Energy, Geneva,* **28**, 331. 1958.

219. TROSTYANSKAYA, E.B.—*Trudy Komissii po analit. khimii,* **6**, 215. 1955.

220. WALDRON, M.B., J. GARSTONE, J.A. LEE, P.G. MARDON, J.A.C. MARPLES, D.M. POOLE, and G.K. WILLIAMSON.—*Second International Conference on Peaceful Uses of Atomic Energy, Geneva,* **6**, 162. 1958.

221. WEINSTOCK, B. and J.G. MALM.—*International Conference on Peaceful Uses of Atomic Energy, Geneva,* **7**, 377. 1955.

222. WEISMAN, S.D., M.L. PERLMAN, and D. LIPKIN.—U.S. Patent 2908563.

223. PHIPPS, T.E., G.W. SEARS, R.L. SEIFERT, and O.C. SIMPSON.—*International Conference on Peaceful Uses of Atomic Energy, Geneva,* **7**, 382. 1955.

224. FLANARY, J.R.—*International Conference on Peaceful Uses of Atomic Energy, Geneva,* **9**, 528. 1955.

225. FLETCHER, J.M.—*International Conference on Peaceful Uses of Atomic Energy, Geneva,* **9**, 459. 1955.

226. FOMIN, V.V. *Khimiya ekstraktsionnykh protsessov (Chemistry of Extraction Processes).*—Moscow, Atomizdat. 1960.

227. FOMIN, V.V., S.P. VOROB'EV, and M.A. ANDREEVA.—*Atomnaya energiya,* **4**, 57. 1958.

228. FOMIN, V.V. and E.P. MAIOROVA.—*Zhurnal neorganicheskoi khimii,* **3**, 1937. 1958.

229. FRIED, S. and V.H. ZACHARIASEN.—*International Conference on Peaceful Uses of Atomic Energy, Geneva,* **7**, 235. 1955.

230. HYDE, E. *Khimiya yadernogo goryuchego (Chemistry of Nuclear Fuel)*, p. 393.—Moscow, Goskhimizdat. 1956.

231. HYDE, I. and G. SEABORG. *Transuranium Elements and Nuclear Energy*.—Univ. of Notre Dame, Indiana. 1948.

232. HAHN, O. *Applied Radiochemistry*.—Ithaca, N.Y., Cornell Univ. Press. 1936.

233. HOWELLS, G.R., T.G. HUGHES, D.R. MCKAY, and K. SADDINGTON.—*Second International Conference on Peaceful Uses of Atomic Energy, Geneva*, **17**, 3. 1958.

234. HAEFFNER, E., G. NILSSON, and A. HULTGREN.—*International Conference on Peaceful Uses of Atomic Energy, Geneva*, **9**, 498. 1955.

235. KHLEBNIKOV, G.I. and E.P. DERGUNOV.—*Atomnaya energiya*, **4**, 376. 1958.

236. KHLOPIN, V.G.—*Trudy Radievogo instituta*, **4**, 34. 1938.

237. HOLLEY, C.E., R.N.R. MULFORD, E.J. HUBER, E.A. HEAD, F.H. ELLINGTON, and C.W. BJORKLUND.—*Second International Conference on Peaceful Uses of Atomic Energy, Geneva*, **6**, 215. 1958.

238. HULTGREN, A. and E. HAEFFNER.—*Second International Conference on Peaceful Uses of Atomic Energy, Geneva*, **17**, 324. 1958.

239. TSALETKA, R. and A.V. LAPITSKII.—*Uspekhi khimii*, **29**, 1489. 1960.

240. CHERNYAEV, I.I., V.A. GOLOVNYA, and G.V. ELLERT.—*Zhurnal neorganicheskoi khimii*, **1**, 2726. 1956.

241. CHANG-WEN-CH'ING. Candidate's Dissertation.—Moscow, Institut geokhimii i analiticheskoi khimii im. V.I. VERNADSKOGO, AN SSSR. 1960.

242. CHMUTOVA, M.K., O.M. PETRUKHIN, and YU.A. ZOLOTOV.—*Zhurnal neorganicheskoi khimii*, **18**, 588. 1963.

243. SHARLOV, L.E. *Analiz mineral'nogo syr'ya (Analysis of Mineral Materials)*.—Moscow–Leningrad, ONTI, p. 164. 1936.

244. SHVEDOV, V.P., S.G. STRIZHOV, and CHIN TSZE-KHOU.—*Radiokhimiya*, **1**, 622. 1959.

245. SHVETSOV, I.K. and A.M. VOROB'EV.—*International Conference on Peaceful Uses of Atomic Energy, Geneva*, **7**, 304, 1955.

246. SHEVCHENKO, V.B. and YU.F. ZHDANOV.—*Radiokhimiya*, **3**, 676. 1961.

247. SHEVCHENKO, V.B., N.S. POVITSKII, and A.S. SOLOVKIN.—*Second International Conference on Peaceful Uses of Atomic Energy, Geneva*, **17**, 46. 1958.

248. SHEVCHENKO, V.B. and V.S. SMELOV.—*Atomnaya energiya*, **5**, 542. 1958.

249. SHEVCHENKO, V.B. and V.S. SMELOV.—*Atomnaya energiya*, **6**, 140. 1959.

250. SHEVCHENKO, V.B., A.S. SOLOVKIN, L.M. KIRILLOV, and A.I. IVANTSOV.—*Radiokhimiya*, **3**, 503. 1961.

251. SHEVCHENKO, V.B., A.S. SOLOVKIN, I.V. SHILIN, L.M. KIRILLOV, A.V. RODIONOV, and V.V. BALANDINA.—*Radiokhimiya*, **2**, 281. 1960.

252. SHEVCHENKO, V.B., V.G. TIMOSHEV, and A.A. VOLKOVA.—*Atomnaya energiya*, **6**, 426. 1959.

253. SHEVCHENKO, V.B. and I.A. FEDOROV.—*Radiokhimiya*, **2**, 6. 1960.

254. SHEVCHENKO, V.B., V.S. SHMIDT, and E.A. MEZHOV.—*Zhurnal neorganicheskoi khimii*, **5**, 1911. 1960.

255. SHRAIBER, G. *Biofizicheskaya radiologiya (Biophysical Radiology)*.—Moscow, Atomizdat. 1960.

256. Yakovlev, G.N. and D.S. Gorbenko-Germanov.—*International Conference on Peaceful Uses of Atomic Energy, Geneva,* **7**, 306. 1955.
257. Yakovlev, G.N., E.P. Dergunov, I.A. Reformatskii, and V.B. Delov.—*International Conference on Peaceful Uses of Atomic Energy, Geneva,* **7**, 57. 1955.
258. Yakovlev, G.N., P.M. Chulkov, V.B. Delov, V.N. Kosyakov, and Yu.P. Sobolev.—*Atomnaya energiya,* **1**, 131. 1956.
259. Yatsimirskii, K.B. and V.P. Vasil'ev. *Konstanty nestoikosti komplesnykh soedinenii (Instability Constant of Complex Compounds).*—Moscow, Izdatel'stvo AN SSSR. 1959.
260. Abraham, B.M., B.B. Brody, N.R. Davidson, F. Hagemann, G. Karle, J.J. Katz, and M.J. Wolf.—*The Transuranium Elements. Nat. Nucl. Energy Ser.,* Div. IV, **14B**, paper 6.7. 1949.
261. Abraham, B.M. and N.R. Davidson.—*Ibid.,* paper 6.10. 1949.
262. Abraham, B.M., N.R. Davidson, and E.E. Westrum.—*Ibid.,* paper 6.18. 1949.
263. Ahrland, S., I. Grenthe, and B. Noren.—*Acta chem. scand.,* **14**, 1059. 1960.
264. Ahrland, S., I. Grenthe, and B. Noren.—*Acta chem. scand.,* **14**, 1077. 1960.
265. Aikin, A.M.—*Chem. Eng. Progress,* **53**, 82. 1957.
266. Alcock, K., G.F. Best, E. Hesford, and H.A.C. McKay.—*J. Inorg. Nucl. Chem.,* **6**, 328. 1958.
267. Alcock, K., S.S. Grimley, T.V. Healy, J. Kennedy, and H.A.C. McKay.—*Trans. Faraday Soc.,* **52**. 39. 1956.
268. Allen, K.A.—*J. Phys. Chem.,* **60**, 239. 1956.
269. Allen, K.A.—*J. Phys. Chem.,* **60**, 943. 1956.
270. Anderson, H.H.—*The Transuranium Elements, Nat. Nucl. Energy, Ser.,* Div. IV, **14B**, paper 6.2. 1949.
271. Anderson, H.H.—*Ibid.,* paper 6.9. 1949.
272. Anderson, H.H.—*Ibid.,* paper 6–14. 1949.
273. Anderson, H.H.—*Ibid.,* paper 6.15. 1949.
274. Anderson, H.H.—*Ibid.,* paper 6.16. 1949.
275. Anderson, H.H.—*Ibid.,* paper 6.17. 1949.
276. Anderson, H.H.—*Ibid.,* paper 6.19. 1949.
277. Anderson, H.H.—*Ibid.,* paper 6.21. 1949.
278. Anderson, H.H.—*Ibid.,* paper 6.220. 1949.
279. Anderson, H.H. and L.B. Asprey.—US Patent 2924506; cited in *Nucl. Sci. Abstr.,* **15**, 24800. 1961.
280. Arselin, F., P. Faugeras, and E. Grison.—*Compt. rend.,* **242**, 1996. 1956.
281. Ayres, G.H., B.L. Tuffly, and J.S. Forrester.—*Anal. Chem.,* **27**, 1742. 1955.
282. Bagnall, K.W., A.M. Deane, T.L. Markin, P.S. Robinson, and M.A.A. Stewart.—*J. Chem. Soc.,* 1611. 1961.
283. Bagnall, K.W., T.L. Markin, D.S. Robertson, P.S. Robinson, and M.A.A. Stewart.—*R. C. T. C.* (p. 128). *At. En. R. E.* No. 14730. 1960; cited in *Nucl. Sci. Abstr.,* **14**, 14730. 1960.
284. Balhis, J.H.—US Patent 2992888; cited in: *Nucl. Sci. Abstr.,* **15**, 22384. 1961.
285. Ball, J.G., J.A. Lee, P.G. Mardon, and J.A.L. Robertson.—*Mem. sci. rev. met.,* **57**, 49. 1960.

286. BALL, J.G. and W.B.H. LORD.—*J. Institute Metals,* **86**, 369. 1958.
287. BALL, J.G., J.A.L. ROBERTSON, P. MARDON, J.A. LEE, and E.T. ADAMS.—*Nature,* **173**, 535. 1954.
288. BARKER, F.—*Nucl. Sci. Abstr.,* **14**, 10460. 1960.
289. BARKER, G.C. and I.L. JENKINS.—*Analyst,* **77**, 685. 1952.
290. BARRICK, I.G. and I.P. MANION.—US Patent 2996352; cited in: *Nucl. Sci. Abstr.,* **15**, 24800. 1961.
291. BEATON, R.H.—US Patent 2942939; cited in: *Nucl. Sci. Abstr.,* **14**, 25575. 1960.
292. BEATON, R.H.—US Patent 2990243; cited in: *Chem. Abstr.,* **55**, 25535. 1961.
293. BERG, VAN DER M. and P.F.A. KLINKENBERG.—*Physica,* **20**, 462. 1954.
294. BERGSTRESSER, K.S.—*Anal. chem.,* **30**, 1630. 1958.
294a. BERGSTRESSER, K.S.—*Anal. chem.,* **29**, 532. 1957.
295. BERNAL, J.D. and R.H. FOWLER.—*J. Chem. Phys.,* **1**, 515. 1933.
296. BEST, G.F., E. HESFORD, and H.A.C. MCKAY.—cited in: *Chemical Abstracts,* **55**, 15088. 1961.
297. BEST, G.F., H.A.C. MCKAY, and P.R. WOODGATE.—*J. Inorg. Nucl. Chem.,* **4**, 315. 1957.
298. BETTS, R.H. and R.M. LEIGH.—*Canad. Res.,* **28**, 514. 1950.
299. BLAEDEL, W. and H. KHIGHT.—*Anal. Chem.,* **26**, 741. 1954.
300. BLUMENTHAL, B. and M.B. BRODSKY. *Plutonium,* p. 171. 1960—London, Cleaver-Hume Press, Ltd., 1961.
301. BOASE, D.G. and J.K. FOREMAN.—*Talanta,* **8**, 187. 1961.
302. BOASE, D.G., J.K. FOREMAN, and J.L. DRUMMOND.—*Talanta,* **9**, 53. 1962.
303. BOAZ, H.E.—*Nucl. Sci. Abstr.,* **10**, 2295. 1956.
304. BOCK, R. and E. BOCK.—*Naturwissenschaft,* **36**, 344. 1949.
305. BOOMEN, G.L.—*Anal. Chem.,* **29**, 213. 1957.
306. BOOMEN, G.L., W.B. HOLBROOK, and J.E. REIN.—*Anal. Chem.,* **29**, 219. 1957.
307. BOYES, G.W. JR., J.B. RAMSAY, and R.T. PHELPS.—*Talanta,* **6**, 209. 1960.
308. BRETT, N.H. and L.E. RUSSELL. *Plutonium,* p. 397. 1960.—London, Cleaver-Hume Press, Ltd., 1961.
309. BREWER, L.—*Chem. Rev.,* **52**, 1. 1953.
310. BREWER L., L. BROMLEY, P. GILLES, and N. LOFGREN.—*The Transuranium Elements, Nat. Nucl. Energy Ser.,* Div. IV, **14B**, paper 6.40. 1949.
310a. BRICKER, C.E. and G.R. WATERBURY.—*Anal. Chem.,* **29**, 558. 1957.
311. British Patent 826910; cited in: *Nucl. Sci. Abstr.,* **14**, 8513. 1960.
312. British Patent 847647; cited in: *Nucl. Sci. Abstr.,* **14**, 25495. 1960.
313. British Patent 852501; cited in: *Chem. Abstr.,* **55**, 10137. 1961.
314. British Patent 855853; cited in: *Nucl. Sci. Abstr.,* **15**, 8912. 1961.
315. British Patent 862352. 1961.
316. BRODY, J.K., J.P. FARIS, and R.F. BUCHANAN.—*Anal. Chem.,* **30**, 1909. 1958.
317. BROTHERS, J.H., R.G. HART, and W.G. MATHERS.—*J. Inorg. Nucl. Chem.,* **7**, 85. 1958.
318. BROWN, F., G.G. GEORGE, D.E. GREEN, and W.D. WATT.—*J. Inorg. Nucl. Chem.,* **13**, 192. 1960.
319. BROWN, F., H.M. OCKENDEN, and G.A. WELCH.—*J. Chem. Soc.,* 3932. 1955.
320. BROWN, F., H.M. OCKENDEN, and G.A. WELCH.—*J. Chem. Soc.,* 4196. 1955.

321. Browne, C.I., D.C. Hoffman, W.T. Crane, J.P. Balagha, G.H. Higgins, J.W. Bornes, R.W. Hoff, H.L. Smith, J.P. Mize, and M.E. Bunker.—*J. Inorg. Nucl. Chem.*, **1**, 254. 1955.
322. Bruce, F.R.—*Progress in Nucl. Energy,* ser. III, **2**, 363. London, Pergamon Press, 1958.
323. Brunstad, A.—*Ind. Eng. Chem.*, **51**, 38. 1959.
324. Brunstad, A.—*Chem. Abstr.*, **51**, 1770. 1957.
325. Bubernak, J., M. Lew, and G. Matlack.—*Talanta,* **6**, 167. 1960.
326. Buchanan, R.F. and J.P. Faris.—*Radioisotopes in the Physical Sciences and Industry,* **2**, 361.—Intern. At. Energy Agency, Vienna. 1962.
327. Buchanan, R.F., J.P. Hughes, and C.A.A. Bloomquist.—*Talanta,* **6**, 100. 1960.
328. Buchanan, R.F., J.P. Hughes, J.J. Hines, and C.A.A. Bloomquist.—*Talanta,* **6**, 173. 1960.
329. Bunce, G.L.—*Referativnyi Zhurnal Khimii.* 7964. 1959.
330. Bunney, L.R., N.E. Ballon, J. Pascual, and S. Foty.—*Anal. Chem.*, **31**, 324. 1959.
331. Burger, L.Z.—*J. Phys. Chem.*, **62**, 590. 1958.
332. Butler, J.P., M. Lounsbery, and J.S. Merritt.—*Can. J. Chem.*, **34**, 253. 1956.
333. Caldwell, C.E., L.F. Grill, R.G. Kurtz, F.G. Miner, and N.E. Moody.—*Anal. Chem.*, **34**, 346. 1962.
334. Campbell, W.M.—*Nucleonics,* **14**, 92. 1956.
335. Campbell, M.H.—*Chem. Abstr.*, **56**, 158c. 1962.
336. Carleson, G.—*Svensk. Kem. tidskr.*, **70**, 55. 1958.
337. Carleson, G.—*Proc. UN Inter. Conf. Peaceful Uses At. Energy, 2nd., Geneva,* 1958; cited in: *Chem. Abstr.*, **54**, 23217. 1960.
338. Carson, W.N.Cr., C.G.W. Vanderwater, and H.S. Gile.—*Anal. Chem.*, **29**, 1417. 1957.
339. Chamberlein, O., G.W. Forwell, and E. Segu.—*Phys. Rev.*, **94**, 1561. 1954.
340. Choppin, G.R. and A. Chatham-Strode.—*J. Inorg. Nucl. Chem.*, **15**, 377. 1960.
341. Clanet, F.—*J. Chromatography,* **6**, 85. 1961.
342. Cohen, D.—*J. Inorg. Nucl. Chem.*, **18**, 207. 1961.
343. Cohen, D.—*J. Inorg. Nucl. Chem.*, **18**, 211. 1961.
344. Coleman, J.S., R.A. Penneman, T.K. Keenan, L.E. La Mar, D.E. Armstrong, and L.B. Asprey.—*J. Inorg. Nucl. Chem.*, **3**, 327. 1956.
345. Connick, R.E.—*J. Am. Chem. Soc.*, **71**, 1528. 1949.
346. Connick, R.E., J.W. Gofman, W.H. McVey, and G.E. Sheline.—*The Transuranium Elements, Nat. Nucl. Energy Ser.*, Div. IV, **14B**, paper 3.14. 1949.
347. Connick, R.E., M. Kasha, W.H. McVey, and G.E. Sheline.—*Ibid.*, paper 3.15. 1949.
348. Connick, R.E., M. Kasha, W.H. McVey, and G.E. Sheline.—*Ibid.*, paper 4.20. 1949.
349. Connick, R.E. and W.H. McVey.—*Ibid.*, paper 3.4. 1949.
350. Connick, R.E. and W.H. McVey.—*Ibid.*, paper 3.9. 1949.
351. Connick, R.E. and W.H. McVey.—*Ibid.*, paper 3.10. 1949.
352. Connick, R.E. and W.H. McVey.—*Ibid.*, paper 4.12. 1949.
353. Connick, R.E. and W.H. McVey.—*J. Am. Chem. Soc.*, **71**, 1534. 1949.
354. Connick, R.E. and W.H. McVey.—*J. Am. Chem. Soc.*, **73**, 1798. 1951.
355. Connick, R.E. and W.H. McVey.—*J. Am. Chem. Soc.*, **74**, 1341. 1952.
356. Connick, R.E. and W.H. McVey.—*J. Am. Chem. Soc.*, **75**, 474. 1953.

357. CONNICK, R.E., W.H. MCVEY, and G.E. SHELINE.—*The Transuranium Elements, Nat. Nucl. Energy Ser.,* Div. IV, **14B**, paper 3.12. 1949.

358. COOK, A.O.—*Ibid.,* paper 3.5. 1949.

359. COOK, G.P., J.K. FOREMAN, and E.F. KEMP.—*Anal. chim. acta,* **19**, 174. 1958.

359a. COOKE, W.D., F. HAZEL, and W.M. MACNABB.—*Anal. chim. acta.,* **3**, 656. 1949.

360. COOPER, G.H.—*Talanta,* **6**, 154. 1960.

361. CORYELL, C.D. and N. SUGARMAN.—*Radiochemical Studies: The Fission Products. Nat. Nucl. Energy Ser.,* Div. IV, **9**, 1951.

362. COWAN, M.R. and J.K. FOREMAN.—*Chem. Industry,* 1583. 1954.

363. CROMER, DON T. and R.B. ROOF.—*Acta. Cryst.,* **12**, 942. 1959.

364. CROUCH, E.A.C. and C.B. COOK.—*J. Inorg. Chem.,* **2**, 223. 1956.

365. CUNNINGHAM, J.G. and C.L. MILES.—*J. Inorg. Nucl. Chem.,* **3**, 54. 1956.

366. CUNNINGHAM, B.B., C. SMITH, and J.S. DIXON.—US Patent 2545606; cited in: *Chem. Abstr.,* **45**, 6508. 1951.

367. CUNNINGHAM, B.B. and L.R. WERNER.—*The Transuranium Elements, Nat. Nucl. Energy Ser.,* Div. IV, **14B**, paper 1.8. 1949.

368. CUNNINGHAM, B. and L.B. WERNER.—*J. Amer. Chem. Soc.,* **71**, 1521. 1949.

369. CVJETICANIN, D.—JENER Rep., No. 57. 1958; cited in: *RZhKh.* 26999. 1959.

370. CVJETICANIN, D. and N. CVJETICANIN.—JENER Rept., No. 54, Kjeller. 1958.

371. DALTON, G.C.—*United Kingdom Atomic Energy Authority, Production Group,* Rept. 284(w). 1962.

372. DAVIDSON, N.R., F. HAGEMANN, E.K. HYDE, J.J. KATZ, and J. SHEFT.—*The Transuranium Elements, Nat. Nucl. Energy, Ser.,* Div. IV, **14B**, paper 6.8. 1949.

373. DAVIES, T.H.—US Patent 2992889; cited in: *Nucl. Sci. Abstr.,* **15**, 22385. 1961.

374. DAWSON, J.K.—*J. Chem. Soc.,* 3393. 1954.

375. DAWSON, J.K.—Ministry of Supply, Hurwell, Berks (Eng). At. Energy Research Estab., Unclassified C/R **861**, 26. 1952. cited in: *Chem. Abstr.,* **46**, 10991. 1952.

376. DAWSON, J.K., R.M. ELLIOT, R. HURST, and A.E. TRUSWELL.—*J. Chem. Soc.,* 558. 1954.

377. DELAHAY, P. and G. MAMONTOV.—*Anal. Chem.,* **27**, 478. 1955.

378. DEMPSEY, E. and A.E. KAY.—*J. Inst. Metals,* **86**, 8, 379. 1958.

379. DIAMOND, H. and R.F. BARNES.—*Phys. Rev.,* **101**, 1064. 1956.

380. DIAMOND, R.M., K. STREET, JR., and G.T. SEABORG.—*J. Am. Chem. Soc.,* **76**, 1461. 1954.

381. DIXON, J.S. and C. SMITH.—*The Transuranium Elements, Nat. Nucl. Energy Ser.,* Div. IV, **14B**, paper 6.39. 1949.

382. DIZDAR, Z.I. and D.H.W. DEN BOER.—*J. Inorg. Nucl. Chem.,* **3**, 923. 1956.

383. DIZDAR, Z., O. TAL, and J. KAJNUAJN.—*Bull. Inst. Nucl. Sci. (Belgrad),* **7**, 43. 1957.

384. DOERNER, H. and W.J. HOSKINS.—*Amer. Chem. Soc.,* **47**, 675. 1955.

385. DREEBEN, A.B.—*Nucl. Sci. Abstr.,* **10**, 4198. 1956.

386. DRUMMOND, J.L., B.J. MCDONALD, H.M. OCKENDEN, and G.A. WELCH.—*J. Chem. Soc.,* 4785. 1957.

387. DRUMMOND, J.L. and G.A. WELCH.—*J. Chem. Soc.,* 2565. 1956.

388. DRUMMOND, J.L. and G.A. WELCH.—*J. Chem. Soc.,* 3781. 1957.

389. DRUMMOND, J.L. and G.A. WELCH.—*J. Chem. Soc.,* 3210. 1958.

390. DUKES, E.K.—*J. Am. Chem. Soc.,* **82**, 9. 1960.

391. DUWELL, E.J. and J.W. SHEPARD.—*J. Phys. Chem.*, **63**, 2044. 1959.
392. ENGELKEMIER, D.W., P.B. FIELDS, S. FRIED, G.L. PYLE, C.M. STEVENS, L.B. ASPREY, C.I. BROWNE, H.L. SMITH, and R.W. SPENCE.—*J. Inorg. Nucl. Chem.*, **1**, 345. 1955.
393. ENGELKEMIER, D.W., P.B. FIELDS, and I.R. HUIZENGA.—*Phys. Rev.*, **90**, 6. 1953.
394. FARIS, B.F.—US Patent 2931701; cited in: *Nucl. Sci. Abstr.*, **14**, 21609. 1960.
395. FARIS, B.F. and C.M. OLSON.—US Patent 2991150; cited in: *Nucl. Sci. Abstr.*, **15**, 22380. 1961.
396. FARIS, B.F. and H.K. STRASSEL.—Cited in: *RZhKh.* 12K35. 1962.
397. FERGUSON, W.S. and D.M. NEWELL.—*Nucl. Sci. Abstr.*, **15**, 1950. 1961.
398. FERRETT, D.G. and G.W.C. MILNER.—*Analyst.* **80**, 132. 1955.
399. FERRETT, D.G. and G.W.C. MILNER.—*Analyst,* **81**, 193. 1956.
400. FERRETT, D.G. and G.W.C. MILNER.—*J. Chem. Soc.*, 1186. 1956.
401. FERRETT, D.G., G.W.C. MILNER, H.L. SHALGOSKY, and L.G. SLEE.—*Analyst,* **81**, 506. 1956.
402. FIELDS, P.R., M.H. STUDIER, A.M. FRIEDMAN, H. DIAMOND, R. SJOBLOM, and P.A. SELLER.—*J. Inorg. Nucl. Chem.*, **1**, 262. 1955.
403. FITCH, F. and D. RUSSELL.—US Patent 2852338; cited in: *RZhKh.* 2722. 1960.
404. FLETCHER, G.M.—*Progr. Nucl. En.*, **3**, 105. 1956.
405. FLORIN, A.E., J.R. TANNENBAUM, and J.F. LEMONS.—*J. Inorg. Nucl. Chem.*, **2**, 368. 1956.
406. FOREMAN, J.K. and W.T. APPLETON.—*Nucl. Sci. Abstr.*, **14**, 19451. 1960.
407. FOREMAN, J.K. and T.D. SMITH.—*J. Chem. Soc.*, 1752. 1957.
408. FOREMAN, J.K. and T.D. SMITH.—*J. Chem. Soc.*, 1758. 1957.
409. FORTUNE, W.B. and M.G. MELLON.—*Ind. End. Chem., Anal. Ed.*, **14**, 869. 1942.
410. FOSTER, E. and G. PHILLIPS.—Cited in: *Chem. Abstr.*, **55**, 16267. 1961.
411. FORWELL, G.W., J.E. ROBERTS, and A.C. WAHL.—*Phys. Rev.*, **94**, 363. 1954.
412. FRED, M., R. NACTRIEB, and F. TOMPKINS.—*J. Opt. Soc. Am.*, **37**, 279. 1947.
413. FRED, M. and F. TOMPKINS.—*J. Opt. Soc. Am.*, **44**, 824. 1954.
414. French Patent 1220061; cited in: *Nucl. Sci. Abstr.*, **15**, 14366. 1961.
415. FRIED, S.—US Patent 2860948; cited in: *RZhKh.* 5243. 1960.
416. FRIED, S. and N.R. DAVIDSON.—*The Transuranium Elements, Nat. Nucl. Energy Ser.*, Div. IV, **14B**, paper 6.11. 1949.
417. FRIED, S. and N.R. DAVIDSON.—*The Transuranium Elements, Nat. Nucl. Energy Ser.*, Div. IV, **14B**, paper 15.5. 1949.
418. FRIED, S. and N.R. DAVIDSON.—*J. Am. Chem. Soc.*, **70**, 3539. 1948.
419. FRIED, S., E.F. WESTRUM, H.L. BAUMBACH, JR., and P.J. KIRK.—*J. Inorg. Nucl. Chem.*, **5**, 192. 1957.
420. FRIED, J.—*Symposium on Rare Metals,* pp. 386–401. Indian Institute of Metals, 1957; cited in: WALDRON, M.B. *Contemporary Physics,* **2**, 385. 1961.
421. FRIES, B.A.—US Patent 2926067; cited in: *Nucl. Sci. Abstr.*, **14**, 15719. 1960.
422. FRITZ, J.S., B.B. CARRALDA, and S.K. KARRAKER.—*Anal. Chem.*, **33**, 882. 1962.
423. FUDGE, A.J., A.G. WOOD, and M.F. BAHNMAN.—*Nucl. Sci. Abstr.*, **14**, 18851. 1960.
424. GARNER, C.S., N.A. BONNER, and G.T. SEABORG.—*J. Am. Chem. Soc.*, **70**, 3453. 1948.
425. GEVANTMAN, D.H. and K.A. KRAUS.—*The Transuranium Elements, Nat. Nucl. Energy Ser.*, Div. IV, **14B**, paper 4.16. 1949.

426. CHIORSO, A., R.A. JAMES, O.O. MORGAN, and G.T. SEABORG.—*Phys. Rev.*, **78**, 427. 1950.
427. GIERST, L.—*Anal. chim. acta*, **15**, 262. 1956.
428. GIERST, L. and A. JULIARD.—*J. Phys. Chem.*, **57**, 701. 1953.
429. GIERST, L. and PH. MECHELYNCK.—*Anal. chim. acta*, **12**, 79. 1955.
430. GLUCKAUF, E. and H.A.C. MCKAY.—*Nature*, **165**, 594. 1950.
431. GOEEKERMAN, R.H.—US Patent 2967395; cited in: *Nucl. Sci. Abstr.*, **15**, 13016. 1961.
432. GORDON, C.L., W.G. SCHLECHT, and E.J. WICHERS.—*Research Nat. Bur. Standards*, **33**, 457. 1944.
433. GRENTH, I. and B. NOREN.—*Acta chem. scand.*, **14**, 2216. 1960.
434. GRIFFIN, P.M. and J.R. MCNALLY, JR.—*J. Opt. Soc. Am.*, **45**, 63. 1955.
435. GRISON, E.—*Progress in Nuclear Energy*, ser. V, **2**, 91. London, Pergamon Press, 1959.
436. GROH, H.J.—Cited in: *Nucl. Sci. Abstr.*, **15**, 14346. 1961.
437. GWOZDZ, R. and S. SIEKIERSKI.—*Nukleonika*, **5**, 671. 1960.
438. TER HAAZ, K. and J. BAZEN.—*Anal. chim. acta*, **9**, 235. 1953.
439. HAEFFNER, E. and A. HULTGREN.—*Nucl. Sci. Eng.*, **3**, 372. 1958.
440. HAGEMANN, F.J.—*Am. Chem. Soc.*, **72**, 768. 1950.
441. HAGEMANN, F., B.M. ABRAHAM, N.R. DAVIDSON, J.J. KATZ, and J. SHEFT.—*The Transuranium Elements, Nat. Nucl. Energy Ser.*, Div. IV, **14B**, paper 6.170. 1949.
442. HAHN, O. and F. STRASSMAN.—*Naturwissenschaft*, **30**, 256. 1942.
443. HALL, G.R. and P.D. HERNIMAN.—*J. Chem. Soc.*, 2214. 1954.
444. HAMAKER, J.W. and C.W. KOCH.—*The Transuranium Elements, Nat. Nucl. Energy, Ser.*, Div. IV, **14B**, paper 6.92. 1949.
445. HAMM, R.E.—*Anal. Chem.*, **30**, 350. 1958.
446. HAMPSON, D.C., G.A. BENNETT, and N.R. CHELLEW.—*Nucl. Sci. Engng.*, **6**, 501. 1959.
447. HANDSHUH, G.W.—*Chem. Abstr.*, **55**, 14093. 1961.
448. HARDY, C.O.—*Progress in Nucl. Energy*, ser. III, **2**, 357. London, Pergamon Press, 1958.
449. HART, R., C.B. BIGHAM, and M. LOUNSBURY.—*Talanta*, **6**, 94. 1960.
450. HARVEY, G.B.—*Nucleonics*, **2**, 30. 1948.
451. HARVEY, G.B., H.G. HEAL, A.G. MADDOCK, and E.L. ROWLEY.—*J. Chem. Soc.*, 1010. 1947.
452. HEAL, H.G.—US Patent 2925322; cited in: *Nucl. Sci. Abstr.*, **14**, 15717. 1960.
453. HEABY, T.V. and H.A.C. MCKAY.—*Trans. Faraday Soc.*, **52**, 633. 1956.
454. HELLY, M.V. and A.W. GARDNER.—*J. Inorg. Nucl. Chem.*, **7**, 245. 1958.
455. HELBIG, W.—*Chemische Technik*, **13**, 514. 1961.
456. HELBIG, W.—*Z. anal. Chem.*, **182**, 15. 1961.
457. HELBIG, W.—*Z. anal. Chem.*, **182**, 19. 1961.
458. HELBIG, W.—*Z. anal. Chem.*, **182**, 84. 1961.
459. HILL, J.D.—*Nature*, **186**, 304. 1960.
460. HINDMAN, J.C.—*The Transuranium Elements, Nat. Nucl. Energy, Ser.*, Div. IV, **14B**, paper 4.4. 1949.
461. HINDMAN, J.C.—*Ibid.*, paper 4.5. 1949.
462. HINDMAN, J.C.—*Ibid.*, paper 4.7. 1949.
463. HINDMAN, J.C. and D.P. AMER.—*Ibid.*, paper 4.2. 1949.

464. HINDMAN, J.C., K.A. KRAUS, J.J. HOWLAND, and B.B. CUNNINGHAM.—*Ibid.*, paper 3.2. 1949.

465. HOFFMAN, D.C.—*Progress in Nucl. Energy*. ser. IX. **2**, 157. London, Pergamon press. 1961.

466. HOFFMAN, D.C., G.P. FORD, and F.O. LAWRENCE.—*J. Inorg. Nucl. Chem.*, **5**, 6. 1957.

467. HOLLANDER, J.M., J. PERLMAN, and G.T. SEABORG.—*Rev. Mod. Phys.*, **25**, 469. 1953.

468. HOLT, B.D.—*Anal. Chem.*, **32**, 124. 1960.

469. HORNER, D.E. and C.F. COLEMAN.—*Nucl. Sci. Abstr.*, **15**, 11055. 1961.

470. HORTON, C.T. and B.J. WARD. *Plutonium*, p. 499. 1960.—London, Cleaver-Hume Press Ltd., 1961.

471. HOWLAND, J.J. JR., J.C. HIDMAN, and K.A. KRAUS.—*The Transuranium Elements, Nat. Nucl. Energy Ser.*, Div. IV, **14B**, paper 3.3. 1949.

472. HOFFORD, D.L. and B.F. SCOTT.—*Ibid.*, paper 16.1. 1949.

473. HULET, E.K., R.G. GUTMACHER, and M.S. COOPS.—*J. Inorg. Nucl. Chem.*, **17**, 350. 1961.

474. HUMMEL, F.C. and H.H. WILLARD.—*Ind. Eng. Chem. Anal. Ed.*, **10**, 13. 1938.

475. HUNTER, G.J. and R.B. CHENLEY.—At. Energy Research Estab. (Gr. Brit.) A.M. 19,5 1959; cited in: *Chem. Abstr.*, **54**, 307. 1960.

476. HURST, R. and G.R. HALL.—*Analyst*, **77**, 790. 1952.

477. JACKSON, N. and J.F. SHORT.—A.E.R.E., N M444 1959; cited in: *RZhKh*, 47007. 1960.

478. JAMES, R.A., A.E. FLORIN, H.H. HOPKINS, JR., and A. GHIORSO.—*The Transuranium Elements, Nat. Nucl. Energy Ser.*, Div. IV, **14B**, paper 22.8. 1949.

478a. JAMES, R.A. and S.G. THOMPSON.—US Patent 2863892; cited in: *RZhKh*, 2721. 1960.

479. JENKINS, E.N.—*J. Inorg. Nucl. Chem.*, **13**, 323. 1960.

480. JENKINS, J.L. and H.A. MCKAY.—*Trans. Faraday Soc.*, **50**, 107. 1954.

480a. JOHNSON, A.J. and E. VEJVODA.—*Anal. Chem.*, **31**, 1643. 1959.

481. JORGENSEN, C.K.—*J. Chem. Phys.*, **23**, 399. 1955.

482. JETTE, E.R.—*J. Chem. Phys.*, **23**, 365. 1955.

483. KABANOVA, O.L., M.A. DANUSCHENKOVA, and P.N. PALEY.—*Anal. chim. acta*, **22**, 66. 1960.

484. KAMBARA, R.—*Bull. Chem. Soc., Japan*, **27**, 523, 527, 529. 1954.

485. KASHA, M. and G.E. SHELINE.—*The Transuranium Elements, Nat. Nucl. Energy Ser.*, Div. IV, **14B**, paper 3.14. 1949.

486. KASHA, M.—*Ibid.*, paper 3.100. 1949.

487. KATZ, J.J. and D.M. GRUEN.—*J. Am. Chem. Soc.*, **71**, 2106. 1949.

488. KEDER, W.E., J.C. SHEPPARD, and A.S. WILSON.—*J. Inorg. Nucl. Chem.*, **12**, 327. 1960.

489. KELLER, C.—*J. Chromatography*, **7**, 535. 1961.

490. KELEY, M.T., H.C. JONES, and D.E. FISHER.—*Anal. Chem.*, **31**, 488, 956. 1959.

491. KELEY, M.T., H.C. JONES, and D.E. FISHER.—*Talanta*, **6**, 185. 1960.

492. KENNEDY, J.W., G.T. SEABORG, E. SEGRE, and A.C. WAHL.—*Phys. Rev.*, **70**, 555. 1946.

493. KENNEDY, J.W., G.T. SEABORG, E. SEGRE, and A.C. WAHL.—*The Transuranium Elements, Nat. Nucl. Energy Ser.*, Div. IV, **14B**, paper 1.2. 1949.

494. KIESS, C.C., C.J. HUMPHREYS, and D.D. LAUN.—*J. Research Nat. Bur. Standards*, **37**, 57. 1946; *J. Opt. Soc. Am.*, **36**, 357. 1961.

495. KING, E.L.—*The Transuranium Elements, Nat. Nucl. Energy Ser.*, Div. IV, **14B**, paper 17.4. 1949.
496. KNOCH, W.—*Naturforsch.*, **16a**, 525. 1961; cited in: *Nucl. Sci. Abstr.*, **15**, 24911. 1961.
497. KNOCH, W. and R.Z. SINDER.—*Electrochem.*, **64**, 1020. 1960.
498. ROY, KO.—*Nucleonics*, **15**, 71. 1957.
499. ROY, KO.—*Anal. Chem.*, **28**, 274. 1956.
500. ROY, KO.—*Appl. Spectroscopy*, **13**, 10. 1959.
501. KOCH, C.W.—*The Transuranium Elements, Nat. Nucl. Energy Ser.*, Div. IV, **14B**, paper 17.4. 1949.
502. KOHMAN, T.P.—*J. Chem. Educ.*, **38**, 73. 1961.
503. KOSHAND, M.E. and A. FRIES.—*Nucl. Sci. Abstr.*, **10**, 5182. 1956.
504. KOSHLAND, D.E., J.C. KRONER, and L. SPECTOR.—*The Transuranium Elements, Nat. Nucl. Energy Ser.*, Div. IV, **14B**, paper 6.6. 1949.
505. KOSHLAND, D.E. and J.E. WILLAND.—US Patent 2995419; cited in: *Nucl. Sci. Abstr.*, **15**, 24916. 1961.
506. KOYAMA, K.—*Anal. Chem.*, **34**, 523. 1960.
507. KRAUS, K.A.—*The Transuranium Elements, Nat. Nucl. Energy Ser., Div. IV*, **14B**, paper 3.16. 1949.
508. KRAUS, K.A.—*Ibid.*, paper 3.17. 1949.
509. KRAUS, K.A.—*Ibid.*, paper 4.17. 1949.
510. KRAUS, K.A. and J.R. DAM.—*Ibid.*, paper 4.14. 1949.
511. KRAUS, K.A. and J.R. DAM.—*Ibid.*, paper 4.15. 1949.
512. KRAUS, K.A. and J.D. DAM.—*Ibid.*, paper 4.18. 1949.
513. KRAUS, K.A. and G.E. MOORE.—*Ibid.*, paper 4.19. 1949.
514. KRAUS, K.A. and F. NELSON.—*J. Am. Chem. Soc.*, **72**, 3901. 1950.
515. KRAUS, K.A., F. NELSON, and G.E. MOORE.—*J. Am. Chem. Soc.*, **77**, 3979. 1955.
516. LAMB, C.E.—*Talanta*, **6**, 20. 1960.
517. LANGER, A. and D.P. STEVENSON.—*Ind. Eng. Chem., Anal. Ed.*, **14**, 770. 1942.
518. LARSEN, R.P. and C.A. SEILS.—*Anal. Chem.*, **32**, 1863. 1960.
519. LEARY, J.A., A.N. MORGAN, and W.J. MARANAN.—*Ind. Eng. Chem.*, **51**, 105. 1959.
520. LEE, J.A. and R.O.A. HALL.—*J. Less Common Metals*, **1**, 356. 1959.
521. LEVINE, C.A. and W.T. SEABORG.—*J. Amer. Chem. Soc.*, **73**, 3278. 1951.
522. LINGANE, G.G.—*Anal. chim. acta*, **2**, 584. 1948.
523. LINGANE, G.G.—*J. Am. Chem. Soc.*, **67**, 1916. 1945.
524. LORD, W.B.H.—*Nature*, **173**, 534. 1954.
525. LORD, W.B.H. and M.B. WALDRON.—*J. Inst. Metals*, **86**, 385. 1958.
526. MADDOCK, A.G. and H. BOOTH.—US Patent 2952511; cited in: *Chem. Abstr.*, **55**, 2302. 1961.
527. MAECK, W.J., G.L. BOOMAN, M.E. KUSSY, and J.E. REIN.—*Anal. Chem.*, **32**, 1874. 1960.
528. MAECK, W.J., G.L. BOOMAN, M.E. KUSSY, and J.E. REIN.—*Anal. Chem.*, **33**, 1775. 1961.
529. MAECK, W.J., M.E. KUSSY, T.D. MORGAN, J.E. REIN, and M.T. LANG.—*Nucleonics*, **20**, N 5, 80. 1926.
530. MAGNUSSON, L.B.—*The Transuranium Elements, Nat. Nucl. Energy Ser.*, Div. IV, **14B**, paper 15.11. 1949.

531. MAGNUSSON, L.B. and T.J. LA CHAPELLE.—*Ibid.*, paper 1.7. 1949.

532. MAGNUSSON, L.B., S.G. THOMPSON, and G.T. SEABORG.—*Phys. Rev.*, **78**, 363. 1950.

533. MAINLAND, E.W.—*Ind. Eng. Chem.*, **53**, No. 9, 685. 1961.

534. MALY, J., H. KURZWEILOVA, R. LENK, and I. REKA.—*Collect. Czechosl. Chem. Comm.*, **25**, 1383. 1960.

535. MANDLEBERG, C.J., H.K. RAE, R. HURST, G. SONG, D. DAVIES, and K.E. FRANCIS.—*J. Inorg. Nucl. Chem.*, **2**, 358. 1956.

536. MARKIN, T.L. and H.A.C. MCKAY.—*J. Inorg. Nucl. Chem.*, **7**, 298. 1958.

537. MCDONALD, B.J. and J.B. FARDON.—*J. Chem. Soc.*, 781. 1956.

538. MCKAY, H.A.C., J.S. NAIRN, and M.B. WALDRON.—*Proc. of the Second Inter. Conf. on the Peaceful Uses of Atomic Energy. Geneva.* 299. United Nations. New York. 1958.

539. MCKENZIE, D.E., Can.—*J. Chem.*, **34**, 749. 1956.

540. MCLANE, C.K., J.S. DIXON, and J.C. HINDMAN.—*The Transuranium Elements, Nat. Nucl. Energy Ser.*, Div. IV, **14B**, paper 4.3. 1949.

541. MCMILLAN, E.M. and P.H. ABELSON.—*Phys. Rev.*, **57**, 1185. 1940.

542. MCNALLY, J.R., Jr.—*J. Opt. Soc. Am.*, **45**, 901. 1955.

543. MEADOWS, J.W. and G.M. MATLACK.—*Anal. chem.*, **32**, 1607. 1960.

544. MEADOWS, J.W.T., G.M. MATLACK, and G.B. NELSON.—*Talanta*, **6**, 246. 1960.

545. MEGGERS, W.F., M. FRED, and F.S. TOMKINS.—*J. Research Nat. Bur. Standards*, **58**, 297. 1957.

546. MERZ, E.Z.—*Anal. Chem.*, **166**, 417. 1959.

547. METZ, C.F.—*Anal. Chem.*, **29**, 1748. 1957.

548. METZ, C.F. and G.R. WATERBURY.—*Talanta*, **6**, 149. 1960.

549. MULLER, H.W. and R.J. BROUNS.—*Anal. Chem.*, **24**, 536. 1952.

550. MILNER, C.W.C. and J.L. WOODHEAD.—*Analyst*, **81**, 427. 1956.

551. MINER, F.J., R.P. DEGRAZIO, C.R. FORREY, Jr., and T.C. JONES.—*Anal. chim. acta*, **22**, 214. 1960.

552. MITCHELL, R.F.—*Anal. Chem.*, **32**, 326. 1960.

553. MONK, R.G.—*Anal. Abstr.*, **5**, 1511. 1958.

554. MOONEY, C.L. and W.H. ZACHARIASEN.—*The Transuranium Elements, Nat. Nucl. Energy Ser.*, Div. IV, **14B**, paper 20.1. 1949.

555. MOORE, F.L.—*Anal. Chem.*, **28**, 997. 1956.

556. MOORE, F.L.—*Anal. Chem.*, **29**, 941. 1957.

557. MOORE, F.L.—*Anal. Chem.*, **30**, 908. 1958.

558. MOORE, F.L.—*Anal. Chem.*, **30**, 1368. 1958.

559. MOORE, F.L.—*Anal. Chem.*, **32**, 1075. 1960.

560. MOORE, G.E. and K.A. KRAUS.—*The Transuranium Elements, Nat. Nucl. Energy Ser.*, Div. IV, **14B**, paper 4.22. 1949.

561. MOORE, F.L. and J.E. HUDGENS.—*Anal. Chem.*, **29**, 1767. 1957.

562. MOORE, F.L. and G.W. SMITH.—*Nucleonics*, **13**, 66. 1955.

563. MORRIS, D.F. and D. SCARGILL.—*Anal. chim. acta*, **14**, 57. 1956.

564. MORRISON, G.H.—*Anal. Chem.*, **22**, 1388. 1950.

565. MORRISON, G.H. and H. FREISER.—*Anal. Chem.*, **30**, 632. 1958.

566. MORRISON, G.H. and H. FREISER. *Solvent Extraction in Analytical Chemistry.*—Wiley, New York. 1957.

567. MULAY, L.N. and P.W. SELWOOD.—*J. Am. Chem. Soc.,* **76**, 6207. 1954.

568. MULFORD, R.N.R. and G.E. STURDY.—*J. Am. Chem. Soc.,* **77**, 3449. 1955.

569. MULFORD, R.N.R. and G.E. STURDY.—*J. Am. Chem. Soc.,* **78**, 3897. 1956.

570. MULLINS, J.L., J.A. LEARY, and C.W. BJORKLUND.—US *At. Energy Comm.,* LAMS 2441. 1960; cited in: *Chem. Abstr.,* **55**, 3240. 1961.

571. MYERS, M.N.—*Nucl. Sci. Abstr.,* **14**, 16680. 1960.

572. NAUMANN, D.—*Z. Chem.,* **1**, 247. 1961.

573. NEBEL, D.—*Chemische Technik,* **13**, 522. 1961.

573a. NEBEL, D. and K. SCHWABE.—*Z. Phys. Chem.,* **220**, 240. 1962.

574. NEBEL, E. and D. NEBEL.—*Kernenergie,* **4**, 15. 1961.

575. NEWTON, T.W.—*J. Phys. Chem.,* **62**, 943. 1958.

576. NEWTON, T.W.—*J. Phys. Chem.,* **63**, 1493. 1959.

577. NEWTON, T.W. and F.B. BAKER.—*J. Phys. Chem.,* **60**, 1417. 1956.

578. NEWTON, T.W. and F.B. BAKER.—*J. Phys. Chem.,* **61**, 934. 1957.

579. NEWTON, T.W. and H.D. COWAN.—*J. Phys. Chem.,* **64**, 244. 1960.

580. NEWTON, T.W. and S.W. RABIDEAU.—*J. Phys. Chem.,* **63**, 365. 1959.

581. NICHOLLS, C.M., J. WELLS, and R. SPENCE.—Canadian Patent 613891; cited in: *Nucl. Sci. Abstr.,* **15**, 20804. 1961.

582. OCKENDEN, D.W. and G.A. WELCH.—*J. Chem. Soc.,* 3359. 1956.

583. ORLEMAN, E.F. and B.B. BRODY.—*The Transuranium Elements, Nat. Nucl. Energy Ser.,* Div. IV, **14B**, paper 3.1. 1949.

584. OSBORN, G.H.—*Analyst,* **73**, 381. 1958.

585. PARKER, G.W., P.M. LANTZ, and W.J. MARTIN. *XVI Congrès intern. de chim. pure et appl., Paris* 1957. *Sect. de chim. minérale,* S. Ed. E. S., Paris, p. 395. 1958.

586. PASTERNAK, A.—*Nukleonica,* **6**, 113. 1961.

587. PATTON, R.L.—*The Transuranium Elements, Nat. Nucl. Energy Ser.,* Div. IV, **14B**, paper 6.33. 1949.

588. PATTON, R.L.—*Ibid.,* paper 6.37. 1949.

589. PASTERNAK, A.—*Nukleonika,* **6**, 113. 1961.

590. PEPPARD, D.F., M.H. STUDIER, M.V. GERGEL, G.W. MASON, J.C. SULLIVAN, and J.F. MECH.—*J. Am. Chem. Soc.,* **73**, 2529. 1951.

591. PERLMAN, J., S.G. THOMPSON, and B.B. CUNNINGHAM.—US Patent 2990245.

592. PIETRI, C.E. and J.A. BAGLIO.—*Talanta,* **6**, 159. 1960.

593. PHILLIPS, G.—*Analyst,* **83**, 75. 1958.

594. PHILLIPS, G. and E.N. JENKINS.—*J. Inorg. Nucl. Chem.,* **4**, 220. 1957.

595. PHILLIPS, G. and E. FOSTER.—AERE, N R2878, 1959; cited in: *RZhKh.* 30538. 1960.

596. PHILLIPS, G., J. WOODHEAD, and E.N. JENKINS.—*Anal. chim. acta,* **19**, 229. 1958.

597. PHIPPS, T.E., G.W. SEARS, and O.C. SIMPSON.—*The Transuranium Elements, Nat. Nucl. Energy Ser.,* Div. IV, **14B**, paper 6.1b. 1949.

598. PHIPPS, T.E., G.W. SEARS, R.L. SIEFERT, and O.C. SIMPSON.—*The Transuranium Elements, Nat. Nucl. Energy Ser.,* Div IV, **14B**, paper 6.1a. 1949.

599. POWELL, R.—*Analyst,* **83**, 252. 1958.

600. Prevot, I. and P. Regnaut.—*Progress Nucl. Energy. ser.* III, **2**, 337. London, Pergamon Press, 1958.
601. Purkayastha, B.C.—*Nucleonics,* **3**, No. 6, 2. 1948.
602. Rabideau, S.W.—*J. Am. Chem. Soc.,* **75**, 798. 1953.
603. Rabideau, S.W.—*J. Am. Chem. Soc.,* **78**, 2705. 1956.
604. Rabideau, S.W.—*J. Am. Chem. Soc.,* **79**, 3675. 1957.
605. Rabideau, S.W.—*J. Phys. Chem.,* **64**, 1491. 1960.
606. Rabideau, S.W. and H.D. Cowan.—*J. Am. Chem. Soc.,* **77**, 6145. 1955.
607. Rabideau, S.W. and R.J. Kline.—*J. Inorg. Nucl. Chem.,* **14**, 91. 1960.
608. Rabideau, S.W. and R.J. Kline.—*J. Phys. Chem.,* **63**, 1502. 1959.
609. Rabideau, S.W. and R.J. Kline.—*J. Phys. Chem.,* **64**, 193. 1960.
610. Rabideau, S.W. and R.J. Kline.—*J. Phys. Chem.,* **64**, 680. 1960.
611. Rabideau, S.W. and J.F. Lemons.—*J. Am. Chem. Soc.,* **73**, 2895. 1951.
612. Rabideau, S.W. and B.J. Masters.—*J. Phys. Chem.,* **65**, 1256. 1961.
613. Rainey, R.H.—USAEC, Rept. No. 59–12–95. 1959.
614. Rasmussen, M.J. and H.H. Hopkins.—*Ind. Eng. Chem.,* **53**, 453. 1961.
615. Reas, W.H.—*The Transuranium Elements, Nat. Nucl. Energy Ser.,* Div. IV, **14B**, paper 4.9. 1949.
616. Reilley, C.N., G.W. Everett, and R. Johns.—*Anal. Chem.,* **27**, 483. 1955.
617. Ritter, D.M. and R.P.S. Black.—Brit. Patent 860458; cited in: *Nucl. Sci. Abstr.,* **15**, 11094. 1961.
618. Rollier, M.A.—*Gazz. chim. ital.,* **84**, 649. 1954; *Z. anal. Chem.,* **147**, 138. 1955.
619. Rost, G.A.—*Anal. Chem.,* **33**, 738. 1961.
620. Runnals, O.S.C.—*Can. J. Chem.,* **34**, 133. 1956.
621. Runnalls, L.J.C.—*Progress in Nuclear Energy, ser.* V, **2**, 98. London, Pergamon Press. 1959.
622. Russell, E.R., A.S. Adamson, J. Schubert, and G.E. Boyd.—US Patent 2859093. 1959.
623. Ryan, J.L.—*J. Phys. Chem.,* **64**, 1378. 1960.
624. Ryan, J.L.—*J. Phys. Chem.,* **65**, 1099. 1961.
625. Ryan, J.L.—*J. Phys. Chem.,* **65**, 1856. 1961.
626. Ryan, J.L. and E.S. Wheelright.—*Ind. Eng. Chem.,* **51**, 60. 1959.
627. Ryan, J.L. and E.J. Wheelwright.—*Secound United Nations Intern. Conf. on the Peaceful Uses of At. Energy, Geneva,* Report No. 1915. USA. 1958.
628. Rydberg, J.—*Acta chim. scand.,* **9**, 1252. 1955.
629. Rydberg, J.—*Acta chim. scand.,* **11**, 201. 1957.
630. Rydberg, J.—*Arkiv för Kemi,* **9**, 109. 1955.
631. Rydberg, J.—*J. Inorg. Nucl. Chem.,* **5**, 79. 1957.
632. Rydberg, J. and B. Bernstrom.—*Acta chem. scand.,* **11**, 86. 1957.
633. Rydberg, J. and L.G. Sillen.—*Acta chem. scand.,* **9**, 1241. 1955.
634. Sackman, J.F. *Plutonium,* 1960, p. 222.—London, Cleaver-Hume Press, Ltd., 1961.
635. Sanchez, E.—*Bol. radioactividad.,* **24**, 42. 1951; cited in: *Chem. Abstr.,* **47**, 67856. 1953.
636. Sandenaw, T.A.—*Phys. and Chem. Solids,* **16**, 325. 1960.

637. SANDERS, S.M.—Cited in: *Nucl. Sci. Abstr.*, **10**, 4012. 1956.

638. SAVVIN, S.B. and A.A. MUK.—*Bull. Inst. Nucl. Sci., (Boris Kidrich)*, **12**, No. 259. 1961.

639. SCHLYTER, R. and L.G. SILLEN.—*Acta chem. scand.*, **4**, 1323. 1950.

640. SCHONFELD, F.W., E.M. CRAMER, W.N. MINER, F.H. ELLINGER, and A.S. COFFINBERRY.—*Progress in Nucl. Energy*, ser. V, **2**, 579. London, Pergamon Press, 1959.

641. SCOTT, F.A. and R.M. PEEKMA.—*Progress in Nucl. Energy*, ser. IX, **1**, 64. London, Pergamon Press, 1959; *Second United Nations Intern. Conf. on the Peaceful Uses of At. Energy, Geneva*, Rep. 914. 1958.

642. SCOTT, F.A. and R.M. PEEKMA.—*Talanta*, **6**, 196. 1960.

643. SCRIBNER, B. and H. MULLIN.—*J. Res. of Nat. Bureau of Stands.*, **37**, 379. 1946.

644. SCHWENDIMAN, L.C. and J. HEALY.—*Progress in Nucl. Energy*, ser. IX, **1**. 357. London, Pergamon Press, 1959.

645. SEABORG, G.T.—*Chem. Eng. News*, **23**, 2190. 1945.

646. SEABORG, G.T.—*Nucleonics*, **5**, No. 5, 16. 1949.

647. SEABORG, G.T. *The Transuranium Elements.*—New Haven, Yale University Press. 1958.

648. SEABORG, G.T.—*The Transuranium Elements, Nat. Nucl. Energy Ser.*, Div. IV, **14B**, paper 21.1. 1949.

649. SEABORG, G.T.—US Patent 2990242; cited in: *Nucl. Sci. Abstr.*, **15**, 19478. 1961.

650. SEABORG, G.T., W.J. BLAEDEL, and M.T. WALLING.—US Patent 2950166. 1960.

651. SEABORG, G.T., R.A. JAMES, and L.O. MORGAN.—*The Transuranium Elements, Nat. Nucl. Energy Ser.*, Div. IV, **14B**, paper 22.1. 1949.

652. SEABORG, G.T., R.A. JAMES, and A. GHIORSO.—*The Transuranium Elements, Nat. Nucl. Energy Ser.*, Div. IV, **14B**, paper 22.2. 1949.

653. SEABORG, G.T., E. MCMILLAN, and J.W. KENNEDY.—*Phys. Rev.*, **69**, 366. 1946.

654. SEABORG, G.T., E.M. MCMILLAN, J.W. KENNEDY, and A.C. WAHL.—*The Transuranium Elements, Nat. Nucl. Energy Ser.*, Div. IV, **14B**, paper 1.1a. 1949.

655. SEABORG, G.T. and J. PERLMAN.—*J. Amer. Chem. Soc.*, **70**, 1571. 1948.

656. SEABORG, G.T. and J. PERLMAN.—*Rev. Mod. Phys.*, **20**, 585. 1948.

657. SEABORG, G.T. and J. PERLMAN.—*The Transuranium Elements, Nat. Nucl. Energy Ser.*, Div. IV, **14B**, paper 1.3. 1949.

658. SEABORG, G.T. and S.G. THOMPSON.—US Patent 2950163; cited in: *Nucl. Sci. Abstr.*, **14**, 25582. 1960.

659. SEABORG, G.T. and A.C. WAHL.—*The Transuranium Elements, Nat. Nucl. Energy Ser.*, Div. IV, **14B**, paper 1.6. 1949.

660. SEABORG, G.T., A.C. WAHL, and J.W. KENNEDY.—*Phys. Rev.*, **69**, 367. 1946.

661. SEABORG, G.T., A.C. WAHL, and J.W. KENNEDY.—*The Transuranium Elements, Nat. Nucl. Energy Ser.*, Div. IV, **14B**, paper 1.1b. 1949.

662. SEABORG, G.T., A.C. WAHL, and J.W. KENNEDY.—*Ibid.*, paper 1.4. 1949.

663. SEGRE, E. and G. WEIGAND.—*Phys. Rev.*, **94**, 157. 1954.

663a. SCHWABE, K. and D. NEBEL.—*Z. phys. Chem.*, **220**, 339. 1962.

664. SCHULTS, W.D.—*Chem. Abstr.*, **55**, 13176. 1961.

665. SCHULTS, W.D.—*Anal. Chem.*, **33**, 15. 1961.

666. SIDDALL, T.H.—*J. Am. Chem. Soc.*, **81**, 4176. 1954.

667. SIDDALL, T.H.—*J. Inorg. Nucl. Chem.*, **13**, 151. 1960.

668. SIDDALL, T.H.—*J. Phys. Chem.*, **64**, 1863. 1960.
669. SIKKELAND, T.—JENER, Rept. No. 38. 1955; cited in: *Chem. Abstr.*, **50**, 5421. 1956.
670. SIKKELAND, T. and G. HAALAND.—JENER, Rept. No. 46, Kjeller. 1956.
671. SLEE, L.J., G. PHILLIPS, and E.N. JENKINS.—*Analyst*, **84**, 596. 1959.
672. SMALES, A.A., J.L. AIREY, G.N. WALTON, and R. BROOKS.—*Atomic Energy Research Establ.* C/R 533. 1955; cited in: *Chem. Abstr.*, **50**, 9490. 1956.
673. SMITH, C.S.—*Metal Progress*, **65**, 81. 1954.
674. SMITH, C.S.—*Phys. Rev.*, **94**, 1068. 1954.
675. SMITH, G.F.—*J. Chem. Educ.*, **7**, 2948. 1930.
676. SMITH, G.F.—*Ind. Eng. Chem., Anal. Ed.*, **6**, 229. 1934.
677. SMITH, H.L.—*J. Inorg. Nucl. Chem.*, **17**, 178. 1961.
678. SMITH, M.E.—*Anal. Chem.*, **30**, 912. 1958.
679. SMITH, M.E.—Cited in: *Nucl. Sci. Abstr.*, **14**, 21396. 1960.
680. SMITH, M.E.—Cited in: *Nucl. Sci. Abstr.*, **14**, 17814. 1960
681. STEINDLER, M.J., D.V. STEIDL, and R.K. STEUNENBERG.—*Nucl. Sci. Engng.*, **6**, No.4, 333. 1959.
682. STOUGHTON, R.W.—US Patent 2970035; cited in: *Chem. Abstr.*, **55**, 11131. 1961.
683. STREET, K. Jr. and G.T. SEABORG.—*J. Am. Chem. Soc.*, **72**, 2790. 1950.
684. STREET, K. Jr. and G.T. SEABORG.—*J. Am. Chem. Soc.*, **72**, 2792. 1950.
685. STROMINGER, D., J. HOLLANDER, and G.T. SEABORG.—*Rev. Mod. Phys.*, **30**, 823. 1958.
686. STUDIER, M.H., P.R. FIELDS, P.H. SELLERS, A.M. FRIEDMAN, C.M. STEVENS, J.F. MECH, H. DIAMOND, J. SEDLET, and J.R. HUIZENGA.—*Phys. Rev.*, **93**, 1433. 1954.
687. STUDIER, M.H. and J.R. HUIZENGA.—*Phys. Rev.*, **96**, 545. 1954.
688. SULLIVAN, J.C., G.Z. PYLE, M.H. STUDIER, P.R. FIELDS, and W.M. MANNING.—*Phys. Rev.*, **83**, 1267. 1951.
689. TAUBE, M.—*J. Inorg. Nucl. Chem.*, **12**, 174. 1959.
689a. TAUBE, M.—*Nukleonika*, **5**, 531. 1960.
690. TAYLOR, R.E.—*Anal. chim. acta*, **21**, 549. 1959.
690a. TAYLOR, C.G.—*Analyst*, **81**, 369. 1956.
691. THOMPSON, S.G., D.R. MULLER, and R.A. JAMES.—US Patent 2989367; cited in: *Nucl. Sci. Abstr.*, **15**, 19473. 1961.
692. THOMPSON, S.G., A. GHIORSO, and G.T. SEABORG.—*Phys. Rev.*, **80**, 781. 1950.
693. THOMPSON, S.G., K. STREET, Jr., A. GHIORSO, and G.T. SEABORG.—*Phys. Rev.*, **80**, 790. 1950.
694. THOMPSON, S.G., K. STREET, Jr., A. GHIORSO, and F.Z. REYNOLDS.—*Phys. Rev.*, **80**, 1108. 1950.
695. THOMPSON, S.G., K. STREET, JR., A. GHIORSO, and F.Z. REYNOLDS.—*Phys. Rev.*, **84**, 165. 1951.
696. TOBER, F.W.—*Second United Nations Intern. Conf. on the Peaceful Uses of At. Energy, Geneva,* Report N 520, USA. 1958.
697. TRENTINION, M. and A. CHESNE.—Cited in: *Nucl. Sci. Abstr.*, **15**, 2695. 1961.
698. TUCK, D.G.—*J. Inorg. Nucl. Chem.*, **6**, 252. 1958.
698a. TUCK, D.G. and G.A. WELCH.—*J. Inorg. Nucl. Chem.*, **9**, 302. 1959.
699. UMEZAWA, H.J.—*Atomic Energy Soc.*, **2**, 478. 1960; *Nucl. Sci. Abstr.*, **14**, 24209. 1960.

700. UKAEA.—PG Report 212 (S). 1961.
701. UKAEA.—PG Report 280 (W). 1961.
702. UKAEA.—PG Report 290 (M). 1962.
703. UKAEA.—PG Report 297 (W). 1962.
704. UKAEA.—PG Report 300 (W). 1962.
705. UKAEA.—PG Report 309 (W). 1962.
706. VAN SMIT, R.J., W. ROBB, and J.J. JACOBS.—*Nucleonics*, **17**, 116. 1959.
707. VOIGT, A., A. KANT, N. SLEIGHT, R. HEIN, J. WRIGHT, F. WOLTER, and H. BROWN.—*The Transuranium Elements, Nat. Nucl. Energy, Ser.*, Div. IV, **14B**, paper 3.8. 1949.
708. VOIGT, A., N. SLEIGHT, R. HEINE, and J. WRIGHT.—*Ibid.*, paper 15.9. 1949.
709. WAHL, C.—US Patent 2994579. 1961.
710. WARD, M. and G.A. WELCH.—*J. Inorg. Nucl. Chem.*, **2**, 395. 1956.
711. WATANABE, K.J.—*At. Energy Soc., Japan*, **3**, 497. 1961.
712. WATERBURY, G.R.—*Anal. Chem.*, **31**, 1138. 1959.
713. WATERBURY, G.R. and C.E. BRICKER.—*Anal. Chem.*, **29**, 129. 1957.
714. WATERBURY, G.R. and C.E. BRICKER.—*Anal. Chem.*, **29**, 1474. 1957.
715. WATERBURY, G.R. and C.E. BRICKER.—*Anal. Chem.*, **30**, 1007. 1958.
716. WATERBURY, G.R., R.M. DOUGLAS, and C.F. MERTZ.—*Anal. Chem.*, **33**, 1018. 1961.
717. WATERBURY, G.R. and C.F. METZ.—*Anal. Chem.*, **31**, 1144. 1959.
718. WATERBURY, G.R. and C.F. METZ.—*Talanta*, **6**, 237. 1960.
719. WEAVER, B. and D.E. HORNER.—*J. Chem. Eng. Data*, **5**, 260. 1960.
720. WEHNER, P. and J.C. HINDMAN.—*J. Am. Soc.*, **72**, 3911. 1950.
721. WEINSTOCK, B. and J.G. MALM.—*J. Inorg. Nucl. Chem.*, **2**, 380. 1956.
722. WEINSTOCK, B., E.E. WEAVER, and J.G. MALM.—*J. Inorg. Nucl. Chem.*, **11**, 104. 1959.
723. WEISS, H.V. and W.H. SHIPMAN.—*Anal. Chem.*, **33**, 37. 1961.
724. WELTON, G.N., E. FURBY, and V. OWEN.—Cited in: *RZhKh*, 22447. 1959.
725. WENDLANDT, W.W.—*Anal. Chem.*, **28**, 499. 1956.
726. WESTRUM, E.F.—*The Transuranium Elements, Nat. Nucl. Energy Ser.*, Div. IV, **14B**, paper 6.57. 1949.
727. WESTRUM, E.F. Jr., J.C. HINDMAN, and R. GREENLEE.—*Ibid.*, paper 22.80. 1949.
728. WESTRUM, E. and J. C. WALLMANN.—*J. Am. Chem. Soc.*, **73**, 3530. 1951.
729. WINCHESTER, R. S. and W. J. MARAMAN.— *Second United Nations, Intern. Conf. on the Peaceful Uses of At. Energy, Geneva*, Report No. 530, USA. 1958.
730. WISE, E.N.—*Nucl. Sci. Abstr.*, **10**, 8776. 1956.
731. WISH, L.—*Anal. Chem.*, **31**, 326. 1959.
732. ZACHARIASEN, W. H.—*Acta Cryst.*, **2**, 388. 1949.
733. ZACHARIASEN, W. H.—*Acta Cryst.*, **5**, 17. 1952.
734. ZACHARIASEN, W. H.—*The Transuranium Elements*, Nat. Nucl. Energy Ser., Div. IV, **14B**, paper 20.2. 1949.
735. ZACHARIASEN, W. H.—*Ibid.*, paper 20.4. 1949.
736. ZACHARIASEN, W. H.—*Ibid.*, paper 20.5. 1949.
737. ZACHARIASEN, W. H.—*Ibid.*, paper 20.6. 1949.
738. ZACHARIASEN, W. H.—*Ibid.*, paper 20.7. 1949.
739. ZOTOV, G. and C. FOWLER.—*At. En. Canada*, Chalk River Project, No. 208. 1955.

EXPLANATORY LIST OF ABBREVIATIONS OF U.S.S.R. INSTITUTIONS AND JOURNALS APPEARING IN THIS TEXT

Abbreviation	*Full name (transliterated)*	*Translation*
AN SSSR	Akademiya Nauk SSSR	Academy of Sciences of the USSR
AN Ukr SSR	Akademiya Nauk Ukrainskoi SSR	Academy of Sciences of the Ukrainian SSR
KhGU	Khar'kovskii Gosudarstvennyi Universitet	The Kharkov State University
LGU	Leningradskii Gosudarstvennyi Universitet	Leningrad State University
ONTI	Ob"edinenie Nauchno-Tekhnicheskikh Izdatel'stv	United Scientific and Technical Publishing Houses
TsITEIN	Tsentral'nyi Institut Tekhniko-Ekonomicheskoi Informatsii	Central Institute of Information on Technology and Economics
RZhKh	Referativnyi Zhurnal, Khimiya	Chemical Abstracts, USSR

SUBJECT INDEX

Printed in Israel
Manufactured at the Israel Program for Scientific Translations, Jerusalem